Photonic
MEMS Devices

Design, Fabrication and Control

OPTICAL SCIENCE AND ENGINEERING

Founding Editor
Brian J. Thompson
University of Rochester
Rochester, New York

1. Electron and Ion Microscopy and Microanalysis: Principles and Applications, *Lawrence E. Murr*
2. Acousto-Optic Signal Processing: Theory and Implementation, *edited by Norman J. Berg and John N. Lee*
3. Electro-Optic and Acousto-Optic Scanning and Deflection, *Milton Gottlieb, Clive L. M. Ireland, and John Martin Ley*
4. Single-Mode Fiber Optics: Principles and Applications, *Luc B. Jeunhomme*
5. Pulse Code Formats for Fiber Optical Data Communication: Basic Principles and Applications, *David J. Morris*
6. Optical Materials: An Introduction to Selection and Application, *Solomon Musikant*
7. Infrared Methods for Gaseous Measurements: Theory and Practice, *edited by Joda Wormhoudt*
8. Laser Beam Scanning: Opto-Mechanical Devices, Systems, and Data Storage Optics, *edited by Gerald F. Marshall*
9. Opto-Mechanical Systems Design, *Paul R. Yoder, Jr.*
10. Optical Fiber Splices and Connectors: Theory and Methods, *Calvin M. Miller with Stephen C. Mettler and Ian A. White*
11. Laser Spectroscopy and Its Applications, *edited by Leon J. Radziemski, Richard W. Solarz, and Jeffrey A. Paisner*
12. Infrared Optoelectronics: Devices and Applications, *William Nunley and J. Scott Bechtel*
13. Integrated Optical Circuits and Components: Design and Applications, *edited by Lynn D. Hutcheson*
14. Handbook of Molecular Lasers, *edited by Peter K. Cheo*
15. Handbook of Optical Fibers and Cables, *Hiroshi Murata*
16. Acousto-Optics, *Adrian Korpel*
17. Procedures in Applied Optics, *John Strong*
18. Handbook of Solid-State Lasers, *edited by Peter K. Cheo*
19. Optical Computing: Digital and Symbolic, *edited by Raymond Arrathoon*
20. Laser Applications in Physical Chemistry, *edited by D. K. Evans*
21. Laser-Induced Plasmas and Applications, *edited by Leon J. Radziemski and David A. Cremers*
22. Infrared Technology Fundamentals, *Irving J. Spiro and Monroe Schlessinger*
23. Single-Mode Fiber Optics: Principles and Applications, Second Edition, Revised and Expanded, *Luc B. Jeunhomme*
24. Image Analysis Applications, *edited by Rangachar Kasturi and Mohan M. Trivedi*
25. Photoconductivity: Art, Science, and Technology, *N. V. Joshi*
26. Principles of Optical Circuit Engineering, *Mark A. Mentzer*
27. Lens Design, *Milton Laikin*

28. Optical Components, Systems, and Measurement Techniques, *Rajpal S. Sirohi and M. P. Kothiyal*
29. Electron and Ion Microscopy and Microanalysis: Principles and Applications, Second Edition, Revised and Expanded, *Lawrence E. Murr*
30. Handbook of Infrared Optical Materials, *edited by Paul Klocek*
31. Optical Scanning, *edited by Gerald F. Marshall*
32. Polymers for Lightwave and Integrated Optics: Technology and Applications, *edited by Lawrence A. Hornak*
33. Electro-Optical Displays, *edited by Mohammad A. Karim*
34. Mathematical Morphology in Image Processing, *edited by Edward R. Dougherty*
35. Opto-Mechanical Systems Design: Second Edition, Revised and Expanded, *Paul R. Yoder, Jr.*
36. Polarized Light: Fundamentals and Applications, *Edward Collett*
37. Rare Earth Doped Fiber Lasers and Amplifiers, *edited by Michel J. F. Digonnet*
38. Speckle Metrology, *edited by Rajpal S. Sirohi*
39. Organic Photoreceptors for Imaging Systems, *Paul M. Borsenberger and David S. Weiss*
40. Photonic Switching and Interconnects, *edited by Abdellatif Marrakchi*
41. Design and Fabrication of Acousto-Optic Devices, *edited by Akis P. Goutzoulis and Dennis R. Pape*
42. Digital Image Processing Methods, *edited by Edward R. Dougherty*
43. Visual Science and Engineering: Models and Applications, *edited by D. H. Kelly*
44. Handbook of Lens Design, *Daniel Malacara and Zacarias Malacara*
45. Photonic Devices and Systems, *edited by Robert G. Hunsberger*
46. Infrared Technology Fundamentals: Second Edition, Revised and Expanded, *edited by Monroe Schlessinger*
47. Spatial Light Modulator Technology: Materials, Devices, and Applications, *edited by Uzi Efron*
48. Lens Design: Second Edition, Revised and Expanded, *Milton Laikin*
49. Thin Films for Optical Systems, *edited by Francoise R. Flory*
50. Tunable Laser Applications, *edited by F. J. Duarte*
51. Acousto-Optic Signal Processing: Theory and Implementation, Second Edition, *edited by Norman J. Berg and John M. Pellegrino*
52. Handbook of Nonlinear Optics, *Richard L. Sutherland*
53. Handbook of Optical Fibers and Cables: Second Edition, *Hiroshi Murata*
54. Optical Storage and Retrieval: Memory, Neural Networks, and Fractals, *edited by Francis T. S. Yu and Suganda Jutamulia*
55. Devices for Optoelectronics, *Wallace B. Leigh*
56. Practical Design and Production of Optical Thin Films, *Ronald R. Willey*
57. Acousto-Optics: Second Edition, *Adrian Korpel*
58. Diffraction Gratings and Applications, *Erwin G. Loewen and Evgeny Popov*
59. Organic Photoreceptors for Xerography, *Paul M. Borsenberger and David S. Weiss*
60. Characterization Techniques and Tabulations for Organic Nonlinear Optical Materials, *edited by Mark G. Kuzyk and Carl W. Dirk*
61. Interferogram Analysis for Optical Testing, *Daniel Malacara, Manuel Servin, and Zacarias Malacara*
62. Computational Modeling of Vision: The Role of Combination, *William R. Uttal, Ramakrishna Kakarala, Spiram Dayanand, Thomas Shepherd, Jagadeesh Kalki, Charles F. Lunskis, Jr., and Ning Liu*
63. Microoptics Technology: Fabrication and Applications of Lens Arrays and Devices, *Nicholas Borrelli*
64. Visual Information Representation, Communication, and Image Processing, *edited by Chang Wen Chen and Ya-Qin Zhang*
65. Optical Methods of Measurement, *Rajpal S. Sirohi and F. S. Chau*

66. Integrated Optical Circuits and Components: Design and Applications, *edited by Edmond J. Murphy*
67. Adaptive Optics Engineering Handbook, *edited by Robert K. Tyson*
68. Entropy and Information Optics, *Francis T. S. Yu*
69. Computational Methods for Electromagnetic and Optical Systems, *John M. Jarem and Partha P. Banerjee*
70. Laser Beam Shaping, *Fred M. Dickey and Scott C. Holswade*
71. Rare-Earth-Doped Fiber Lasers and Amplifiers: Second Edition, Revised and Expanded, *edited by Michel J. F. Digonnet*
72. Lens Design: Third Edition, Revised and Expanded, *Milton Laikin*
73. Handbook of Optical Engineering, *edited by Daniel Malacara and Brian J. Thompson*
74. Handbook of Imaging Materials: Second Edition, Revised and Expanded, *edited by Arthur S. Diamond and David S. Weiss*
75. Handbook of Image Quality: Characterization and Prediction, *Brian W. Keelan*
76. Fiber Optic Sensors, *edited by Francis T. S. Yu and Shizhuo Yin*
77. Optical Switching/Networking and Computing for Multimedia Systems, *edited by Mohsen Guizani and Abdella Battou*
78. Image Recognition and Classification: Algorithms, Systems, and Applications, *edited by Bahram Javidi*
79. Practical Design and Production of Optical Thin Films: Second Edition, Revised and Expanded, *Ronald R. Willey*
80. Ultrafast Lasers: Technology and Applications, *edited by Martin E. Fermann, Almantas Galvanauskas, and Gregg Sucha*
81. Light Propagation in Periodic Media: Differential Theory and Design, *Michel Nevière and Evgeny Popov*
82. Handbook of Nonlinear Optics, Second Edition, Revised and Expanded, *Richard L. Sutherland*
83. Polarized Light: Second Edition, Revised and Expanded, *Dennis Goldstein*
84. Optical Remote Sensing: Science and Technology, *Walter Egan*
85. Handbook of Optical Design: Second Edition, *Daniel Malacara and Zacarias Malacara*
86. Nonlinear Optics: Theory, Numerical Modeling, and Applications, *Partha P. Banerjee*
87. Semiconductor and Metal Nanocrystals: Synthesis and Electronic and Optical Properties, edited by *Victor I. Klimov*
88. High-Performance Backbone Network Technology, *edited by Naoaki Yamanaka*
89. Semiconductor Laser Fundamentals, *Toshiaki Suhara*
90. Handbook of Optical and Laser Scanning, *edited by Gerald F. Marshall*
91. Organic Light-Emitting Diodes: Principles, Characteristics, and Processes, *Jan Kalinowski*
92. Micro-Optomechatronics, *Hiroshi Hosaka, Yoshitada Katagiri, Terunao Hirota, and Kiyoshi Itao*
93. Microoptics Technology: Second Edition, *Nicholas F. Borrelli*
94. Organic Electroluminescence, *edited by Zakya Kafafi*
95. Engineering Thin Films and Nanostructures with Ion Beams, *Emile Knystautas*
96. Interferogram Analysis for Optical Testing, Second Edition, *Daniel Malacara, Manuel Sercin, and Zacarias Malacara*
97. Laser Remote Sensing, *edited by Takashi Fujii and Tetsuo Fukuchi*
98. Passive Micro-Optical Alignment Methods, *edited by Robert A. Boudreau and Sharon M. Boudreau*
99. Organic Photovoltaics: Mechanism, Materials, and Devices, *edited by Sam-Shajing Sun and Niyazi Serdar Saracftci*
100. Handbook of Optical Interconnects, *edited by Shigeru Kawai*
101. GMPLS Technologies: Broadband Backbone Networks and Systems, *Naoaki Yamanaka, Kohei Shiomoto, and Eiji Oki*

102. Laser Beam Shaping Applications, *edited by Fred M. Dickey, Scott C. Holswade and David L. Shealy*
103. Electromagnetic Theory and Applications for Photonic Crystals, *Kiyotoshi Yasumoto*
104. Physics of Optoelectronics, *Michael A. Parker*
105. Opto-Mechanical Systems Design: Third Edition, *Paul R. Yoder, Jr.*
106. Color Desktop Printer Technology, *edited by Mitchell Rosen and Noboru Ohta*
107. Laser Safety Management, *Ken Barat*
108. Optics in Magnetic Multilayers and Nanostructures, *Štefan Višňovský*
109. Optical Inspection of Microsystems, *edited by Wolfgang Osten*
110. Applied Microphotonics, *edited by Wes R. Jamroz, Roman Kruzelecky, and Emile I. Haddad*
111. Organic Light-Emitting Materials and Devices, *edited by Zhigang Li and Hong Meng*
112. Silicon Nanoelectronics, *edited by Shunri Oda and David Ferry*
113. Image Sensors and Signal Processor for Digital Still Cameras, *Junichi Nakamura*
114. Encyclopedic Handbook of Integrated Circuits, *edited by Kenichi Iga and Yasuo Kokubun*
115. Quantum Communications and Cryptography, *edited by Alexander V. Sergienko*
116. Optical Code Division Multiple Access: Fundamentals and Applications, *edited by Paul R. Prucnal*
117. Polymer Fiber Optics: Materials, Physics, and Applications, *Mark G. Kuzyk*
118. Smart Biosensor Technology, *edited by George K. Knopf and Amarjeet S. Bassi*
119. Solid-State Lasers and Applications, *edited by Alphan Sennaroglu*
120. Optical Waveguides: From Theory to Applied Technologies, *edited by Maria L. Calvo and Vasudevan Lakshiminarayanan*
121. Gas Lasers, *edited by Masamori Endo and Robert F. Walker*
122. Lens Design, Fourth Edition, *Milton Laikin*
123. Photonics: Principles and Practices, *Abdul Al-Azzawi*
124. Microwave Photonics, *edited by Chi H. Lee*
125. Physical Properties and Data of Optical Materials, *Moriaki Wakaki, Keiei Kudo, and Takehisa Shibuya*
126. Microlithography: Science and Technology, Second Edition, *edited by Kazuaki Suzuki and Bruce W. Smith*
127. Coarse Wavelength Division Multiplexing: Technologies and Applications, *edited by Hans Joerg Thiele and Marcus Nebeling*
128. Organic Field-Effect Transistors, *Zhenan Bao and Jason Locklin*
129. Smart CMOS Image Sensors and Applications, *Jun Ohta*
130. Photonic Signal Processing: Techniques and Applications, *Le Nguyen Binh*
131. Terahertz Spectroscopy: Principles and Applications, *edited by Susan L. Dexheimer*
132. Fiber Optic Sensors, Second Edition, *edited by Shizhuo Yin, Paul B. Ruffin, and Francis T. S. Yu*
133. Introduction to Organic Electronic and Optoelectronic Materials and Devices, *edited by Sam-Shajing Sun and Larry R. Dalton*
134. Introduction to Nonimaging Optics, *Julio Chaves*
135. The Nature of Light: What Is a Photon?, *edited by Chandrasekhar Roychoudhuri, A. F. Kracklauer, and Katherine Creath*
136. Photonic MEMS Devices: Design, Fabrication and Control, *Ai-Qun Liu*

Photonic
MEMS Devices

Design, Fabrication and Control

Ai-Qun Liu

Co-Authors
Xuming Zhang
Jing Li
Selin Hwee Gee Teo
Frank L. Lewis
Bruno Borovic

CRC Press
Taylor & Francis Group
Boca Raton London New York

CRC Press is an imprint of the
Taylor & Francis Group, an **informa** business

CRC Press
Taylor & Francis Group
6000 Broken Sound Parkway NW, Suite 300
Boca Raton, FL 33487-2742

First issued in paperback 2019

© 2009 by Taylor & Francis Group, LLC
CRC Press is an imprint of Taylor & Francis Group, an Informa business

No claim to original U.S. Government works

ISBN-13: 978-1-4200-4568-0 (hbk)
ISBN-13: 978-0-367-38694-8 (pbk)

Visit the Taylor & Francis Web site at
http://www.taylorandfrancis.com

and the CRC Press Web site at
http://www.crcpress.com

Dedication

This book is dedicated to my late father

and my beloved mother

Contents

1   MEMS Optical Switches and Systems...1
    Jing Li and Ai-Qun Liu

2   Design of MEMS Optical Switches 27
    Jing Li and Ai-Qun Liu

3   MEMS Thermo-Optic Switches...89
    Jing Li, Tian Zhong, and Ai-Qun Liu

4   PHC Microresonators Dynamic Modulation Devices 129
    Selin Hwee Gee Teo and Ai-Qun Liu

5   MEMS Variable Optical Attenuators..................................173
    Xuming Zhang, Hong Cai, and Ai-Qun Liu

6   MEMS Discretely Tunable Lasers.....................................237
    Xuming Zhang, Hong Cai, and Ai-Qun Liu

7   MEMS Continuously Tunable Lasers 273
    Xuming Zhang and Ai-Qun Liu

8   MEMS Injection-Locked Lasers 317
    Xuming Zhang and Ai-Qun Liu

9   Deep Etching Fabrication Process...................................353
    Jing Li and Ai-Qun Liu

10 Deep Submicron Photonic Bandgap Crystal Fabrication Processes.............393
    Selin Hwee Gee Teo and Ai-Qun Liu

11 Control Strategies for Electrostatic MEMS Devices.........................429
    Bruno Borovic and Frank L. Lewis

12 Control of Optical Devices...447
    Bruno Borovic and Frank L. Lewis

Index..467

Preface

Optics is conventionally used in the general science of the study of light. Photonics, on the other hand, was originally introduced as a counterpart of electronics that uses photons instead of electrons for signal processing and transfer of information, implying a wider range of applications and possibilities. Photonic micro-electro-mechanical systems (MEMS) is a significant branch of MEMS technology that originated from optical MEMS technology. Its wide potential is very closely related to photonics applied in different research fields such as photonic bandgap crystals, plasmonics, biophotonics, etc. Photonic MEMS devices may be passive or active devices spinning from single devices to large integrated MEMS systems with electronic control circuits. Photonic MEMS covers a broad spectrum of applications ranging from fiber optical communication to optofluidic or micro-optical-fluidic-systems (MOFS) for biomedical applications. The purpose of this book *Photonic MEMS Devices: Design, Fabrication and Control* is to collect, edit, and summarize different aspects of photonic MEMS research, providing as much background and foundational information as possible on the essential theories for the analysis and design of the photonic MEMS devices.

This is the first research textbook written to meet the research, development, and academic needs at undergraduate and graduate levels for this burgeoning field of MEMS technology. *Photonic MEMS Devices* is not reviewing a narrow subject. The topic spans photonic and optical MEMS technology as well as MEMS device design, fabrication, modeling, control, and all the variations in between. It also deals with MEMS electronic, mechanical, and optical design; materials and their different level of micro and nanofabrication processes; experimental techniques; actuator control methods; and so on.

The book describes a broad range of optical and photonic MEMS devices, from MEMS optical switches to photonic bandgap crystal switches, and from variable optical attenuators (VOA) to injection locked tunable lasers. The book deals rigorously and completely with all these technologies at a fundamental level to define and illustrate the key concepts in each topic and introduce critical nomenclature in a systematic fashion. Each chapter includes a fundamental introduction, analysis techniques with equations, research case studies, discussions, and conclusions.

I am grateful to all my Ph.D. students, Xuming, Jing, and Selin, and Prof. Frank Lewis's Ph.D. student, Bruno Borovic, for their excellent research work and contributions in writing this book. Frank is not only my research collaborator in MEMS control, but also my friend who initiated the idea of writing this book. I am also grateful to Jessica Vakili at Taylor & Francis Group, who met every one of our needs in the process of bringing this book to final fruition.

I am sure that I have not remembered all who contributed to the creation of this book, but I extend my sincerest appreciation to those above and any I have regrettably forgotten to name. My final thanks are to my research group members, both past and present, for contributing their efforts and full support to me.

Although I have made my best effort in writing this book, I have undoubtedly made errors or omissions and sincerely apologize for them. I welcome hearing from readers about such errors, if any key concepts have been omitted, and format suggestions. Updates, corrections, and other information will be available at my Web site, http://nocweba.ntu.edu.sg/laq_mems.

Ai-Qun Liu
Nanyang Technological University
Singapore

About the Author

Ai-Qun Liu (A. Q. Liu) received his Ph.D. degree from National University of Singapore in 1994. He obtained his M.Sc. degree from Beijing University of Posts and Tele-communications in 1988, and B.Eng. degree from Xi'an Jiaotong University in 1982. Currently, he is associate professor at the Division of Microelectronics, School of Electrical and Electronic Engineering, Nanyang Techno-logical University. He was the recipient of the IES Presti-gious Engineering Achievement Award in 2006 and the University Scholar Award in 2007. He is also as an associ-ate editor for the *IEEE Sensor Journal* and a guest editor for *Sensors & Actuators A: Physical.*

Co-Authors

Xuming Zhang   University of Maryland, Baltimore

Jing Li   Institute of Microelectronics, Singapore

Selin Hwee Gee Teo   Institute of Microelectronics, Singapore

Frank L. Lewis   University of Texas, Arlington

Bruno Borovic   InvenSense, Inc., Santa Clara, California

1

MEMS Optical Switches and Systems

Jing Li and Ai-Qun Liu

CONTENTS

1.1 Introduction to MEMS Optical Switch...2
    1.1.1 Different Types of Optical Switches..2
    1.1.2 MEMS Optical Switches...4
        1.1.2.1 2-D MEMS Optical Switches 4
        1.1.2.2 3-D MEMS Optical Switches...5
1.2 Optical Switch Matrix...6
1.3 Optical Switching Systems 6
    1.3.1 Reconfigurable Optical-Switching OADM..8
        1.3.1.1 Construction of the Optical-Switching OADM...............................9
        1.3.1.2 MEMS Optical Switch and Tunable FBG 10
        1.3.1.3 Experimental Results and Discussion... 13
    1.3.2 Tunable Laser-Integrated OADM 16
        1.3.2.1 Configuration of Tunable Laser Integrated OADM...................... 16
        1.3.2.2 Experiment Results and Discussions 18
1.4 Summary... 23
References 23

With the rising demand for information processing, storage, and transfer in all fields of engineering today, reliable high-speed broadband communication systems have become indispensable. The so-called "information superhighway" is coming into its own as a requirement, not a luxury, for modern-day engineered systems. This concept can be realized by improvements in the techniques used for radio frequency (RF) technology, which is the common platform of WiFi, Bluetooth, and all other communications standards, wireless or wired. A powerful method for speeding up communications, enhancing security, and improving performance is the development of an all-optical infrastructure for RF. To meet these needs, high-performance optical components, subsystems, and systems are necessary as they hold promise for eliminating current communication bottlenecks, thanks to the properties that stem from the parallelism and reliability of processing inherent in optical systems. Among optical components, the optical switch is a key device, as it plays multiple roles in control, monitoring, protection, management, and elsewhere in RF communication networks.

In this chapter, various optical switches are reviewed, and their performances are compared, with a focus on the micro-electro-mechanical system (MEMS) optical switch, which has the following advantages: low loss, low power consumption, middle switching speed, small size, and low cost due to batch fabrication. Then, applications in the fiber optical communication networks are discussed with examples of several different systems.

1.1   Introduction to MEMS Optical Switch

The optical switch is nowadays playing a significant role in optical communication networks. For example, in an all-optical network (AON), optical switches select the directions of the signal, adding or dropping information, protecting networks, and so on. These functions can be realized with traditional electrical switches after converting the optical signal to an electrical one, which is then converted back to an optical signal for further transmission.

Just as electrical switching replaced mechanical relays of the past, optical switching is on the verge of replacing electrical switching modules. Optical switching is more and more important [1] because the conversion of optical to electronic signals is a critical bottleneck in AONs when a huge amount of information needs to be switched through various nodes [2]. Therefore, optical switching without electrical-based traffic protocols is a promising development, and a keystone in dynamically reconfigurable optical networks.

1.1.1   Different Types of Optical Switches

The optical switch is a significant component in a fiber-optical communication system. It has been developed using different types of techniques such as optical MEMS [2–5], thermo-optical methods [6–8], semiconductor optic amplifiers (SOAs) [9], liquid crystals [10,11], electroholography [12], electro-optical LiNbO$_3$ [13], electronically switchable waveguide Bragg grating [14], and acousto-optical [15] and opto-optical switches [16].

Optical MEMS switch: An optical MEMS refers to the integration of optical, electrical, and mechanical functionalities onto a single chip. By using a mirror or membrane to physically alter the propagation of a light beam, MEMS optical switches show a tremendous advantage in achieving low cross talks, high extinction ratios, independence of data rate and data format, wavelength insensitivity, and easy large-scale integrations. However, mechanical stability and millisecond-scale response time are challenges of the optical MEMS switches for practical optical communications networks.

Thermal optical switch: This category of switch is based on the thermo-optic effect of waveguides or other materials. Their main advantages are low insertion loss, polarization insensitivity, with switching speed in sub-milliseconds or even microseconds. Recently, thanks to the advancement of the silicon-on-insulator (SOI) planar lightwave circuit (PLC) technology [8], the overall performance of waveguide-based thermal optical switches has significantly improved, and now they are ready to be marketed. In fact, thermal optical switches present a viable solution both technologically and economically: They are free of severe disadvantages in performance and easy to fabricate through the standard integrated circuit (IC) process. The most popular types of switches in this category include digital optical switches (DOSs) [17] and interferometric switches [18,19]. A more detailed investigation of these devices will be undertaken in Chapter 3.

Electro-optic switch: As its name implies, this category of devices operates on electro-optic effects. The main types are LiNbO$_3$ switches [13], SOA-based switches [9], liquid crystal switches [10,11], and electroholographic (EH) optical switches [12]. The LiNbO$_3$ switch uses the large electro-optic coefficient of LiNbO$_3$ [13]. A typical device is an interference-based directional coupler, whose coupling ratio is regulated by changing the refractive index of the material in the coupling area. The disadvantages of this switch are high insertion loss and high cross talk. The DOSs based on mode evolution are also built utilizing the $LiNbO_3$ electro-optic effect. The DOS device has a step-like switch response to the applied voltage, and thus exhibits a relatively high switching speed in nanoseconds. SOA-based switches refer to current-controlled optical switches. While the SOA is turned off or on by controlling the bias currents, the transmission of light is modulated. Liquid crystal (LC) switches are based on control of the polarization of light. The electro-optic coefficient in LC is much higher than in $LiNbO_3$, so LC switches are more efficient. Lastly, electroholographic optical switches are based on control of the reconstruction process of volume holograms by external application of an electric field. No matter what type is used, electro-optic switches generally require special materials other than common semiconductors. This is because the electro-optic coefficient of semiconductors is relatively small, and thus the devices fabricated on semiconductors are less efficient. As a result, electro-optic switches have a high cost and pose difficulties for on-chip integration.

Acousto-optic switches: This type of switch utilizes the acousto-optic effect in crystals such as TeO_2. Ultrasonic waves are used to deflect the light. These switches could have good performance in terms of high speed, low cross talk, and polarization independence; however, the major drawback lies in the lack of scalability. The device usually has a millimeter-scale size with limited potential for miniaturization, thus posing a hurdle to large-scale integration. Besides, the requirement of special crystal materials makes the acousto-optic switches very expensive and hardly compatible with the semiconductor fabrication process.

All these mechanisms have their own pros and cons in terms of characteristics, including basic performances (insertion loss, switching speed, cross talk, polarization-dependent loss, bit rate and protocol transparency, and operation bandwidth), network requirements (multicast, switching device dimensions, scalability, and nonblocking), and system requirements (stability, repeatability, power consumption, and cost), where the performances depend on each other. Table 1.1 lists the major performance characteristics associated with each of the optical switches mentioned. For the purpose of a general comparison, the average performance of each type is considered. Nevertheless, the combination of the characteristics of different switches determines the potential application of that type of switch. It can be clearly seen that optical MEMS and thermal optical switches are best

TABLE 1.1

Performance Comparisons of Different Optical Switches

Performance Switch Type	Switching Speed	Cross Talk	Size	Actuation Voltage/Power Dissipation	Scalability
Optical MEMS (actuated mirrors)	Sub-ms to ms	≤−30 dB	mm × mm	≤50 V	Good
Thermal optical switch (MZI interferometer)	A few μs to ms	>−30 dB	mm × μm	90 mW	Good
Thermal optical switch (DOS)	<5 ms	≤−30 dB	mm × mm	250 mW	Limited
Electro-optic switch ($LiNbO_3$)	5 ns	<−45 dB	3-in.	18 V	Limited
SOA-based switch	200 ps	≤−12 dB	mm × mm × mm	200 mA	Medium
Liquid crystal switch	ms	≤−35 dB	mm × mm × mm	Very low	Medium
Electroholographic optical switches	<10 ns	Very low	mm × mm	High voltage	Good
Acousto-optic switch	300 ns	32 dB	cm length	200 mW	Poor

suited for low-cost portable devices, optical cross connects (OXCs) and optical add/drop multiplexers (OADMs), whereas electro-optic and acousto-optic switches have the potential for highly reliable high-speed packet switching, although the device might be bulky and expensive. In this book, we will focus on the development of small-size, low-cost, and low-power optical-switching technology for OXC and OADM applications. Therefore, optical MEMS and thermal optical switches are of greater interest than their electro-optic and acousto-optic counterparts. The MEMS optical switch and its subsystems will be the focus of this chapter, and the thermo-optical switch will be discussed in Chapter 3.

1.1.2 MEMS Optical Switches

According to the type of movable element, MEMS optical switches can be divided into two main categories, namely, switches with moving fibers and switches with moving mirrors or lenses. The former transfers light into different fibers by moving the fibers, whereas the latter shifts light between different output fibers by moving mirrors or lenses. However, the former type suffers from poor expandability to large matrixes, and the switching speed is another critical limitation due to the thermal actuation mechanism.

Switches with moving mirrors or lenses have a micromirror that directs the input signal into the targeted output. For this purpose, their structure employs either a 2-D mirror, in which light signals are routed within a plane, or a 3-D one, in which light travels in 3D space according to its design architecture.

1.1.2.1 2-D MEMS Optical Switches

The 2-D MEMS optical switch has its own advantages over the 3-D mirror switch, such as low cost and simple fabrication. Additionally, reliability and repeatability are better. Although scalability is a problem, the 2-D MEMS optical switch has the advantage of small and medium matrices. The market is dominated by small- (2×2 and 4×4) and medium-sized (such as 16×16) matrices; the 2-D switch can be used in matrices with a size up to 64×64.

For the torsional 2-D MEMS switch cell, a high reflective mirror is driven to realize a torsional motion about a certain axis (or several axes). The micromirror is suspended over the substrate by microbeams along the axial direction, and these in turn supported by anchors at the ends. The opposite electrodes lie under the micromirror. In this way, the micromirror can be driven to rotate around the beam axis by electrostatic force.

As most of the previously mentioned 2-D torsional switches are fabricated by surface micromachining, the air gap between the bottom of the micromirror and the substrate cannot be too large because of the thickness limitation of the sacrificial layers leading to a very limited rotation angle. Worse still, the flatness of the mirror is degraded by the stress of mismatched materials, and a warp may occur as a result of the nonuniform distribution of the electrostatic attractive force and the elastic balancing force. The other critical weakness of this type of switch is the stiction problem caused by either wet release or the increased electrostatic attractive force between the mirror and substrate.

Compared to 2-D rotational mirrors, the translation 2-D-mirror-based optical switches, schematically shown in Figure 1.1a, are easier to fabricate, especially when SOI substrates are used, because the micromirror and the actuator are fabricated by deep reactive ion etching (DRIE) process together with the fiber grooves. The performance of this switch is

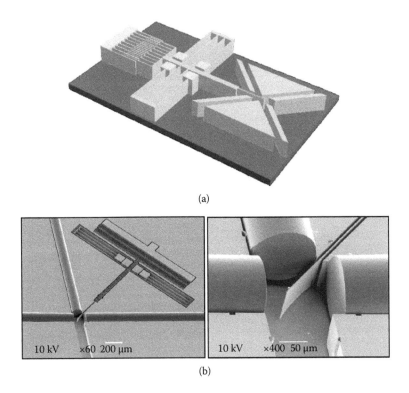

(a)

(b)

FIGURE 1.1
Configuration of the 2-D switch: (a) overview, (b) scanning electron microscope (SEM) picture of an assembled optical switch, and (c) SEM picture of micromirror and fiber ends.

inherently better because the movement of the mirror is translational, where no deformation can occur to the mirror. Additionally, there is no stress induced by the fabrication process as the mirror is coated on both sides, and the built-in stress is canceled out. This type of optical switch is developed by using either an SOI [20] or a plane silicon substrate [2]. There are two different states. When the mirror is in the beam path, the incident light is reflected into the orthogonal output, realizing the cross state. When the mirror is driven out of the beam path, the incident light can travel through to the opposite output fibers, realizing the bar state. Figures 1.1(b) and (c) are SEM pictures of a 2 × 2 optical switch.

1.1.2.2 3-D MEMS Optical Switches

The 3-D MEMS optical switch is best exemplified by Lucent OXCs, which can scale to thousands of ports with low loss and high uniformity. Its micromirror can be tilted along two orthogonal axes when a proper driving voltage is applied to the stationary and movable electrodes. There are clear and critical drawbacks: (1) complex feedback mechanisms are necessary to position the mirrors; (2) bundles of input and output fibers are to be aligned; and (3) the production processes are complicated and costly. Manufacturing this type of MEMS optical switch is extremely complicated as it requires sophisticated analog control mechanisms and intricate packaging to connect the wires onto the electrodes.

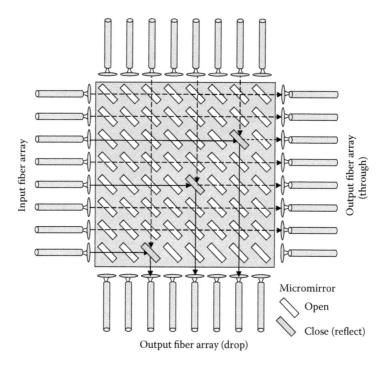

Input fiber array

Output fiber array (through)

Micromirror
◇ Open
◆ Close (reflect)

Output fiber array (drop)

FIGURE 1.2
Schematic of in-plane 8 × 8 switch matrix.

1.2   Optical Switch Matrix

The small switch cell can be expanded to an $N \times N$ matrix. First, one out-plane 4×4 switch matrix was developed. Then, an in-plane 8×8 MEMS optical switch matrix, as reported in Lin, L. Y.; Goldstein, E. L., Military communications conference proceedings, 1999, MILCOM IEEE, Vol. 2, pages 954–957, was developed with the configuration shown in Figure 1.2, using tilted mirrors.

The number of mirrors is N^2 for a normal $N \times N$ switch matrix, so that 64 micromirrors are employed in the previously mentioned 8×8 optical switch matrix. To reduce this number, an L-shape configuration is proposed, as shown in Figure 1.3. Here, the number of mirrors is reduced by 25% [21].

By combining two $N \times N$ optical switch matrices and two N^2 optical fiber bundles, an $N \times N$ OXC was demonstrated by Lucent as schematically shown in Figure 1.4.

As a Gaussian beam diverges with propagation, a collimator should be used, or a waveguide [22] is introduced to confine the beam size. At the same time, an actuator with the capability of providing a large displacement is required to control the direction of the beams.

1.3   Optical Switching Systems

Integration plays a key role in the development of many technologies such as integrated circuits (IC) and planar lightwave circuits. The integration of various components into a system can significantly improve functionality, compactness, and reliability while reducing

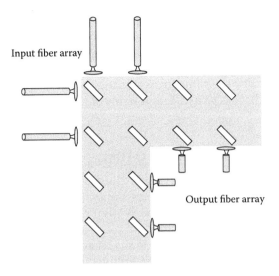

FIGURE 1.3
Configuration of L-shaped switch matrix.

cost. Various devices have already been developed using MEMS technology. However, most of them work independently and function only at component level. Much effort has been made for the integration of single MEMS components with other components such as optical fibers [23,24], microlenses [25], IC control circuits, and planar lightwave circuits. The process of fabricating MEMS components has become technologically mature, and the integration of MEMS components into systems represents the latest trend and has attracted considerable research interest. For example, an optical disk pickup subsystem has been demonstrated by integrating several surface-micromachined components onto single chips [26,27]; various microspectrometers have been prototyped by integrating gratings, micromirrors, and photodetectors within a small box [28], and a hybrid tunable laser has been formed by combining a MEMS-actuated mirror and a laser gain chip with other optical components [29].

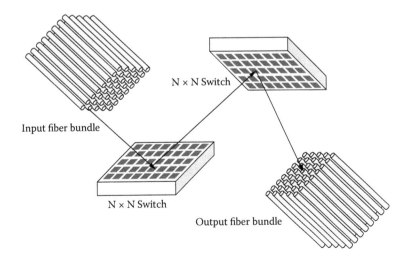

FIGURE 1.4
Schematic of an N × N optical cross connect (OXC) switch.

Not only does the MEMS optical switch play an increasingly important role in high-speed, broadband optical communication systems but it is also a key element in optical subsystems. Thus, its applications become wider and wider. On the other hand, the merit of microbatch fabrication is embodied and employed well; because the different MEMS elements can be realized by the same process, time-consuming assembly can be avoided. In this section, we will discuss two different kinds of optical subsystems that integrate the optical switch cell.

1.3.1 Reconfigurable Optical-Switching OADM

With the deployment of dense wavelength division multiplexing (DWDM) systems, the efficient use of fiber bandwidth becomes imperative [30,31]. The OADM, which has the ability to selectively drop or add an individual channel or a subset of wavelengths from the transmission system without full opto-electronic regeneration of all wavelengths, makes it possible to manipulate the traffic on the basis of wavelength at the optical layer.

To design an optimal system, the OADM architecture as well as the specific requirements of the network must be considered. The OADM should have more accurate and narrower filtering characteristics than the channel spacing, and relatively high speed for DWDM systems. Other critical issues in the design are insertion loss, cross talk, and component count [32]. The OADMs can be divided into two categories, namely, static (fixed) and dynamic (reconfigurable). The former enables predetermined channels to be added or dropped, whereas the latter can provide much more flexibility as it allows the channels to be added or dropped dynamically. Cost is a critical consideration for the implementation of OADMs with improved performance.

Various types of OADMs based on different optical devices have been proposed and developed. N. A. Riza proposed a reconfigurable OADM [33] in which fiber Bragg gratings (FBGs) and ferroelectric liquid crystal (FLC) switches were cascaded in series, and the FLC switches provided fast selection and switching of add/drop channels. A couple of references [34,35] show the applications of FBGs in various OADMs and OXCs, in which the integration of FBGs, optical switches, and optical circulators (OCs) is a huge technological challenge. Waveguide technology presents a significant improvement in the integration technology for OADM [36,37], as it is difficult to integrate the different substrate components into a monolithic chip.

MEMS optical switch [38] and liquid crystal (LC) switch [39] represent two new technologies for the integration of reconfigurable OADM. The MEMS optical switch is very small; hundreds of them can be fitted in the same space as one single macro device that performs the same function. The precision is improved, and they are much less expensive than their counterparts because hundreds of devices can be fabricated on a single wafer in an IC process. In 1999, tilting micromirrors [40] and light valves [41] were developed and applied to the reconfigurable OADM. The 2×2 MEMS optical switch is a significant component that improves the performance of the reconfigurable OADM [42]. In 1999, MEMS actuator-based 2×2 fiber-optical modules were fabricated on a SOI substrate by Marxer and Rooij [4]. The fabrication process is simple owing to the SOI technology, but the wafer is expensive. In 2003, a new DRIE fabrication process on a silicon wafer [2] was developed. It prevents the stiction and notching effect of the SOI-based process and hence reduces the unit fabrication cost of the 2×2 MEMS optical switch.

In this section, a serial-type dynamic OADM based on a 2×2 MEMS optical switch, a tunable FBG, and a OC is discussed. It is insensitive to polarization and bitrate. The structure of the serial-type OADM is described in detail in the following text. The fabrication and properties of the two key components, the MEMS optical switch and thermally tuned

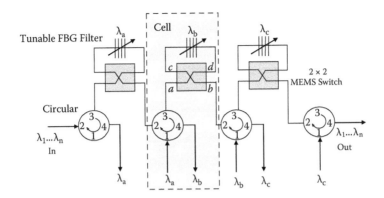

FIGURE 1.5
Construction of the serial dynamic OADM.

FBG, are explained. The performance of this type of OADM is discussed and investigated experimentally.

1.3.1.1 Construction of the Optical-Switching OADM

The serial dynamic OADM consists of n cells cascaded in sequence as shown in Figure 1.5. Each cell is composed of a piece of a 2×2 MEMS optical switch, a four-port OC, and a tunable FBG whose central Bragg wavelength, matching the DWDM signal λ_i, is adjustable. The tunable FBG is inserted into the fiber grooves of the switch. The details will be described in the next subsection. In this way, the fiber-to-fiber splicing or fiber-to-waveguide coupling is not necessary.

To illustrate the working principle of the OADM an individual cell, which is enclosed by the dashed line in Figure 1.5, is described. When the multichannel beam launches to the cell through port 2 of the four-port OC, they will go to branch a of the 2×2 switch via port 3 of the OC. When the switch is in the bar state, all signals pass through the switch to branch b, and no channel will be dropped at this cell. When the switch is in the cross state, the incident beam from branch a is routed to the tunable FBG through branch d of the switch. This beam is fed to the tunable filter, and one wavelength is selected by the tunable FBG. The selected channel is reflected back to port 3 of the OC, and it then emerges at port 4 of the OC. The other channels mismatching the center wavelength of the FBG pass through the grating to branch c and they are directed by the 2×2 optical switch to the subsequent cell through branch b. The wavelength that is dropped by the previous cell is used at this cell by adding a new signal carried by the same wavelength from port 1 of the OC. These wavelengths passing through and the channel added will continue their propagation to the subsequent cells for further adding and dropping.

The first and last cells have the same structure as described previously, but there is a slight difference in the implementation. The first cell has no responsibility to add any channel. Therefore, port 1 of the OC is unused. The last four-port OC is only responsible for adding a wavelength channel, which is dropped at the previous cell. Hence, port 3 is unused. However, it provides more flexibility to cascade more cells.

Each of these cells can handle four wavelengths thanks to the large tuning range of the FBG. However, only one of them can undergo an add/drop process at a time. This type of OADM is more flexible and reconfigurable than the conventional static OADM, even

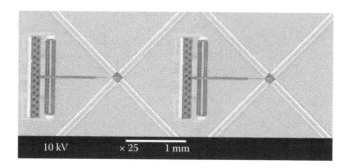

FIGURE 1.6
SEM micrograph of the integrated MEMS optical-switching OADM.

though the former is still not fully flexible. Upgrading the system is realized by cascading more cells without replacing the existing components. Another notable merit of this structure is that the MEMS switches can be fabricated on the same chip as shown in Figure 1.6, in which two cells are integrated together. Therefore, size and cost can be reduced. Because the switch is bit-rate independent, various data formats, such as gigabit Ethernet, synchronous digital hierarchy (SDH), or Internet protocol (IP) can be dropped or added by this dynamic OADM. The coupling loss between the switch and FBG is much lower compared with the Mach–Zehnder switch because the two ends of the FBG represent the two branches of the switch. The fibers are inserted into the grooves, which are fabricated by high-resolution lithography and pattern-transferring technology. The main problem is that the loss will be accumulated due to the serial structure, and it is impossible to drop or add all channels at the same time. However, high insertion loss can be compensated by employing erbium-doped fiber amplifier (EDFA).

The properties of the OADM depend on the components and their construction. The fabrication and characteristics of the MEMS optical switch and tunable FBG will be described in the next subsection.

1.3.1.2   MEMS Optical Switch and Tunable FBG

In this subsection, the fabrication of the MEMS optical switch and tunable FBG is explained, and experimental results are also discussed.

1.3.1.2.1   2 × 2 MEMS Optical Switch

An in-plane 2-D MEMS optical switch, as shown in Figure 1.7a, consists of three main parts: a comb-drive actuator, a folded suspension beam, and a close-up micromirror as shown in Figure 1.7b. The three key parts are connected by a backbone. The fiber grooves, for example, groove *b* in this figure, are fabricated for optical fiber assembly. As shown in this figure, the optical fibers *a*, *c*, and *d* are inserted and fixed in these grooves.

The comb-drive actuator is used to switch the position of the micromirror in and out of the beam path. The folded suspension beam is used to balance the electrostatic force induced by the comb drive when external voltage is applied to the comb-drive actuator. The elastic force generated by the beam will drive the switch to its original position when the voltage applied is released. The micromirror is another key component of the switch that is used to direct the incident beam. When a proper voltage is applied to the actuator, the micromirror is driven out of the beam path by the comb drive, and the incident beam will pass through from input fiber *a* to output fiber *b* directly. When the voltage is released,

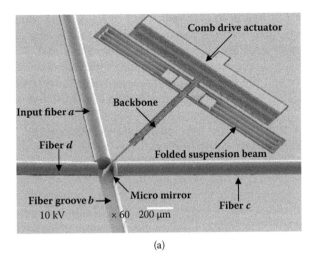

(a)

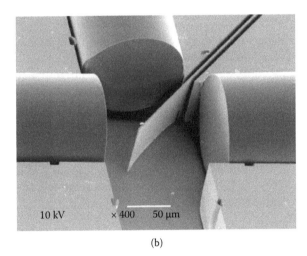

(b)

FIGURE 1.7
SEM micrograph of an in-plane 2-D 2 × 2 MEMS optical switch: (a) overview of the MEMS switch and (b) close-up of micromirror with assembled fibers.

the electrostatic force is unloaded while the elastic force of the suspension beam continues to exist and drive the micromirror to the original place, changing the switch to the cross state. The incident beams from branches *a* and *c* will be reflected by the micromirror to *d* and *b,* respectively, as a result of high reflectivity of both sides of the micromirror.

The MEMS optical switch is fabricated using deep dry etching technology. After patterning all structures by lithography, DRIE is employed to etch the switch, including the three important components mentioned earlier and the backbone, followed by the release of movable components. The electrodes (anchors) are kept unreleased. The optical-fiber grooves are etched in this process, too. The width of the grooves matches the diameter of the fiber and the depth of the grooves is 75 μm. This helps to confine the fibers in their position so that good alignment between inputs and outputs can be achieved.

The reflectivity of the mirror is a critical parameter of OADM at the fiber communication wavelength range. The effects of the type and thickness of the metal will be discussed in Chapter 2. Aluminum (Al) is sputter-deposited in argon (Ar) plasma to get better step coverage and more uniform thickness along the sidewalls. During this metal sputtering, both sides of the micromirror are coated by high-reflective metal. Therefore, the stress is balanced to prevent mirror curvature. Also, the interference problem, caused by the multireflection effect between the silicon/air and silicon/metal interfaces, which exists when the devices are fabricated by surface micromachining is eliminated with both sides of the mirror used.

The characteristics of the switch were examined experimentally. The driving voltage is 30 V, which is relatively low compared with the published result of 60 V. The power consumption is in the order of milliwatts; the current required to produce the electrostatic force is quite low. The insertion loss at the bar state and cross state is about 1.3 dB and 2.0 dB, respectively. This loss comes from free space transmission and mirror reflection. Cross talk is lower than −60 dB for each of the states. The polarization-dependent loss is less than 0.1 dB. The switching speed is greater than 0.1 ms, as shown in Figure 1.8. The ripple of the response can be reduced by increasing damping.

1.3.1.2.2   *Thermally Tuned FBG*

The tuning of the diffracted wavelength of the FBG can be implemented by temperature tuning or by applying stress to the fiber [43]. The fixed FBG is coated with a uniform high-resistance metal layer by using a modified sputter. When current is injected into the metal layer, electrical power converts to thermal power and the center wavelength shifts as a result of varied temperature.

To obtain sufficient heat efficiency, a high-resistivity metal titanium was chosen. The length of the FBG is 2.0 cm, and the thickness of the coating is 1000 Å, resulting in the resistance of 5.27 KΩ. This kind of tunable FBG makes use of the thermal effect of the grating instead of piezoelectric transducers (PZTs) tuning. It is easy to shift the center wavelength by adjusting the injected current. This method is relatively insensitive to excitation frequency, compared with piezoelectrically tuned FBGs [43].

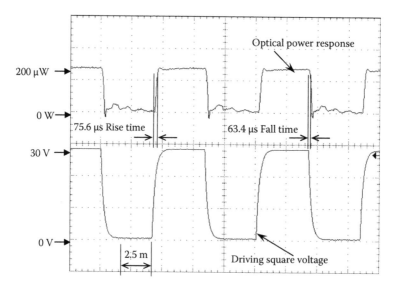

FIGURE 1.8
Dynamic response of the MEMS optical switch.

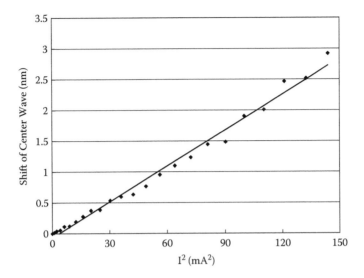

FIGURE 1.9
Center wavelength shift versus injected current square.

Figure 1.9 shows the tuning performance of the thermally tuned FBG. Obviously, the center wavelength of the FBG shifts to longer wavelength when the injected current is increased. The tuning range is 2.9 nm with a maximum current of 12 mA. The shift of the center wavelength is proportional to the square of the injected current because the wavelength shift is proportional to temperature change, which is in turn proportional to the current square. In this range, the degradation of the characteristic of the FBG is negligible, and the isolation is maintained at −30 dB. The dots represent the test result, and the solid line represents the theoretical curve. They match.

The disadvantage of this tunable FBG is the tuning speed, several hundreds of milliseconds, which varies with the tuning gap and direction of shifting. For instance, the switching time is 392 ms, including 162 ms rise time and 230 ms delay time, when the wavelength is increased by 0.12 nm with an injected current of 3.0 mA from the original center wavelength, as shown in Figure 1.10a. The returning time is 325 ms, consisting of 165 ms fall time and 160 ms delay time. When the wavelength is tuned from 1550.95 nm of 1.2 mA to 1551.38 nm of 5 mA (the response is shown in Figure 1.10b), it has a rise time of 230 ms with 500 ms delay and 137 ms fall time with 100 ms delay. The dependence of the tuning time on the shifting gap is due to the square relationship between wavelength shift and injected current. The difference between wavelength increasing time and decreasing time is due to the different heating and cooling mechanisms. This slow tuning speed could be improved using PZT actuator to realize the center wavelength shift. The setting time of less than 2.0 ms has been obtained.

1.3.1.3 Experimental Results and Discussion

The insertion loss between any two adjacent ports of the OC is 0.6 dB in this setup. Depending on the structure and properties of the components, the optical switch and OC contribute mainly to insertion loss. The insertion losses of various signals are listed in Table 1.2. The loss is about 1.9 dB for a passing-by signal and 5.2 dB for a dropped channel consisting of a 4.0 dB loss induced by double reflection and 1.2 dB from passing through

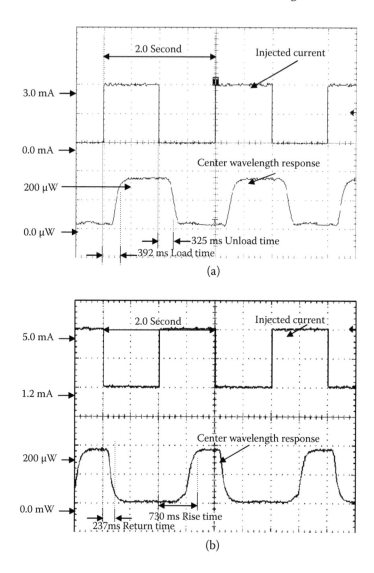

FIGURE 1.10
Dynamic response of the tunable FBG: current change from (a) 0.0 mA to 3.0 mA and (b) 1.2 mA to 5.0 mA.

TABLE 1.2

Insertion Loss for Different States of the Reconfigurable OADM

State of the Reconfigurable OADM		Insertion Loss (dB)
Pass state		1.9
Drop state		5.2
Add state	Without wavelength drop	2.5
	With drop (any wavelength of the cell)	5.2

Note: OADM = optical add/drop multiplexer.

the adjacent ports of a four-port OC. The loss from FBG can be neglected. The loss for an added channel is 2.5 dB when no wavelength is dropped at one cell or 5.2 dB when the switch is at the cross state to selectively drop one of the wavelengths. These losses come from individual components the switch and circulator. Therefore, one way to reduce the loss is to improve the performance of the optical switches using low-loss circulator components. Because the cross talk of an MEMS optical switch is less than −60 dB, the cross talk of the OADM depends on a 99.9% reflection of the FBG. Hence, the isolation of each cell is as high as 30 dB. The tuning range of each cell is 2.9 nm, which corresponds to four channels of the 100G DWDM system with a channel spacing of 0.8 nm.

The performance of one cell of this reconfigurable serial OADM is also experimentally investigated. The four adjacent input channels have wavelengths of 1552.50 nm, 1553.31 nm, 1554.13 nm, and 1554.94 nm with a spacing of 0.8 nm at 1550 nm and 0 dBm power level. For the first characterization, the micromirror is actuated so that all traffic goes through. Figure 1.11a shows the spectrum passing through without a channel drop. The loss is about 1.9 dB, comprising 0.6 dB from the OC and 1.3 dB from the switch.

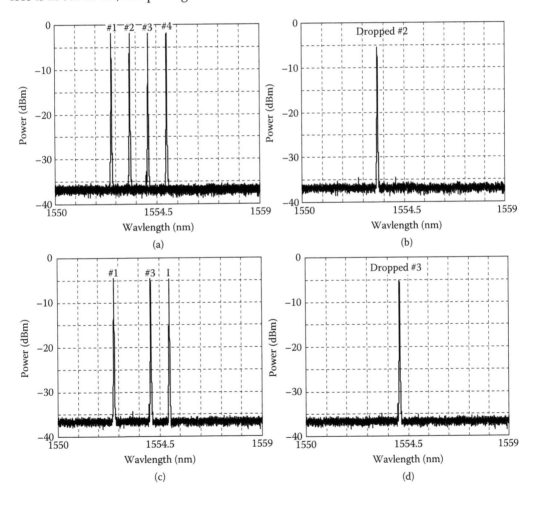

FIGURE 1.11
Experimental results on a single cell of optical add/drop multiplexer (OADM): (a) spectrum of through traffic without channel drop, (b) spectrum of dropped channel 2, (c) output traffic with channel 2 dropped, and (d) dropped channel when injected current is 9.3 mA.

The OADM is then configured to drop the second channel, which means that the switch is changed to the cross state and a current of 6.6 mA is injected to the metal-coated thermally tuned FBG. Figure 1.1b shows the dropped channel obtained from the optical spectrum analyzer. The loss of this channel is about 5.18 dB due to the loss of the switch and OC, as discussed earlier. The spectrum of the other three channels passing through is shown in Figure 1.11c, in which the loss of each channel is about 4.6 dB as a result of reflection by the micromirror of the switch two times and a 0.6 dB loss between port 2 and port 3 of the OC. The isolation is as high as 30 dB and can be improved further by increasing the reflection of the FBG. Figure 1.11d shows the dropped channel when the injected current is adjusted to 9.3 mA.

The insertion loss of the modularized serial OADM will accumulate as a result of the cascade structure. The cross talk of this type of OADM is less than −30 dB. The use of MEMS technology realizes a compact system, and the size is reduced significantly. The size of a single switch is 1.6 × 1.2 mm, and that of a two-switch chip is 4.0 × 1.6 mm. Furthermore, the tunable FBG and circulator are packaged with the same substrate of the switch chip. Similarly, the electronic control circuit can be integrated onto a single chip. The metal sputter is a simple but effective method to realize thermal tuning for the FBG. The reliability this OADM is guaranteed, and cost is reduced. The power consumption is in the order of hundreds of milliwatts, which is low compared with other macro systems.

1.3.2 Tunable Laser-Integrated OADM

A single-chip photonic subsystem with a dimension of 3.5 × 3.0 × 0.6 mm is developed by the integration of a tunable laser with an in-plane 2-D optical switch using MEMS technology. This subsystem has potential usage in niche applications such as reconfigurable OADM wavelength converters. In addition to the compact size and high tuning speed, MEMS integration also brings in other advantages, such as more functionality, high reliability, batch fabrication, and low cost. More significantly, the prototype demonstrates MEMS integration successfully. This is, to our knowledge, the first realization of a single-chip MEMS subsystem by integrating different functional MEMS components.

1.3.2.1 Configuration of Tunable Laser Integrated OADM

The configuration and application of the MEMS tunable laser-integrated OADM is illustrated in Figure 1.12.

The multiple-wavelength channels from an input fiber are first separated into different links by a demultiplexer. At switching state, as shown in Figure 1.12a, the wavelength channel to be dropped is connected to the desired destination through a 2 × 2 optical switch. At the same time, the tunable laser provides a local channel and adds it through the optical switch. The added channel and the other pass-through channels are then combined into an output fiber using a wavelength-division-multiplexing (WDM) combiner [44], which can merge several fiber links into one fiber at a very low insertion loss and wavelength-dependence loss (WDL). Compared with a common multiplexer, whose input ports work at a fixed wavelength, the combiner allows the input wavelength to change within a certain range and, therefore, can receive the added channel at a different wavelength, which is provided by the tunable laser. To avoid conflict, the added channel should not use any of the wavelengths that have already been used by the other channels. The local data signal can be added to the tunable laser by direct or external modulation. The optical switch can also be set in the idle state as shown in Figure 1.12b. It allows the incoming channel to pass through directly to the output end without adding or dropping any channel.

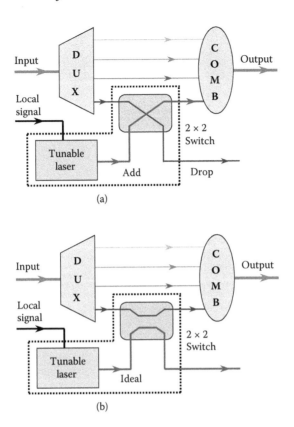

FIGURE 1.12
Configuration of the MEMS tunable laser integrated OADM: (a) the switching state and (b) the idle state. DUX and COMB represent a demultiplexer and a combiner, respectively. The dotted frame indicates the scope of the subsystem.

Compared with the fixed-wavelength lasers, the tunable laser has a particular advantage that the wavelength of the added channel can be changed intentionally. The local signal can use the wavelength of the dropped channel or any other idle wavelength even though the dropped channel may have a fixed wavelength. This property is very useful for the wavelength-based switching and routing networks. For example, in a passively routing network [45], each user is assigned a fixed wavelength for incoming signals but is able to send out one of many wavelengths, only the path and the destination being determined by the signal wavelength through the passive router. By tuning the wavelength of the added channel, one end user can deliver the signal directly to the desired recipient without any intermediate processing, switching, or reconfiguration at intermediate nodes. In addition, the tunable laser in the reconfigurable OADM makes it possible to reuse the idle wavelengths, which can save wavelength conversion significantly. It would help to cut costs because the wavelength is a rare and expensive source in optical networks. Moreover, a multichannel OADM can be implemented by using an array of these subsystems.

The MEMS system can also be used as a wavelength converter as shown in Figure 1.13. The tunable laser provides the targeted wavelengths whereas the optical switch selects the state in which the wavelength converter works, that is, an idle state or a conversion state. In the idle state, as shown in Figure 1.13b, the optical switch directly forwards the input to the output, and the tunable laser is idle. In the conversion state, as shown in Figure 1.13a,

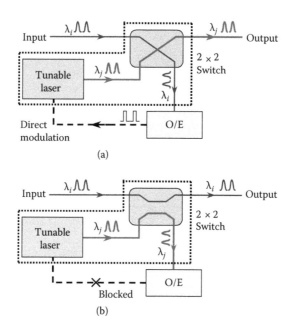

FIGURE 1.13

Application of the MEMS subsystem as a tunable wavelength converter: (a) the conversion state and (b) the idle state. O/E stands for optical-to-electrical converter. The dotted frame indicates the scope of the subsystem.

the input signal at a wavelength of λ_i is switched by the optical switch to a receiver for O/E conversion. The electrical signal is then used to directly modulate the tunable laser, whose wavelength can be adjusted to the desired wavelength of λ_j independently. It is noted that the tunable laser in the photonic subsystem is able to work at a continuous wave (CW) as well as direct modulation. In addition to the direct modulation, external modulation is also possible. The tunable laser provides CW light to act as the signal carrier while the electrical signal is applied to an external modulator to code the laser light on the output side of the optical switch. Compared with the conventional wavelength converters, which have a bulky size and fixed-output wavelength, the MEMS subsystem is compact which can be batch-fabricated, and is capable of handling wide ranges of input and output wavelengths.

1.3.2.2   Experiment Results and Discussions

An overview of a fabricated MEMS subsystem is shown in the SEM in Figure 1.14, consisting of a 2×2 optical switch, a tunable laser, and a microlens. In the MEMS subsystem, the laser works in the wavelength domain to provide the light source for different wavelength channels, whereas the optical switch works in the spatial domain to switch the laser light to different light paths. The optical switch is similar to the previous 2-D optical switch. The microlens is used to couple the laser output to the optical switch. It is designed to have a focus length of 160 μm and a magnification of 1.14. A metal layer (0.2-μm-thick gold) is coated onto the flat and curved mirrors to improve the reflectivity, whereas the microlens is protected from coating. After MEMS fabrication, the laser chip and optical fibers are integrated and packaged. The optical axis of the system is chosen at 45 μm depth from the top surface. The system has a dimension of only $3.5 \times 3.0 \times 0.6$ mm (excluding the full length of the optical fibers), which is quite small compared with the conventional devices.

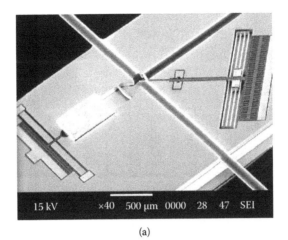

(a)

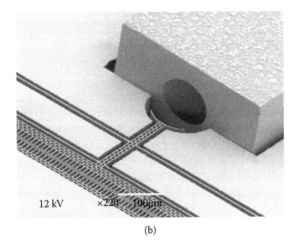

(b)

FIGURE 1.14
The on-chip MEMS subsystem: (a) overview, (b) close-up of the tunable laser, and (c) close-up of the optical switch with microlens.

The tunable laser uses a curved mirror as an external reflector to form an external cavity to the Fabry–Pérot (FP) semiconductor laser. The laser output can be tuned to different wavelengths by actuating the mirror to change cavity length. This configuration is an improved version of the simple configuration that uses a flat mirror as the reflector [46]. The position of the curved mirror is controlled by applying an electrostatic voltage to a comb-drive microactuator. When the driving voltage is greater than 2.5 V, the mirror displacement increases with higher driving voltage and reaches 4 μm at 18 V. The laser facets are not coated (i.e., $r_2 = 0.56$). The effective reflectance of the curved mirror is estimated to be $r_3 = 0.03$ with the threshold current variation measured while the curved mirror is moved. The effective reflectance takes into account the losses due to the curved mirror reflectance, the dispersion of the laser beam, the misalignment, and the coupling of the reflected light into the laser. The laser chip can respond to 2.5 GHz small-signal square-wave-driving current over the threshold, making it possible for direct modulation.

The spectra of all the output states during the wavelength tuning are shown in Figure 1.15. The injection current is kept at about 28% over the threshold current of 18.2 mA to

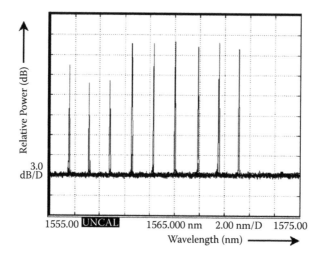

FIGURE 1.15
Spectra of all the output wavelengths of the MEMS subsystem.

assure that the laser can be operated in a continuous wave without additional heat sink. The output power is about 1 mW and varies with the wavelength tuning (variation <6 dB). The suppression ratio of the side mode is about 20 dB.

Figure 1.16 shows the change of the laser wavelength with the mirror displacement. The wavelength is initially 1570.04 nm and remains constant for very small mirror displacements. Further movement of the curved mirror makes it drop abruptly to 1556.56 nm. After this, the wavelength appears at the positions of the laser modes and increases in constant steps of 1.69 nm each. A tuning range of 13.5 nm is obtained. Further displacement of the curved mirror does not produce a higher wavelength. Instead, the wavelength

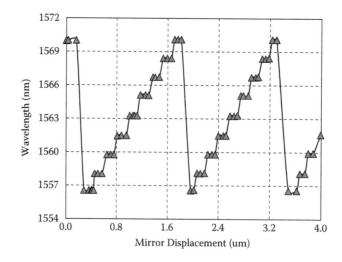

FIGURE 1.16
Wavelength variation with mirror displacement.

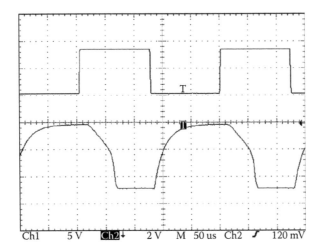

Ch1   5 V   Ch2↓   2 V   M 50 us Ch2 ∫   120 mV

FIGURE 1.17
Dynamic light path switching of the MEMS subsystem under a quasi-square-wave driving signal.

changes periodically. The mirror shift corresponding to one period is about one wavelength. The abrupt drop and stepwise increase of the wavelength are evidence of discrete wavelength tuning, which is different from those tunable lasers whose wavelengths can be tuned continuously [47,48]. The output wavelength is well locked to the positions of the FP laser modes such that the wavelength difference between the output and the nominal laser mode is less than 0.04 nm. It is observed that, at any wavelength, the output is a stable single mode when the mirror is displaced within a certain range (stable region about 0.11–0.14 μm).

The optical switch in the MEMS system gives a good performance, as discussed previously. Figure 1.17 shows the dynamic response of the optical switch under a 400 Hz quasi-square-wave driving signal. The rise time is about 84 μs, and the fall time is about 52 μs. Although the switch structure experiences some oscillation before it settles at the on or off state, this oscillation is not reflected in the optical response because the mirror displacement is large enough to keep the light path fully open or switched even during oscillation. In this way, the ripple of the light power during switching is avoided.

The polarization-dependence loss (PDL) is shown in Figure 1.18 when the optical switch is at the cross state (i.e., the mirror reflects the light). The PDL falls in a range between −0.06 dB and 0.05 dB when the polarization angle is rotated 90° using a polarization controller. At the bar state (i.e., the mirror does not intercept the light), the PDL is less than 0.1 dB. The insertion loss of the optical switch is 0.64 dB at the cross state and 0.55 dB at the bar state. The optical switch has been run for 25 million cycles without operational degradation.

The WDL of the MEMS subsystem is shown in Figure 1.19. When the optical switch is in the cross state, the laser light is reflected by the optical switch and coupled to the fiber by the microlens. The coupling ratio varies slightly with the laser wavelength. The influence of laser power variation during tuning is normalized by monitoring the power change. When the wavelength sweeps from 1556.56 to 1570.04 nm, the WDL varies from −0.05 to 0.12 dB (in total, 0.17 dB). The WDL tends to decrease with longer input wavelength.

The coupling loss from the tunable laser to the output fibers is quite large because the deeply etched microlens cannot focus light in the vertical direction, which accounts for

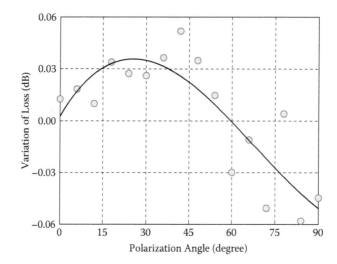

FIGURE 1.18
Polarization-dependence loss of the optical switch in the cross state. The circles are measured data and the solid line is the trend line.

about −13 dB loss. The reflection in the air-silicon surfaces of the microlens causes a loss of about −3 dB. Although the low coupling efficiency may prevent practical uses of the MEMS subsystem, it demonstrates MEMS integration. To reduce the losses to acceptable levels, hybrid lenses or tapered silicon rib waveguide can be used.

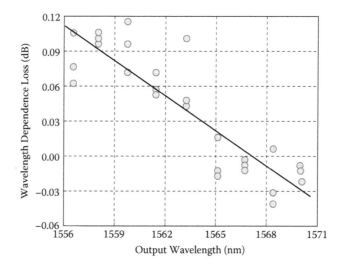

FIGURE 1.19
Wavelength-dependence loss of the MEMS subsystem. The circles are measured data and the solid line is the trend line.

1.4   Summary

With the development of optical communication, optical switches and switch matrices are gaining their significance with the rising demand for low-cost, small-footprint, and high-performances optical devices MEMS technology-based optical switch has begun to attract great interest. After comparing the various types of optical switches in this chapter, the MEMS optical switch is reviewed in detail—including 2-D and 3-D switches. Then, different configurations of switch matrices on a single cell are listed together with their characteristics. A significant benefit of MEMS technology is realized by the integration of the same type (or with other kinds) of optical components. First, the in-plane 2-D MEMS optical switch is assembled with a four-port OC and thermally tunable FBG to realize a reconfigurable OADM by cascading the cells. Then, the in-plane 2-D MEMS optical switch is integrated on a single chip with a tunable laser and a microlens using the same technology. This specific subsystem has the capability to switch, simultaneously, between different wavelength channels and different light paths, and has applications in optical networks such as reconfigurable OADMs and tunable wavelength converters. The photonic subsystems and even the whole optical communication system draw on MEMS technology and integration demonstrate the merits of compact size, low weight, batch fabrication, high mechanical reliability, and easy integration with the IC circuits.

References

1. Tomsu, P. and Schmutzer, C., *Next Generation Optical Networks: The Convergence of IP Intelligence and Optical Technologies*, Prentice Hall, Upper Saddle River, NJ, 2002.
2. Li, J., Zhang, Q. X., and Liu, A. Q., Advanced fibre optical switches using deep RIE (DRIE) fabrication, *Sens. Actuators A.*, 102, 286, 2003.
3. Lin, L. Y. and Goldstein, E. L., Opportunities and challenges for MEMS in lightwave communications, *IEEE J. Sel. Top. Quantum Electron.*, 8, 1, 163, 2002.
4. Marxer, C. and Rooij, N. F. de, Micro-optomechanical 2×2 switch for single-mode fibers based on a plasma-etched silicon mirror and electrostatic actuation, *J. Lightwave Technol.*, 17, 2, 1999.
5. Chai, T. Y., Cheng, T. H., Bose, S. K., Lu, C., and Shen, G., Array interconnection for rearrangeable 2-D MEMS optical switch, *J. Lightwave Technol.*, 21, 5, 1134, 2003.
6. Kasahara, R., Yanagisawa, M., Goh, T., Sugita, A., Himeno, A., Yasu, M., and Matsui, S., New structure of silica-based planar lightwave circuits for low-power thermooptic switch and its application to 8×8 optical matrix switch, *J. Lightwave Technol.*, 20, 6, June, 993, 2002.
7. Dainesi, P., Kung, A., Chabloz, M., Lagos, A., Fluckiger, P., Ionescu, A., Fazan, P., Declerq, M., Renaud, P., and Robert, P., CMOS compatible fully integrated Mach-Zehnder interferometer in SOI technology, *IEEE Photon. Technol. Lett.*, 12, 6, 660, 2000.
8. Espinola, R. L., Tsai, M. C., Yardley, J. T., and Osgood, R. M., Fast and low-power thermo-optic switch on thin silicon-on-insulator, *IEEE Photon. Technol. Lett.*, 15, 10, 1366, 2003.
9. Dorgeuiele, F., Noirie, L., Faure, J.-P., Ambrosy, A., Rabaron, S., Boubal, F., Schilling, M., and Artigue, C., 1.28 Tbit/s throughput 8×8 optical switch based on arrays of gain-clamped semiconductor optical amplifier gates, *Optical Fiber Communication Conference 2000*, 4, 221, 2000.
10. Riboli, F., Daldosso, N., Pucker, G., Lui, A., and Pavesi, L., Design of an integrated optical switch based on liquid crystal infiltration, *IEEE J. Quantum Electron.*, 41, 9, 1197, 2005.

11. Zhang, A., Chan, K. T., Demokan, M. S., Chan, V. W. C., Chan, P. C. H., Kwok, H. S., and Chan, A. H. P., Integrated liquid crystal optical switch based on total internal reflection, *Appl. Phys. Lett.*, 86, 211108, 2005.

12. Didosyan, Y. S., Hauser, H., Fiala, W., Nicolics, J., and Toriser, W., Latching type optical switch, *J. App. Phys.*, 91, 10, 7000, 2002.

13. R. hl, Howerton, M. M., Dubinger, J., and Greenblatt, A. S., Performance and modeling of advanced Ti: LiNbO$_3$ digital optical switches, *J. Lightwave Technol.*, 20, 1, 92, 2002.

14. Domash, L. H., Chen, Y. M., Haugsjaa, P., and Oren, M., Electronically switchable waveguide Bragg gratings for WDM routing, *Digest of the IEEE/LEOS Summer Topical Meetings*, 34, 11–15 August 1997.

15. Ivtsenkov, G., Narver, V., Magdich, L., and Solodovnikov, N., Acousto-optical switch for fiber optic lines, *US Patent 6539132*.

16. Fischer, U., Zinke, T., and Petermann, K., Integrated optical waveguide-switches in SOI, *Proc. 1995 IEEE Int. SOI Conf.*, 141, 2–5 October 1995, Tucson, Arizona.

17. Siebel, U., Hauffe, R., Bruns, J., and Petermann, K., Polymer digital optical switch with an integrated attenuator, *IEEE Photon. Technol. Lett.*, 13, 9, 957, 2001.

18. Sakamoto, T. and Kikuchi, K., Analyses of all-optically regenerated transmission system using nonlinear interferometric switches, *IEEE Photon. Technol. Lett.*, 13, 9, 1020, 2001.

19. Kang, K. I., Chang, T. G., Glesk, I., and Prucnal, P. R., Comparison of Sagnac and Mach-Zehnder ultrafast all-optical interferometric switches based on a semiconductor resonant optical nonlinearity, *Appl. Opt.*, 35, 417, 1996.

20. Li, J., Zhang, Q. X., Liu, A. Q., Goh, W. L., and Ahn, J., Technique for preventing stiction and notching effect on silicon-on-insulator microstructure, *J. Vacuum Sci. Technol.*, B, 21, 6, 2530, 2003.

21. Li, C. Y., Li, G. M., Li, V. O. K., Wai, P. K. A., Xie, H., and Yuan, X. C., Using 2 × 2 switching modules to build large 2-D MEMS optical switches, *Global Telecommunications Conference*, 5, 2798, 1–5 December 2003.

22. Agranat, A. J., Electroholographic waveguide selective crossconnect, 61–62, *Dig. LEOS Summer Topical Mtgs.*, 1999.

23. Corning® SMF-28™ optical fiber product information, Corning Inc., 1–4, April 2001.

24. Corning® OptiFocus™ collimating lensed fiber product information, Corning Inc., August 2006.

25. Yuan, S. and Riza, N. A., General formula for coupling-loss characterization of single-mode fiber collimators by use of gradient-index rod lenses, *Appl. Opt.*, 38, 15, 3214, 1999.

26. Lin, L. Y., Shen, J. L., Lee, S. S., and Wu, M. C., Realization of novel monolithic free-space optical disk pickup heads by surface micromachining, *Opt. Lett.*, 21, 2, 155, 1996.

27. Lee, S. S., Lin, L. Y., and Wu, M. C., Surface-micromachined free-space micro-optical systems containing three-dimensional micrograting, *Appl. Phys. Lett.*, 67, 15, 2135, 1995.

28. Wolffenbuttel, R. F., Microresonator array for high-resolution spectroscopy, *IEEE Trans. Instrument. Meas.*, 53, 1, 197, 2004.

29. Berger, J. D. and Anthon, D., Tunable MEMS devices for optical networks, *Opt. Photonics News*, 42, March 2003.

30. Pu, C., Lin, L. Y., Goldstein, E. L., and Tkach, R. W., Client-configurable eight-channel optical add/drop multiplexer using micromachining technology, *IEEE Photon. Technol. Lett.*, 12, 1665, 2000.

31. Tran, A. V., Zhong, W. D., Tucker, R. S., and Song, K., Reconfigurable multichannel optical add-drop multiplexers incorporating eight-port optical circulators and fiber Bragg gratings, *IEEE Photon. Technol. Lett.*, 13, 1100, 2001.

32. Okamoto, K., Okuno, M., Himeno, A., and Ohmori, Y., 16-channel optical add/drop multiplexer consisting of arrayed-waveguide gratings and double-gate switches, *Electron. Lett.*, 32, 1471, 1996.

33. Riza, N. A., Ferroelectric liquid crystal polarization switching-based high speed multi-wavelength add/drop filters using fiber and array waveguide gratings, *International Optics in Computing 98 Conference (SPIE 3490)*, 335–338, 1998.

34. Liaw, S.-K., Ho, K.-P., and Chi, S., Multichannel add/drop and cross-connect using fiber Bragg gratings and optical switches, *Electron. Lett.*, 34, 1601, 1998.
35. Chen, Y.-K. and Lee, C.-C., Fiber Bragg grating-based large nonblocking multiwavelength cross-connects, *J. Lightwave Technol.*, 16, 1746, 1998.
36. Albert, J., Bildeau, F., Johnson, D. C., Hill, K. O., Hattori, K., Ktagawa, T., Hibino, Y., and Abe, M., Low-loss planar lightwave circuit OADM with high isolation and no polarization dependence, *IEEE Photon. Technol. Lett.*, 11, 346, 1999.
37. Doerr, R., Stulz, L. W., Cappuzzo, M., Laskowski, E., Paunescu, A., Gomez, L., Gates, J. V., Shunk, S., and White, A. E., 40-wavelength add drop filter, *IEEE Photon. Technol. Lett.*, 11, 1437, 1999.
38. Riza, N. A. and Yuan, S., Reconfigurable wavelength add-drop filtering based on a banyan network topology and ferroelectric liquid crystal fiber-optic switches, *J. Lightwave Technol.*, 17, 1575, 1999.
39. Ford, J. E., Aksyuk, V. A., Bishop, D. J., and Walker, J. A., Micromechanical wavelength add/drop switching: From device to network architecture, *J. Lightwave Technol.*, 17, 904, 1999.
40. Monteverde, R. J., MEMS-based light valves manage wavelengths for DWDM optical networks, http://www.eetimes.com/story/OEG20020606S0079.
41 Riza, N. A. and Polla, D. L., Micro-dynamical fiber-optic switch, U.S. Patent No. 5,208,880, May 4, 1993.
42. Marxer, C. and Rooij, N. F. de, Micro-opto-mechanical 2×2 switch for single-mode fibers based on plasma-etched silicon mirror and electrostatic actuation, *J. Lightwave Technol.*, 17, 2, 1999.
43. A., Limberger, H. G., and Salathe, R. P., Bragg grating fast tunable filter, *Electron. Lett.*, 33, 2147, 1997.
44. Datasheet of WDM combiner DW603 series, http://www.furukawaamerica.com.
45. Zouganeli, E., Tunable lasers route optical signals, *WDM Solutions*, 2, 3, 26 2000.
46. Liu, A. Q., Zhang, X. M., Murukeshan, V. M., and Lam, Y. L., A novel integrated micromachined tunable laser using polysilicon 3-D mirror, *IEEE Photon. Technol. Lett.*, 13, 5, 427, 2001.
47. Liu, K. and Littman, M. G., Novel geometry for single-mode scanning of tunable lasers, *Opt. Lett.*, 6, 3, 117, 1981.
48. Syms, R. A. and Lohmann, A., Tuning mechanism for a MEMS external cavity laser, *Proc. Optical MEMS*, 183, 2002.

2

Design of MEMS Optical Switches

Jing Li and Ai-Qun Liu

CONTENTS

2.1 Introduction 28
2.2 Optical Design 29
    2.2.1 Insertion Loss 29
        2.2.1.1 Fiber Coupling Loss 29
        2.2.1.2 Fresnel Loss 32
    2.2.2 Optical Loss Related to Micromirror 33
        2.2.2.1 Surface Roughness 33
        2.2.2.2 Surface Material 36
    2.2.3 PDL of Metal Coating..41
2.3 Electromechanical Design 47
    2.3.1 Lateral-Mode of the Comb-Drive Actuator.................................47
    2.3.2 Parallel Plate Capacitor 47
    2.3.3 Comb-Drive Actuator..49
        2.3.3.1 Capacitor Analysis..49
        2.3.3.2 Lateral Electrostatic Force 50
    2.3.4 Stability of the Comb Drive..51
    2.3.5 Displacement of Actuator 53
2.4 Fabrication Tolerance Effect..55
    2.4.1 Effects of Etch Profiles...55
        2.4.1.1 Profile of Comb Fingers 55
        2.4.1.2 Profile of Folded Beam 56
        2.4.1.3 Profiles of Comb Fingers and Folded Beam 58
    2.4.2 Native Oxidation on Comb Fingers 60
    2.4.3 Tapered Microstructures and Oxidation.......................................60
    2.4.4 Uneven Depth Comb Fingers...61
    2.4.5 Undercut...64
    2.4.6 Nonuniform Gap..67
    2.4.7 Overall Effects of Various Tolerances 69
2.5 Large-Displacement Actuator 70
    2.5.1 Design and Theoretical Analysis 71
        2.5.1.1 Configuration of a Stable and Large Displacement Actuator 71
        2.5.1.2 Theoretical Analysis of Linear Displacement Amplification.....72
        2.5.1.3 Self-Latching by Bifurcation Effect 79
    2.5.2 Experimental Results and Discussions 83
2.6 Summary 85
References 86

As a result of more stringent bandwidth requirements, MEMS optical-fiber switches are gaining in significance in optical communication. It is thus necessary to realize a high-performance switch. At the same time, the design of these MEMS devices poses many challenges as a number of properties need to be considered and traded off against one another. This chapter discusses the design issues in relation to two major characteristics of the optical switch: optical performance and electromechanical properties.

For optical design, the insertion loss induced by fiber misalignment; Fresnel reflection; and micromirror-related issues, including coating material, thickness, and roughness of the surface, are investigated. The effect of the mirror on different wavelengths will then be presented. Polarization-dependent loss (PDL) is also be discussed in detail.

For electromechanical design, the actuation of the optical switch is elaborated. After a brief comparison among different actuating mechanisms, the electrostatic comb-drive actuator is discussed from the viewpoints of force, stiffness, displacement, and stability. Besides, an actuator with large displacement amplification is introduced to produce a latched output.

2.1 Introduction

As discussed in Chapter 1, a MEMS optical switch consisting of an actuator, an optical guiding element, and an optical mirror, is shown schematically in Figure 2.1 [1]. The detailed design and analysis of the optical and electromechanical properties of this MEMS optical switch are discussed in this chapter.

The optical signal transmits from one of the input fibers to one of the outputs through free space. Also, light reflected by the optical mirror is affected due to imperfect reflection and the interfaces between two different materials. Therefore, the optical design of the switch is of importance to optical insertion loss and different insertion loss for different polarization states, which is called PDL. All of these losses are discussed in this chapter.

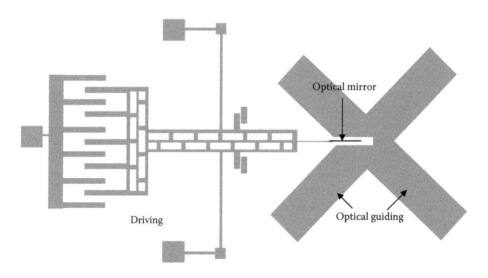

FIGURE 2.1
Schematic of 2-D MEMS optical switch.

TABLE 2.1

Comparison of Various Types of Actuators

Type of Actuation	Force	Speed	Range	Power Source	Compatibility	Repeatability	Function
Piezoelectric	High	High	Low	Voltage	Low	High	AC
Magnetic	High	Low	High	Amp	Low	High	AC
Thermal	High	Low	High	Amp	High	Low	Assembly
Surface tension	High	Low	High	Temp	High	Low	Assembly
Electrostatic	Low	High	High/low	Voltage	High	High	AC

The actuator is used to determine the translation or rotation of the micromirrors so that the incident optical signals can be directed to the expected outputs. There are numerous kinds of microactuators, which have been extensively analyzed and widely used, such as comb drives, thermal actuators [2–4], magnetic actuators [5,6], and scratch drive actuators (SDA) [7,8]. Compared to comb drives, thermal actuators are too sensitive to ambient temperature. Magnetic and SDA actuators are much more complicated in terms of fabrication as they require special materials that are not compatible with CMOS fabrication. For this reason, the comb-drive actuator is the most attractive with broad applications in MEMS systems [1,9–14], such as optical switch, inertial gyroscope, accelerometer, RF, etc. The advantages and disadvantages of the various actuators are compared in Table 2.1.

2.2 Optical Design

The optical properties are discussed in this section. First, we present a detailed analysis of the optical power loss including its sources, quantitative calculations, and approaches to control. Then, the cross talk of the alternative output channels is discussed. Another important specification of PDL is analyzed in relation to the various materials that can be used for the surface coating of the mirror. These are the guidelines for the design and fabrication of a high-performance optical switch.

2.2.1 Insertion Loss

The optical beam is always transmitted along the path shown in Figure 2.2 as long as a micromirror is used, whether it is an in-plane, an up-down, or a rotational MEMS optical switch. The signal comes from the input fiber to the micromirror and goes out to either output fiber 1 or output fiber 2, depending on the position of the micromirror. In the bar state, output channel 1 is selected without changing the beam direction. Alternatively, to redirect the input signal to channel 2, the mirror redirects the path of the beam, at the same time, inducing an extra loss. Therefore, the beam power distribution, mirror surface roughness, mirror verticality, and the properties of the mirror material are the main causes of optical power loss in the system.

2.2.1.1 *Fiber Coupling Loss*

The fundamental mode field distribution of single-mode fiber (SMF) can be well approximated by a Gaussian function. The following empirical expression describes the waist w_0

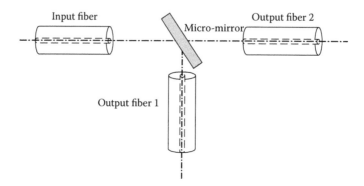

FIGURE 2.2
Schematic of the beam path of an optical switch.

of the Gaussian beam [15]:

$$w_0 = \left(0.65 + \frac{1.619}{V^{3/2}} + \frac{2.879}{V^6} \right) \cdot a \qquad (2.1)$$

where w_0 is the waist of the Gaussian beam, a is the core radius, and V is the waveguide parameter given in [15]. The beam will diverge when it leaves the fiber enter into a free space because there is no total reflection. The beam profile is shown in Figure 2.3.

The Gaussian beam coming from the fiber propagates in free space and its beam size is given as follows [16]:

$$w = w_0 \left[1 + \left(\frac{\lambda z}{\pi w_0^2} \right)^2 \right]^{\frac{1}{2}} \qquad (2.2)$$

For $w_0 = 5.2\,\mu m$, the beam size in free space is shown in Figure 2.4. The beam waist increases to $w_1 = 7.88\,\mu m$ and $w_2 = 12.95\,\mu m$ after propagates for of 62.5 and 125 μm, respectively.

When two SMFs are coupled with each other, three different kinds of misalignment, namely, longitude, lateral, and angular, can appear based on the relative position of the two fibers, as shown in Figure 2.5. Power is lost because of the divergence of the beam.

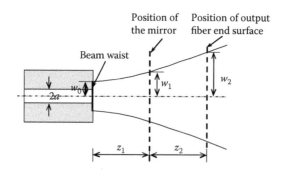

FIGURE 2.3
The divergence of the Gaussian beam.

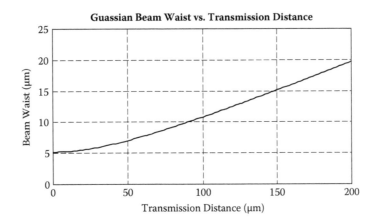

FIGURE 2.4
Gaussian beam waist versus transmission distance.

Nemoto and Makimoto developed the Gaussian approximation analytical model to describe insertion losses in fiber splices, coupling loss due to misalignments, as well as the difference between the mode field radii of the two fibers [17]. The total insertion loss can be calculated from:

$$L = -10 \log \left\{ 4\frac{D}{B} \exp(-A \cdot C / B) \right\} dB \qquad (2.3)$$

where

$$A = (k \cdot w_T)^2/2, \quad k = 2\pi n_0/\lambda \qquad (2.3a)$$

$$B = G^2 + (D+1)^2 \qquad (2.3b)$$

$$C = (D+1) \cdot F^2 + 2DFG \cdot \text{Sin}(\Delta\theta) + D(G^2 + D + 1)\text{Sin}^2(\Delta\theta) \qquad (2.3c)$$

$$D = (w_R / w_T)^2 \qquad (2.3d)$$

$$F = 2 \cdot \frac{\Delta x}{k \cdot w_T^2} \qquad (2.3e)$$

$$G = 2 \cdot \frac{\Delta z}{k \cdot w_T^2} \qquad (2.3f)$$

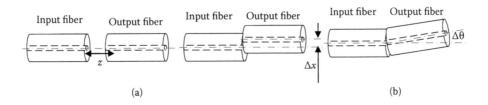

FIGURE 2.5
Misalignment between the input and output fibers: (a) longitudinal, (b) lateral, and (c) angular misalignment.

and w_T and w_R represent the Gaussian mode field radii of the transmitting and receiving fibers, respectively. To analyze the different effects on the losses, these three kinds of misalignment are treated separately.

Longitude loss: Let $\Delta x = 0$ and $\Delta\theta = 0$, so that, the relationship between the insertion loss and the distance between the ends of the two fibers can be written in the form of:

$$L = -10\log\left\{4 \cdot \frac{1}{M^2 + 4}\right\}$$

(2.3g)

where $M = \frac{\lambda z}{\pi n w_0^2}$; and n is the refractive index of the media, that is, 1.0 in a free space; and $w_T = w_R = w_0$. The relation is shown in Figures 2.6a,b. Obviously, the separation between the two fibers introduces a great loss, which increases very fast with the distance. It is impossible to realize a large matrix with normal SMF and free-space transmission. To ensure that the loss is less than 1.0 dB, the space between the two adjacent fibers must be less than 125 μm. When this requirement is not met, the design must be modified to solve this problem.

Lateral misalignment loss: Let $\Delta z = 0$ and $\Delta\theta = 0$, so that the relationship between the insertion loss and the lateral distance between the ends of the two fibers has the form of:

$$L = -10\log\left\{\exp\left(-\frac{\Delta x^2}{w_0^2}\right)\right\}$$

(2.3h)

where Δx is the lateral misalignment distance, and $w_T = w_R = w_0$. The relation between power loss and lateral misalignment is shown in Figure 2.7. The loss is not infinite, although the lateral displacement is larger than the radius of the fiber because the beam waist is greater than the radius of the fiber.

Angular misalignment loss: Let $\Delta z = 0$ and $\Delta\theta = 0$, so that the relationship between the insertion loss and the angular misalignment between the ends of the two fibers has the form of:

$$L = -10\log\left\{\exp\left(-\frac{\pi n_0 w_0 \sin \Delta\theta}{\lambda}\right)^2\right\}$$

(2.3i)

where $\Delta\theta$ is the misalignment angle, and $w_T = w_R = w_0$. Figure 2.8 shows the relationship between the fiber-coupling loss and the angular misalignment.

2.2.1.2   Fresnel Loss

Back reflection (also known as Fresnel Loss) is another intrinsic loss noted when the two fibers are coupled. The structure fiber–air–fiber produces this loss. Fresnel loss due to this double interface can be calculated as:

$$L_{BR} = -10\log\left\{1 - \frac{(n_F - n_0)^2}{(n_F + n_0)^2}\right\} dB$$

(2.4)

where n_F and n_0 are the refractive indexes of the fiber core and the air, which are 1.4638 and 1, respectively. Therefore, the back-reflection loss is equal to 0.1567 dB per interface.

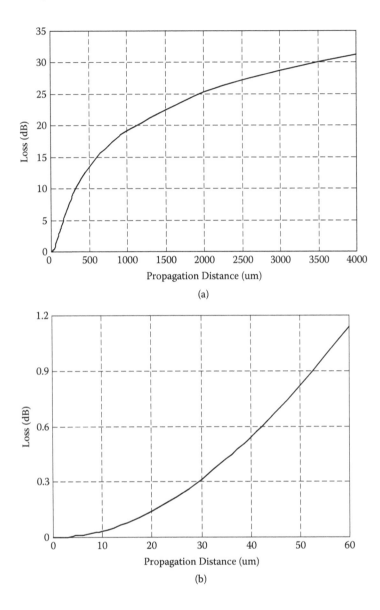

FIGURE 2.6
Coupling loss as a function of the transmission distance over a range of (a) 0 to 4000 μm and (b) 0 to 60 μm.

2.2.2 Optical Loss Related to Micromirror

For the in-plane optical switch structure, the micromirror is the directing beam component. Therefore, not only by the quality of the coupling between the input and output fibers, but the final performance of the switch and the optical system is determined also the optical properties of the mirrors.

2.2.2.1 Surface Roughness

Scattering loss at the mirror surface is related to surface roughness. The total integrated scatter is used to measure the fractional scattered power from an ideal smooth, clean,

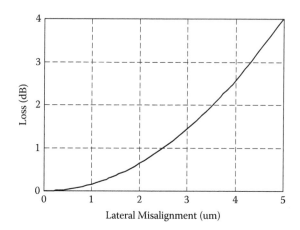

FIGURE 2.7
Coupling loss versus lateral misalignment.

conducting surface. The scattering power due to surface roughness is expressed as [18]:

$$\eta = 1 - \exp\left[-\left(\frac{4\pi\sigma\cos\theta_i}{\lambda}\right)^2\right] \qquad (2.5)$$

where η is the percentage of scattering loss, σ the root-mean-square (RMS) roughness of the mirror surface, θ_i the incident angle, and λ the light wavelength.

The scattering power percentage due to the mirror surface roughness at different wavelengths of the S, C, and L bands is shown in Figure 2.9. To obtain a scattering loss of less than 10%, the roughness of the mirror should be less than 57 nm for 1550 nm wavelength. However, the scattering power is only slightly dependent on the wavelength of the beams. For instance, it is 1.4% for 1520 nm, and 1.2% for 1620 nm when the incident angle is 45° and the RMS roughness is 20 nm. Thus, one of the merits of this in-plane switch is that the loss is wavelength independent. However, the wavelength-dependent

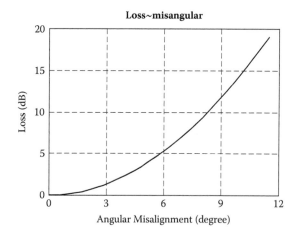

FIGURE 2.8
Coupling loss versus misalignment angle.

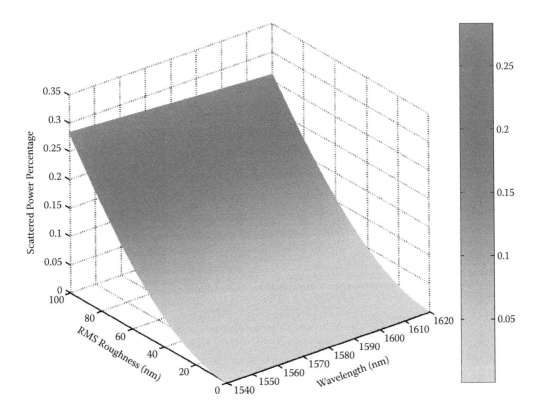

FIGURE 2.9
Scattering power percentage due to root mean square (RMS) roughness.

loss is dependent on the roughness of the mirror surface. For example, the scattered power will be 5.3% for a wavelength of 1520 nm and 4.7% for that of 1620 nm when the RMS roughness is 40 nm, as compared to 1.4% and 1.2%, respectively, when the RMS roughness is 20 nm.

Another variable that affects the optical properties of the optical MEMS device is the incident angle. Although the designed value is 45°, any fabrication error will induce some extra loss. The effect is illustrated by Figure 2.10, assuming that the RMS roughness is 20 nm and the beam to be switched has a wavelength of 1550 nm. It indicates that the smaller the incident angle, the higher the scattered power. Even though the fiber-to-output beam coupling is perfect, 1.3% power is scattered when the incident angle is exactly 45°, but it increases to 1.5% and decreases to 1.1% when the incident angle is 40° and 50°, respectively.

Hence, the loss in dB, induced by the micromirror is expressed as:

$$L_{rough} = -10\log\left\{\exp\left[-\left(\frac{4\pi\sigma\cos\theta_i}{\lambda}\right)^2\right]\right\} \text{ dB} \tag{2.6}$$

The wavelength λ is determined by the fiber-optic communication system, and the incident angle θ_i is designed to be 45°. The surface roughness is a key issue to be considered during the fabrication process for a low-loss MEMS optical switch.

FIGURE 2.10
Scattered power percentage difference due to the different incident angle.

2.2.2.2 Surface Material

The reflection of the mirror is another important factor that strongly influences the loss of the MEMS optical switch. If the mirror is made of single-crystal silicon (SCS), the reflection is defined by the Fresnel equation as:

$$R = \left(\frac{n_{Si} - n_0}{n_{Si} + n_0} \right)^2 \tag{2.7}$$

where n_{Si} and n_0 are the refractive indexes of the SCS and the air, which is 3.5 and 1.0, respectively. Therefore, the reflection is only 30.9%, a value that is too low because the loss is 1.6 dB even if the surface of the mirror is perfectly smooth (i.e., the RMS roughness is 0).

Some effective approaches to increase the reflection of the mirror are multilayer dielectric films of alternative high-low refractive index coating or a proper metal coating on the mirror surface. However, dielectric materials is not compatible with the CMOS process. Therefore, proper metal coating is generally used to obtain the desired high reflectivity, which also depends on the type of materials and the thickness of the metal.

2.2.2.2.1 Minimum Thickness of Metal for Highest Reflectivity

The space and time derivatives of the electromagnetic wave are related by Maxwell's equations, which hold at every point in a neighborhood where the physical properties of the medium are continuous:

$$\nabla \times \vec{H} - \frac{1}{c} \frac{\partial \vec{D}}{\partial t} = \frac{4\pi}{c} \vec{J} \tag{2.8a}$$

$$\nabla \times \vec{E} + \frac{1}{c} \frac{\partial \vec{B}}{\partial t} = 0 \tag{2.8b}$$

$$\nabla \cdot \vec{D} = 4\pi\rho \tag{2.8c}$$

$$\nabla \cdot \vec{B} = 0 \tag{2.8d}$$

Four basic quantities of the electric field $\vec{E}$ (i.e., displacement $\vec{D}$, magnetic induction [magnetic flux density] $\vec{B}$, the magnetic field $\vec{H}$, and the current density $\vec{J}$) are connected by the preceding equations. The variable ρ is the density of electric charge. To allow a unique determination of the field vectors from a given distribution of currents and charges, these equations must be supplemented by the relations that describe the behavior of substances under the influence of the electromagnetic field. These relations are known as material equations or constitutive relations, and are listed as:

$$\vec{D} = \varepsilon_0 \vec{E} + \vec{P} \tag{2.9a}$$

$$\vec{H} = \vec{B} / \mu_0 - \vec{M} \tag{2.9b}$$

$$\vec{J} = \sigma \vec{E} \tag{2.9c}$$

where ε_0 is the permittivity of free space, $\vec{p}$ the polarization, μ_0 the permeability of free space, $\vec{M}$ the magnetization, and σ the specific conductivity. Although an ordinary piece of metal is a crystalline aggregate, it consists of small crystals of random orientation. Therefore, it can be regarded as a homogeneous isotropic medium with dielectric constant ε, permeability μ, and conductivity σ. The material equations become:

$$\vec{D} = \varepsilon \vec{E} \tag{2.10a}$$

$$\vec{B} = \mu \vec{H} \tag{2.10b}$$

$$\vec{J} = \sigma \vec{E} \tag{2.10c}$$

Diverging from Equation 2.8a, we obtain:

$$-\frac{\varepsilon}{c}\left(\nabla \cdot \frac{\partial \vec{E}}{\partial t}\right) = \frac{4\pi\sigma}{c}\frac{4\pi}{\varepsilon}\rho$$

Also, differentiation of Equation 2.8c with respect to time gives:

$$\nabla \cdot \frac{\partial \vec{E}}{\partial t} = \frac{4\pi}{\varepsilon}\frac{\partial\rho}{\partial t}$$

Therefore, we obtain:

$$\frac{\partial\rho}{\partial t} + \frac{4\pi\sigma}{\varepsilon}\rho = 0 \tag{2.11}$$

Integrating on time t, we have:

$$\rho = \rho_0 e^{-t/\tau}, \text{ where } \tau = \frac{\varepsilon}{4\pi\sigma} \tag{2.12}$$

Obviously, the electric density ρ falls off exponentially with time t. For a medium with appreciable conductivity, the relaxation time τ is extremely small. For metals, it is on the order of 10^{-18} s, which is three orders less than the periodic time of the infrared light of 1.55 µm. Hence, $\rho \approx 0$ and $\nabla \cdot \vec{D} = 0$. Equation 2.8b becomes:

$$\nabla^2 \vec{E} = \frac{\mu\varepsilon}{c^2}\frac{\partial^2 \vec{E}}{\partial t^2} + \frac{4\pi\mu\sigma}{c^2}\frac{\partial E}{\partial t} \tag{2.13}$$

If $\vec{E}$ and $\vec{H}$ are in the form of $\vec{E} = \vec{E}_0 e^{-i\omega t}$ and $\vec{H} = \vec{H}_0 e^{-i\omega t}$, that is, if the field is strictly monochromatic with angular frequency ω, Maxwell's equations can be rewritten as:

$$\nabla \times \vec{H} + \frac{i\omega}{c}\left(\varepsilon + i\frac{4\pi\sigma}{\omega}\right)\vec{E} = 0 \qquad (2.14a)$$

$$\nabla \times \vec{E} - \frac{i\omega\mu}{c}\frac{\partial \vec{H}}{\partial t} = 0 \qquad (2.14b)$$

and Equation 2.13 becomes:

$$\nabla^2 \vec{E} + \hat{k}^2 \vec{E} = 0 \qquad (2.15)$$

where

$$\hat{k}^2 = \frac{\omega^2 \mu}{c^2}\left(\varepsilon + i\frac{4\pi\sigma}{\omega}\right) = \frac{\omega^2 \mu}{c^2}\hat{\varepsilon} \qquad (2.16)$$

Then, the complex dielectric constant $\hat{\varepsilon}$ is defined as:

$$\hat{\varepsilon} = \varepsilon + i\frac{4\pi\sigma}{\omega} \qquad (2.17)$$

Hence,

$$\hat{v} = \frac{c}{\sqrt{\mu\hat{\varepsilon}}}, \quad \hat{n} = \frac{c}{\hat{v}} = \sqrt{\mu\hat{\varepsilon}} = \frac{c}{\omega}\hat{k} \qquad (2.18)$$

Let

$$\hat{n} = n\left(1 + i\kappa\right) \qquad (2.19)$$

where n and κ are real, and κ is called the *attenuation index*. Squaring Equation 2.19, gives:

$$\hat{n}^2 = n^2(1 - \kappa^2 + 2i\kappa) = \mu\hat{\varepsilon} = \mu\left(\varepsilon + i\frac{4\pi\sigma}{\omega}\right) \qquad (2.20)$$

Equating the real and imaginary parts, the following relations are obtained:

$$n^2(1 - \kappa^2) = \mu\varepsilon \qquad (2.20a)$$

$$n^2\kappa = \frac{2\pi\mu\sigma}{\omega} = \frac{\mu\sigma}{v} \qquad (2.20b)$$

From these equations it follows that:

$$n = sqrt\left[\frac{1}{2}\left(\sqrt{\mu^2\varepsilon^2 + \frac{4\mu^2\sigma^2}{v^2}} + \mu\varepsilon\right)\right] \qquad (2.21a)$$

$$n\kappa = sqrt\left[\frac{1}{2}\left(\sqrt{\mu^2\varepsilon^2 + \frac{4\mu^2\sigma^2}{v^2}} - \mu\varepsilon\right)\right] \qquad (2.21b)$$

Therefore, the electrical field can be expressed as

$$\vec{E} = \vec{E}_0 \exp[i(k\hat{r} \cdot \vec{s} - \omega t)] = \vec{E}_0 \exp\left(-\frac{\omega}{c} n\kappa \vec{r} \cdot \vec{s}\right) \exp\left[i\omega\left(\frac{n}{c}\vec{r} \cdot \vec{s} - t\right)\right] \tag{2.22}$$

This shows that the plane wave with a wavelength of $\lambda = 2\pi c/\omega n$ is attenuated exponentially. The energy density w is proportional to the time average of E^2, so w decreases in accordance with the relation:

$$w = w_0 \exp(-\chi\vec{r} \cdot \vec{s}) \tag{2.23}$$

where the absorption coefficient is given by

$$\chi = \frac{2\omega}{c} n\kappa = \frac{4\pi\nu}{c} n\kappa = \frac{4\pi}{\lambda_0} n\kappa = \frac{4\pi}{\lambda} \kappa \tag{2.24}$$

where λ_0 denotes the wavelength in vacuum and λ the wavelength in medium. The skin depth d, defined as the distance where the energy density falls to $1/e$ of its value, is

$$d = \frac{1}{\chi} = \frac{\lambda_0}{4\pi n\kappa} = \frac{\lambda}{4\pi\kappa} \tag{2.25}$$

Equation 2.21 shows that $n = \sqrt{\mu\varepsilon}$ and $\kappa = 0$ when $\sigma = 0$, or the medium is formed of dielectric materials. For metals, $\sigma \neq 0$, and it is in fact so large that ε may be neglected in comparison to $2\sigma/\mu$. Equation 2.21a and 2.21b can be rewritten as:

$$n \approx n\kappa \approx \sqrt{\frac{\mu\sigma}{\nu}} \tag{2.26}$$

The skin depth becomes

$$d = \frac{\lambda_0}{4\pi n\kappa} = \frac{\lambda_0}{4\pi}\sqrt{\frac{\nu}{\mu\sigma}} = \frac{1}{4\pi}\sqrt{\frac{c\lambda_0}{\mu\sigma}} \tag{2.27}$$

The skin depths of various metals are listed in Table 2.2.

The power transmitted in percentage is $P = \exp(-\frac{s}{d})$, where s is the transmission distance. Hence, the reflection percentage is expressed as:

$$R = 1 - P = 1 - \exp\left(-\frac{s}{d}\right) \tag{2.28}$$

TABLE 2.2

Skin Depth of Selected Metals

Metal	σ (*$10^7(\Omega m)^{-1}$)	d (nm)
Al	3.771	5.891
Au	4.521	5.380
Ni	0.1431	30.242
Pt	0.7741	13.002

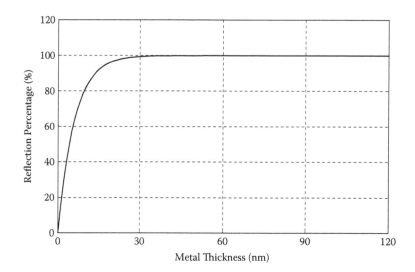

FIGURE 2.11
Relative reflection in percentage versus thickness of aluminum coating.

Considering an ideal mirror surface, the reflection in terms of the thickness of the metal layer is illustrated in Figure 2.11. Obviously, a specific thickness is required to realize the maximum reflection.

The reflection depends not only on the thickness of the coating but also on the coating metal. Figure 2.12 shows the reflectivity function for different coating metals with various thicknesses at normal incidence. An aluminum-coated mirror reaches its maximum reflectivity of 97% when the film is thicker than 40 nm. Gold also provides high reflectivity of 97.5% with a thickness of more than 60 nm. Both metals can yield better reflection than nickel and chrome, which has reflectivity of 72 and 63%, respectively.

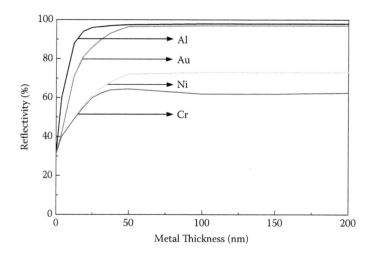

FIGURE 2.12
Reflectivity of metal-coated silicon surface as a function of metal thickness.

2.2.2.2.2 Reflectivity of Various Wavelengths

Reflectivity also varies with different signal wavelengths. This relation can be written as:

$$R = \frac{2\frac{\sigma}{v} + 1 - 2\sqrt{\frac{\sigma}{v}}}{2\frac{\sigma}{v} + 1 + 2\sqrt{\frac{\sigma}{v}}} \tag{2.29}$$

The value of σ/v is sufficiently small when the light is in the infrared band that 1 may be neglected in comparison with the other terms, and we obtain:

$$R = 1 - 2\sqrt{\frac{v}{\sigma}} + \cdots \tag{2.30}$$

Therefore,

$$\frac{1 - R_1}{1 - R_2} = \frac{2\sqrt{v_1/\sigma}}{2\sqrt{v_2/\sigma}} = \sqrt{\frac{v_1}{v_2}} = \sqrt{\frac{\lambda_2}{\lambda_1}} \tag{2.31}$$

The reflection of the metals at the specific wavelength of 1.55 μm could be attained when those at other wavelengths are known. Table 2.3 lists the highest reflection of four different metals at this wavelength.

2.2.2.2.3 Reflectivity of Various Metals and Wavelengths

Combining Equation 2.28 and the highest reflection, the minimum thickness and the highest possible reflection could be obtained as shown in Figure 2.13. It indicates that the highest reflectivity is different when different metals are deposited and the minimum thickness to realize it varies from one metal to another. Table 2.4 provides a comparison. Among the four metals shown aluminum and gold are the best two because they provide the highest reflectivity with relatively a thin metal layer.

2.2.3 PDL of Metal Coating

PDL is another important specification for a high-quality mirror of an optical switch. It depends on the mirror materials.

When an electric field is incident from medium 1 to medium 2 at an angle of θ_i, as shown in Figure 2.14, it will be reflected and transmitted along a specific direction. Let A be the amplitude of the electric vector of the incident field, where it is a complex number, with the phase

TABLE 2.3

Highest Reflection of Some Metals at 1.55 μm

Metal	R_1 ($\lambda_1 = 5893$Å)	R_2 ($\lambda_2 = 15500$Å)
Al	0.83	0.90
Au	0.82	0.89
Ni	0.66	0.79
Pt	0.59	0.75

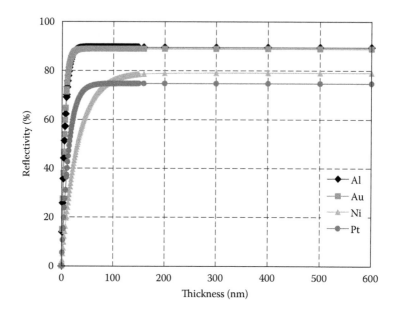

FIGURE 2.13
Reflectivity of various metals with different thickness.

TABLE 2.4

Minimum Thickness for Highest Reflection of Some Metals

Metal	R_2 $(\lambda_2 = 15500\text{Å})$	Minimum Thickness (nm)
Al	0.90	30
Au	0.89	30
Ni	0.79	200
Pt	0.75	100

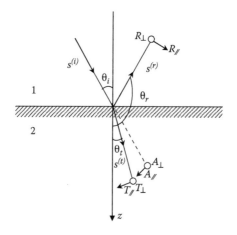

FIGURE 2.14
Reflection and refraction of a plane wave.

equal to the constant part of the argument of the wave function and the variable part being:

$$\tau_i = \omega\left(t - \frac{\vec{r} \cdot \vec{s}^{(i)}}{v_1}\right) = \omega\left(t - \frac{x\sin\theta_i + z\cos\theta_i}{v_1}\right) \tag{2.32}$$

The vector is decomposed into two components parallel (denoted by subscript //) and perpendicular (denoted by subscript ⊥) to the plane of incidence as shown in Figure 2.14. The sign of the parallel components, too, is indicated in this figure, in which the perpendicular components must be visualized at right angles to the plane of the figure. Therefore, the electric vector components are:

$$E_x^{(i)} = -A_{//}\cos\theta_i e^{-i\tau_i}, \quad E_y^{(i)} = -A_\perp e^{-i\tau_i}, \quad E_z^{(i)} = A_{//}\sin\theta_i e^{-i\tau_i} \tag{2.33}$$

and, with the assumption of $\mu = 1$, the components of the magnetic vector are:

$$H_x^{(i)} = -A_\perp \cos\theta_i \sqrt{\varepsilon_1} e^{-i\tau_i}, \quad H_y^{(i)} = A_{//}\sqrt{\varepsilon_1} e^{-i\tau_i}, \quad H_z^{(i)} = A_\perp \sin\theta_i \sqrt{\varepsilon_1} e^{-i\tau_i} \tag{2.34}$$

For the transmitted field, these components are:

$$E_x^{(t)} = -T_{//}\cos\theta_t e^{-i\tau_t}, \quad E_y^{(t)} = T_\perp e^{-i\tau_t}, \quad E_z^{(t)} = T_{//}\sin\theta_t e^{-i\tau_t} \tag{2.35}$$

$$H_x^{(t)} = -T_\perp \cos\theta_t \sqrt{\varepsilon_2} e^{-i\tau_{ti}}, \quad H_y^{(t)} = T_{//}\sqrt{\varepsilon_2} e^{-i\tau_t}, \quad H_z^{(t)} = T_\perp \sin\theta_t \sqrt{\varepsilon_2} e^{-i\tau_t} \tag{2.36}$$

with

$$\tau_t = \omega\left(t - \frac{\vec{r} \cdot \vec{s}^{(t)}}{v_2}\right) = \omega\left(t - \frac{x\sin\theta_t + z\cos\theta_t}{v_2}\right) \tag{2.37}$$

For the reflected field:

$$E_x^{(r)} = -R_{//}\cos\theta_r e^{-i\tau_r}, \quad E_y^{(r)} = R_\perp e^{-i\tau_r}, \quad E_z^{(r)} = R_{//}\sin\theta_r e^{-i\tau_r} \tag{2.38}$$

$$H_x^{(t)} = -R_\perp \cos\theta_r \sqrt{\varepsilon_1} e^{-i\tau_r}, \quad H_y^{(r)} = R_{//}\sqrt{\varepsilon_1} e^{-i\tau_r}, \quad H_z^{(r)} = R_\perp \sin\theta_r \sqrt{\varepsilon_1} e^{-i\tau_r} \tag{2.39}$$

with

$$\tau_r = \omega\left(t - \frac{\vec{r} \cdot \vec{s}^{(r)}}{v_1}\right) = \omega\left(t - \frac{x\sin\theta_r + z\cos\theta_r}{v_1}\right) \tag{2.40}$$

Because of the boundary conditions, the tangential components of E and H should be continuous across the boundary. Hence,

$$E_x^{(i)} + E_x^{(r)} = E_x^{(t)}, \quad E_y^{(i)} + E_y^{(r)} = E_y^{(t)}, \quad H_x^{(i)} + H_x^{(r)} = H_x^{(t)}, \quad H_y^{(i)} + H_y^{(r)} = H_y^{(t)} \tag{2.41}$$

Substituting these into Equation 2.40 for all the components, and using $\cos\theta_r = -\cos\theta_i$, we obtain four relations:

$$\cos\theta_i(A_{//} - R_{//}) = \cos\theta_t T_{//} \tag{2.42a}$$

$$A_\perp + R_\perp = T_\perp \tag{2.42b}$$

$$\sqrt{\varepsilon_1}\cos\theta_i(A_\perp - R_\perp) = \sqrt{\varepsilon_2}\cos\theta_t T_\perp \tag{2.42c}$$

$$\sqrt{\varepsilon_1}(A_{//} + R_{//}) = \sqrt{\varepsilon_2}T_{//} \tag{2.42d}$$

It is noted that these four equations could be divided into two independent groups, one of which contains only the components parallel to the incidence plane, whereas the other contains only those perpendicular to the plane of incidence. Solving them in combination with the refractive law $\sin\theta_t = \frac{1}{n}\sin\theta_i$, we express the amplitude of the reflection and the refraction fields as:

$$T_{//} = \frac{2\sin\theta_t\cos\theta_i}{\sin(\theta_i + \theta_t)\cos(\theta_i - \theta_t)}A_{//} \tag{2.43a}$$

$$T_\perp = \frac{2\sin\theta_t\cos\theta_i}{\sin(\theta_i + \theta_t)}A_\perp \tag{2.43b}$$

$$R_{//} = \frac{\tan(\theta_i - \theta_t)}{\tan(\theta_i + \theta_t)}A_{//} \tag{2.43c}$$

$$R_\perp = -\frac{\sin(\theta_i - \theta_t)}{\sin(\theta_i + \theta_t)}A_\perp \tag{2.43d}$$

For the reflection electric field, the amplitude of the reflections along the parallel and perpendicular directions are:

$$r_{//} = \frac{R_{//}}{A_{//}} = \frac{\tan(\theta_i - \theta_t)}{\tan(\theta_i + \theta_t)} \tag{2.44a}$$

$$r_\perp = \frac{R_\perp}{A_\perp} = -\frac{\sin(\theta_i - \theta_t)}{\sin(\theta_i + \theta_t)} \tag{2.44b}$$

The power or energy is more widely used to measure the reflectivity and transmissivity of an interface. The Poynting vector, whose magnitude represents the light intensity and direction, is the propagation direction of the light, which is defined as:

$$\vec{S} = \frac{c}{4\pi}\sqrt{\frac{\varepsilon}{\mu}}E^2\vec{s} \tag{2.45}$$

Obviously, the power is proportional to the square of the amplitude. Therefore, the reflectivity of the intensity is:

$$R_{//P} = (r_{//})^2 = \frac{\tan^2(\theta_i - \theta_t)}{\tan^2(\theta_i + \theta_t)} \tag{2.46a}$$

$$R_{\perp P} = (r_\perp)^2 = \frac{\sin^2(\theta_i - \theta_t)}{\sin^2(\theta_i + \theta_t)} \tag{2.46b}$$

where the refractive angle and incident angle meet the relation decided by Snell's law, $\sin\theta_t = \frac{1}{n}\sin\theta_i$, where n is the relative refractive index between the two media. In these equations, all variables are real when the two media are both of zero conductivity, perfectly transparent, and have a magnetic permeability that differs from unity by a negligible amount. However, the situation changes when one of the media is a conductive metal because the refractive index is no longer real. Therefore, by analogy with the law of refraction, $\sin\theta_t = \frac{1}{\hat{n}}\sin\theta_i$, where $\hat{n}$ and θ_t are complex and the latter quantity no longer has the simple significance of an angle of refraction. The first medium is a dielectric material, and the reflected wave is an ordinary (homogeneous) wave with a real phase factor. The amplitude reflections expressed by Equation 2.35 are still applicable except that the refractive angle is complex.

$$r_{//} = \frac{R_{//}}{A_{//}} = \frac{\tan(\theta_i - \hat{\theta}_t)}{\tan(\theta_i + \hat{\theta}_t)} = \rho_{//}e^{i\varphi_{//}} \tag{2.47a}$$

$$r_{\perp} = \frac{R_{\perp}}{A_{\perp}} = -\frac{\sin(\theta_i - \theta_t)}{\sin(\theta_i + \theta_t)} = \rho_{\perp}e^{i\varphi_{\perp}} \tag{2.47b}$$

It indicates that characteristic phase changes occur on reflection. Thus, the incident linearly polarized light will in general become elliptically polarized on reflection at the metal surface. The absolute reflections are represented by $\rho_{//}$, $\rho_{\perp}$, and the phase changes are $\varphi_{//}$ and $\varphi_{\perp}$, respectively, when the incident light is linearly polarized in azimuth α_i, that is:

$$\tan\alpha_i = \frac{A_{\perp}}{A_{//}} \tag{2.48}$$

Let α_r be the azimuth angle of the reflected light, which is generally complex. Then,

$$\tan\alpha_r = \frac{R_{\perp}}{R_{//}} = -\frac{\cos(\theta_i - \theta_t)}{\cos(\theta_i + \theta_t)} = Pe^{-i\Delta}\tan\alpha_i \tag{2.49}$$

where

$$P = \frac{\rho_{\perp}}{\rho_{//}} = \left|\frac{\cos(\theta_i - \theta_t)}{\cos(\theta_i + \theta_t)}\right| \text{ and } \Delta = \varphi_{//} - \varphi_{\perp} \tag{2.50}$$

As mentioned previously, α_r is complex in most cases except the following two:

(i) When the incident angle $\theta_i = 0$ (normal incidence), $P = 1$, and $\Delta = -\pi$, so that $\tan\alpha_r = \tan\alpha_i$. The normal incidence results in an opposite reflection; thus, the negative sign implies that the azimuth of the linearly polarized light is unchanged in its absolute direction in space.

(ii) When the incident angle $\theta_i = \pi/2$ (grazing incidence), $P = 1$ and $\Delta = 0$, so that $\tan\alpha_r = \tan\alpha_i$. The absolute direction remains unchanged when the incidence is grazing.

When the incident angle is between the two values mentioned above, the polarization state, including the amplitude and phase, will change, which induces the PDL.

$$\sin\theta_t = \frac{1}{\hat{n}}\sin\theta_i = \frac{1}{n + in\kappa}\sin\theta_i \tag{2.51}$$

$$\cos\theta_t = \sqrt{1 - \sin^2\theta_t} = \sqrt{1 - \left(\frac{1}{n + in\kappa}\sin\theta_i\right)^2} \tag{2.52}$$

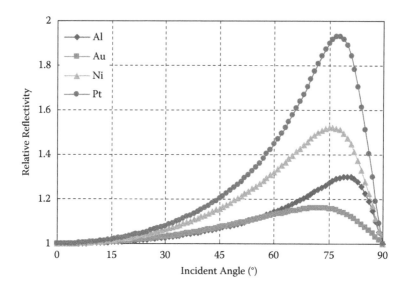

FIGURE 2.15
Relative reflectivity between perpendicular and parallel components of various incident angles with different metals.

Obviously, the reflection amplitude and phase shift depend on the incident angle θ_i and the complex refractive index $\hat{n}$ of metals. The reflection ratio between the perpendicular and parallel components of the four different kinds of metals are calculated and plotted in Figure 2.15. It indicates that the polarization-dependent reflectivity increases with the increase in the incident angle for any metals. When the incident angle is fixed, aluminum and gold provide the smaller difference. The incident angle of interest is 45° as it is used in the MEMS optical switch. The ratio of the two orthogonal components is very similar at this point for aluminum and gold, that is, 1.07, whereas it is 1.15 for nickel and 1.20 for platinum. Figure 2.16 shows the phase difference between the two orthogonal components.

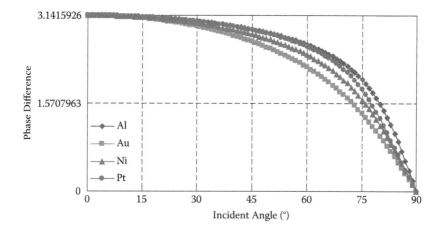

FIGURE 2.16
Phase difference between two orthogonal components.

TABLE 2.5

Polarization Dependence Reflection
of Selected Metals

Metal	Polarization Dependence
Al	6.9%
Au	7.3%
Ni	15.9%
Pt	20.3%

According to Equation 2.37, the reflectivity of the intensity is the square of amplitude reflectivity. Thus, the relative power reflectivity ratio is expressed as:

$$\Gamma = \left(\frac{R_\perp}{R_{//}}\right)^2 = \left(-\frac{\cos(\theta_i - \theta_t)}{\cos(\theta_i + \theta_t)}\right)^2 = P^2 \qquad (2.53)$$

When the incident angle is less than 15°, no obvious difference is observed between the various metals. With the increase in the incident angle θ_i, the reflection of different polarization lights deviate further for different metals. In the optical switch, an incident angle of 45° is most important. In such cases, divergence varies with the types of metal as listed in Table 2.5.

2.3 Electromechanical Design

The electromechanical actuator is used to provide translation or rotation to the micromirrors so that the incident optical signals are directed to the expected outputs. There are numerous kinds of microactuators (for a comparison see Section 2.1). Because the comb-drive actuator is the most attractive structure with broad applications in MEMS systems, the design related to this structure and its characteristics are discussed in this section.

2.3.1 Lateral-Mode of the Comb-Drive Actuator

The comb-drive actuator consists of two series of comb-like structures with the fingers interdigitating with each other. One set of comb fingers is free to move to determine the motion of the functional component, whereas the other is fixed in a stationary position. The device can actuate along lateral, transversal, or vertical directions. The rotary movement (in plane and out of plane) can be achieved by this actuator as well. For more clarity, the conventional direction is standardized as defined in Figure 2.17, where b and H is the width and height, respectively, of the finger, g is the gap between the adjacent overlapped fingers, and L_1 is the initial overlap length of the fingers. The mechanical support is provided by various suspension beams, among which the folded beam is the most widely accepted design whose width, length, and height is denoted by b_f, l_f, and t_f, respectively. Irrespective of the direction of the movement, the actuation is obtained on the basis of the plane capacitor, which is analyzed in the following sections.

2.3.2 Parallel Plate Capacitor

The plate capacitor, shown in Figure 2.18, consists of two plates, A and B, with overlapping area S (product of plate depth e and length y_0) and the gap d (which is d_0 at the initial

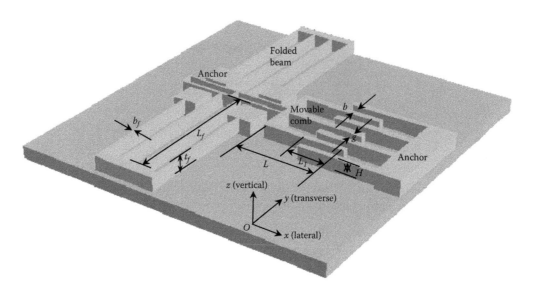

FIGURE 2.17
Schematic of a comb drive.

position). Thus, the capacitance is expressed as $C = S\varepsilon\varepsilon_0/d$. The electric energy stored in the capacitor is $E_e = S\varepsilon\varepsilon_0 V^2/2d$.

If an external force is applied to the system, the position and energy of the system will change. The force can be normal or tangential to the plane.

For normal force (F_n): The energy conservation equation is $F_n\Delta x + \frac{dE_e}{dx}\Delta x + \frac{dE_s}{dx}\Delta x = 0$, where E_s is the internal source energy and E_e is the electrical energy. Therefore, the normal force is:

$$F_n = -\frac{dE_e}{dx} - \frac{dE_s}{dx} = -\frac{S\varepsilon\varepsilon_0}{2x^2}V^2 \tag{2.54}$$

The negative indicates that the force applied to move the plate is an attractive force from the stationary electrode.

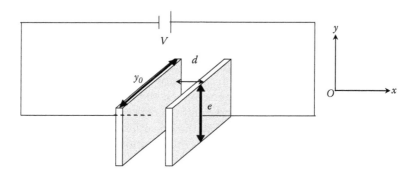

FIGURE 2.18
Parallel plate capacitor.

For tangential force (F_t): When a tangential force is applied, the equation $F_t \Delta y + \frac{dE_e}{dy} \Delta y + \frac{dE_s}{dy} \Delta y = 0$ should be satisfied. Hence, the tangential force is:

$$F_t = -\frac{e\varepsilon_0}{2d_0} V^2 \tag{2.55}$$

Therefore, the ratio of the two forces at the initial position is expressed as:

$$\frac{F_n}{F_t} = \left(\frac{S\varepsilon\varepsilon_0}{2d_0^2} V^2 \right) \Big/ \left(\frac{e\varepsilon\varepsilon_0}{2d_0} V^2 \right) = \frac{y_0}{d_0} \tag{2.56}$$

This ratio is often much greater than 1.0 because the overlapping length y_0 is usually much larger than the gap d_0 between the two plates; thus, the normal force is dominant. However, when either of the two plates is moving in the normal direction, the normal force F_n varies with the motion, whereas the tangential force F_t is independent of the tangential movement of the plate.

2.3.3 Comb-Drive Actuator

The comb-drive actuator makes use of the tangential force for driving, which is independent of the tangential displacement. Although the normal force is much larger than the tangential force as discussed previously, its effect is eliminated by arranging the fixed fingers symmetrically on both sides of each movable finger, so that the normal forces from both sides are cancelled out in pairs as shown in Figure 2.19. However, obtaining an accurate force generated by the comb-drive actuator is still more complicated than in the plate capacitor case because of the fringe effect.

2.3.3.1 *Capacitor Analysis*

When a voltage is applied to the comb fingers, the electrical field has a distribution as shown in Figure 2.19a. All dielectric/air interfaces are modeled as equivalent magnetic walls as is usually done in the partial-capacitance technique [19–21]. The electric walls are assumed at the planes of symmetry of the electric field distribution, where the field lines are normal to the electric wall. The capacitance C of the comb structure is approximated as the sum of the contribution of the field capacitance C_u on the vertical sides of the comb fingers (see Figure 2.19b) and the contribution of the fringe field capacitance [22,23], C_f due to the top and bottom sides of the comb fingers (see Figure 2.19c).

The height of the comb fingers obtained by the deep reactive ion etching (DRIE) technique is so large that the field capacitance C_u on the vertical sides of the comb fingers is the dominant part of the total capacitance. This is similar to the parallel plate capacitor except for the extra capacitor between the finger tip and the comb finger. The capacitance is expressed as:

$$C = \frac{2n\varepsilon_0 Hx}{g} + \frac{\left(1 + \frac{\pi}{2}\right)[(2n-1)b + 2w]\varepsilon_0 H}{L - x} \tag{2.57}$$

where n is the number of movable comb fingers, $n+1$ is the number of fixed comb fingers, b is the width of the middle comb fingers, w is the width of the two side fixed comb fingers, L is the total length of the finger, and x is the overlap length. The first term comes from the overlapping finger parts, and the second term is the contribution of the capacitor between the tip of the comb fingers and the beam.

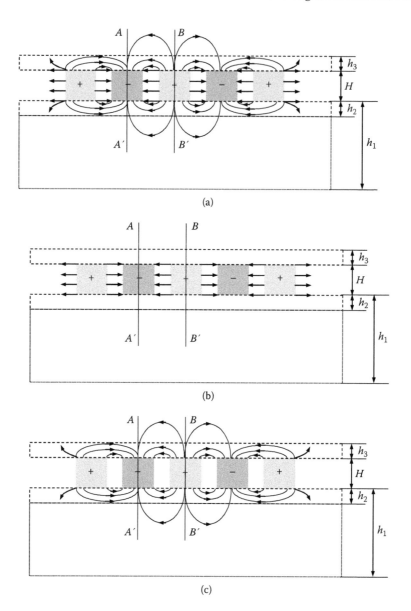

FIGURE 2.19
Electrical field of a comb drive: (a) electrical field distribution of comb drive, (b) field distribution on the vertical sides, and (c) fringe field due to the top and bottom sides of the comb fingers.

2.3.3.2 Lateral Electrostatic Force

The electrostatic force is derived by differentiating the potential energy with respect to any direction of the axis. From Equation 2.4, the lateral electrostatic force is determined as:

$$F_x = \frac{1}{2}\frac{\partial C}{\partial x}V^2 = \left\{ \frac{n\varepsilon_0 H}{g} + \frac{\left(1+\frac{\pi}{2}\right)[(2n-1)b+2w]\varepsilon_0 H}{2(L-x)^2} \right\}V^2 \tag{2.58}$$

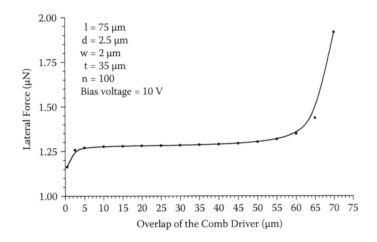

FIGURE 2.20
Relation of lateral force with overlap of the comb drive.

The relationship between the electrostatic force and the overlapping distance of the comb drive is plotted in Figure 2.20. It is clearly shown that this relation is nonlinear at the initial and final overlapping distances due to the fringing electric field effect and the dominant parasitic capacitor. The driving force is constant and independent of the lateral displacement, within a range of $0.3L < x < 0.7L$, and the expression of the electrostatic force is simplified to:

$$F_x = \frac{1}{2}\frac{\partial C}{\partial x}V^2 = \frac{n\varepsilon_0 H}{d}V^2 \tag{2.59}$$

2.3.4 Stability of the Comb Drive

Though the normal forces on both sides of the movable finger are cancelled out as a result of the symmetric structure, they may still cause instability to the system when the voltage is sufficiently high. In Figure 2.21, the fingers (or plates) A and B are fixed, and C, located exactly between A and B, is movable; ky is the elastic constant along the y direction, and g is the initial gap between the neighboring fixed and movable fingers.

When the movable plate C has a displacement y along the transversal direction, the capacitance of the system is expressed as:

$$C(y) = \frac{A\varepsilon\varepsilon_0}{g+y} + \frac{A\varepsilon\varepsilon_0}{g-y} = \frac{2A\varepsilon\varepsilon_0 g}{g^2 - y^2} \tag{2.60}$$

whereas the initial capacitance is $C_0 = \frac{2A\varepsilon\varepsilon_0}{g}$. Let $\tilde{y} = \frac{y}{g}$ estimate the asymmetry; the change of the capacitance is expressed as

$$\Delta C(y) = \frac{2A\varepsilon\varepsilon_0 g}{g^2 - y^2} - C_0 = C_0 \frac{\tilde{y}^2}{1-\tilde{y}^2} \tag{2.61}$$

FIGURE 2.21
Schematic of stability analysis.

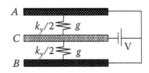

With reference to the energy of the initial position (without displacement in the transverse direction, or $y = 0$), the system energy changes to:

$$E(\tilde{y}) = \frac{1}{2}k_y g^2 \tilde{y}^2 - \frac{1}{2}\Delta C(\tilde{y})V^2 \tag{2.62}$$

By differentiating the system energy at displacement $\tilde{y}$ (and let it be zero) to find the balance positions, we obtain:

$$k_y g^2 \tilde{y} - \frac{\tilde{y}}{(1-\tilde{y}^2)^2}C_0 V^2 = 0 \tag{2.63}$$

There are three solutions for this equation:

$$\tilde{y}_1 = 0 \quad \text{and} \quad \tilde{y}_{2,3} = \pm\sqrt{1 - \sqrt{\frac{C_0 V^2}{k_y g^2}}} \tag{2.64}$$

The second-order derivative of the energy at displacement $\tilde{y}$ is used to analyze the stability of these solutions:

$$\frac{\partial^2 E(\tilde{y})}{\partial \tilde{y}^2} = k_y g^2 - \frac{1+3\tilde{y}^2}{(1-\tilde{y}^2)^3}C_0 V^2 \tag{2.65a}$$

For $\tilde{y}_1 = 0$,

$$\frac{\partial^2 E(\tilde{y})}{\partial \tilde{y}^2} = k_y g^2 - C_0 V^2 \tag{2.65b}$$

If $V < g\sqrt{\frac{k_y}{C_0}}$, $\frac{\partial^2 E(\tilde{y})}{\partial \tilde{y}^2} > 0$ the system energy has a minimum value at this position, which implies that it is a stable state. In this case, $\tilde{y}_2$ and $\tilde{y}_3$ are two real solutions. The second-order derivative of the energy evaluated at the two states becomes:

$$\frac{\partial^2 E(\tilde{y}_{2,3})}{\partial \tilde{y}^2} = 4k_y g^2 \left(1 - \sqrt{\frac{k_y g^2}{C_0 V^2}}\right) < 0 \tag{2.66}$$

The negative value means that these two solutions are unstable corresponding to the two positions on both the sides of the movable finger. Thus, if the movable plate or finger C moves a little farther from either of these two unstable positions, $\tilde{y}_{2,3}g$, to the adjacent fixed fingers A or B, then it will continue moving until it makes contact with either fixed finger A or B, and it will remain there forever. On the contrary, if finger C moves a little bit back to middle position, it will move back to the stable state at $\tilde{y} = y = 0$. Therefore, the region where finger C can return to the stable state at $\tilde{y} = y = 0$ is between $\left(g\sqrt{1-\sqrt{\frac{C_0 V^2}{k_y g^2}}}\right)$ and $\left(-g\sqrt{1-\sqrt{\frac{C_0 V^2}{k_y g^2}}}\right)$. If the position of the middle finger is outside this range, then it will move toward and stick to the nearest fixed finger permanently. According to the expression of $\tilde{y}_{2,3}$ in Equation 2.64, the larger the applied voltage, the smaller is $|\tilde{y}_{2,3}|$, and thus the more stable the region will be.

If $V > \sqrt{\frac{k_y g^2}{C_0}}$ and $\frac{\partial^2 E(\tilde{y})}{\partial \tilde{y}^2} < 0$, the energy has a maximum value at this position, which indicates that $y = 0$ is no longer a stable position, in which case, the other two solutions are not real-valued, and the active movable finger C will always be pulled into stiction to one of the two fixed fingers next to it.

The stability of the actuator depends on the initial capacitance C_0 (defined by the overlap length, height of the fingers, and gap between the adjacent fingers), the external applied voltage, and the stiffness along the transverse direction. When the design is fixed, the stability varies with the driving voltage. The actuator can provide a stable motion along the lateral direction when the voltage is within the range of 0 to $\sqrt{k_y g^2/C_0}$. The pull-in and side stiction will occur with an increase in the driving voltage. However, the critical voltage relies on the comb drive that generates the electrostatic force and the suspension that provides the support for the movable structures and the elastic force to balance the deformation. Because the displacement provided by the comb-drive actuator depends on the stiffness along the lateral direction, the stability of the actuator is evaluated by the ratio of the stiffness along the transverse direction k_y to that along the motion direction k_x. To accomplish a large, stable displacement, the ratio k_y/k_x must be much greater than 1.

2.3.5 Displacement of Actuator

In the previous sections, the electrostatic force along the lateral direction was analyzed, and the stability of the actuator was gauged by the stiffness ratio between the transverse direction and the lateral direction. To simulate the displacement under this electrostatic force, the suspension is described and analyzed in this section. The suspensions should provide freedom of movement along the lateral direction (x) and constrain the traveling along the orthogonal directions (y and z) to prevent instability of the system. The suspension should also be able to relieve the built-in stress and the axial stress induced by a large displacement.

The folded beam fulfills the requirements of the suspension [21]. The stiffness along the x direction, k_x, and along the y direction, k_y, can be expressed as:

$$k_x = \frac{24EI_z}{L_f^3} = \frac{24E}{L_f^3} \frac{t_f b_f^3}{12} = \frac{2Et_f b_f^3}{L_f^3} \tag{2.67}$$

$$k_y = \frac{8AE}{L_f} = \frac{8bt_f E}{L_f} \tag{2.68}$$

The factor eight of k_y is due to the eight individual beams with the assumption that the outer connecting trusses are rigid, which is the recommended option. Therefore, the ratio between the spring constant along the y direction and that along the x direction is:

$$\frac{k_y}{k_x} = \frac{8b_f t_f E / L_f}{2Et_f b_f^3 / L_f^3} = \frac{4L_f^2}{b_f^2} \tag{2.69}$$

It is clear that the stability of the actuator strongly depends on the relative dimensions of the length and width of the folded beams. Fortunately, in most cases, the length of the beam is several hundred times the width. Therefore, the motion in the y direction due to beam extension and compression is unlikely as long as the applied voltage is not beyond the critical value of $\sqrt{k_y g^2/C_0}$.

When half or more of the eight parallel beams buckle under a normal force F_y in the y direction, the structure will move sideways. In the worst scenario, Euler's simple buckling criterion can be used to evaluate the critical force F_y required to buckle a fixed-fixed beam with length L_b [24]:

$$F_y = \frac{\pi^2 E I_z}{L_f^2} \tag{2.70}$$

Here, I_z is the moment of inertia along the z direction, that is, $I_z = \frac{t_b b_b^3}{12}$. Thus, the force ratio along the two orthogonal directions is:

$$\frac{F_y}{F_x} = \frac{F_y}{k_x x} = \frac{\pi^2 E I_z / L_f^2}{24 E I_z x / L_f^3} = \frac{\pi^2 L_f}{24 x} \tag{2.71}$$

For a typical structure with 600 μm long folded beam, the force F_y required to buckle the four supporting beams is about 50 times higher than the force F_x required to pull the structure for a displacement of 20 μm in the x direction. Another possible cause for the structure to move in the y direction is the comb-finger misalignment, which can be ignored because all of the features are patterned with one mask and etched by the same DRIE process. Therefore, the folded beam is used along with a dense comb drive thanks to its high stability and large possible displacement.

As the displacement along the lateral direction (x direction) is small compared to the length of the folded beam, the elastic force F_e generated by deformation of the folded beam can be regarded as a linear spring and is analyzed with the help of Hook's law $F_e = k_x \cdot \Delta x$. The elastic force is balanced by the electrostatic force generated by the two sets of comb fingers. Therefore, the displacement along the lateral direction, regardless of the damping, is expressed as:

$$\Delta x = \frac{F_e}{k_x} = \frac{F_{es}}{k_x} = \frac{n\varepsilon\varepsilon_0 H V^2 / g}{2 E t_f b_f^3 / L_f^3} = \frac{n\varepsilon\varepsilon_0 H V^2 L_f^3}{2 E t_f b_f^3 g} \tag{2.72}$$

In most cases, the thickness of the comb finger is equal to that of the suspension beam, and the dielectric material is air, implying that the relative permittivity ε is 1.0. Thus, the displacement along the x direction is simplified to:

$$\Delta x = \frac{n\varepsilon_0 V^2 L_f^3}{2 E b_f^3 g} \tag{2.73}$$

The displacement is proportional to the number of fingers n, the square of applied voltage V, and the cube of the length of the folded beam L_f; and in inverse proportion to the finger gap g and cube of folded beam width b_f. However, to prevent the sideway movement and stiction, the voltage cannot be too high. The overall size of the switch cannot be too large, which means that the number of fingers must be limited. To ensure high quality factor, the stiffness along the x direction should not be too small. Therefore, optimizing the whole system involves trade-offs.

Similarly, the comb-drive actuator for the other directions can be accomplished by employing the energy stored in the comb-drive structure and the stiffness along the directions of motion.

2.4 Fabrication Tolerance Effect

All preceding analyses are theoretical results regardless of the fabrication tolerances, which affect the performances of the actuators. The tolerances due to the fabrication include (1) misalignment between different mask layers [24], (2) tapered profiles of the microbeams, (3) various materials of the beam, (4) uneven depth of the individual beams, (5) undercut, and (6) different gaps of the adjacent fingers between the overlap and nonoverlap sections [25,26]. These tolerances depend significantly on the process methods and conditions. In the planar comb-drive actuator, there is no misalignment between the structures as they are patterned and deep etched together in the entire process. However, the remaining five other types of tolerances, are significant when the DRIE fabrication technique is used. Their effects on the performances of the comb-drive actuator, in terms of electrostatic force, mechanical stiffness, stability, and displacement, are discussed in the following sub-sections. These analytical results can be used to compensate for the fabrication tolerances at the design stage and allow the actuators to provide more predictable performance.

2.4.1 Effects of Etch Profiles

The etched profile is described by an angle α, as shown in the schematic of Figure 2.22 (anti-clockwise from the vertical line to the sidewall), which depends on the process conditions and designs. Generally, the top is wider than the bottom of narrow trenches, whereas the reverse is true of wide trenches, and trenches with different widths have different slope profiles. Three types of profiles are considered. These are, positive, negative, and vertical ($\alpha = 0$) profiles.

   In the comb-drive actuator, the trenches of the comb fingers, the folded beams, and the key element—the micromirror—have different widths, thus inducing various profiles, which in turn affecting the electrostatic force, mechanical stiffness, stability, and displacement.

2.4.1.1 *Profile of Comb Fingers*

The cross-sectional view of the fabricated comb finger with slight taper is schematically shown in Figure 2.23a, whereas Figure 2.23b is the SEM micrograph of the deep-etched comb fingers, in which the slope angle is denoted as α, and the gap on the top, as g_0.

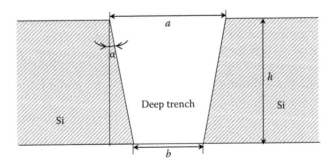

FIGURE 2.22
Schematic of the DRIE-etched slanted trench.

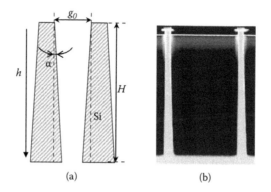

FIGURE 2.23
Sloped comb fingers: (a) schematic of the sloped comb fingers and (b) SEM micrograph of the sloped comb-finger set.

The gap between two adjacent comb fingers is a function of the finger height and is expressed as:

$$g(h) = g_0 - 2h \tan \alpha \tag{2.74}$$

Accordingly, the capacitance of the comb sets changes to:

$$C_{slope} = \int_0^C 2n dC = \frac{n\varepsilon(L_1 + \Delta x)}{\tan \alpha} \ln\left(\frac{g_0}{g_0 - 2H \tan \alpha}\right) \tag{2.75}$$

where dC is the micro capacitance of a small section with a height dh. Therefore, when the same voltage is applied, the ratio of the electrostatic force generated by the sloped comb fingers to that of the vertical fingers calculated by Equation 2.46 is

$$\frac{F_{slope}}{F_{design}} = \frac{(\partial U/\partial x)_{slope}}{(\partial U/\partial x)_{design}} = \frac{g_0}{2H \tan \alpha} \ln\left(\frac{g_0}{g_0 - 2H \tan \alpha}\right) \tag{2.76}$$

Figure 2.24 shows that the relative electrostatic force is a function of the slanted angle α under the conditions of $H = 50$ μm and $g_0 = 2.5$ μm, where α varies from $-1.0°$ to $+1.0°$. The electrostatic force generated by the negatively sloped comb finger is smaller, as opposed to positively sloped comb finger, which is more significant. For instance, only 75.8% of the electrostatic force is generated when the comb finger has an angle of $-1.0°$, but this force increases by 1.716 times when the comb finger has an angle of $+1.0°$. The angle cannot be too large because an unwanted force along the z direction will then be generated, and in the case that the negatively slope angle is large, this force can even be comparable with lateral force [27].

2.4.1.2   Profile of Folded Beam

An imperfect vertical profile can be observed for the folded suspension beam, inducing deviation in the stiffness and stability of the actuator. The top width and height of the trapezoidal cross section of the beam is denoted by b_0 and H, respectively, whereas the cross

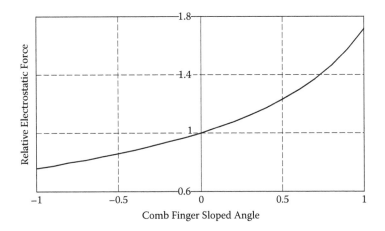

FIGURE 2.24
Relative electrostatic force as a function of the finger profile.

profile has an angle of β, as shown in Figure 2.25a and b. The second moment of inertia I_z can be rewritten as:

$$I_{zslope} = \frac{H}{12}\left(b_0^3 + 3Hb_0^2 \tan \beta + 4H^2 b_0 \tan^2 \beta + 2H^3 \tan^3 \beta\right) \tag{2.77a}$$

The stiffness ratio along the lateral direction between the slanted and the vertical beam is:

$$\frac{k_{xslope}}{k_{xdesign}} = \frac{b_0^3 + 3Hb_0^2 \tan \beta + 4H^2 b_0 \tan^2 \beta + 2H^3 \tan^3 \beta}{b_0^3} \tag{2.77b}$$

Figure 2.26 shows the normalized stiffness along the lateral direction with tapered profile under the conditions of $b_0 = 2.5$ μm and $H = 50$ μm. The stiffness of the negatively sloped folded beam decreases, whereas, for the positively sloped beam, it increases nonlinearly. The stiffness decreases to only 36% for the folded beam with −1.0°, and rises 2.62 times for the folded beam with an angle of +1.0°.

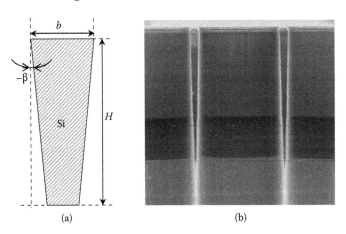

(a)   (b)

FIGURE 2.25
Sloped folded beam: (a) schematic cross-sectional view and (b) SEM micrograph of sloped beam.

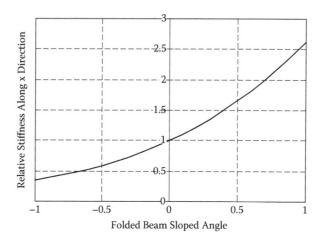

FIGURE 2.26
Normalized lateral stiffness of the nonideal folded beam.

The stability of the comb-drive actuator evaluated by the ratio of stiffness along the transversal and lateral direction is also influenced by the tapered deep-etched folded beam. The relative stiffness along the transverse direction is expressed as:

$$\frac{k_{yslope}}{k_{ydesign}} = \frac{b_0 + H\tan\beta}{b_0} \tag{2.78}$$

The stiffness along the lateral and transverse directions decreases with the positively sloped beam at a different speed, and this affects the stability of the system. The relative stiffness ratio for the transverse direction to the lateral direction between sloped and vertical designed folded beam is expressed as:

$$\left(\frac{k_y}{k_x}\right)_{slope} \bigg/ \left(\frac{k_y}{k_x}\right)_{design} = \frac{k_{yslope}}{k_{ydesign}} \cdot \frac{k_{xdesign}}{k_{xslope}} = \frac{b_0^2}{b_0^2 + 2b_0 H\tan\beta + 2H^2\tan^2\beta} \tag{2.79}$$

and is used to evaluate the effect of the sloped profile on the system stability as shown in Figure 2.27, which indicates that the actuator can provide more stable actuation when the deep-etched folded beams are negatively sloped.

2.4.1.3 Profiles of Comb Fingers and Folded Beam

As the comb fingers and the folded beams have various trenches, the profile may be different after being etched under the same process conditions. As a result, the displacement along the lateral direction obtained by the slanted comb fingers and the sloped folded beam deviates from the designed value when the same voltage is applied. The normalized displacement under the applied voltage V can be expressed as:

$$\frac{\Delta x_{slope}}{\Delta x_{ndesign}} = \frac{F_{slope}/k_{xslope}}{F_{design}/k_{xdesign}} = \frac{F_{slope}}{F_{design}}\frac{k_{xdesign}}{k_{xslope}}$$

$$= \frac{g_0}{H\tan\alpha}\, In\!\left(\frac{g_0 + 2H\tan\alpha}{g_0}\right)\frac{b_0^3}{b_0^3 + 3Hb_0^2\tan\beta + 4H^2 b_0\tan^2\beta + 2H^3\tan^3\beta} \tag{2.80}$$

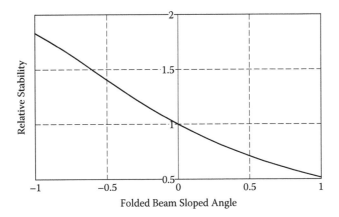

FIGURE 2.27
Relative stiffness along the y direction of the nonideal folded beam.

The relative displacement as a function of the slope of the comb fingers, α, and the folded beam, β, is illustrated in Figure 2.28, which shows that the outward comb fingers, combining with the inward folded beams, give a larger displacement. For example, the displacement increases 4.832 times when the comb finger is $+1.0°$ sloped and the folded beam is $-1.0°$ sloped. The slanted profile also influences the properties of the transverse, vertical, and rotational actuators, which can be analyzed in a similar way.

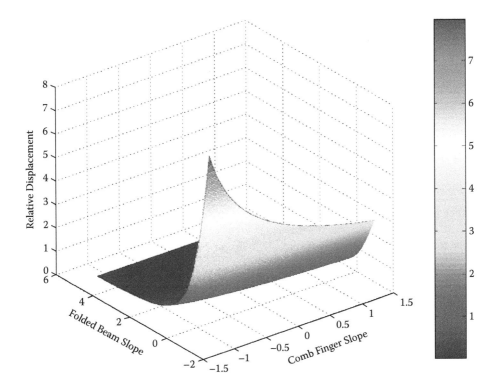

FIGURE 2.28
Relative displacement versus profile of comb finger and folded beam.

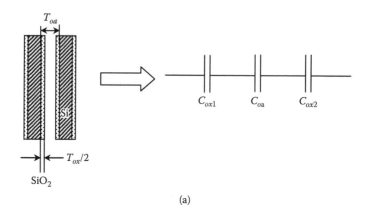

(a)

FIGURE 2.29
The oxidized comb finger and equivalent circuit: (a) schematic of dioxide-covered comb finger and (b) equivalent electrical circuit.

2.4.2 Native Oxidation on Comb Fingers

The sidewall of the microstructure is apt to be oxidized to a thickness of Tox/2, and the properties change after fabrication, unless vacuum-packaging is carried out. This model can be used to explain the measurement difference at various periods. The oxidized comb fingers and the equivalent circuit is shown in Figures 2.29a, and b, respectively, where the total capacitance is the result of these three capacitors in the series. Two of them come from the oxide layer denoted by C_{ox1} and C_{ox2}, and the other one is the generated air gap, C_{oa}. The change of these capacitances correspondingly results in the alteration of the actuator characteristics. The capacitance of the oxide-covered comb drive is changed to:

$$C_{oxideCover} = 2n\left(\frac{\varepsilon_a \varepsilon_{ox} \varepsilon_0 H(L_1 + \Delta x)}{\varepsilon_{ox}(T_{oa} - T_{ox}) + \varepsilon_a T_{ox}}\right) \qquad (2.81)$$

and the relative electrostatic force is expressed by:

$$\frac{F_{oxideCover}}{F_{design}} = \frac{\partial\left(\frac{1}{2}C_{oxideCover}V^2\right)/\partial x}{\partial\left(\frac{1}{2}C_0 V^2\right)/\partial x} = \frac{\varepsilon_{ox} T_{oa}}{\varepsilon_{ox}(T_{oa} - T_{ox}) + \varepsilon_a T_{ox}} \qquad (2.82)$$

The effect of the oxide layer on the driving force under these conditions is shown in Figure 2.30. Obviously, the thicker the oxide layer, the larger the force generated as a result of higher relative permittivity of the oxide layer and narrower air gap.

2.4.3 Tapered Microstructures and Oxidation

When the comb fingers have a sloped profile and are oxidized, the capacitance will be different from that in the previous two cases discussed. In general, the oxide layer thickness is uniform from top to bottom. The schematic of these two tolerances is plotted in Figure 2.31. The final capacitance becomes:

$$C_{total} = 2n\frac{C_{ox} \cdot C_a}{C_{ox} + C_a} = \frac{2n\varepsilon_{ox}\varepsilon_0\varepsilon_a H(L_1 + x) \cdot \ln\left(\frac{g_0 + 2Htg\alpha}{g_0}\right)}{2\varepsilon_{ox}Htg\alpha + \varepsilon_a T_{ox}\cos\alpha \ln\left(\frac{g_0 + 2Htg\alpha}{g_0}\right)} \qquad (2.83)$$

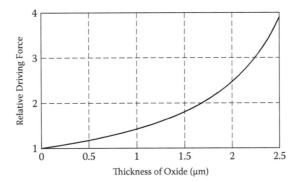

FIGURE 2.30
Relative driving force with various thicknesses of oxide.

The relative electrostatic force of the sloped comb finger with a thin oxide layer to the designed vertical silicon comb is expressed as:

$$\frac{F_{slopedoxide}}{F_{design}} = \frac{\varepsilon_{ox}(g_0 + T_{ox}) \cdot \ln\left(\frac{g_0 + 2Htg\alpha}{g_0}\right)}{2\varepsilon_{ox}Htg\alpha + \varepsilon_a T_{ox} \cos\alpha \ln\left(\frac{g_0 + 2Htg\alpha}{g_0}\right)} \tag{2.84}$$

Let $H = 50$, $g_0 = 2.0$, $T_{ox} = 0.5$, and $\varepsilon_{ox} = 3.9$; the normalized force is a function of the slope angle and the oxide-layer thickness. This is shown in Figure 2.32. Obviously, the positive angle decreases the electrostatic force, whereas the oxide layer increases it; however, the effect of the oxide layer thickness on the electrostatic force is weaker. Thus, the performance of the comb-drive actuator depends significantly on the profiles of the comb fingers and the folded beam.

2.4.4 Uneven Depth Comb Fingers

The reactive ion etching (RIE) lag or microloading effect [28,29], whereby the etching rate decreases when the trench width increases, is commonly observed in fluorinated plasma etching, and the etch parameters are contributory factors to the microloading effect. Usually, monolithic silicon or a silicon-on-insulator (SOI) substrate is used to fabricate MEMS devices. During the fabrication process, uneven-depth beams occur as a result of the different wide trench beside the beams [30], where the finger gap of the overlapping part is narrower than that of the nonoverlapping part.

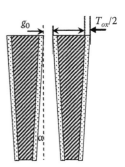

FIGURE 2.31
Schematic of tapered comb finger with oxide coverage.

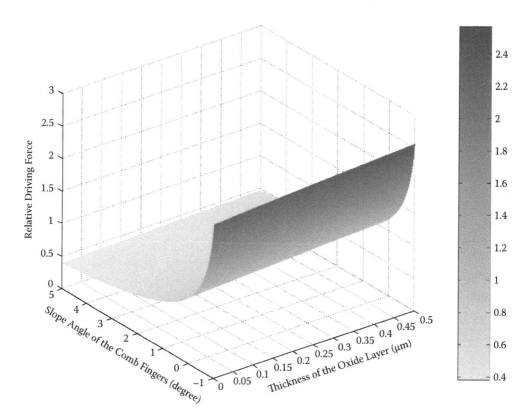

FIGURE 2.32
The effect of oxide on sloped comb fingers.

Figure 2.33a is the SEM micrograph of uneven-depth comb fingers, and Figure 2.33b is the schematic graph, in which the overlapping part with a gap of g_1 has a depth H_1 and length L_1, whereas the nonoverlapping part has a depth and gap of H_2 and g_2, respectively. Two types of uneven-depth comb fingers exist because of different substrates, which correspond to the two cases $H_2 > H_1$ and $H_2 < H_1$.

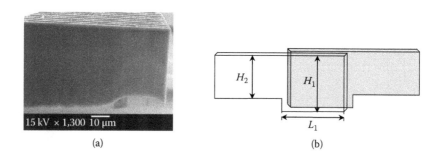

(a)   (b)

FIGURE 2.33
Comb drive with various depth parts: (a) SEM micrograph of damaged comb finger and (b) schematic of nonuniform-depth comb.

The capacitance, electrostatic force, and the corresponding displacement differ from that of the even-depth finger design. The initial capacitance is expressed as:

$$C_0 = \frac{2n\varepsilon_a\varepsilon_0 H_1 L_1}{g_0} \tag{2.85}$$

When $H_1 < H_2$ and the displacement along lateral direction is Δx, the capacitance is

$$C_1 = \frac{2n\varepsilon_a\varepsilon_0}{g_0}(L_1 + \Delta x)H_1 \qquad (0 \le \Delta x \le L_1) \tag{2.86a}$$

$$C_2 = \frac{2n\varepsilon_a\varepsilon_0}{g_0}[(\Delta x - L_1)H_2 + 2L_1 H_1] \qquad (L_1 \le \Delta x) \tag{2.86b}$$

When $H_1 > H_2$,

$$C_1' = \frac{2n\varepsilon_a\varepsilon_0}{g_0}[2\Delta x H_2 + (L_1 - \Delta x)H_1] \qquad (0 \le \Delta x \le L_1) \tag{2.86c}$$

$$C_2' = \frac{2n\varepsilon_a\varepsilon_0}{g_0}(L_1 + \Delta x)H_2 \qquad (L_1 \le \Delta x) \tag{2.86d}$$

When $0 \le \Delta x \le L_1$, the electrostatic force under the applied voltage V is:

$$F_1 = \frac{\partial}{\partial(\Delta x)}\left(\frac{1}{2}C_1 V^2\right) = \frac{n\varepsilon\varepsilon_0 V^2}{g_0}H_1 \qquad (H_1 < H_2) \tag{2.87a}$$

$$F_1' = \frac{\partial}{\partial(\Delta x)}\left(\frac{1}{2}C_1' V^2\right) = \frac{n\varepsilon\varepsilon_0 V^2}{g_0}[2H_2 - H_1] \qquad (H_1 > H_2) \tag{2.87b}$$

When $\Delta x \ge L_1$,

$$F_2 = F_2' = \frac{\partial}{\partial(\Delta x)}\left(\frac{1}{2}C_0 V^2\right) = \frac{n\varepsilon\varepsilon_0 V^2}{g_0}H_2 \tag{2.87c}$$

By substitution of the designed force as expressed in Equation 2.46, the relative forces can be expressed as:

$$\frac{F_1}{F_0} = \frac{H_1}{H_0} \qquad (H_1 < H_2) \tag{2.88a}$$

$$\frac{F_1'}{F_0} = \frac{2H_2 - H_1}{H_0} \qquad (H_1 > H_2) \tag{2.88b}$$

when the displacement is less than the initial overlapping length $(0 \le \Delta x \le L_1)$. The driving force changes abruptly when the displacement is over the initial overlap such that:

$$\frac{F_2}{F_0} = \frac{F_2'}{F_0} = \frac{H_2}{H_0} \tag{2.88c}$$

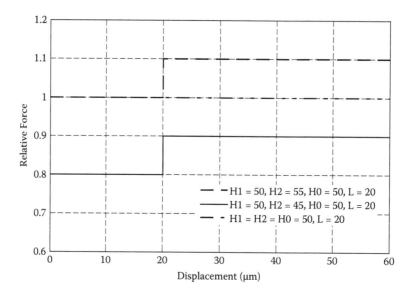

FIGURE 2.34
Electrostatic force versus the displacement of the uneven-deep comb drive.

Figure 2.34 shows the relative electrostatic force as a function of the displacement of the comb-drive actuator. The electrostatic force increases abruptly when the displacement is equal to the initial overlapping length, no matter which part is deeper. The force is weaker than the designed value when the initial nonoverlapped section is shallower than the overlapped part whose depth is equal to the designed value, whereas the opposite fabrication result generates a greater electrostatic force when the displacement is over the initial overlapped length. Consequently, the displacement can be expressed as:

$$\Delta x = \frac{F_1}{k_x} = \frac{n\varepsilon\varepsilon_0 V^2}{g_0 k_x} H_1 \qquad (0 \le \Delta x \le L_1 \,\&\, H_1 < H_2) \tag{2.89a}$$

$$\Delta x' = \frac{F_1'}{k_x} = \frac{n\varepsilon\varepsilon_0 V^2}{g_0 k_x}[2H_2 - H_1] \qquad (0 \le \Delta x' \le L_1 \,\&\, H_1 > H_2) \tag{2.89b}$$

$$\Delta x_2 = \frac{F_{x2}}{k_x} = \frac{n\varepsilon\varepsilon_0 H_2}{g_0 k_x} V^2 \qquad (0 \le \Delta x \le L_1) \tag{2.89c}$$

Let $k_x = 1\mu N/\mu m$, $n = 150$, $g_0 = 2.5\mu m$, and $\varepsilon = 1.0$; the displacement which is a function of the applied voltage is illustrated in Figure 2.35, where there is a jitter at the critical displacement, which is equal to the initial overlapping length. It is more significant when the nonoverlapped portion is shallower than the designed depth as the displacement obtained is always less than the designed value.

2.4.5 Undercut

The microstructure is fabricated by lithographically pattern and hard mask etching to deep silicon etching. During these pattern transfers, some deviations from the soft mask to the hard mask can be observed because of imperfect anisotropic etching. This problem

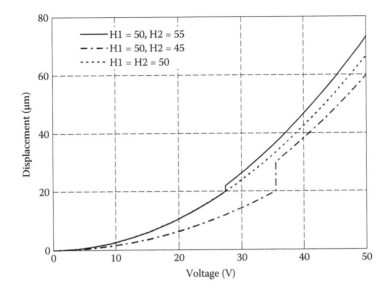

FIGURE 2.35
Displacement versus voltage of the uneven comb drive.

appears to be more serious when the pattern is transferred to the high-aspect-ratio silicon microstructures, especially when the process starts from the etching cycle with SF_6 because it is an isotropic chemical process. This can be observed from Figure 2.36, in which the width of the silicon beam is less than the upper hard mask. When this process continues, this undercut is more serious and is similar to the case of cryogenic enhancement deep etching, in which an 18 μm undercut is observed for a 275 μm deep etching, and an 8 μm undercut is observed for a 125 μm deep etching [31].

Let Δa and Δb denote the undercuts at each side of the comb finger and the folded beams, respectively. The gap of the consequent finger increases to $g + 2\Delta a$ and a smaller driving force is generated. When the folded beam narrows to $b_f - 2\Delta b$, the stiffness along the lateral direction decreases. The displacement is affected by the two undercuts. Substituting the

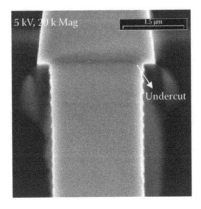

FIGURE 2.36
SEM micrograph of undercut of deep reactive ion etching (DRIE).

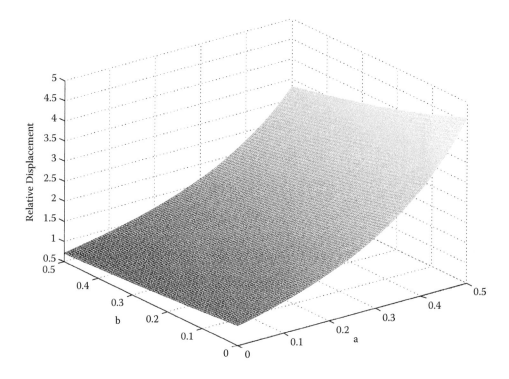

FIGURE 2.37
Relative displacement generated by undercut microbeams.

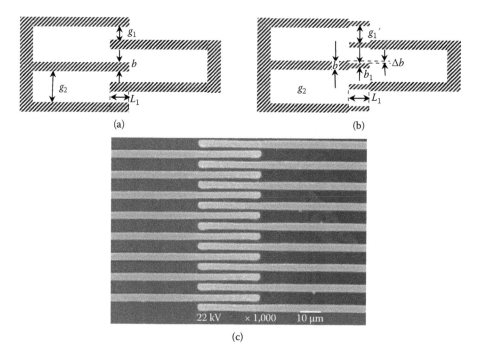

FIGURE 2.38
Comparison between the designed pattern and the fabricated structure: (a) designed uniform comb fingers, (b) schematic of a nonuniform gap comb set, and (c) SEM micrograph of a fabricated nonuniform gap comb set.

gap and beam width in Equation 2.59, the relative displacement can be expressed as:

$$\frac{\Delta x_{undercut}}{\Delta x} = \frac{g_0}{(g_0 + 2\Delta a)}\left(\frac{b_f}{b_f - 2\Delta b}\right)^3$$ (2.90)

Figure 2.37 illustrates the effects on the displacement of the undercuts of the comb finger and the folded beam. It shows that the displacement decreases when the undercut is observed on the comb fingers only. However, the displacement increases when the undercut is observed on the folded beam.

The stability changes as a result of the variation of stiffness along lateral and transverse directions, and the relative stability is expressed as:

$$\left(\frac{k_y}{k_x}\right)_{undercut} \bigg/ \left(\frac{k_y}{k_x}\right)_{design} = \left(\frac{b_f}{b_f - 2\Delta b}\right)^2$$ (2.91)

The relative stability increases with a rise in the undercut because the stiffness along the lateral direction decreases faster than that along the transverse direction.

2.4.6 Nonuniform Gap

The finger width and the gap between the two adjacent fingers are designed to be constant along the whole length, as shown in Figure 2.38a. The gap at the overlap section can be expressed as:

$$g = g_1 = \frac{g_2 - b}{2}$$ (2.92)

where b is the width of the fingers, g_1 and g_2 are the gaps of the overlap and nonoverlap, respectively, and L_1 is the initial overlap length. However, the width of the finger is nonuniform, as shown in Figure 2.38b, due to different undercuts. The finger width of the overlap part is changed to $b_1 = b - 2\Delta b$, and the gap becomes $g_1' = g_1 + 2\Delta b$, whereas the finger width and gap of the nonoverlap part are maintained at b and g_2, respectively. Figure 2.38c shows the SEM micrograph of the nonuniform movable comb set. The driving force under the same voltage deviates from the designed value during the movement. On the assumption that the height of the whole fingers is uniform at H, the capacitance is:

$$C_0 = 2n\varepsilon\varepsilon_0 H\left[\frac{2\Delta x}{g_1 + \Delta b} + \frac{L_1 - \Delta x}{g_1 + 2\Delta b}\right] \quad (0 \le \Delta x \le L_1)$$ (2.93a)

$$C_0 = 2n\varepsilon\varepsilon_0 H\left[\frac{2L_1}{g_1 + \Delta b} + \frac{(\Delta x - L_1)}{g_1}\right] \quad (L_1 \le \Delta x)$$ (2.93b)

Therefore, the electrostatic force can be changed as:

$$F_x = n\varepsilon\varepsilon_0 H\left[\frac{2}{g_1 + \Delta b} - \frac{1}{g_1 + 2\Delta b}\right]V^2 \quad (0 \le \Delta x \le L_1)$$ (2.94a)

$$F_x = \frac{n\varepsilon\varepsilon_0 H}{g_1}V^2 \quad (L_1 \le \Delta x)$$ (2.94b)

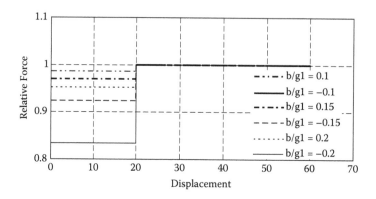

FIGURE 2.39
The relative force with different deflection gaps.

The relative force compared to the ideal comb drive of even width and gap is plotted in Figure 2.39, where a jump occurs at the critical point when the displacement is equal to the initial overlap length. The force is less than in the ideal case, whether the overlap part is wider or narrower than the nonoverlap part. It decreases more when $\Delta b/g_1 < 0$, or when the gap of the overlap part is narrower than the nonoverlap part. However, the drop is not symmetric. The electrostatic force decreases faster on the left-hand side when $\Delta b/g_1 < 0$ as shown in Figure 2.40. Therefore, the displacement under the same applied voltage and suspended by the same folded beam is different.

The static displacement is formulated as:

$$\Delta x = \frac{F_x}{k_x} = \frac{n\varepsilon\varepsilon_0 h}{k_x}\left[\frac{2}{g_1+\Delta b} - \frac{1}{g_1+2\Delta b}\right]V^2 \qquad (0 \le \Delta x \le L_1) \tag{2.95a}$$

$$\Delta x = \frac{F_x}{k_x} = \frac{n\varepsilon\varepsilon_0 h}{g_1 k_x}V^2 \qquad (L_1 \le \Delta x) \tag{2.95b}$$

Let $k_x = 1N/m$, $n = 150$, $g_1 = 2.5\mu m$, and $\varepsilon = 1.0$; the displacement varies with the applied voltage as illustrated in Figure 2.41. The displacement is still proportional to the square of the voltage, but the coefficient is no longer a constant. An abrupt change in the displacement

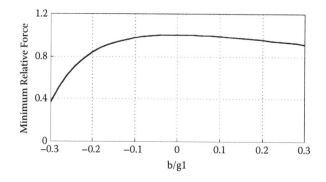

FIGURE 2.40
The minimum relative force versus deflection when $\Delta x \le L_i$.

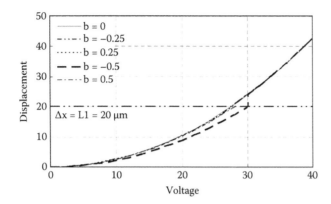

FIGURE 2.41
Displacement of the different deformed gap versus applied voltage.

is observed when the motion is equal to the initial overlap length. The corresponding voltage is different as a result of different deformations. The jumping distance changes according to the gaps. For example, when $\Delta b = -0.50\,\mu m$, the key voltage is 30.083 V and the jumping displacement is 4.036 μm. When $\Delta b = 0.25\,\mu m$, the key voltage is 27.650 V and the jumping displacement is 0.306 μm.

2.4.7 Overall Effects of Various Tolerances

The effects of each individual DRIE fabrication tolerance on the performance of the lateral comb-drive actuator were discussed. A positive slanted comb finger generates greater force as a result of the larger capacitance of a narrower gap, whereas a negative sloped folded beam decreases the stiffness along the direction of motion. These two effects provide larger displacement than the designed corresponding value. The uneven depth of the different parts of the individual comb finger makes the force and displacement jump when the displacement reaches the overlap length. The last type of tolerance, undercut, reduces the electrostatic force and the stiffness along the direction of movement as well. These two effects decide the displacement of the actuator. However, in actual experiments, a combination of these fabrication tolerances is observed.

To investigate the effects of these fabrication tolerances on the lateral comb-drive actuator, one deep-etched planar comb-drive actuator with comb finger gap g of 2.5 μm, depth of the released beams H of 30 μm, and folded beam width b_0 of 2.0 μm was fabricated. The slope of the comb finger, α, is $+1.0°$, and the slope of the folded beam, β, is $-1.0°$. The undercut for the comb finger, Δa, is 0.025 μm, and the undercut for the folded beam, Δb, is 0.075 μm. Substituting all the values in Equations 2.67 and 2.77, the theoretical relative displacement compared to the ideal vertical microstructures is:

$$\frac{\Delta x_{slope-undercut}}{\Delta x_{ndesign}} = 3.023 \tag{2.96}$$

The experimental results are plotted in Figure 2.42, and the designed curve is also plotted according to $\Delta x_{design} = 0.011V^2$ which is modified to $\Delta x_{design(tolerance)} = 0.033V^2$, considering the fabrication tolerances for comparison. It is observed that the modified designed curve and experimental results are consistent with each other except at a higher driving

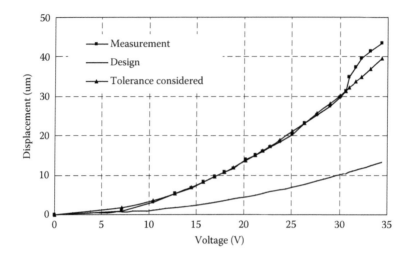

FIGURE 2.42
Displacement of the undercut sloped comb-drive actuator.

voltage range. This is the result of a more significant fringe effect because the distance between the fingertip and the frame of the fixed comb set is reduced with the motion of the movable comb set. In this case, an additional term for the capacitance of the comb sets, expressed as:

$$C_{fringe} = \frac{(1+\frac{\pi}{2})(2n+1)b\varepsilon_0 H}{L-x}$$

has to be considered. Hence, the electrostatic force becomes:

$$F_x = \left\{ \frac{n\varepsilon_0 H}{g} + \frac{(1+\frac{\pi}{2})(2n+1)b\varepsilon_0 H}{2(L-x)^2} \right\} V^2$$

When the overlap length of the two comb sets is over 70% of the total length of L, the electrostatic force is no longer quadratic to the applied voltage, but is dependent on the displacement as well.

2.5   Large-Displacement Actuator

A self-latched micromachined mechanism with a large displacement ratio is discussed in this section. The large output displacement is obtained by amplifying the small input motion through the elastic deformation of the compliant configuration, which realizes the self-latched output by local bifurcation effect. The design theory and synthesis of compliant microstructures are presented. The numerical and analytical simulations are implemented to the linear amplification, which shows that the displacement magnification is as high as 50 times. Self-latching is realized by the bifurcation effect to maintain fixed output displacement even though the input goes further as indicated by the large displacement simulation.

2.5.1 Design and Theoretical Analysis

In this section, a stable self-locked actuator with large displacement ratio (output to input) is designed and theoretically analyzed. This actuator is a linear displacement device with adjustable displacement ratio ranging from 5 to 100 through the elastic deformation of the compliant microstructure. Self-locking is achieved by the bifurcation effect, whereby the output displacement remains stable even when the input displacement changes. Therefore, the drawbacks of the previous actuators in terms of instability and small displacement are eliminated. A stable output is ensured by the self-locking features, which consequently eliminates the effects of vibration. Moreover, many benefits, such as prevention of assembly, avoidance of stiction, elimination of joint friction, precision, accuracy, repeatability, absence of backlash, and [32,33] can be achieved by this design. Additionally, fabrication is entirely compatible with surface and bulk micromachining.

2.5.1.1 Configuration of a Stable and Large Displacement Actuator

The configuration of the self-latched micromachined actuator with large displacement ratio is schematically shown in Figure 2.43. In this self-latched micromachined actuator system, the driving force is generated by a comb drive, and the displacement amplification is realized by the compliant configuration consisting of six beams that can receive the input force, store the energy as strain energy by the deformation of the beams, and finally release the energy with predetermined displacement. Also, self-latching is obtained by the compliant structures when the loading force is over the critical value. The folded suspension beam provides a very stiff flexure along the y direction to ensure the stability of the entire system. Besides, the stiffness along direction of motion of the folded beam is small compared to that of the compliant microstructure such that most of the force generated by the comb-drive actuator falls on the compliant microstructure.

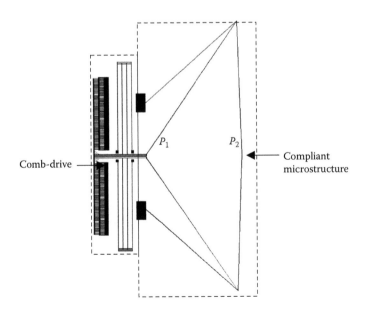

FIGURE 2.43
The schematic of a self-latched micromachined actuator.

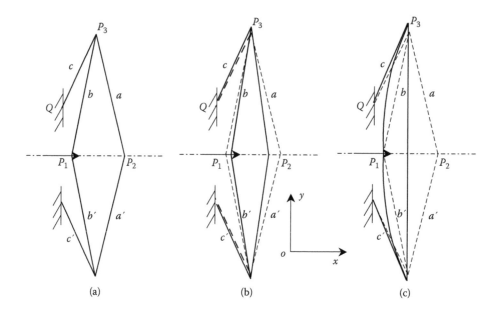

FIGURE 2.44
Schematic of a compliant structure of the actuator: (a) original configuration, (b) linear amplification, and (c) self-locking position.

In the linear amplification phase, the input displacement of point P_1 is amplified by the compliant microstructure to achieve a large output displacement at point P_2. With a rise in the driving force, the bifurcation phenomenon, leading to the self-latching of the output point P_2 through the input point P_1, goes further, and a stable output displacement is achieved. When the driving voltage is reduced, all of the microbeams release back to their original position as the deformation is elastic.

The compliant structure is schematically shown in Figure 2.44. It consists of six beams with axial symmetry along the x axis. Beams c and c' are fixed to the substrate at Q and Q', respectively. Beams a, b, and c are connected at point P_3. The electrostatic force is applied to P_1 along the positive x direction, whereas the amplified output displacement is achieved at point P_2 along either the positive or the negative x direction. In the linear amplification range, the input displacement of point P_1 is amplified by the compliant microstructure to achieve a large output displacement at point P_2, as shown in Figure 2.44b. With a rise in the driving force, bifurcation occurs, leading to the self-locking of the output displacement at point P_2 (see the schematic view in Figure 2.44c). Afterward, the output point P_2 remains fixed even though the input point P_1 goes further, and thus a stable output displacement is achieved. When the driving voltage is unloaded, all the microbeams are released back to their original positions, proving that the deformation is elastic. The linear displacement amplification and bifurcation effect of the compliant microstructure are analyzed in the following subsections.

2.5.1.2 Theoretical Analysis of Linear Displacement Amplification

As the compliant microstructure is plane-symmetric along the x-axis to ensure motion stability, only half of the model, as shown in Figure 2.45, is used to analyze the amplification. Half of every compliant microstructure is composed of three beams denoted as a, b, and c.

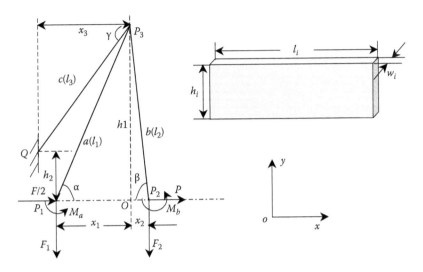

FIGURE 2.45
Analytical schematic of an actuator.

The left end of beam a is connected to the driving part, which provides the input displacement at point P_1. The final amplified output point P_2 is at the right end of beam b. The left end of beam c is fixed to the substrate at point Q. The other ends of the three beams are connected together at point P_3. In the compliant microstructure, the cross-sectional areas, lengths, and inertial moments of the three beams are A_i (product of thickness t_i and width w_i), L_i, and I_i ($i = 1$, 2, and 3 representing beams a, b, and c), respectively. The configuration of the system is further represented by angles α, β, and γ.

When point P_1 is actuated and obtains an input displacement of δ_1, the final output displacement at point P_2 reaches δ_2. The displacement ratio is defined as:

$$R = \frac{\delta_2}{\delta_1} \tag{2.97}$$

The sign of ratio R depends on the relative directions of motion of the input and output points. If they move in the opposite directions, R is negative, and vice versa. The amplitude of the displacement ratio is the absolute value of R.

In general, such a compliant design is governed by the relationship between the input and output forces and displacements. A multicriterion optimization-based method can be used for the topological synthesis of the compliant structure, taking into account both the kinematical requirements and the structural requirements while maximizing the mechanical efficiency [34]. The advantage of such an approach is that precise control over the mechanical and geometric merits of the structure can be enforced during the optimization process to obtain a quantitative determination of an optimized design for the actuator. In this study, the criterion for the geometric design is how to maximize the amplitude of the stable displacement ratio defined in Equation 2.84. As illustrated in Figure 2.45, the driving force applied to the half structure along the displacement direction is half of the total driving force F according to the symmetric characteristics of the structure system. The generalized reacting forces and moments that come from the other half are F_1, F_2, M_a, and M_b. The internal forces N_i and momentums M_i (i is equal to 1, 2, and 3 representing beams

a, b, and c) of the individual beams in the linear amplification range can be expressed as:

$$N_1 = F_1 \sin \alpha - \frac{F}{2} \cos \alpha \tag{2.98a}$$

$$M_1 = M_a + F_1 \cos \alpha \cdot x + \frac{F}{2} \sin \alpha \cdot x \tag{2.98b}$$

Similarly,

$$N_2 = F_2 \sin \beta \tag{2.99a}$$

$$M_2 = M_b - F_2 \cos \beta \cdot x \tag{2.99b}$$

$$N_3 = \frac{F}{2} \cos \gamma - F_1 \sin \gamma - F_2 \sin \gamma \tag{2.100a}$$

$$M_3 = M_a + M_b + \frac{F}{2} l_1 \sin \alpha - F_2 l_2 \cos \beta + F_1 l_1 \cos \alpha - \frac{F}{2} \sin \gamma \cdot x - (F_1 + F_2) \cos \gamma \cdot x \tag{2.100b}$$

The elastic strain energy of the beam system can be expressed as:

$$U = \sum_{i=1}^{3} \left(\int_{l_i} \frac{N_i^2}{2EA_i} dx_i + \int_{l_i} \frac{M_i^2}{2EI_i} dx_i \right) \tag{2.101}$$

where E is Young's modulus of the structure material that is equal to 150 GPa for single-crystalline silicon.

On the basis of the Castigliano theory, the input and output displacements of points P_1 and P_2 can be deduced under the loading force F. The well-known Castigliano's theorem [35] states that, if a structure is subjected to system-external forces $f_1, f_2, ..., f_n$, and if only one virtual displacement $\Delta\delta_i$ is applied in the direction of the i-th displacement δ_i, the expression for δ_i is:

$$\delta_i = \frac{\partial U}{\partial f_i} \tag{2.102}$$

in which δ_i denotes the generalized displacement.

According to the boundary conditions of the compliant structure, the relative generalized translational displacements in the y direction and rotational displacements at points P_1 and P_2 should be zero. Therefore, for the compliant microstructure, the strain energy and generalized force has the following relations:

$$\frac{\partial U}{\partial F_m} = 0 \quad \text{and} \quad \frac{\partial U}{\partial M_n} = 0 \tag{2.103}$$

where m can take values 1 and 2, indicating the reacting forces, and n can be a or b, representing the moments, which come from the other half of the structure. Thus, the relationship between the forces and displacements is described as:

$$\begin{bmatrix} C_{11} & C_{12} & C_{13} & C_{14} \\ C_{21} & C_{22} & C_{23} & C_{24} \\ C_{31} & C_{32} & C_{33} & C_{34} \\ C_{41} & C_{42} & C_{43} & C_{44} \end{bmatrix} \cdot \begin{bmatrix} F_1 \\ F_2 \\ M_a \\ M_b \end{bmatrix} = \begin{bmatrix} D_1 \\ D_2 \\ D_3 \\ D_4 \end{bmatrix} \tag{2.104}$$

where the elements of the coefficient matrix [C] are the functions of the geometric configurations and the structure stiffness of the beam system, whereas the column matrix [D] depends on the external loading force F. These elements can be deduced from Equations 2.85 to 2.91, and the detailed formulations are listed in the appendix. Therefore, the general reacting forces F_m and the momentums M_n can be expressed by the loading force F, the structure configuration, and the individual stiffness of beams. The displacements of point P_1 can be expressed as:

$$\delta_1 = \frac{l_1}{EA_1}\left(-\frac{F_1}{2}\sin\alpha\cos\alpha + \frac{F}{4}\cos^2\alpha\right) + \frac{l_3}{EA_3}\left(\frac{F}{4}\cos^2\gamma - \frac{F_1}{2}\sin\gamma\cos\gamma\right.$$

$$\left. -\frac{F_2}{2}\sin\gamma\cos\gamma\right) + \frac{1}{EI_1}\left(\frac{M_1}{4}l_1^2\sin\alpha + \frac{F_1}{6}l_1^3\cos\alpha\sin\alpha + \frac{F}{12}l_1^3\sin^2\alpha\right)$$

$$+\frac{l_3}{EI_3}\left(\frac{M_1}{2}l_1\sin\alpha + \frac{M_2}{2}l_1\sin\alpha + \frac{F}{4}l_1^2\sin^2\alpha - \frac{F_2}{2}l_1l_2\sin\alpha\cos\beta\right.$$

$$+\frac{F_1}{2}l_1^2\sin\alpha\cos\alpha - \frac{F}{8}l_1l_3\sin\alpha\sin\gamma - \frac{F_1}{4}l_1l_3\sin\alpha\cos\gamma - \frac{F_2}{4}l_1l_3\sin\alpha\cos\gamma$$ (2.105)

$$-\frac{M_1}{4}l_3\sin\gamma - \frac{M_2}{4}l_3\sin\gamma - \frac{F}{8}l_1l_3\sin\alpha\sin\gamma + \frac{F_2}{4}l_2l_3\sin\gamma\cos\beta$$

$$\left. -\frac{F_1}{4}l_1l_3\sin\gamma\cos\alpha + \frac{F}{12}l_3^2\sin^2\gamma + \frac{F_1}{6}l_3^2\sin\gamma\cos\gamma + \frac{F_2}{6}l_3^2\sin\gamma\cos\gamma\right)$$

To obtain the displacement of point P_2, a virtual force P is assumed to point P_2 as shown in Figure 2.44. Hence, the internal forces and momentums are the functions of the virtual force. The displacement of point P_2 is expressed as:

$$\delta_2 = \frac{l_2}{EA_2}\left(\frac{F_2}{2}\sin\beta\cos\beta\right) + \frac{1}{EI_2}\left(\frac{M_2}{4}l_2^2\sin\beta - \frac{F_2}{6}l_2^3\cos\beta\sin\beta\right)$$

$$+\frac{1}{EA_3}\left(\frac{F}{4}l_3\cos^2\gamma - \frac{F_1}{2}l_3\sin\gamma\cos\gamma - \frac{F_2}{2}l_3\sin\gamma\cos\gamma\right)$$

$$+\frac{l_3}{EI_3}\left(\frac{M_1}{2}l_1\sin\alpha + \frac{M_2}{2}l_1\sin\alpha + \frac{F}{4}l_1^2\sin^2\alpha - \frac{F_2}{2}l_1l_2\sin\alpha\cos\beta\right.$$

$$+\frac{F_1}{2}l_1^2\sin\alpha\cos\alpha - \frac{F}{8}l_1l_3\sin\alpha\sin\gamma - \frac{F_1}{4}l_1l_3\sin\alpha\cos\gamma - \frac{F_2}{4}l_1l_3\sin\alpha\cos\gamma$$ (2.106)

$$-\frac{M_1}{4}l_3\sin\gamma - \frac{M_2}{4}l_3\sin\gamma - \frac{F}{8}l_1l_3\sin\alpha\sin\gamma + \frac{F_2}{4}l_2l_3\sin\gamma\cos\beta$$

$$\left. -\frac{F_1}{4}l_1l_3\sin\gamma\cos\alpha + \frac{F}{12}l_3^2\sin^2\gamma + \frac{F_1}{6}l_3^2\sin\gamma\cos\gamma + \frac{F_2}{6}l_3^2\sin\gamma\cos\gamma\right)$$

Substituting the aforementioned relationship between F_m, M_n, and load F, the displacements of the input point P_1 and output point P_2 are obtained. The displacement ratio is obtained using Equation 2.84.

TABLE 2.6

Dimensions of the Compliant Mechanical Mechanism

Parameters	x_1	x_2	x_3	h_1	h_2	w_i	t_i
Dimensions (μm)	1002.5	51.5	1014.0	1432.5	296.5	3.0	25

The configuration can also be represented by the coordinates as x_1, x_2, x_3, h_1 and h_2 (see Figure 2.44); this representation makes it easier to understand the design parameters and their influence on the amplification properties. The effects of the individual parameter on the amplification are analyzed on the basis of the dimensions listed in Table 2.6.

Figure 2.46a shows the effect of x_2 on the displacement ratio where it sweeps from −100.0 μm to +400.0 μm, whereas the other parameters and loading force remain fixed. The most

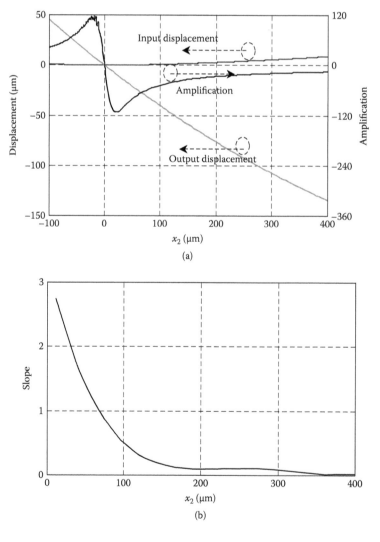

FIGURE 2.46

The effect of x_2 on amplification: (a) displacement and amplification, and (b) deviation of amplification.

significant phenomenon is that the ratio R changes from positive to negative when $x_2 = 0.0$, which means that the output point P_2 alters the direction of motion at this critical point even though the input displacement keeps its moving direction. Therefore, the output displacement direction can be adjusted by the position of the output point when the driving force is fixed. When the output point P_2 is located on the right side of the original point O, x_2 is positive and the ratio is negative. The maximum negative ratio is attained when x_2 is 20 μm, and it decreases with the rise of x_2. The amplitude of the displacement ratio decreases significantly from 150.0 to 26.5 when x_2 increases from 20.0 μm to 200.0 μm, and it remains nearly constant (at about 14.6) afterward. This relationship can be illustrated by the deviation of the displacement ratio to the distance x_2 (see Figure 2.46b). The deviation remains nearly zero when the position of P_2 is farther away from 200.0 μm, and this verifies the fact that the displacement ratio remains constant after this particular location.

Figures 2.47a,b illustrate the effect of lateral positions of the input point and the joint point, represented by x_1 and x_3, on the performance of the amplifier. The magnification

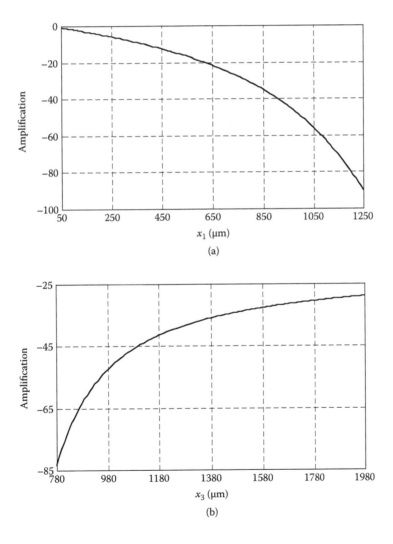

FIGURE 2.47
The effect of the design parameters: (a) x_1 and (b) x_3.

is larger when the input point P_1 is farther away from the original point in the lateral direction as a result of the softer compliant structure. The amplitude of the amplification increases from 10 to 90 when x_1 increases from 380 μm to 1250 μm as shown in Figure 2.47a. The amplification decreases with an increase of x_3, which is opposite to the effect of x_1. Figure 2.47b shows that the amplification decreases from 80 to 30 when x_3 increases from 790 μm to 1850 μm.

The orthogonal coordinates of the joint point and fixed point represented by h_1 and h_2 also affect the displacement ratio. The dependence of the displacement ratio on the joint point position h_1 is plotted in Figure 2.48, which illustrates that the ratio is linear with respect to the joint position h_1 as a result of less stiffness along the x direction. The influence of the fixed point denoted by h_2 is shown in Figure 2.48b, implying that the amplification decreases when the fixed points are farther away in the vertical direction.

The analytical solutions are verified by ANSYS version 8.0 numerical simulation. Case I and case II, corresponding to a fixed x_1 at 1000 μm while x_2 is equal to 50 and 80 μm,

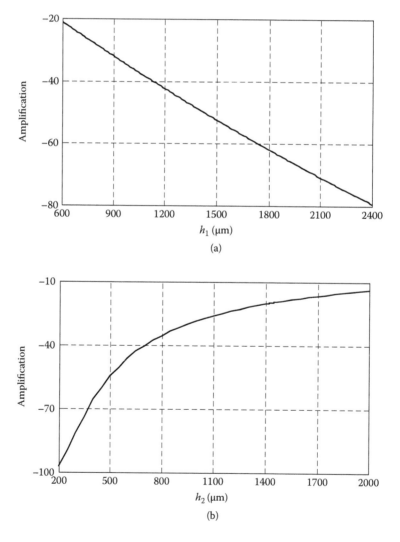

FIGURE 2.48
Amplification versus vertical coordinates: (a) joint point h_1 and (b) fixed point h_2.

respectively, are simulated, which shows that the amplification is −49.7 and −32.6, respectively. This particular difference comes from the different location of the output point P_2. Also, the motion along the y direction of case I is 5.4 μm, which is larger than that of case II of 3.9 μm, and this will determine a greater amplification when the input displacement is equal to 1.0 μm.

2.5.1.3 Self-Latching by Bifurcation Effect

The local bifurcation effect occurs for this compliant microstructure with an increased input displacement under a higher loading. To find the critical load of the system, the bifurcation analysis of the beams system of the micromachined actuator is carried out using the finite element method (FEM).

Because the beams of this actuator are slender, the Euler–Bernoulli hypothesis is valid. The beam warp is negligible and the strains are small, whereas the displacements and rotations are moderately large in the analysis. The element stiffness matrix of beams can be derived employing the principle of stationary potential energy Π_e which can be expressed as:

$$\Pi_e = U_e - V_e \tag{2.107}$$

where U_e is the strain energy stored in the element, and V_e is the external work done to the element; the subscript e represents the element level. For a beam element, the strain energy can be expressed as:

$$U_e = \frac{1}{2} \int_0^L \left[EA \left(\frac{\partial u}{\partial x'} \right)^2 + EI \left(\frac{\partial^2 v}{\partial x'^2} \right)^2 \right] dx' +$$

$$\int_0^L \left[\frac{F_{x'}}{2} \left(\frac{\partial v}{\partial x'} \right)^2 - F_{y'} \left(\frac{\partial u}{\partial x'} \frac{\partial v}{\partial x'} \right) \right] dx' \tag{2.108}$$

where E is Young's modulus, L is the length of element, A is the cross-sectional area, I is the second moment of area, u is the axial displacement, and v is the lateral displacement. $F_{x'}, F_{y'}$, and M are the nodal forces and moments relative to the local coordinate system of the beam element in which x' is the axial coordinate and y' is the lateral coordinate. The first term of Equation 2.96 leads to the linear stiffness matrix, and the second term relates to the geometric stiffness matrix, which can accurately predict lateral bifurcation. The element nodal force vector is constrained in this study to a 2-D beam element, and hence is expressed as:

$$\{F\} = [F_{x'1}, F_{y'1}, M_1, F_{x'2}, F_{y'2}, M_2] \tag{2.109}$$

In Equation 2.96, the bending moment is assumed to be distributed linearly. Therefore, the forces $F_{x'}, F_{y'}$, and M at the internal cross section with the coordinate of x' can be expressed in terms of those at the element ends by using:

$$F_{x'} = F_{x'2} = N \tag{2.110a}$$

$$F_{y'} = -(M_1 + M_2)/L \tag{2.110b}$$

$$M = M_1(1 - x'/L) + M_2(x'/L) \tag{2.110c}$$

In the updated Lagrangian formulation [36], let $\{f\}_e$ and $\{u\}_e$ represent the incremental nodal force and incremental nodal displacement vectors at the two ends of the element. The work done by the element nodal force under nodal displacement increments is:

$$V_e = \{u\}_e^T \{f\}_e \qquad (2.111)$$

Linear interpolation functions can be adopted for the axial displacement u. Cubic interpolation functions can be employed for the lateral displacement v. Substituting these functions into the preceding equations, the expression for the total potential energy may be defined in terms of the incremental nodal displacements at the two ends of the element.

The beam element secant equilibrium equation can be formulated according to the principle of stationary potential energy:

$$\delta \Pi_e = \delta U_e - \delta V_e \qquad (2.112)$$

which leads to the element secant stiffness matrix in the updated Lagrangian formulation, which can be written as:

$$\{f\}_e = ([K_L]_e + [K_G]_e)\{u\}_e \qquad (2.113)$$

where $[K_L]_e$ is the linear stiffness matrix, and $[K_G]_e$ is the geometric stiffness matrix, both of which are discussed in standard textbooks [37].

For this beam system of the micromachined actuator, FEM equations can be obtained by assembling the element stiffness matrix. Hence, the FEM system equation of the micromachined actuator can be written as:

$$([K]_E + \lambda[K]_G)\{\Delta q\} = \lambda\{\Delta F\} \qquad (2.114)$$

where $\{\Delta q\}$ and $\{\Delta F\}$ represent the incremental generalized displacement and load incremental vectors, respectively, and λ is the scaling factor of the corresponding load. The matrix of $[K]_E$ and $[K]_G$ represent the classical first-order elastic stiffness matrix and the geometric stiffness matrix of the system, respectively.

The nonlinear large displacement bifurcation analysis is carried out using commercial software, MSC/NASTRAN, and the critical load of the system is determined. The result is schematically shown in Figure 2.49. When the load reaches the critical value, local bifurcation takes place where a stable output is achieved. Before bifurcation, the stiffness of the system is governed, and contributed to, by beams a, b, and c. For the present beam system, local bifurcation of beam b takes place when the critical load of the system is reached. Figure 2.49 compares the relative position and shape of the compliant structure before and after bifurcation. The thick line shows the beams (with the subscript b) before bifurcation, whereas the thin line indicates the bifurcated deformation (with the subscript a). It shows that beam b deforms most and maintains the position after local bifurcation occurs, even if the driving force increases further, whereas the deformations of other beams along the lateral direction are relatively insignificant, such that they cannot be clearly seen in this figure. Therefore, Figure 2.49b shows the bending of other beams, when all deformations are elastic and the beams will return to their original position on release of the load.

Although local bifurcation takes place, the overall performance of the beam system is still stable. In such a stable state, further horizontal displacement at P_2 is so small compared to those of beams a and c that it can be ignored. At this stage, beam b makes a big contribution of a very large longitude or tension stiffness to the system, which means that the driving force F does not contribute to the transversal deformation of beam b after

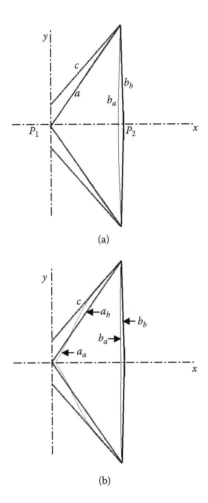

(a)

(b)

FIGURE 2.49
Comparison of the compliant structure: (a) before bifurcation and (b) after bifurcation.

bifurcation. Four models are built with the ramping of h_2 from 317.5 μm to 2053.5 μm (see Table 2.7), and the other parameters have the same values as listed in Table 2.6. It must be noted that the fixed point is on the outside of the joint point P_3 for model 4. The individual amplification under the ramping loading and the resonant frequency for different models are simulated and listed in Table 2.7. Their amplification properties are plotted in Figure 2.50a, and these results coincide with the analytical results. More important, this figure also illustrates clearly that the universal bifurcation effect occurs to all the models when the loading exceeds their corresponding individual critical value.

TABLE 2.7

Models Definition, Resonant Frequency, and Linear Displacement Ratio

Model Number	M1	M2	M3	M4
h_2 (μm)	317.5	567.5	1067.0	2053.0
Resonant frequency (Hz)	2398	2480	2632	2739
Linear displacement ratio	−25.0	−52.0	−29.9	−14.6

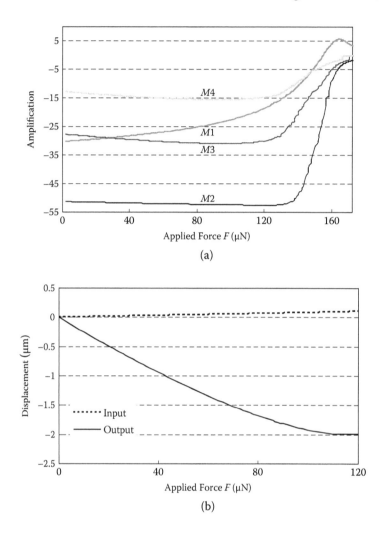

FIGURE 2.50
Numerical simulation results of the bifurcation effect of the amplifier: (a) amplifications for the four models and (b) displacement for model 1.

In normal mechanical systems, bifurcation should be avoided to prevent the system from becoming unstable. However, the bifurcation phenomenon is significant for this micromechanical actuator, so that it can provide a stable self-latched output displacement. In Figure 2.50b, the displacement ratio of every model drops dramatically when the bifurcation effect takes place because the input motion is continuously increasing, whereas the output displacement remains a constant afterward. For example, the displacement ratio is fixed at about 52.3 in the linear amplification range for model 2, whereas it decreases to 50 when the loading reaches its critical value and down to 1 rapidly. Similarly, for model 1, the input displacement increases from 0 to 0.088 μm continuously, whereas the output displacement increases to 1.95 μm when the loading reaches its critical value, and this displacement remains fixed although the loading and input displacement are further increasing, as shown in Figure 2.50b.

Therefore, the self-latched micromachined mechanism with a large displacement ratio is realized by using the linear amplification of the compliant configuration combined with

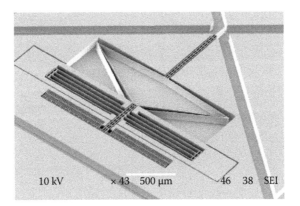

FIGURE 2.51
SEM micrograph of fabricated actuator.

the local bifurcation effect. The linear amplification obtains a large output displacement as it amplifies the small input displacement effectively. Thereafter, the local bifurcation effect works and realizes the self-latching mechanism. Thus, the value of the output displacement remains unchanged even though the input displacement generated by the driving force fluctuates.

2.5.2 Experimental Results and Discussions

In this study, the micromachined actuator is fabricated by using a DRIE process on an SOI wafer. The architecture of the compliant microstructure is optimized according to the theoretical analysis as listed in Table 2.6. The SEM micrograph of a fabricated self-locking actuator with a large displacement ratio is shown in Figure 2.51. Figures 2.52a,b are the SEM micrographs of the input point P_1, and the joint point of the three beams, P_3, respectively. The solid joint point of the three beams serves to prevent the beam from breaking as the stress is high at this particular point. Figure 2.53 shows the deformation of the compliant beams when external voltage is applied to the comb drive, with a comparison of the

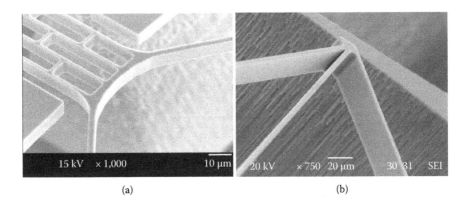

(a)   (b)

FIGURE 2.52
SEM micrographs of the elements: (a) linking input point and (b) joint point.

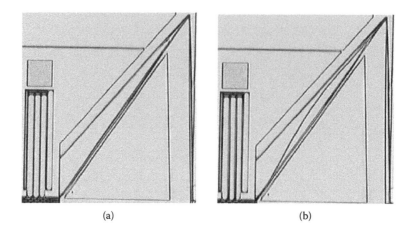

(a)   (b)

FIGURE 2.53
Comparison of the compliant amplifier with and without deformation: (a) original configuration and (b) deformation.

original configuration. It is clearly shown that microbeams *a* and *b* are considerably deformed, and the bifurcation occurs where the output point P_2 is self-locked.

The features of the compliant microstructure are examined and the output displacement versus input displacement is plotted (see Figure 2.54). The input displacement ranges from

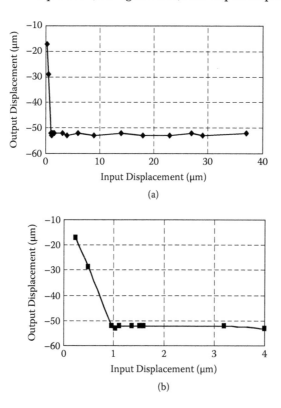

FIGURE 2.54
Measurement results of the output displacement versus input displacement: (a) input ranging from 0 to 38.0 μm and (b) input ranging from 0 to 4.0 μm.

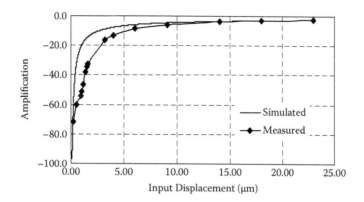

FIGURE 2.55
Measurement results of amplification versus input displacement.

0 to 38.0 μm (see Figure 2.54a). A close-up of the input and output motion relationship, with the input displacement ranging from 0 to 4.0 μm, is shown in Figure 2.54b.

The experiment shows that a stable 52.0 μm output displacement is obtained as a consequence of the efficient amplification of the original input displacement. The output displacement is stabilized as a result of the bifurcation effect within the testing tolerance, as indicated in Figure 2.54a. The figure illustrates that the maximum output displacement is achieved when the input displacement reaches 0.96 μm and the output displacement remains constant even when the input displacement increases further to 38.0 μm. Figure 2.54b shows the linear amplification process, in which the output displacement depends linearly on the input displacement before bifurcation. For example, the output is −17.16 μm when the input displacement is 0.24 μm, resulting in a magnification of 71.5.

The amplification as a function of input displacement is plotted in Figure 2.55, with the simulation curve plotted in the same figure for comparison. Within the motion range the output and input move in opposite directions due to the design, in which the output point P_2 is located at the right side of the joint point P_3 in the lateral direction. Before the bifurcation point, where the critical input displacement is 0.96 μm, the displacement ratio of the output to input is as high as −100 and the magnification decreases with a rise in input displacement. After the critical input value, the magnification of displacement ratio drops as a result of the self-locked output displacement, which means that the output displacement remains at a fixed maximum value even though the input displacement increases continuously. These experimental results confirm the theoretically simulated results.

This character is repeatable in the entire motion range, implying that all the deformation is elastic. All the beams can return to their initial positions when the loading is released. The amplification occurs due to the linear elastic deformation, whereas self-latching is obtained from the bifurcation effect.

2.6 Summary

The optical and electromechanical properties of the MEMS optical switch are discussed in this chapter from the design perspective. Various types of optical coupling loss are discussed, including fiber-to-fiber coupling, mirror-coating material, mirror-coating thickness, and verticality. The PDL is analyzed based on the coupling efficiency between p and s light. This can be used to design and evaluate the optical performance of the optical MEMS devices. After reviewing the principle of the comb-drive actuator, a self-locked micromachined actuator with a large displacement ratio, which has been practically realized, is introduced. The compliant microstructure consists of six symmetrical beams along the direction of movement. A theoretical model is built and used to obtain the optimum design. The analysis shows that the proposed actuator can realize a large displacement ratio because of the elastic deformation of the compliant microstructure. The bifurcation effect is successfully used to attain the self-locked stable output displacement. The experimental results confirm the analytical and numerical simulation results. This self-locked micromachined actuator with a large displacement ratio can be used not only in an optical switch but also in many other applications that require an effective displacement amplification and a stable, large output displacement.

References

1. Li. J., Zhang, Q. X., and Liu, A. Q., Advanced fiber optical switches using deep RIE (DRIE) fabrication, *Sens. Actuators A*, 286, 102, 2003.
2. Popa, D. O., Kang, B. H., Wen, J. T., Stephanou, H. E., Skidmore, G., and Geisberger, A., Dynamic modeling and input shaping of thermal bimorph MEMS actuators, *Proceedings. ICRA '03. IEEE International Conference*, 1, 14–19 September, 1470, 2003.
3. Pichonat-Gallois, E. and Labachelerie, M. de, Thermal actuators used for a micro-optical bench: application for a tunable Fabry-Perot filter, *Transducers, Solid-State Sensors, Actuators and Microsystems, 12th International Conference*, 2, 8–12 June, 14191, 2003.
4. Girbau, D., Lazaro, A., and Pradell, L., RF MEMS switches based on the buckle-beam thermal actuator, *Microwave Conference, 33rd European*, 2, 7–9 October, 651, 2003.
5. Houlet, L., Reyne, G., Bourouina, T., Yasui, M., Hirabayashi, Y., Ozawa, T., Okano, Y., and Fujita, H., Design and realization of magnetic actuators for optical matrix micro-switches, *Optical MEMs, IEEE/LEOS*, 20–23 August, 113, 2002.
6. Lagorce, L. K., Brand, O., and Allen, M. G., Magnetic microactuators based on polymer magnets, *J. Microelectromech. Syst.*, 8, 1, 2, March 1999.
7. Akiyama, T., Collard, D., and Fujita, H., Scratch drive actuator with mechanical links for self-assembly of three-dimensional MEMS, *J. Microelectromech. Syst.*, 6, 1, 10, March 1997.
8. Lai, Y. J., Lee, C., Wu, C. Y., Lin, Y. S., Tasi, M. H., Huang, R. S., and Lin, M. S., Fabrication of corrugated curved beam type electrostatic actuator and SDA driven self-assembling mechanism for VOA applications, *Microprocesses and Nanotechnology Conference*, 29–31, 286, October 2003.
9. Marxer, C. and de Rooij, N. F., Micro-opto-mechanical 2 × 2 switch for single-mode fibers based on plasma-etched silicon mirror and electrostatic actuation, *J. Lightwave Technol.*, 17, 1, 2, January 1999.
10. Akimoto, K., Uenishi, Y., Hunma, K., and Nagaoka, S., Evaluation of comb-drive nickel micromirror for fiber optical communication, *IEEE Annual International Workshop on MEMS*, 26–30, 66, January 1997.
11. Bernstein, J., Cho, S., King, A. T., Kourepenis, A., Maciel, P., and Weinberg, M., A micromachined comb-drive tuning fork rate gyroscope, *MEMS '93, IEEE Proceedings 'An Investigation of Micro Structures, Sensors, Actuators, Machines and Systems'*, 7–10 February, 143, 1993.
12. Xie, H. and Fedder, G. K., A DRIE CMOS-MEMS gyroscope, *Sensors, 2002, Proceedings of IEEE*, 2, 12–14 June, 1413, 2002.

13. Iliescu, C. and Miao, J., One-mask process for silicon accelerometers on Pyrex glass utilising notching effect in inductively coupled plasma DRIE, *Electron. Lett.*, 39, April, 658, 2003.
14. Xie, H., Erdmann, L., Zhu, X., Gabriel, K. J., and Fedder, G. K., Post-CMOS processing for high-aspect-ratio integrated silicon microstructures, *J. Microelectromech. Syst.*, 11, April, 93m 2002.
15. Born, M. and Wolf, E., *Principle of Optics Electromagnetic Theory of Propagation, Interference and Diffraction of Light*, 7th edition, Cambridge University Press, Cambridge.
16. Ghatak, A. K. and Thyagarajan, K., *Optical Electronics*, Cambridge University Press, Cambridge, 410–411, 1989.
17. Nemoto, S. and Markimoto, T., Analysis of splice loss in single-mode fibers using a Gaussian field approximation, *Opt. Quantum Electron.*, 11, 447, 1979.
18. Beckmann, P. and Spizzichino, A., *The Scattering of Electromagnetic Waves from Rough Surfaces*, Artech House, New York, 1987.
19. Bao, M. H., Micro Mechanical Transducers, Vol. 8, Handbook of Sensors and Actuators, Elsevier Science, New York, 2000.
20. Joanson, W. A. and Warne, L. K., Electrophysics of micromechanical comb actuators, *J. Microelectromech. Syst.*, 4, 1 March, 49, 1995.
21. Tang, W. and Chi-Keung, Electrostatic comb drive for resonant sensor and actuator applications, Ph.D. thesis, University of California, Berkeley, 1990.
22. Gere, J. M. and Timoshenko, S. P., Mechanics of Material, 2nd Edition, Belmont, CA: Wadsworth, 1984.
23. Keller, C. G. and Howe, R. T., Nickel-filled HEXSIL thermally actuated tweezers, *8th International Conference on Solid-State Sensors and Actuators (Transducers '95)*, 2, 376, 1995.
24. Avdeev, I. V., Lovell, M. R., and Oipede, Jr. D., Modeling in-plane misalignment in lateral combdrive transducers, *J. Micromech. Microeng.*, 13, 809, 2003.
25. Lee, K. W. and Holmes, A. S., Electrostatic actuation in electroformed Ni microstructures, *IEE Seminar on Microengineering, Modelling and Design* (Ref. No. 1999/052), 4/1, 4 March 1999.
26. Williams, K., A silicon microvalve for the proportional control of fluids, *10th International Conference on Solid-State Sensors and Actuators (Transducers'99)*, 1804, 1999.
27. Li, J., Liu, A. Q., and Zhang, Q. X., Tolerance analysis for comb-drive actuator using DRIE fabrication, *Sens. Actuators A*, 125, 494, 2006.
28. Kiihamaki, J., Kattelus, H., Karttunen, J., and Franssila, S., *Sens. Actuators A*, 82 (1), 234, 2000.
29. Jansen, H., Boer, M. de, and Elwenspoek, M., The black silicon method. VI. High aspect ratio trench etching for MEMS applications, *MEMS '96 Proc. IEEE*, 250, 1996.
30. Li, J., Zhang, Q. X., Liu, A. Q., Goh, W. L., and Ahn, J., Technique for preventing stiction and notching effect on silicon-on-insulator microstructure, *J. Vac. Sci. Technol. B*, 21 (6), 2530, 2003.
31. Singh, A., Mehregany, M., Phillips, S. M., Harvey, R. J., and Benjamin, M., Micromachined silicon fuel atomizor for gas turbine engines, *Proc. IEEE 9th Int. Workshop on MEMS*, 473, 1996.
32. Su, X.-P. S. and Yang, H. S., Design of compliant microleverage mechanisms, *Sensors and Actuators A* 87, 146, 2001.
33. Kota, S., Compliant systems using monolithic mechanisms, *Smart Materials Bulletin*, 7, March 2001.
34. Canfield, S. and Frecher, M., Topology optimization of compliant mechanical amplifiers for piezoelectric actuators, *Struct. Multidisc. Optim.*, 20, 269, 2000.
35. Przemieniecki, J. S., *Theory of Matrix Structural Analysis*. New York: McGraw-Hill, 1968.
36. Bathe, K. J., *Finite Element Procedures*, Prentice-Hall, Englewood Cliffs, 1995.
37. Liu, A. Q., Li, J., Liu, Z. S., Lu, C., Zhang, X. M., and Wang, M. Y., A micromachined self-locked actuator with large displacement ratio, *J. Microelectromech. Syst.*, 15 (6), 1576, 2006.

3

MEMS Thermo-Optic Switches

Jing Li, Tian Zhong, and Ai-Qun Liu

CONTENTS

3.1 Introduction 90
    3.1.1 Mach–Zehnder Interferometer Thermo-Optic Switch 90
    3.1.2 Multimode Interference Thermo-Optic Switch 91
    3.1.3 Spatial Modulation Thermo-Optic Switch 92
3.2 Physics of Thermo-Optic Effect 93
    3.2.1 Thermo-Optic Effect of Silicon Material 94
    3.2.2 Total Internal Reflection 97
    3.2.3 Thermo-Optic Effect and TIR of Silicon 98
3.3 Triangular Thermo-Optic Switch 100
    3.3.1 Configuration of Triangular Prism Thermo-Optic Switch 100
    3.3.2 Modeling and Analysis of the Triangular Prism Switch 102
        3.3.2.1 Gaussian Beam Divergence 102
        3.3.2.2 Fabry–Pérot Effect and Polarization Effect 104
        3.3.2.3 Sources of Optical Losses 107
3.4 Double Cylindrical Prism Thermo-Optic Switch 109
    3.4.1 Configuration of Double Cylindrical Prism Thermo-Optic Switch ...109
    3.4.2 Modeling and Analysis of the Double Cylindrical Prism Switch ...112
        3.4.2.1 Analysis of Optical Effects 112
        3.4.2.2 Simulation of Localized Heating 116
3.5 Implementation of MEMS Thermo-Optic Switches 118
    3.5.1 Fabrication and Experiment on the Triangular Prism Thermo-Optic Switch....118
    3.5.2 Fabrication and Experiment on Double Cylindrical Prism Thermo-Optic Switch 122
        3.5.2.1 Static Switching Characterization 123
        3.5.2.2 Dynamic Switching Characterization 124
3.6 Summary 125
References 126

Thermo-optic switch is another popular optical switching technology for future optical networks. In this chapter, a new type of optical switching device structure—MEMS thermo-optic switch—is explored. It combines the strengths of both MEMS and a waveguide-based thermo-optic device, realizing a low-power, high-speed, and ultra-compact silicon-based optical switch. The new structure is able to break down the design tradeoffs existent in the traditional thermo-optic switches, and utilizes MEMS to enable tuning of the device's working conditions. This chapter begins with a review of current thermo-optic switching technologies. Then, the principles and mechanisms of such MEMS thermo-optic switch are explained by introducing two examples. The design, modeling, fabrication, and characterization of each device are discussed in detail. The chapter concludes with a summary of the advantages of the device with respect to its MEMS and waveguide thermo-optic switch counterparts.

3.1 Introduction

As explained in Chapter 1, the thermo-optic switch is one of the major and most matured optical switching technologies that play a vital role in creating an optical network. It has the significant merits of low-cost and relatively simple implementation. Unlike electro-optic or acousto-optic effects, the thermo-optic effect is found in many materials, such as semiconductors, polymers, dielectrics, etc. So the device could be made using standard IC process. However, the main problem it faces now is most likely its moderate switching speed, ranging from tens to hundreds of microseconds. With the advanced fabrication technologies and careful optimizations of device structures, the latest devices have achieved a response time of a few seconds to even submicroseconds. In this section, the progress of thermo-optic switches is briefly discussed in three of the most popular device structures: Mach–Zehnder interferometer, multimode interference (MMI), and spatial modulation type thermo-optic switch.

3.1.1 Mach–Zehnder Interferometer Thermo-Optic Switch

Mach–Zehnder (MZ) interferometer is a commonly used switching structure. It consists of two 3 dB MZ couplers, two phase shifting branches, and a thin film heater over one branch, which is schematically shown in Figure 3.1. The beam splitter first divides the beam into two beams I_1 and I_2, each passing through one of the two branches and merging as an output, I_{out}. By heating one branch by the thin film heater, the refractive index of this branch is changed. Thus, a phase difference of $\Delta\Phi$ is introduced between these two beams. When they are in phase, constructive interference occurs corresponding to the on-state of the switch; when the phase difference is π, destructive interference occurs, which is the off-state. By modulating the phase difference, the switching effect is realized. The optical output of such devices is expressed as

$$I_{out} = \frac{I_1 + I_2}{2} + \sqrt{I_1 I_2} \cos(\Delta\Phi) \tag{3.1}$$

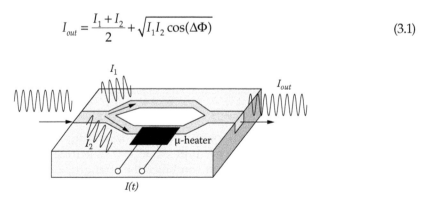

FIGURE 3.1
Schematics of a Mach–Zehnder interferometer thermo-optic switch.

It can be inferred that the optimal contrast (extinction) ratio happens when the power in each branch is equal, for example, $I_1 = I_2$. Therefore, the device performance depends very much on the characteristics of the 3 dB coupler or the splitter. The more the power is evenly distributed in the two branches, the better the switching contrast is. In real practice, the extinction ratio could reach up to 20–30 dB.

Over the last decade, high-speed, low-power, substantial miniaturization has been demonstrated feasible in such an MZ configuration. In 1998, Takashi Goh et al. [1] demonstrated a high extinction ratio thermo-optic matrix switch. It used a double MZ interferometer structure to obtain a 55 dB extinction ratio by blocking the optical leakage first from MZ interferometer. The power consumption of each switching unit is slightly more than 1 W. However, the response speed of such device was unrevealed. In 2003, Espinola et al. [2] improved the MZ switch to a <3.5 μs switching rise time. Such enhanced response time should be attributed to the thin silicon-on-insulator (SOI) technology whose material system exhibits a large change in the refractive index of the silicon-guiding layer with temperature variation.

In 2004, Harjanne et al. [3] successfully fabricated an MZ thermal switch with submicrosecond switching time, which indeed renewed the impression of high-speed thermal photonic devices. In that work, an intensive study was made on the control methods for realizing ultrafast heat removal to push the speed limits of an MZ thermal switch. Bias heating and differential heating were analyzed and compared in terms of switching response time. At the expense of power consumption, the latter method allowed the device to operate at a speed of about 700 ns per switching—the fastest speed ever reported for thermo-optic switches.

With the aid of advancing nanofabrication technology, Chu et al. [4], in 2005, came out with a truly nanosized $1 \times N$ thermo-optic switch using silicon nanowire waveguides. The device achieved a 30 dB extinction ratio, 90 mW power, and about 100 μs switching time with an ultra compactness. Unfortunately, the size of the heater still remains in micrometers scale, which in effect offset the effort of using nanowires to miniaturize the device.

3.1.2 Multimode Interference Thermo-Optic Switch

The multimode interference (MMI) thermo-optic switch actually shares very much the same structure with MZ switches. The difference is that instead of using conventional 3 dB MZ couplers, MMI coupler is adopted. This allows 2×2 switching to be realized with an even more compact configuration than normal MZ switches (only 1×1 switching), thus making easier large-scale integration. Other advantages of MMI thermo-optic switch include wide-band operations and low polarization dependence, which are highly favored by dense wavelength division multiplexing (DWDM) systems. MMI coupler uses the self-imaging principle; in multimode waveguide, as the wave propagates the images of the input field periodically will form in space. Based on this phenomenon, an MMI thermo-optic switch splits the incident light into two equal phase and amplitude beams, modulates the phase difference between two beams, and combines them to form two outputs.

The basic structure is shown in Figure 3.2. In this configuration, a signal can be routed to one of the two outputs by heating up one waveguide branch. In 1995, Fisher et al. [5] fabricated a 2×2 SOI MMI thermo-optic switch. It obtained a switching power as low as 85 mW by etching an SiO_2 thermal isolation trench outside the rid waveguide. The whole device insertion loss was 2.4 dB, approaching the practical requirement. Another similar work was done by Lai et al. [6] in 1998, who also used a deep trench on the sidewall of the waveguide to enhance the thermal isolation, thus reducing the heating power. However,

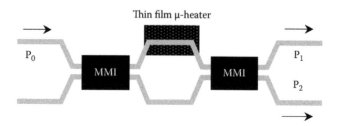

FIGURE 3.2
Schematics of a multimode interferometer (MMI) thermo-optic switch.

as the trench prevents the heat loss, it will slow down the cooling process when the switch is turned off. Consequently, the switching times obtained are slightly longer: 150 μs rise time and 180 μs fall time.

3.1.3 Spatial Modulation Thermo-Optic Switch

As its name implies, the spatial modulation thermo-optic switch controls light by changing its spatial field distribution. Unlike the interference type switches (such as MZ), spatial light modulators make more abrupt transitions between on–off states. The device usually has only two stable switching states, and the output could be only in an on or off state with no other intermediate values. (MZ switches do have intermediate states between on and off.) Because of this property, such a switch is also called a *digital optical switch* (digital implies only 0 and 1 states). It can be inferred that these digital switches have a more sensitive switching characteristics than their interference-type counterparts, exhibiting greater potential for fast- and low-power switching. In addition, digital switches are less polarization-dependent and less sensitive to fabrication tolerance, which is not the case for MZ and MMI switches.

The common designs of this type include *Y* branch, *X* branch, and multiple branch structures. The schematics of the first two are shown in Figures 3.3 and 3.4, respectively.

In a *Y* branch device, without heating, output 1 has a much stronger coupling with the input waveguide than output 2 so that the input light is transmitted to output 1. When

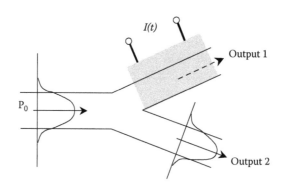

FIGURE 3.3
Schematics of *Y* branch digital thermo-optic switch.

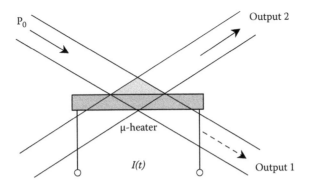

FIGURE 3.4
Schematics of *X* branch total internal reflection thermo-optic switch.

heating is applied to output 1, its refractive index will change, making its coupling to the input waveguide minimal. As a result, the input light goes to output 2. Interestingly, almost all *Y* branch thermo-optic switch are made from polymer waveguides. This might be because, compared to semiconductors, polymer materials have lower refractive index so that the percentage change of index can be much larger. In 1996, Moosburger et al. [7] demonstrated their polymer rid waveguide thermo-optic switch. The branching angle was 0.12°, switching power was around 120 mW, and the temperature increase was 78°C. In 2000, Siebel et al. [8] proposed a similar structure with an extra center branch between two *Y* branches. The advantage of doing so was to obtain a low cross talk of −30 dB, a value independent of the increasing heating power.

For the type of *X* branch devices, the structure consists of two crossing waveguides. At the intersection region, there is a heater for changing the refractive index. When no power is applied, the input light propagates to output 1, whereas an index contrast at the intersection region forms a total internal reflection (TIR) interface, causing input light to be reflected to output 2 when heating is applied.

In 2005, Wang et al. [9] fabricated a polymer-based "X branch" thermo-optic switch for true-time delay application. The crossing angle between waveguides was 6°, the corresponding switching power was 130 mW, and the lowest cross talk was −42 dB; this switch can achieve a switching speed of 2 *μ*s. In 2006, the same authors [10] went on to demonstrate their latest work. With an increased crossing angle, the new switch obtained a low power of 66 mW without worsening the cross talk.

It might have been noticed that the angles between adjacent waveguides are limited to very small values for both *X* and *Y* branch devices. This is because an increase in the branching angle will significantly raise the power consumption. However, by keeping the angles small, the device dimension would be very large to avoid interferences between waveguides. Normally, these kinds of devices have sizes in the millimeter scale, marking a serious disadvantage for on-chip large-scale integration.

3.2 Physics of Thermo-Optic Effect

In this section, the principles of the MEMS thermal-optic switch are discussed in detail to achieve fast switching. Two important mechanisms are involved for this specific switch: thermo-optic effect of single crystal silicon (CSS) and total internal reflection (TIR).

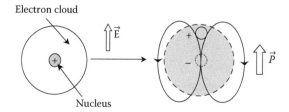

FIGURE 3.5
Schematic of the polarization induced by an external electric field.

3.2.1 Thermo-Optic Effect of Silicon Material

The thermo-optic effect refers to the dependence of the optical refractive index of a material on temperature. This effect is present in all transparent materials including dielectrics, semiconductors, and organic materials, and it has wide application in various optical devices [11]. The thermo-optic coefficient can be positive or negative, depending on the material property. Polymers such as acrylates and polyimides have negative thermo-optic coefficients [12], whereas semiconductor materials such as silicon have positive coefficients. These materials are extensively used in tunable devices. In general, the sign of the coefficient is dependent on the polarizability and the density of a particular material because these two properties are functions of temperature.

The polarizability is a material property that characterizes the ability of distortion of the electron cloud of a molecular entity of that respective material by an electric field [13]. As schematically shown in Figure 3.5, the dipole moment $\vec{p}$ is induced by the external electric field $\vec{E}$.

The relative distance between the positive and negative center is represented by $\vec{x}$. The dipole moment $\vec{p}$ is expressed as

$$\vec{p} = q\vec{x} = \rho\vec{E} \tag{3.2}$$

where ρ is the polarizability of the medium given by $\rho = q\vec{x}/\vec{E}$. A higher polarizability leads to a higher refractive index.

Another term affecting the thermo-optic coefficient is the alteration of the density of the medium induced by a temperature change. This is indicated by the coefficient of thermal expansion (CTE), which is an intrinsic property of the material. Assuming that the medium is homogeneous and isotropic, the material commonly expands uniformly in all directions upon heating as shown in Figure 3.6. Because the number of atoms or molecules in the material remains constant, the density decreases as the volume increases.

The total polarization is expressed as

$$\vec{P} = N\vec{p} = N\rho\vec{E} = \varepsilon\chi_e\vec{E} \tag{3.3}$$

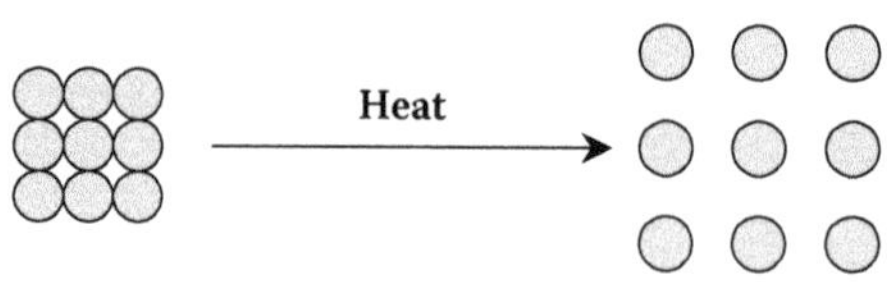

FIGURE 3.6
Schematic of the thermal expansion of the material.

where N is the density of the atom or molecule, ε is the electric permittivity, and χ_e is the electric susceptibility, which is expressed using the refractive index n of the material as:

$$\chi_e = \sqrt{n^2 - 1} \tag{3.4}$$

Thus,

$$N\rho = \varepsilon(n^2 - 1)^{\frac{1}{2}} \tag{3.5}$$

Therefore, the refractive index n can be expressed as:

$$n = \left(1 + \left(\frac{N\rho}{\varepsilon}\right)^2\right)^{\frac{1}{2}} \tag{3.6}$$

This equation shows that the refractive index is directly related to the density N and the polarizability ρ of the material.

The thermo-optic effect of silicon is explained by the single oscillator model [14]. For the critical points of the interband transitions in the first Brillouin zone, the transition energies decrease with higher temperature. Coupling among various types of oscillators to the electromagnetic radiation field is often employed to explain the optical properties of the material. The real part of the dielectric constant ε in the transparent range can be written as [14]:

$$\varepsilon = n^2 = 1 + E_p^2 \sum_k \frac{f_{cv}(k)}{E_{cv}^2(k) - E^2} \tag{3.7}$$

where E_p is the electronic plasma energy and is expressed as $E_p = \sqrt{4\pi N\hbar^2 e^2 / m_e}$ with the effective mass of m_e and using the reduced Planck constant $\hbar$ and single electron charge e. E is the photon energy, k is the reciprocal lattice vector, and $E_{cv}(k)$ and $f_{cv}(k)$ are the transition energy and the interband oscillator strength between the valence and conduction band, respectively.

When a single oscillator occurs, Equation 3.7 can be simplified to:

$$n = \sqrt{1 + \frac{E_p^2}{E_g^2 - E^2}} = \sqrt{1 + \frac{4\pi N\hbar^2 e^2}{m_e\left(E_g^2 - E^2\right)}} \tag{3.8}$$

where E_g is the optical band gap. Thus,

$$\frac{dn}{dT} = \frac{dn}{dN}\frac{dN}{dT} + \frac{dn}{dE_g}\frac{dE_g}{dT} \tag{3.9}$$

$$\frac{dn}{dN}\frac{dN}{dT} = \frac{1}{2}\left(1 + \frac{4\pi N\hbar^2 e^2}{m_e\left(E_g^2 - E^2\right)}\right)^{-\frac{1}{2}} \frac{4\pi\hbar^2 e^2}{m_e\left(E_g^2 - E^2\right)}\frac{dN}{dT} = -\frac{3k_{ex}(n^2 - 1)}{2n} \tag{3.10}$$

where T is the temperature. k_{ex} is the thermal expansion coefficient of silicon, which is expressed as [15]:

$$k_{ex}(T) = 3.725 \times 10^{-6}\{1 - \exp[-5.88 \times 10^{-3}(T - 124)]\} + 5.548 \times 10^{-10}T \tag{3.11}$$

The optical band gap is affected by the temperature as given by:

$$\frac{dn}{dE_g}\frac{dE_g}{dT} = -\frac{n^2-1}{2nE_g}\frac{2}{1-(\frac{E}{E_g})^2}\frac{dE_g}{dT}$$ (3.12)

According to the empirical relationship given by Varshni [16–18], $E_g(T)=E_g(0)-\frac{\alpha_g T^2}{(T+\beta_g)}$, where $E_g(0)$ is the average band gap at 0 K and the constants α_g and β_g are related to the electron–phonon interaction and Debye temperature of silicon. With this relation, Equation 3.12 can be further expressed as:

$$\frac{dn}{dE_g}\frac{dE_g}{dT} = \frac{n^2-1}{2nE_g}\frac{2\alpha_g}{1-(\frac{E}{E_g})^2}\left[\frac{2T}{T+\beta_g} - \frac{T^2}{(T+\beta_g)^2}\right]$$ (3.13)

Therefore, the thermo-optic coefficient is expressed as:

$$\frac{dn}{dT} = \frac{n^2-1}{2n}\left\{-3k_{ex} + \frac{2\alpha_g}{E_g}\frac{1}{1-(E/E_g)^2}\left[\frac{2T}{T+\beta_g} - \frac{T^2}{(T+\beta_g)^2}\right]\right\}$$ (3.14)

When the constants take the values of $E_g(0) = 4.03(eV)$, $E_p = 13.31(eV)$, $\alpha_g = 3.41 \times 10^{-4}$ $(eV \cdot K^{-1})$, $\beta_g = 439(K)$, [14] and $n = 3.42$, the thermo-optic coefficient can be expressed as:

$$\frac{dn}{dT} = 8.919 \times 10^{-5} + 3.156 \times 10^{-7}T - 1.615 \times 10^{-10}T^2 (K^{-1})$$ (3.15)

By neglecting the higher order term of the thermo-optic coefficient, the relationship between the temperature change and the refractive index change is calculated and plotted in Figure 3.7. It shows that the refractive index of single crystal silicon increases linearly with the temperature.

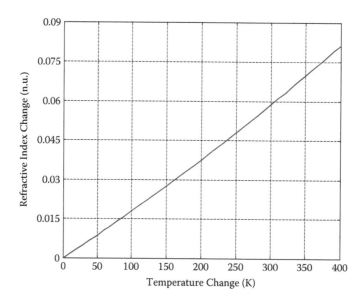

FIGURE 3.7
Relationship between refractive index change and temperature change.

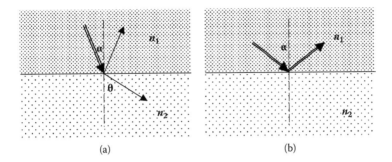

FIGURE 3.8
Optical beam propagation between two media: (a) transition and reflection, and (b) total internal reflection (TIR).

3.2.2 Total Internal Reflection

To make use of the thermo-optic effect for optical switching, the TIR phenomenon plays an important role and thus needs a detailed study, including the basic relations and wavelength dependence.

When one light is incident on the interface of two media, for most cases, major part is transmitted into medium 2 and the rest is reflected into the original medium as shown in Figure 3.8a. If, however, the index of refraction of medium 1 (n_1) is greater than the index of medium 2 (n_2) and the angle of incidence exceeds the critical angle α_c, TIR occurs. The incident light is reflected completely into medium 1 as schematically shown in Figure 3.8b.

According to Snell's law [19], the relationship between the incident angle α and the refractive angle θ is governed by:

$$n_1 \sin\alpha = n_2 \sin\theta \tag{3.16}$$

where n_1 and n_2 are the refractive indices of the two media, respectively. In this study, the two media are silicon and air, Equation 3.16 is written as:

$$n \sin\alpha = \sin\theta \tag{3.17}$$

where n is the refractive index of silicon and $n_2 = 1$. At room temperature and for 1550 nm wavelength, it has $n \approx 3.42$ [20]. Although single crystalline silicon is anisotropic, it has inversion symmetry implying a constant refractive index. When the incident angle α meets the condition of $\sin\alpha \leq n_0/n = 1/n$, the refraction angle θ is a real value. At the onset of TIR, the refractive angle is equal to 90°, in which the critical incident angle α_c is given as:

$$\alpha_c = \sin^{-1}(n_1/n_2) = \sin^{-1}(1/n) \tag{3.18}$$

For silicon–air interface, it has $\alpha_c = \sin^{-1}(1/3.42) = 17.0°$. Once the incident angle is larger than this critical one, the TIR would take place.

For silicon-based prism, it has $\alpha_c = \arcsin(1/3.42) = 17.0°$.

There are two methods to switch the incident light between transmission and total internal reflection. One is to change the incident angle between less and beyond the critical angle, so that the light can be changed between the TIR state and the transmission state. But it is a practical way to realize the switching because the input light is normally fixed.

The second is to change the critical TIR angle. According to the analysis of thermo-optic effect, the refractive index of the material can be tuned by adjusting the temperature while the incident angle is fixed. Compared with the method of adjusting the incident angle, this approach is more efficient and reliable because there is no mechanical movement required. The stability and reliability are therefore assured. Additionally, the strong dependence of the refractive angle on the optical properties of the medium makes the switch more sensitive to the control parameters.

3.2.3 Thermo-Optic Effect and TIR of Silicon

Equation 3.18 shows that the critical angle α_c is the function of the refractive index of single crystal silicon assuming that the refractive index of the other medium, air maintains a constant, 1.0. Hence, if the incident angle is fixed at a value slightly shifted from the critical angle, it is possible to tune the direction of the output light by adjusting the refractive index of silicon. For example, the incident angle is slightly less than the critical angle. Then, it partially transmits to the other medium, which is air. When the refractive index is increased and the incident angle maintains the initial value, the critical TIR angle will decrease. If the increase of the refractive index of silicon is proper, the TIR will happen due to the change of the critical TIR angle. Thus the transmitted light will experience a sudden drop in intensity and the incident light will be reflected to medium 1, silicon, as a result of TIR. The relationship between the refraction index change Δn required and the original refractive angle θ is expressed as:

$$\Delta n = \frac{1}{\sin \alpha} - n_{ini} = \frac{3.42}{\sin \theta} - 3.42 \tag{3.19}$$

Figure 3.9 shows the required refractive index change Δn for different original refractive angle from 86° to 90°. It depicts the change of the refractive index needed to realize the total internal reflection when the initial refractive angle θ varies from 86° to 90°. It is observed that the output selection is so sensitive to the refractive index change that only a 5.2×10^{-4} increase of the refractive index is enough to switch the transmitted light at 89° to the TIR state. Therefore, changing the refractive index of silicon is an effective way to switch the input signal between the transmission state and the TIR state.

By considering the thermo-optic effect of silicon, the relationship between the temperature change required and the initial refraction angle of the transmitted light is simulated

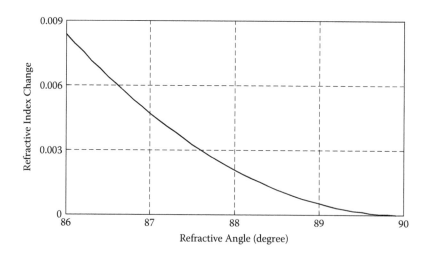

FIGURE 3.9
Required refractive index versus the refractive angle shift from TIR.

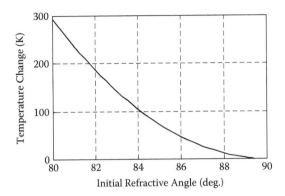

FIGURE 3.10
Relationship between the required temperature change and the initial refractive angle.

and plotted in Figure 3.10. The horizontal axis represents the initial refractive angle, whereas the vertical axis represents the required temperature change to make the refractive angle increase to 90° (i.e., TIR).

In this way, the TIR may occur and the incident light will be reflected out to the alternative output from the transmitted output when the temperature of the optical component is increased accordingly. On the other hand, the TIR reflected light can be switched back to the transmission output by lowering the temperature. However, the optical refractive index is also wavelength-dependent due to the dispersion effect. The refractive index of silicon at the room temperature (293 K) was derived by Edwards et al. [21] as:

$$n^2(\lambda) = \varepsilon_0 + \frac{A}{\lambda^2} + \frac{B\lambda_1^2}{\lambda^2 - \lambda_1^2} \tag{3.20}$$

where $\lambda_1 = 1.1071\ \mu m$, $\varepsilon_0 = 11.6858$, $A = 0.939816$, and $B = 8.10461 \times 10^{-3}$. As most of the applications in optical communications utilize the S, C, and L bands (corresponding to a wavelength range from 1400 nm to 1650 nm), the analysis below will focus on this wavelength range. The deviation of the refractive index is plotted in Figure 3.11 (the right-side

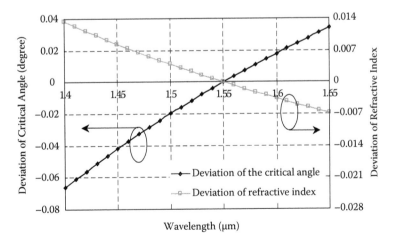

FIGURE 3.11
Relationship between silicon's refractive index and wavelength.

y axis). It is clearly shown that longer incident wavelength results in smaller refractive index. However, the absolute deviation value is less than 0.013 over this wavelength range.

According to Equation 3.18, the corresponding critical angle varies with different wavelength, as expressed by:

$$\theta(\lambda) = \arcsin\left[\frac{1}{n(\lambda)}\right] = \arcsin\left(1\bigg/\sqrt{\varepsilon + \frac{A}{\lambda^2} + \frac{B\lambda_1^2}{\lambda^2 - \lambda_1^2}}\right) \tag{3.21}$$

This relationship is also illustrated in Figure 3.11 (the primary left-side y axis). In applications where multiple wavelengths are used, the switching should satisfy all of three bands or even wider for wavelengths. In addition, this transition is abrupt as the transmitted light vanishes instantaneously at the onset of TIR. The situation here is indeed a perfectly sharp "1" to "0" switching, meaning that the switching contrast ratio (power ratio of "1" to "0" state) is infinitely high in theory. Therefore, the ultrasensitive light transition of TIR could potentially lead to very low power consumption and high-speed switching.

Two types of thermo-optic switches with different configurations are discussed in the following sections, which makes use of the above TIR and thermo-optic effect of single crystal silicon.

3.3 Triangular Thermo-Optic Switch

Triangular prism thermo-optic switch is discussed in this section with the configuration, modeling of the optical properties, and various optical losses.

3.3.1 Configuration of Triangular Prism Thermo-Optic Switch

The architecture of the triangular prism thermo-optic switch based on TIR is in Figure 3.12. It consists of a silicon prism, three optic fibers, and a high reflection mirror. The output direction of the incident beam is determined by the controlled refractive index of

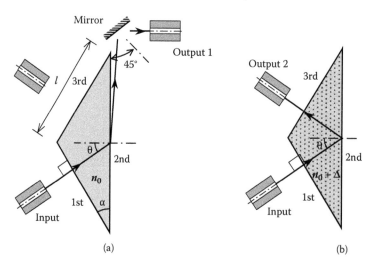

FIGURE 3.12
Architecture and working principle of triangular prism optical switch: (a) transmission state at lower temperature with smaller refractive index; and (b) TIR state at higher temperature and larger refractive index.

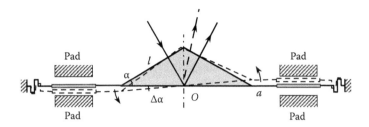

FIGURE 3.13
Rotation of the triangular prism by a pair of bidirectional parallel plate capacitor actuators.

the prism. The high reflection mirror is used to change the direction of the transmitted light into the output fiber 1. The mirror is orientated in such a way that its normal makes a 45° angle with the 2nd surface of the triangular prism. This is to ensure a good coupling of light into the output fiber. The optical axis of the input beam is perpendicular to the incident surface of the prism. Therefore, the propagation direction remains unchanged at the first interface. Hence, the incident angle at the second interface is equal to the prism angle α. The two switching states of the optical switch are schematically shown in Figures 3.12a and 3.12b, respectively.

At low temperature, the refractive index of the prism is small resulting in a larger critical angle. If the incident angle (fixed value) is smaller than the critical angle, the input light is refracted to the output 1 at the transmission state (see Figure 3.12a). On the other hand, the input light will be totally reflected to output 2 when the temperature of the prism is sufficiently high so that the increased refractive index causes the critical angle to be smaller than the incident angle (see Figure 3.12b).

The initial incident angle should be selected sufficiently close to the critical angle at room temperature to minimize the requirement for the heating power. In this design, a pair of bidirectional parallel plate capacitor actuators is used to finely adjust the incident angle by rotating the prism as shown in Figure 3.13. The electro-static force induces displacements of the supporting beam at two ends of the prism, and these symmetric displacements rotate the prism without affecting its center point O. The side length l of the prism is selected at 108.5 μm for the ease of positioning the input and output fibers. The prism angle is chosen to be 17°, exactly the room temperature critical angle for silicon–air interface. Other design parameters are listed in Table 3.1.

From the architecture of this optical switch, a few advantages over the other technologies can be identified. First, the required change in refractive index is not as critical as the waveguide-based switch because the switching function can be realized as long as the TIR condition is satisfied. Second, the mechanical instability issue is no longer a concern as the switching is realized by changing the temperature instead of mechanical movement. Last, the effective size of the device is small because the TIR switching mechanism

TABLE 3.1

Design Parameters of the Silicon Prism Thermo-Optic Switch

Prism side length l	108.5 μm
Prism angle α	17.0°
Actuator maximum displacement s	3.0 μm
Connecting beam length a	470 μm
Incident angle tuning range $\Delta\alpha_{max}$	±0.3°
Optical wavelength	1550 nm

requires only an interface instead of whole waveguides as in the case of current thermo-optic switches.

3.3.2 Modeling and Analysis of the Triangular Prism Switch

In the following sections, a complete analysis of the light switching characteristics will be presented to account for the influence of Gaussian beam divergence, F-P cavity effect, polarization effect, and various optical losses that are associated with the triangular prism design. The divergence of beam requires a safe angle to obtain high isolation (i.e., low cross talk) between the output ports; the F-P cavity effect determines the best combination of the parameters; the polarization affects the achievable specifications; and the loss factors cause insertion losses.

3.3.2.1   *Gaussian Beam Divergence*

The input light from the single mode optic fiber is a Gaussian beam, which exhibits beam divergence along its propagation in the air and inside the prism as illustrated in Figure 3.14. Therefore, the rays in the input light, when striking the 2nd surface of the prism, will see a spread of angles of incidence. Such beam divergence would affect the choice of the incident angle and would broaden the range of refractive index and the corresponding temperature change over which a significant modulation of output power could be achieved. As illustrated by an extreme case in Figure 3.14 in which the primary axis of the incident light is right at the critical TIR angle in the 2nd surface, the upper half of the light is beyond the critical angle and thus will be reflected, whereas the lower half is refracted (e.g., at point B). This causes a low isolation (i.e., large cross talk) between the two outputs. For a high-isolation optical switching from the transmission to the TIR state, the incident light should initially be below the critical angle by a certain value (i.e., safe angle) to make sure the majority of rays (i.e., the optical power) is transmitted, and then the temperature change should be high enough to let the majority of rays be reflected.

For a simple estimation of the safe angle and the corresponding additional amount of refractive index change caused by the Gaussian beam divergence, it is assumed that the prism is high enough in the depth direction and the rays at the 2nd surface have approximately the same curvature and beam waist. For practical uses, it requires an isolation >40 dB. That is, in the transmission state, the reflected power is less than 10^{-4} of the incident power, and in the TIR state the refracted power is also less than 10^{-4} as well. The isolation of 40 dB is thus set as the target for choosing the working conditions of the prism switch.

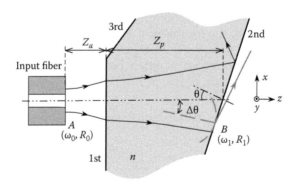

FIGURE 3.14
Spreading of incident angle due to Gaussian beam divergence.

According to the ABCD law of Gaussian optics [22], the equivalent traveling distance of the incident Gaussian beam from the input fiber facet to the 2nd surface of the prism is $Z = Z_p/n + Z_a$ (see Figure 3.14); therefore, the beam radius ω and the curvature radius R of phase wavefront can be expressed as $\omega = \omega_0\sqrt{1+(Z/z_R)^2}$ and $R = Z[1+(z_R/Z)^2]$, where ω_0 is the waist radius of Gaussian beam from the input fiber, z_R is the Rayleigh range given by $z_R = \pi\omega_0^2/\lambda$, and λ is the wavelength. The field distribution of the incident light at the 2nd surface can be expressed as:

$$U(x,y) = u_0 \exp\left\{-\frac{x^2+y^2}{\omega^2}\right\} \tag{3.22}$$

where $u_0 = \sqrt{2/\pi\omega^2}$ is the normalization factor, x is along the surface of the prism and y is in the vertical direction. The divergent angle $\Delta\theta$ in the prism horizontal plane is related to x by $\Delta\theta = x/nR$. Therefore, the power distribution over the divergent angle is also a Gaussian distribution. If the prism can steer all the angles $\leq\Delta\theta$ to one output (e.g., the transmitted output 1), the rays with divergent angle $>\Delta\theta$ will be reflected to output 2. This part of power $P(\Delta\theta)$ is considered as leaked optical power, which can be expressed as:

$$P(\Delta\theta) = \int_{nR\Delta\theta}^{+\infty} dx \int_{-\infty}^{+\infty} dy\, |U(x,y)|^2 = \frac{1}{2}\left[1 - \mathrm{Erf}\left(\frac{\sqrt{2}nR}{\omega}\Delta\theta\right)\right] \tag{3.23}$$

where Erf is the error function defined by $\mathrm{Erf}(x) = \frac{2}{\sqrt{\pi}}\int_0^x \exp(-t^2)dt$. Here, the leaked power actually represented the cross talk between the two outputs, i.e., the isolation of one output to the other.

The calculated relationship between the leaked power (in dB) and the divergent angle is plotted in Figure 3.15. The parameters are $\lambda = 1.55\ \mu m$, $Z_a = 10\ \mu m$, and $\omega_0 = 15\ \mu m$, which are the same as those in the experiment later. According to Figure 3.13 and Table 3.1,

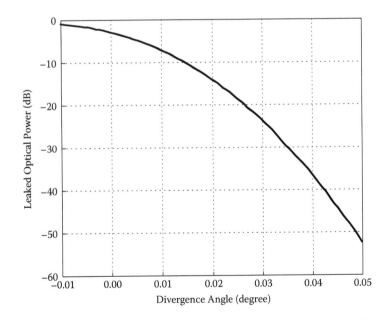

FIGURE 3.15
Leaked optical power beyond a certain divergent angle.

$Z_p = l \cos \alpha \sin \alpha = 33.34$ μm. Therefore, at the 2nd surface of the prism, $\omega = 15.01$ μm and $R = 11039$ μm. As observed in Figure 3.15, most of the power will be leaked at $\Delta\theta = -0.01°$. With the increase of $\Delta\theta$, the leaked power quickly drops. At $\Delta\theta = 0.05°$, it is already only -52.5 dB. Therefore, a safe angle of $0.05°$ is large enough to guarantee the targeted isolation of 40 dB. Here, the incident Gaussian beam should have a large radius so as to reduce the divergence. If the normally cleaved Corning SMF-28 (SMF—single mode fiber) is used [23], which has $\omega_0 = 5.4$ μm, a safe angle of $0.05°$ can only obtain an isolation of approximately 2.5 dB, too low for practical uses. Such beam size can be readily achieved by use of lensed fibers [24] or thermal-diffusion-expanded core fibers.

3.3.2.2 Fabry–Pérot Effect and Polarization Effect

Light passing through any air–silicon interface will generally undergo a partial reflection. When there are two or more interfaces present in the path of the beam, multiple reflections will be generated, causing interference. A special case will be met for normal incidence when the switch works in the TIR state, in which the 1st and 3rd surfaces of the prism form a Fabry–Pérot (F-P) cavity via the reflection at the 2nd surface as shown in Figure 3.16. When the refractive index and the corresponding optical path length are tuned, the output light intensity is anticipated to vary significantly. To take into account such F-P cavity effect, the optical powers at two outputs need to be formulated based on the Fresnel's equations [25].

The analytical model of the F-P cavity effect is illustrated in Figure 3.16, which also defines various terminologies to be used in the formulation. Here r and t represent the amplitude reflection coefficient and transmission coefficient for light incident from air into silicon, respectively; and r' and t' refer to the counterparts when light incident from silicon into air. In addition, r_2 and t_2 are used for the reflection and transmission coefficients at the 2nd surface of prism. For different polarizations, the expressions of t and t' are the same as given by $t = \frac{2}{1+n}$ and $t' = \frac{2n}{1+n}$, but r, r', r_2 and t_2 are different [25]. For the s-polarization (TE polarization, i.e., the electric field is perpendicular to the incident plane), they are $r = -r' = \frac{1-n}{1+n}$, $r_2 = \frac{n \cos \alpha - \cos \phi}{n \cos \alpha + \cos \phi}$ and $t_2 = \frac{2n \cos \alpha}{n \cos \alpha + \cos \phi}$. For the p-polarization (TM polarization, i.e., the electric field lies in the incident plane), they become $r = -r' = \frac{n-1}{1+n}$, $r_2 = \frac{\cos \alpha - n \cos \phi}{\cos \alpha + n \cos \phi}$ and

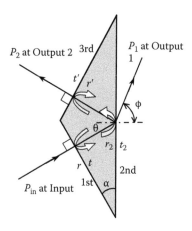

FIGURE 3.16
Influence of Fabry–Pérot cavity effect on the output powers due to the interference of multiple reflections at the air–silicon interfaces.

$t_2 = \frac{2n\cos\alpha}{\cos\alpha + n\cos\phi}$. Here the expressions are for normal incidence at the 1st surface. The powers P_1 and P_2 of two outputs can be calculated by adding up the complex amplitudes of all the light resulted from multiple reflections, yielding:

$$P_1 = \frac{t^2 t_2^2}{1 + r_2^4 r^4 - 2r_2^2 r^2 \cos\delta} \cdot \frac{\cos\phi}{n\cos\alpha} \tag{3.24}$$

$$P_2 = \frac{r_2^2 t^2 t'^2}{1 + r_2^4 r^4 - 2r_2^2 r^2 \cos\delta} \tag{3.25}$$

where ϕ refers to the refraction angle at the 2nd surface, and δ denotes the phase difference between any two consecutively reflected lights as given by $\delta = 2\pi n l \sin 2\alpha / \lambda$, where l and α are the geometry of the prism defined in Figure 3.13, and n being the refractive index, respectively. The term $\cos\phi / n\cos\alpha$ in Equation 3.23 is to take into account the cross-sectional area, which is necessary for converting the flux to optical power. It is noted that Equations 3.21 and 3.22 are derived for normal incidence at the 1st surface. The expressions will be more complicated for non-normal incidence but can be derived in the similar way.

The influence of the polarization on the switching function is studied in Figure 3.17, in which the transmitted optical power at output 1 and the reflected power at output 2 are

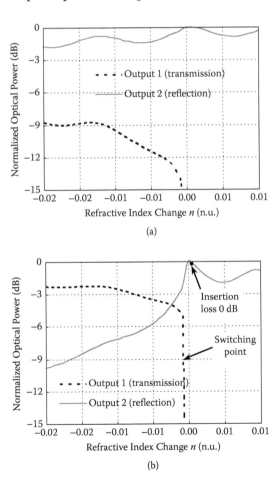

(a)

(b)

FIGURE 3.17
Optical power of different outputs for different polarization states: (a) s-polarization; and (b) p-polarization.

plotted against the refractive index change for the s-polarization and the p-polarization cases. The parameters for calculation are normal incidence (i.e., $\Delta\alpha = 0$), $\lambda = 1550$ nm, $n_0 = 3.42$, and $\alpha = \theta_{c0} = 17.0°$. The choice of parameters is to reach the critical TIR state for $\Delta n = 0$. For the s-polarization in Figure 3.17a, the reflection is always greater than -1.8 dB, whereas the transmission is smaller than -8.8 dB. Here, the powers are normalized relative to the incident power. With Δn approaches 0 from the negative side, the transmission is turned off but the reflection does not change obviously. This is not preferable for optical switching as otherwise the transmission suffers large insertion loss, although the reflection does not exhibit a sharp transition. The ripple on the curves comes from the F-P cavity effect due to the change of effective cavity length in response to the refractive index variation. For the p-polarization as shown in Figure 3.17b, the reflection and transmission at $\Delta n = -0.02$ are -9.8 and -2.2 dB, respectively. With the increase of Δn, the reflection is gradually increased when the transmission is reduced.

Over the critical TIR state, the reflection goes up to 0 dB whereas the transmission suddenly drops to very low (<-50 dB). Such properties are very useful for the optical switch to obtain very low insertion loss, abrupt power change, and high isolation between the two outputs. For this reason, the discussion and experiment henceforward are restricted to the p-polarized light. In Figure 3.17, the reflected and transmitted power may not add up to 1.0 because there are two other outputs. One is the back reflection on the incident side and the other is the refracted light out of the 2nd surface due to the reflection from the 3rd surface (symmetric to the position of output 1). The presence of F-P cavity effect causes the complexity of power fluctuation with respect to the changes of wavelength and refractive index. However, it brings in an important advantage that the insertion loss can be reduced to 0 by proper combination of the parameters as indicated in Figure 3.17b. Without this F-P cavity effect, the insertion loss cannot be 0 since the back reflection at the 1st surface always wastes approximately 30% incident power. It can also be observed from Figure 3.17b that a refractive index change of 0.010 (from -0.005 to $+0.005$) is large enough for realizing the optical switching function. This value is the same as that required by the Gaussian beam divergence.

The study above is for normal incidence. In real practice, however, perfect normal incidence is least likely ensured. Through the angular adjustment by the MEMS actuators, the amount of deviation $\Delta\alpha$ might take a very small but nonzero value (expectedly within $\pm 0.1°$). As the switching characteristics are also functions of $\Delta\alpha$, it is important to take into account such incident angle deviation. The output power is contoured in Figure 3.18 with respect to two variables: the refractive index change and the incident angle deviation. The bright part represents high output power, and the dark part for low power. For the transmitted power of output 1 as shown in Figure 3.18a, the high-power region and the low-power region are separated by a diagonal straight line, which represents the critical TIR state. The dark half above the diagonal line corresponds to the TIR state and thus output 1 has nearly no power. At a given angle deviation, it requires a minimum amount of refractive index change to switch from high output to low output (i.e., from the transmission state through the critical TIR state to finally the TIR state). For example, if $\Delta\alpha = -0.05°$, it should have $\Delta n \geq 0.010$ (taking $\Delta n = 0.013$ to have a safe margin). For the reflected power of output 2 as shown in Figure 3.18b, high-power region is only a small stripe, i.e., high-power output can only be reached by proper combination of the incident angle deviation and the refractive index change. For example, when $\Delta\alpha = -0.05°$ as given by the requirement of Gaussian beam divergence, it should have $\Delta n = 0.013$ (correspondingly, $\Delta T = 69$ K) to switching the light from the initial low-power position to the brightest part. Otherwise, it will end with low output power. The fluctuation in the right-top half is also due to the F-P cavity effect. The value of 0.013 for the non-normal incidence is a bit

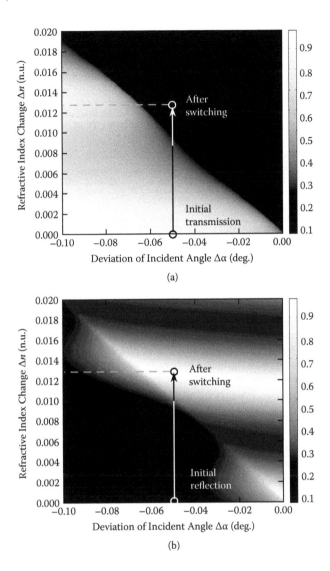

FIGURE 3.18
Contours of output power as functions of incident angle and refractive index change: (a) transmission at output 1; and (b) reflection at output 2.

larger than the value of 0.010 as required by the beam divergence and F-P cavity effect in the normal incidence.

To obtain an effective optical switching from output 1 to 2 (i.e., transmission to reflection), it would require the incident angle to be 0.05° below the critical TIR angle at room temperature, and then an increase of refractive index from 3.420 to 3.433 by heating up the prism by 69 K.

3.3.2.3   Sources of Optical Losses

In this optical switch, there are various loss mechanisms associated with each output. For output 1 in the transmission state, the loss is caused by the Fresnel reflections, fiber coupling loss, and the mirror scattering loss of the 2nd surface. For output 2 in the TIR state,

TABLE 3.2

Optical Losses at Two Switching States

Loss Factors	Output 1 (Transmission)	Output 2 (Reflection)
Fresnel reflections	2.6 dB	0.4 dB (best) 2.4 dB (worst)
Fiber coupling	0.6 dB	0.4 dB
Surface roughness	10.1 dB	1.9 dB
Mirror scattering	0.5 dB	—
Total loss	13.8 dB	2.7 ~ 4.7 dB

the optical loss comes from the F-P cavity effect, fiber coupling loss due to the Gaussian beam divergence, and scattering loss due to surface roughness at the 2nd surface. In this section, the various optical loss factors will be evaluated for an estimation of the device performance as summarized in Table 3.2.

Loss of Fresnel Reflection

Fresnel reflections are universally present when an optical beam traverses through boundaries of two or more homogeneous media with different refractive indices. In this switch, such losses mainly occur at the surfaces of the prism by single and multiple reflections, as already analyzed in studying the F-P cavity effect. The subsequent amount of losses can be extracted directly from Figure 3.18. For the *p*-polarized light as shown in Figure 3.18b, output 1 (transmission output) has an insertion loss of approximately 2.2 dB when the incident light is not close to the critical TIR angle; although output 2 (reflection output) can have no loss at its peak value, it is always <2 dB thereafter. The interface of fiber facet and free space also contributes to the reflection loss, with a value of 0.2 dB per interface. So in total, there is 2.6 dB Fresnel reflection loss associated with output 1, whereas for output 2 it ranges between 0.4 and 2.4 dB.

Loss of Fiber Coupling

In this design, the input beam is injected from an SMF whereas outputs are collected by the SMF of the same type. The coupling losses can be associated with three kinds of fiber misalignments: longitudinal separation, lateral shift, and angular misalignment. In this optical switch, lateral shift and angular misalignment can be well reduced thanks to the high-resolution lithography and etching technology. Thus, the insertion loss for each output is solely determined by the optical path length traveled by the input beam. The light from the input fiber to any output should pass through certain lengths in air and inside the prism. The longitudinal separation ΔZ can be expressed as $\Delta Z = Z_{air} + Z_{prism}/n$, where Z_{air} and Z_{prism} are the physical distances of the light path in air and prism, respectively. For output 1 at room temperature, it has $n = 3.42$, $Z_{prism} = 30.3$ μm and $Z_{air} = 140$ μm, which includes 10 μm from input fiber to the 1st surface plus 130 μm from the 2nd surface via the mirror to the output fiber (see Figure 3.13). As a result, it has the separation $\Delta Z_1 = 148.9$ μm. For output 2 at the TIR state, it has $Z_{prism} = 60.7$ μm, $Z_{air} = 20$ μm (10 μm from the input to the 1st surface and 10 μm from the 3rd surface to the output fiber) and $n \approx 3.433$; hence, the separation $\Delta Z_2 = 37.7$ μm. According to the formulae presented by Yuan and Riza [26], the total coupling losses are as small as 0.1 dB and nearly 0 for output 1 and output 2, respectively. In practical application, the lateral and angular alignment cannot be ideal. Assuming it has a lateral shift of 2 μm and an angular misalignment of 0.5°, the coupling losses rise to 0.6 and 0.4 dB for output 1 and output 2, respectively. It should be noticed that after the refraction at the 2nd surface, the beam transmitted towards output 1 does not possess a symmetric Gaussian distribution. Thus, the method used here might underestimate the coupling loss of output 1.

Loss of Prism Surface Scattering

As the prism is fabricated using deep reactive ion etching (DRIE) process, the prism surfaces are not as smooth as an ideal mirror and thus would introduce scattering loss. Such loss is especially large in the 2nd surface since the transmission state has an incident angle very close to the critical TIR angle and the refractive light grazes over the 2nd surface. Therefore, the transmitted power is very sensitive to the surface roughness. The roughness of a surface is described by its root mean square (RMS) deviation and the correlation distance. In the etched surface, two types of roughness are present. The first is the high frequency component of the surface roughness with a correlation length of 0.5 μm. The second is called bumpiness with a correlation length of 2.5 μm [27]. The RMS roughness of the 2nd surface is about 60 nm, and could be as fine as 20 nm in the upper part of the prism after the DRIE process. Such roughness makes the reflection and transmission powers deviate from those of perfect surface at the TIR condition due to enhanced backscattering and dispersed transmission [28]. By the extinction theory [29], the losses with respect to surface roughness of 60 nm are estimated to be 10.1 and 1.9 dB for output 1 and output 2, respectively. The scattering at the 2nd surface also causes the higher cross talk since the scattering power provides a strong background.

Loss of Mirror Scattering

Output 1 has to experience an additional loss in comparison to output 2. It comes from the scattering of the mirror, which is used to direct output 1 into the output fiber as illustrated in Figure 3.13. This configuration facilitates the positioning of output 1 fiber but at the expense of scattering loss. In real device, the mirror is fabricated by deep etching and metal coating, yielding certain surface roughness that serve as light scatterers. Such scattering depends on the surface roughness, incident wavelength, and incident angle [30] as given by:

$$\eta = 1 - \exp\left[-\left(\frac{4\pi\sigma\cos\theta_i}{\lambda}\right)^2\right] \tag{3.26}$$

where η is the scattering percentage, σ is the RMS roughness of the mirror surface, and θ_i is the incident angle. In the designed structure, the transmitted beam is assumed to impinge on the mirror at approximately 45°. For a surface roughness of 60 nm, therefore, the mirror scattering loss particularly for output 1 is calculated to be 0.5 dB. The respective losses analyzed above are listed in Table 3.2.

3.4 Double Cylindrical Prism Thermo-Optic Switch

Double cylindrical prism thermo-optic switch is another example of the implementation of the TIR and thermo-optic effect of silicon.

3.4.1 Configuration of Double Cylindrical Prism Thermo-Optic Switch

This double cylindrical design differs from the previous triangular prism switch: it has a pair of prisms, instead of one. The mechanism of the double cylindrical prism thermal-optic

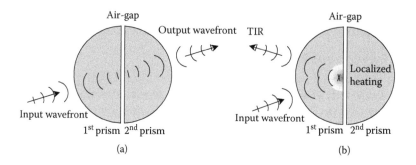

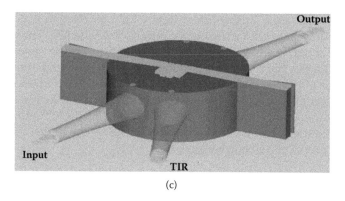

FIGURE 3.19
Working principle of the designed thermo-optics switch: (a) on-state: at room temperature, input wave transmits through the air gap; (b) off-state: upon localized heating (shaded area), FTIR of input wave occurs at the air gap; and (c) 3-D view of the device structure.

switch is illustrated in Figure 3.19, which is similar to its triangular prism counterpart. It is made up of two identical micromachined silicon cylindrical prisms separated by a thin air gap. The width of air gap is comparable to the wavelength of the input light. In the initial state (see Figure 3.19a), the input light is chosen to make a sub-critical angle θ (slightly less than the critical angle) at the 1st prism–air gap interface. Due to the narrow gap, the light penetrates through the air gap and propagates forward into the output (e.g., on-state). Thereafter, the temperature of a small area near the 1st prism–air gap interface is raised (Figure 3.19b). The 1st prism refractive index increases owing to the thermo-optic effect of silicon material [31,32], which in turn causes frustrated total internal reflection (FTIR) [33]. Under the condition of FTIR, almost all the input light is reflected by the air gap, leaving nearly zero power to the previous output (e.g., off-state). The switching from the on-state in Figure 3.19a to the off-state in Figure 3.19b is realized by localized heating, which is implemented by an aluminum micro-heater deposited right on top of the 1st prism, as shown in the 3-D view of the device in Figure 3.19c. When the heating stops, the 1st prism cools down and the device goes back to the on-state in Figure 3.19a. The double prisms are hemispherical in shape with a radius of 80 μm. They are also used as micro-lenses to collimate the light from input fiber such that the lightwave has exactly the plane wavefronts when hitting the silicon–air gap interfaces. This ensures the desired refractions and reflections for both switching states. In addition, due to structural symmetry, the double prisms focus the input light into the output fiber, resulting in a minimized coupling loss.

Due to fabrication tolerance and uncertainty in material properties, an adjustment of incident angle and air-gap width is needed for the device to operate properly. Figure 3.20

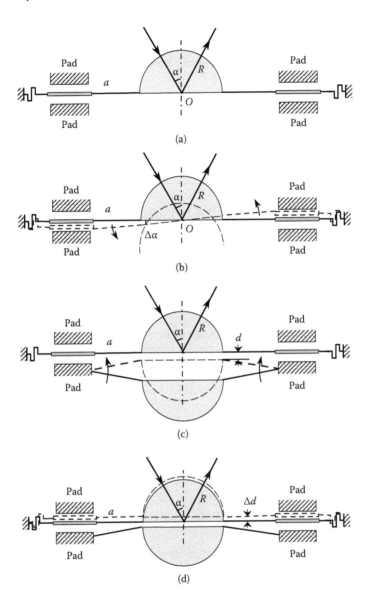

FIGURE 3.20
Design of MEMS structures for fine tuning of the device working conditions: (a) prism at the equilibrium position; (b) rotation of the prism; (c) bending of the prism to form thinner air gap; and (d) translation of the prism to adjust the air-gap width.

shows the design of MEMS structure for fine tuning of the device's working conditions. Various terminologies are labeled in Figure 3.20: a refers to the length from the prism to the edge of the parallel plate capacitance actuator, R is the radius of the hemisphere, d is the width of the air gap, and α is the incident angle of input light. In Figure 3.20a, the double prism structure is at equilibrium position. Figure 3.20b illustrates the rotation of the prism by applying a same voltage on the actuators at opposite sides of the prism. This is the same as the incident angle adjustment for the triangular prism switch. In Figure 3.20c, the 2nd prism is bent towards the 1st prism to form the thinner air gap. The initial position of the 2nd prism (dotted line) is far away from the 1st prism to ensure a good surface roughness

TABLE 3.3

Design Parameters of the Silicon MEMS Thermo-Optic Switch

Hemispherical prism radius R	80 μm
Incident angle α	17.0°
Width of aluminum micro-heater filament	2 μm
Equilibrium air-gap width d	1 μm
Actuator maximum displacement Δd	±5.0 μm
Connecting beam length a	470 μm
Incident angle tuning range $\Delta\alpha_{max}$	± 0.2°
Targeted optical wavelength λ	1550 nm

during the dry etching process. Then the 2nd prism is flipped to another mechanically stable position that is much closer to the 1st prism. In this way a thin air gap of about a wavelength width could be obtained. Figure 3.20d depicts the translation of the 1st prism to adjust the air-gap width. By applying a same voltage on actuators at the same side of the beam, the air-gap width could be either increased or decreased by Δd. The values of various device design parameters are summarized in Table 3.3.

3.4.2 Modeling and Analysis of the Double Cylindrical Prism Switch

The modeling of the double cylindrical prism thermo-optic switch is multiphysics in nature. The operation of the device consists of optical and thermal effects, and simulations on these two effects are carried out separately. Basically, the optical analysis leads us to an expression of optical power as a function of material refractive index $P(n)$, *although* the thermal analysis should end up with a temperature distribution function in space as well as time $T(x, y, z, t)$. As the thermo-optic effect relationship $n(T)$ is known, two aspects of the simulation could be linked and a final model for the device switching characteristics, either static or dynamic, could be obtained.

3.4.2.1 *Analysis of Optical Effects*

Optical Switching Characteristics

The switching characteristics are mainly determined by the transmission of wave through the air gap. Figure 3.21 represents the model of our design, in which the input wave propagates through a thin homogeneous film (i.e., air gap) surrounded by two semi-infinite dielectric media (i.e., two prisms). Here, we use λ to denote the input wavelength, d the

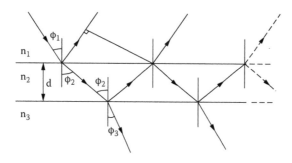

FIGURE 3.21
Reflection and transmission at a thin, homogeneous film representing the air gap. The upper and lower dielectric medium represent the double prisms.

air-gap width, and ϕ_1 the incident angle, respectively. n_1, n_2 (=1), *and* n_3 refer to the refractive index of the 1st prism, air gap, and 2nd prism, respectively. Note that n_1 is the temperature-dependent silicon index in the heated area. ϕ_2, ϕ_3 refer to the refractive angles at the 1st prism–air gap interface and the 2nd prism–air gap interface. By following the Fresnel relations [29], and adding up the multiple reflections, the total amplitude reflection coefficient can be expressed as

$$re^{i\theta} = (E^{(r)}/E^{(i)}) = \left[\left(r_{12} + r_{23} e^{i\delta} \right) / \left(1 + r_{12} r_{23} e^{i\delta} \right) \right] \tag{3.27}$$

where r_{12}, r_{23} are the amplitude reflection coefficients for the 1st prism–air gap interface and the 2nd prism–air gap interface, respectively. $\delta = (4\pi d/\lambda)(n_2^2 - n_1^2 \sin^2 \phi_1)^{1/2}$ is the phase difference between consecutively reflected waves. For both the parallel and perpendicular polarization components, the transmission coefficient is

$$T = 1 - |r|^2 \tag{3.28}$$

An interesting physical phenomenon, FTIR, occurs when ϕ_1 exceeds the critical angle of the 1st prism–air gap interface. It refers to the penetration of the wave (evanescent wave) into medium 2 when total internal reflection occurs. Such penetration have a finite depth; so, when medium 3 is brought close enough to medium 1, the wave could enter into medium 3. Later, the discovery of barrier penetration in quantum mechanics brought the insight that it was a phenomenon rather analogous to FTIR. Hence, the name "optical tunneling" for FTIR is used fairly often as an illustration of quantum mechanical tunneling [34–36]. Following Equations 3.24 and 3.25, the mathematical expression for the transmission coefficient T of FTIR [33] is given as

$$1/T = \alpha \sinh^2 y + \beta \tag{3.29}$$

where

$$y = (2\pi d/\lambda)\left(n_1^2 \sin^2 \phi_1 - n_2^2 \right)^{1/2} \tag{3.30}$$

And α and β take the following forms for different polarizations [33]:

$$\alpha_\perp = \frac{\left(n_1^2 - 1 \right)\left(n_3^2 - 1 \right)}{4n_1^2 \cos\phi_1 \left(n_1^2 \sin^2 \phi_1 - 1 \right)\left((n_3/n_1) - \sin^2 \phi_1 \right)^{1/2}} \tag{3.31}$$

$$\beta_\perp = \frac{[(n_3/n_1)^2 - \sin^2 \phi_1)^{1/2} + \cos\phi_1]^2}{4\cos\phi_1 \left(n_3^2/n_1^2 - \sin^2 \phi_1 \right)^{1/2}} \tag{3.32}$$

$$\alpha_\parallel = \left(\alpha_\perp n_1^2/n_3^2 \right)\left[\left(n_1^2 + 1 \right)\sin^2 \phi_1 - 1 \right] \times \left[\left(n_3^2 + 1 \right)\sin^2 \phi_1 - n_3^2/n_1^2 \right] \tag{3.33}$$

$$\beta_\parallel = \frac{[(n_3/n_1)^2 - \sin^2 \phi_1)^{1/2} + (n_3/n_1)\cos\phi_1]^2}{4(n_3/n_1)^2 \cos\phi_1 \left(n_3^2/n_1^2 - \sin^2 \phi_1 \right)^{1/2}} \tag{3.34}$$

For this specific structure, $n_1 = n_3 = 3.42$ for $\lambda = 1550$ nm, ϕ_1 is varied from 17° to 18.6°, where 17° is the critical angle for silicon–air interface. Figure 3.22 plots T against (d/λ) for the incident field of both polarizations. It is noted that for the perpendicular polarization, the transmission is minimized and the transmission drop is almost independent of the incident angles. This means there is a good isolation of optical output at the off-state. On the other hand, for the parallel polarization, the leakage is significant even when the gap

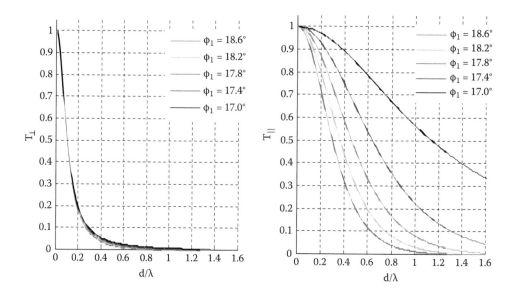

FIGURE 3.22

Transmission coefficient against (d/λ) for different polarization light: (a) for perpendicular polarization under different incident angles; and (b) transmission coefficient against (d/λ) for parallel polarization under different incident angles.

is 1.6 λ wide. Therefore, it concludes that the device should operate for perpendicularly polarized field input. It is for this reason that all the simulations, experiments henceforth assume a perpendicularly polarized input light and the air-gap width is set to be 1.6 λ. Substituting in the values λ = 1550 nm and d = 1.6 λ, the optical output power contour can be plotted as a function of both the change in refractive index and the angle deviation from the ideal 17° as shown in Figure 3.23.

It is interesting to see that the output characteristics now completely deviate from the conventional TIR case: the transmission is almost zero even though the incident angle is less than the critical angle. In fact, only in a very narrow strip (bright area in Figure 3.23) in the $\Delta\alpha$–Δn map, the maximum light transmission can be found. The bright region in Figure 3.23 corresponds to the condition when multiple reflected waves are all added up in phase. In other words, this means the optical path difference between consecutively reflected waves is multiple integers of wavelength.

The optimal on–off switching could be envisioned on the contour map as shown by the arrow. The starting point is at 0.1° below 17°, which signifies an on-state. When the prism index is subsequently increased due to heating, the transmission drops rapidly, thus resulting in an off-state. To achieve a contrast ratio of at least 25 dB, the required increase in refractive index should be 0.0076. From the thermo-optic coefficient given in Equation 3.15, it can be calculated that the required increase of temperature is 35.4 K.

Optical Losses

The total insertion loss can be decomposed into contributions from Fresnel reflections, fiber coupling loss, and scattering loss of surface roughness.

a.   Loss of Fresnel Reflection   Loss of Fresnel reflections occurs at the surfaces of the prisms by single and multiple reflections. When multiple reflections take place with constant phase delay, F-P cavity effect arises. At the on-state, transmission through the air gap reaches unity,

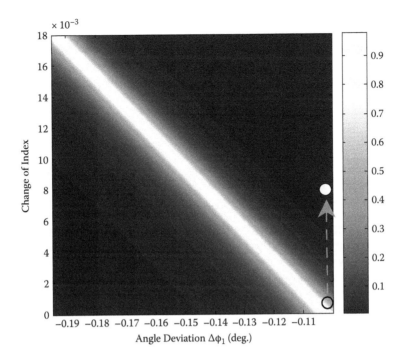

FIGURE 3.23
Contours of optical power as a function of deviant incident angle $\Delta\phi_1$ and change of refractive index Δn_1.

and the double prisms essentially form an F-P cavity whose physical length is just the diameter of the circle when two hemispheres are combined. The two curved surfaces of silicon prisms are equivalent to partial reflecting mirrors at the ends of a cavity, with a reflectance of 0.3. The worst case loss happens when the transmission coefficient of the F-P cavity falls to the minimum. According to the expression for F-P etalon [25], the loss for the silicon–air "cavity" is

$$T_{F-P}\big|_{min} = \frac{(1-R_{si})^2}{1+R_{si}^2+2R_{si}} = -5.38 dB \tag{3.35}$$

where R_{si} is the reflectance of a silicon–air interface. For the best case, an F-P cavity has 0.0 dB loss, but this only occurs at very special conditions. In addition, the interface of fiber facet and free space also contributes to the reflection loss, with a value of 0.2 dB per interface. Totally there is 0.4 ~ 5.78 dB Fresnel reflection loss.

b. Loss of Fiber Coupling In this design, the input beam is incident from an SMF, whereas outputs are collected by the SMF of the same type. The coupling loss in this design is solely determined by the optical path length traveled by the input beam. The light from the input fiber to the output should pass through certain lengths in air and inside the prisms. The longitudinal separation ΔZ can be expressed as

$$\Delta Z = Z_{air} + Z_{prism}/n \tag{3.36}$$

where Z_{air} and Z_{prism} are the physical distances of the light path in air and prisms, respectively. The fiber coupling loss on the transverse direction (parallel to the plane of incidence) should be eliminated due to the collimation and focusing effects mentioned earlier. But the absence of coupling mechanism in the vertical direction (along the height of the prisms) results in some loss. Given, $R_{prism} = 80\ \mu m$, $\lambda = 1.55\ \mu m$, $n_{si} = 3.42$, and $\omega_0 = 5.6\ \mu m$ for SMF, both input and output fibers are to be positioned 149.5 μm away from their respective prisms' surface in order to achieve the desired collimation and focusing effects. As a result, the total separation $\Delta Z_1 = 2\,Z_a + 2\,R_{prism}/3.42 = 345.8\ \mu m$. According to the formulae presented by Yuan and Riza [26], the total coupling loss in this case is

$$L = -10\log\left\{4\Big/\left[\left(\frac{\lambda z}{\pi n w_0^2}\right)^2 + 4\right]\right\} = 10.38\ \text{dB} \tag{3.37}$$

This equation, however, assumes two-dimensional divergences. In the case of only one dimensional divergence along the vertical direction, the loss value is half. Therefore, the overall divergence loss due to fiber coupling difficulty is 5.19 dB.

c. Loss of Prism Surface Scattering   As the prism is to be fabricated using DRIE process, the prism surfaces are not perfectly smooth and thus would introduce a scattering loss. By the extinction theory [29], the losses with respect to surface roughness are estimated to be 1.9 dB for transmission. The scattering at the silicon–air interface also causes some power leakage to the output, introducing a kind of noise background.

Table 3.4 lists the various loss factors and the estimated total insertion loss. In future implementations, the total loss can be significantly improved by either applying antireflection coating on the prism surfaces (to remove the Fresnel reflections) or using 3-D focusing system [37] to improve the coupling along the vertical direction.

3.4.2.2   Simulation of Localized Heating

The temperature distribution of the prism upon ohmic heating is simulated by finite element method using the DC conduction and heat transfer modules of Comsol Multiphysics V. 3.1. The boundary conditions are as follows.

$$\frac{\partial T}{\partial s} = -h(T_s - T_A) \quad \text{on top surface and 1st prism–air gap interface} \tag{3.38}$$

$$\frac{\partial T}{\partial s} = 0 \quad \text{on other lateral surfaces} \tag{3.39}$$

$$T = 20°\,C \quad \text{on bottom surface} \tag{3.40}$$

where $h = 2$ mW/cm²·K is the heat transfer coefficient for free convection [38]. Due to localized heating, the temperature rise is confined to a small volume near the top surface and 1st prism–air gap interface; therefore, other surfaces are assumed adiabatic. For the initial condition, the whole structure is assumed to be at room temperature, and the current for the microheater is 1.5A and it is to be turned on at $t = 0$. Figure 3.24 shows a snapshot of the temperature distribution over the 1st prism–air gap interface at a point when the

TABLE 3.4

Optical Insertion Loss and its Components of the Switch

Loss Factors	Loss Values
Fresnel reflections	0.4–5.78 dB
Beam divergence	5.19 dB
Surface roughness	1.9 dB
Total insertion loss	7.09–12.87 dB

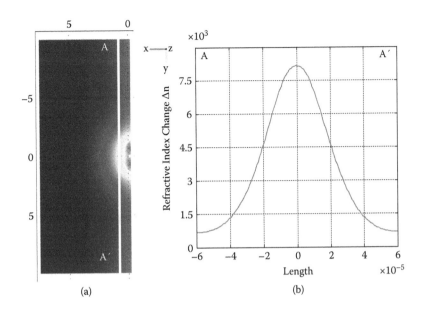

FIGURE 3.24
Thermal effect and corresponding refractive index change: (a) temperature distribution on the 1st prism–air gap interface when the region input light strikes on reaches 35.4 K temperature rise; and (b) refractive index change distribution along section A–A′.

temperature rise reaches 34.5 K at 12.5 μm below the center of the top surface. This region is of particular interest as center of the input light beam pass through it (the fiber core is 62.5 μm, and the prisms are 75 μm above the substrate). It can be clearly seen that the temperature rise, and thus the refractive index increase, is exactly confined to a small area for FTIR to occur. A plot of effective refractive index along the section A–A′ is also shown in Figure 3.24.

The dynamic thermal response of the device is simulated by pumping in a 1.5 A current pulse of 2 μs duration. The resultant time evolving temperature at the point of 12.5 μm below the center of the top surface is shown in Figure 3.25. The temperature rise time is

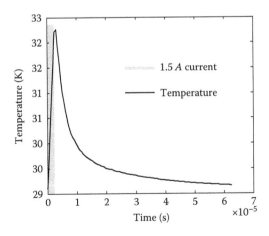

FIGURE 3.25
Transient thermal response of the prism driven by a 1.5 A, 2 μs current pulse.

2.5 μs and fall time is 51 μs. The rise time is short as only the large slope part of an exponential transient response is used. It is noted that the fall time is significantly longer than the rise time, and follows a well-defined exponential decay.

3.5 Implementation of MEMS Thermo-Optic Switches

The above two types of MEMS thermo-optic switches are fabricated, assembled, and tested. The results are listed and compared with the simulation corresponding values. The physics of the thermo-optic effect and TIR effect are verified by real devices.

3.5.1 Fabrication and Experiment on the Triangular Prism Thermo-Optic Switch

The scanning electron micrograph (SEM) of a packaged optical switch is shown in Figure 3.26. The device consists of a silicon prism, two bidirectional actuators, a high-reflection mirror, and four optic fibers (three for input and outputs, and one for alignment monitoring). The device has an overall footprint of 5×4 mm.

The fabrication employs deep reactive ion etching (DRIE) process on silicon-on-insulator (SOI) wafers. The device silicon layer has a 75 μm depth. Shadow mask is used to coat the mirror surface while keeping the prism untouched. The close-ups of the prism and one of the actuators are shown in Figure 3.27. As the prism is quite wide for dry release or wet release, an etching process called backside opening has to be used to remove the part of substrate that is right under the prism [39]. As a result, the prism is hung in free space as can be seen from Figure 3.27a. The removal of the substrate eliminates the possible stiction problem due to the dust particles and the substrate's static charge, which happens easily in the conventional SOI devices that remain the substrate and have the movable parts only about 2 μm above the substrate. As observed from the sidewalls of the prism in Figure 3.27a, the upper part (from top to about 40 μm depth) has quite a smooth surface whereas the lower part is rougher. The light spot striking onto the prism from input fiber should be aligned to the smooth range of depth by properly burying the fiber into the etched grooves. To have enough space to place the fibers for input and output 2, the fiber ends have to have a large distance (here 210 μm) from the prism surfaces. To reduce the

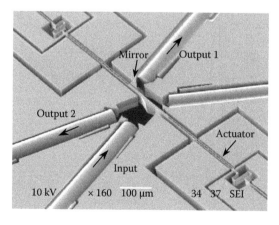

FIGURE 3.26
SEM micrograph of a packaged triangular prism optical switch.

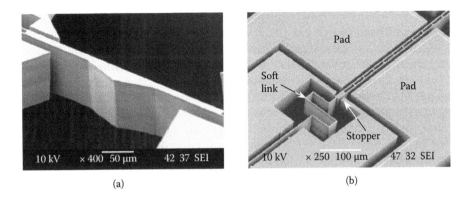

(a)   (b)

FIGURE 3.27
Close-ups of the optical switch: (a) micromachined silicon prism; and (b) bidirectional electrostatic actuator for fine angular adjustment of the prism.

coupling loss, lensed fibers that have the Gaussian beam waist (radius 15 μm) at 200 μm away from the fiber end are used for the input and outputs.

The bidirectional actuator is shown in Figure 3.27b. The central part is a beam connected to the prism and is generally electrically grounded. It will be attracted to displace when a potential is applied at any one of the two pads. The maximum displacement in any direction is 5 μm for a driving potential of 6.8 V. The stopper is to avoid the contact and the shorting of the central beam and the pad at too high voltage. A serpentine beam acts as a soft link to connect the actuator to the anchor, which allows a certain amount of extension (or compression) in the axial direction when the beam is subject to fabrication stress and temperature change. This feature helps maintain the prism position during the heating up and cooling down. The actuation relationship between the adjustment angle and the driving voltage is shown in Figure 3.28. With the increase of the voltage from 0 to 6.4 V, the rotational angle rises up to 0.15° continuously. The angle can be adjusted within the range of ±0.3° with an accuracy of approximately 0.02°/V.

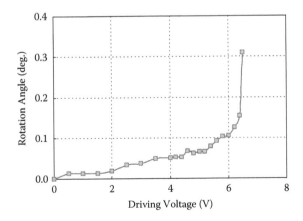

FIGURE 3.28
Adjustment angle of the prism with respect to the actuation voltage applied on the electrostatic parallel-plate actuator.

This optical switch is experimentally characterized using a laser light source at a power of 0 dBm at wavelength of 1550 nm (Ando AQ4321D). The *p*-polarization state of the input beam is ensured by a polarization controller. Two optical power meters (Newport 2832-C) are employed to detect the powers at the two outputs. A thermo-electric cooler (TEC) is stuck beneath the optical switch by a thermal conductive tape for precise temperature control. This TEC provides temperature modulation over the range from 5 to 150°C by pumping in a DC voltage with proper intensity and polarity. By accurate calibration against the injected current intensity, the actual temperature of the TEC and thus the switch is monitored by a thermometer. For visualization, a microscope with an infrared camera (Hamamatsu C5332) mounted is used to monitor the power distribution by capturing the scattered light.

The optical switching can be realized by choosing either the TIR state or the transmission state at room temperature as the initial condition. The former would require a decrease in temperature so as to switch the light from the TIR to the transmission state. However, such cooling is difficult to obtain large temperature decrease in open environment. It is also subject to condensation of water vapors on the prism surfaces, especially in a high humidity environment. To avoid such problems, the device is initially set at the transmission state and then it is heated up to achieve the TIR state. The initial orientation of the prism is optimized with the aid of bidirectional rotational actuators. In experiment, the prism is heated up to about 92°C, then DC voltage is applied to the actuators to finely rotate the prism in the direction of increasing incident angle until the first peak is observed in the output 2. This corresponds to the peak power due to the F-P cavity effect as discussed earlier. After this adjustment, the DC voltage is maintained but the temperature is returned to room temperature. As mentioned above, such adjustment makes it convenient to build up the initial condition regardless of the fabrication error and uncertain factors. The choice of temperature 92°C (i.e., $\Delta T = 69$ K for room temperature, 23°C in the experiment) is to match the designed working condition as listed in Table 3.1 and to make sure the safe angle is approximate to 0.05°.

The switching is observed by the infrared camera as illustrated by the snapshots in Figure 3.29. The scattering at the interfaces of fiber/air and prism/air indicates qualitatively the light paths, power distribution, and surface roughness. At room temperature, the incident light is transmitted to output 1 as shown in Figure 3.29a. It has strong scattering at the 2nd surface of the prism and the mirror, indicating quite rough surface quality.

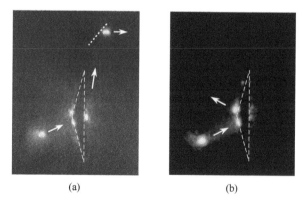

(a)   (b)

FIGURE 3.29
Infrared snapshots of the different switching states: (a) initial transmission state at room temperature; and (b) TIR state after heating up by 69 K.

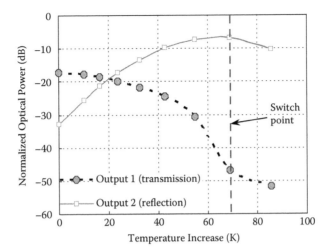

FIGURE 3.30
Measured changes of the transmitted and reflected powers with respect to the increase of temperature.

At the same time, a residue reflection can be observed at the 3rd surface to output 2. Upon heating, the final state in Figure 3.29b shows almost no power present at output 1 whereas the light at output 2 is significantly intensified. The experiment was repeated several times and yielded precisely the same results provided that it is given sufficient time for the switch to be cooled down to the initial condition after every heating [40].

The real characterization is carried out by slowly raising the temperature of the prism from room temperature. The power meter readings for both outputs are recorded by varying the TEC currents. Through the earlier calibration of TEC, an experimental result in the form of optical powers against temperature change can be plotted as shown in Figure 3.30. At room temperature (i.e., transmission state), the transmitted power at output 1 is −17.3 dB (normalized relative to the input power) whereas the reflected power at output 2 is −32.9 dB. With the increase of temperature, the power of output 1 gradually drops as that of output 2 rises. When the prism is heated up to 69 K, it comes to an obvious switching point. The reflected power of output 2 rises up to −6.7 dB as that of output 1 decreases dramatically to −46.8 dB. Further increase of temperature results in lower power at output 1; however, it also reduces the power at output 2 due to the F-P cavity effect. According to Figure 3.30, the transmission state and the TIR state have isolations of 15.6 and 40.1 dB, respectively, and insertion losses of 17.3 and 6.7 dB, respectively. The TIR state has better specifications. The measured insertion losses agree reasonably with the analysis of optical losses 13.8 and 4.7 dB. The excess losses might be attributed to the fiber shift induced by the thermal expansion of the MEMS substrate. For output 1, the distorted Gaussian profile in the transverse direction also contributes more loss as pointed out in previous analysis. It is observed in Figure 3.30 that the power transition at output 1 exhibits a more smoothened instead of an abrupt change as predicted by the calculation. This might mostly be due to the scattering by the roughness of the 2nd surface of the prism, as the literature has confirmed that with the presence of surface roughness, the distribution of the evanescent components of the scattered transmitted field is broadened and hence results in a smooth transition of reflectivity at a dielectric–vacuum interface [29]. The power of output 2 also experiences a slower change than analyzed value. This could be mainly due to the divergence of the incident Gaussian beam, which causes the intermediate state that the power distributed on some angle region is transmitted whereas the other is reflected. Based on

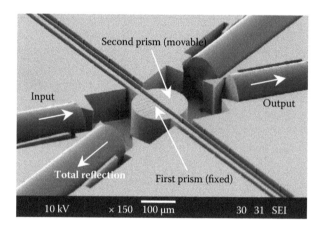

FIGURE 3.31
SEM micrograph of the packaged MEMS thermo-optic switch.

the experimental experience, it is found that the surface roughness is the main cause of the specification degradation in the aspects of high insertion loss, smoothened power transition, and large scattering loss and high cross talk in the TIR state. Further improvement will focus on the fabrication process for better sidewall smoothness. As it is slow and inconvenient to heat up the whole prism, localized heating is under development by patterning electrical heating wires right on top of the prism or by coating electrical conductive transparent films (such as indium-tin oxide) on the prism sidewalls.

3.5.2 Fabrication and Experiment on Double Cylindrical Prism Thermo-Optic Switch

The SEM of a packaged optical switch similar to that of Figure 3.26, having double cylindrical prisms instead of a single triangular one is shown in Figure 3.31. The device consists of double silicon prisms, two bidirectional actuators, and four optic fibers (two for input and output, one for TIR, and one for alignment monitoring). Also, similar to that in 3.5.1, the fabrication employs DRIE process on SOI wafers, with etched silicon layer having a depth of 75 μm. Shadow mask is used to coat the conducting beam while keeping the prisms untouched. The magnified view of the prisms and one of the actuators are shown in Figure 3.32. The bidirectional actuator is actually a parallel plate as shown in Figure 3.32b. The central part is a beam connected to the prism and is generally electrically grounded. It will be attracted to displace when a potential is applied at any one of the two pads. The maximum displacement in any direction is 5 μm for a driving potential of 6.8 V. A serpentine beam acts as a soft link to connect the actuator to the anchor, which allows a certain amount of extension (or compression) in the axial direction when the beam is subjected to fabrication stress and temperature change. This feature helps maintain the prism position during the heating up and cooling down [41].

The fabricated optical switch was experimentally investigated at a wavelength of 1547 nm using a laser light source (Ando AQ4321D) at a power of 0 dBm. A polarization controller was used to keep the input light at perpendicular polarization. A power meter was employed to detect the output power. First, the movable prism was manually bent towards the fixed prism using a probe under a microscope. The initial air-gap width was about 1 μm. Then, DC and AC voltage were injected into the heater to characterize the switching characteristics.

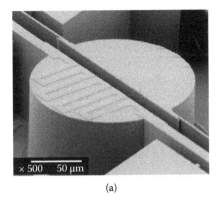

(a)

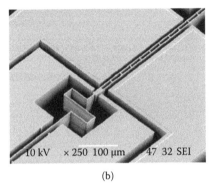

(b)

FIGURE 3.32

Close-ups of the MEMS thermo-optic switch components: (a) micromachined double prisms with deposited aluminum micro-heater on top surface; and (b) parallel plate capacitor MEMS actuators.

3.5.2.1 Static Switching Characterization

The static characterization was carried out by measuring the transmitted output as the electrical heating power is increased. The incident angle is first adjusted to 0.1° below the critical angle 17° by applying 5.9 V to the opposite pair of MEMS actuators. Furthermore, the air gap was adjusted to be 1.6 times wavelength, and the input wavelength was 1547 nm. A DC voltage from a source measurement unit (SMU) was applied to the microheater. Figure 3.33 plots the measured output optical power verses electrical power. The optical power is normalized with respect to the input power. It can be seen that when the heating power reaches 119.2 mW, the optical output is only −39.5 dB, signifying the off-state. The contrast ratio at this power is 30.2 dB. Contrary to the response of an F-P cavity, this transition has a much larger modulation depth (the ratio between maximum and minimum powers) than the maximum allowed modulation depth for a silicon–air F-P cavity (5.38 dB as calculated from Equation 3.34). Thus we could exclude the probability of F-P effect in the measurement and verify the working of FTIR.

The insertion loss is 9.5 dB. Though not ideal, it has an excellent agreement with our previous loss analysis: the total insertion loss between 7.09 and 12.87 dB. Unlike other waveguide switches, this device directly interfaces with fiber optics; so, by nature the loss is greater than others' by the amount of fiber coupling loss. According to the analysis, major loss contributors are the Fresnel reflections and the beam divergence. These losses

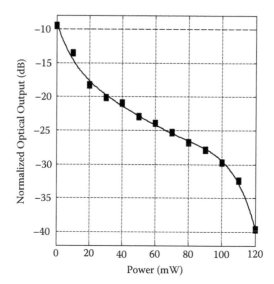

FIGURE 3.33
Static switching characteristics: normalized optical output against electrical heating power.

can be readily minimized in future work by antireflection coating on the prism surfaces or using a 3-D focusing system [37] to improve the fiber coupling. The falling of the optical power does not follow exactly an exponential decay. Especially at higher heating powers (>100 mW), the output drops suddenly. This discrepancy might be explained by the enhanced thermal disturbance at high temperatures. When ohmic heating is strong, a large amount of heat is transferred to the air in the thin air gap. This hot air causes a microscale thermal current, which could be significant enough to widen the air gap. As a result, the coupling of two prisms drops and the output power becomes less than what the simulated value under constant air-gap width assumption. It is noted that this disturbance is actually favorable. It brings down the output further so that switching contrast is improved. This is the reason why the measured contrast 30.2 dB is larger than the simulation value 25 dB. Nevertheless, such phenomenon might only occur during static heating. When the device is switching dynamically at a microsecond speed, this thermal current-induced mechanical movement is suppressed as the switching frequency is far beyond the mechanical vibration frequency (normally in microseconds for this scale structure).

The static switching power 119.2 mW of this device is comparable to other works of low power thermo-optic switches. However, the device reduces the total size of the working components to only $160 \times 160 \ \mu m^2$ compared to other millimeters long thermo-optic switches. This means, owing to localized heating, the current design is much more efficient in terms of the power consumed per unit area. Nevertheless, the power consumption could be improved even further if the heating could be applied directly to the sidewall of the prism using some transparent conducting materials, such as indium-tin oxide.

3.5.2.2 Dynamic Switching Characterization

To measure the optical transient response in terms of the on and off switching time, a fast photodetector was used instead of a power meter. The electrical signal and the photodetector signal were fed into an oscilloscope. Here a square current pulse of 1.8 mA, with 5 μs duration was used to drive the device. Between pulses, the device was cooled down

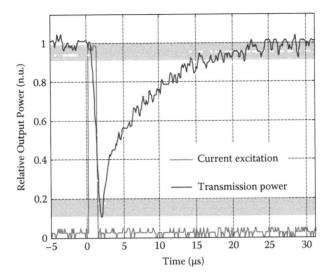

FIGURE 3.34
Dynamic switching characteristics: time evolution of output optical power when a 1.8 mA, 2 μs pulse is applied.

given sufficiently long time intervals between consecutive pulses. The rise and fall times are measured as 10%–90% power transitions. Figure 3.34 shows the measured transient response of the MEMS thermo-optic switch. The electrical pulse is also plotted in Figure 3.34. The gray strips indicate the 10% and 90% portion of the optical power transitions. The fall time (on-state to off-state) is measured as 2.2 μs and the rise time (off-state to on-state) is 22.9 μs. Figure 3.34 shows that the rise time is significantly larger than the fall time by approximately 20 times. This phenomenon coincides with our earlier thermal analysis, which also reveals a long tail of temperature drop due to the slow heat dissipation process in bulk silicon. The reason for this slow cooling is the absence of a heat sink. The speed of heat removal is solely determined by the heat transfer rate at silicon–air surfaces, which is proportional to the temperature difference between the two media, as shown in Equation 3.37. So the heat transfer rate drops as the prism is cooled down close to room temperature. A handy method to curtail this rise time has been proposed by some others' work [3,38]. They push up the initial temperature of the device above room temperature so that there are virtually heat sinks on all device surfaces. The method is usually named thermal bias control, and is able to reduce the rise time to be comparable with the fall time.

3.6 Summary

In this chapter, various thermo-optic switching technologies are described as the potential direction for the development of portable, low-cost, highly scalable optical switches in the future optical network. MEMS thermo-optic switch, which is a new technology that hybrids the strength of both MEMS and waveguide thermo-optic switch, is discussed in detail in this chapter. The two main physical mechanisms, thermo-optic effect of silicon and TIR, are described in detail. Two device examples, the triangular prism and double cylindrical prism switch, are given in terms of their design, theoretical analysis, fabrication, and experiments. The devices exhibit high extinction ratio (>30 dB) and fast switching speed (2.2 µs). Unlike MZ, Y branch or X cross type switch, which achieves a more effective switching by increasing the waveguide lengths, the TIR- or FTIR-induced enhancement of sensitivity of switching to small refractive index change leads to low-power and high-speed switching for MEMS thermo-optic switch design. Such enhancement can be implemented by localized heating, so that the improvements in power and speed come together with a reduction in device size. The trade-offs between power, speed, and size found in previous thermo-optic switches could thus be overcome. This new MEMS thermo-optic switch might become another promising solution in turning future optical network into reality.

References

1. Takashi, G., Mitsuho, Y., Kuninori, H., Akira, H., Masayuki, O., and Yasuji, O., Low-loss and high-extinction ratio silica-based strictly nonblocking 16×16 thermooptic matrix switch, *IEEE Photon Technol. Lett.*, 10(6), 810, 1998.
2. Espinola, R.L., Tsai, M.C., Yardley, J.T., and Osgood, R.M., Fast and low-power thermo-optic switch on thin silicon-on-insulator, *IEEE Photon Technol. Lett.*, 15(10), 1366, 2003.
3. Harjanne, M., Kapulainen, M., Aalto, T., and Heimala, P., Sub-µs switching time in silicon-on-insulator Mach–Zehnder thermo-optic switch, *IEEE Photon Technol. Lett.*, 16(9), 2039, 2004.
4. Chu, T., Hirohito, Y., Satomi, I., and Yasuhiko, A., Compact $1 \times N$ thermo-optic switches based on silicon photonic wire waveguides, *Opt. Express*, 13(25), 10109, 2005.
5. Fischer, U., Zinke, T., and Petermann, K., Integrated optical waveguide-switches in SOI, *Proc. 1995 IEEE Int. SOI Conf.*, Oct. 2–5, 1995, Tucson, AZ, pp. 141–142.
6. Lai, Q., Hunziker, W., and Melchior, H., Low-power compact 2×2 thermooptic silicon-on-silicon waveguide switch with fast response, *IEEE Photon Technol. Lett.*, 10(5), 681, 2003.
7. Moosburger, R., Brose, E., Fischbeck, G., Kostrzewa, C., Schuppert, B., and Petermann, K., Robust digital optical switch based on a novel patterning technique for 'oversized' polymer rib waveguides, *22nd Euro. Conf. Opt. Comm. (ECOC'96)*, 2(15–19), 67, Oslo, Sept. 15–19, 1996.
8. Siebel, U., Zinke, T., and Petermann, K., Crosstalk-enhanced polymer digital optical switch based on a W-shape, *IEEE Photon. Technol. Lett.*, 12(1), 40, 2000.
9. Wang, X., Howley, B., Chen Maggie, Y., Zhou, Q., Chen, R., and Basile, P., Polymer-based thermo-optic switch for optical true time delay, *Proc. SPIE Photon West Conf.*, 5728, 60–67, San Jose, CA, 2005.
10. Wang, X., Howley, B., Chen Maggie, Y., and Chen, R., Crosstalk-minimized polymeric 2×2 thermooptic switch, *IEEE Photon. Technol. Lett.*, 18(1), 16, 2006.
11. Tomsu, P. and Schmutzer, C., *Next-Generation Optical Networks: The Convergence of IP Intelligence and Optical Technologies*, Prentice-Hall, Upper Saddle River, NJ, 2002.
12. Rosch, O.S., Bernhard, W., Muller-Fiedler, R., Dannberg, P., Brauer, A., Buestrich, R., and Popall, M., High-performance low-cost fabrication method for integrated polymer optical devices, *Proc. SPIE*, 3799, 214, Bellingham, 1999.
13. http://notes.chem.usyd.edu.au/course/masters/Chem1/Special%20Topics/Polarizability/.
14. Corte, F.G.D., Montefusco, M.E., Moretti, L., Rendina, I., and Cocorullo, G., Temperature dependence analysis of the thermo-optic effect in silicon by single and double oscillator models, *J. Appl. Phys.*, 88(12), 7115, Dec. 15, 2000.
15. Okada, Y. and Tokumaru, Y., Precise determination of lattice parameter and thermal expansion coefficient of silicon between 300 and 1500 K, *J. Appl. Phys.*, 56(2), 314, July 15, 1984.
16. Varshni, Y.P., Temperature dependence of the energy gap in semiconductors, *Physica* (Utrecht), 34, 149–154, 1976.
17. Thurmond, C.D., The standard thermodynamic function of the formation of electrons and holes in Ge, Si, GaAs and GaP, *J. Electrochem. Soc.*, 122(8), 1133, 1975.

18. Bludau, W., Onton, A., and Heinke, W., Temperature dependence of the band gap of silicon, *J. Appl. Phys.*, 45, 1846, 1974.
19. Born, M. and Wolf, E., *Principles of Optics, Electromagnetic Theory of Propagation, Interference and Diffraction of Light*, 7th ed., Cambridge University Press, Cambridge, 1999.
20. http://www.ioffe.rssi.ru/SVA/NSM/Semicond/SiGe/optic.html.
21. http://www.irfilters.reading.ac.uk/library/technical_data/infrared_materials/si_dispersion.htm.
22. Kogelnik, H.W. and Li, T., On the propagation of Gaussian beams of light through lenslike media including those with a loss or gain variation, *Appl. Opt.*, 41(2), 1562, 1965.
23. Corning® SMF-28™ Optical Fiber Product Information, Corning Inc., New York, April 2001, pp. 1–4.
24. Corning® OptiFocus™ Collimating Lensed Fiber Product Information, Corning Inc., New York, August 2006.
25. Hecht, E., *Optics*, 4th ed., San Francisco, Addison-Wesley, 2002.
26. Yuan, S. and Riza, N.A., General formula for coupling-loss characterization of single-mode fiber collimators by use of gradient-index rod lenses, *Appl. Opt.*, 38(15), 3214, 1999.
27. Lee, S.M., Chew, W.C., Moghaddam, M., Nasir, M.A., Chuang, S.L., Herrick, R.W., and Balestra, C.L., Modeling of rough-surface effects in an optical turning mirror using the finite-difference time-domain method, *IEEE J. Lightwave Technol.*, 9(11), 1471–1480, 1991.
28. Nieto-Vesperinas, M. and Sanchez-Gil, J.A., Light scattering from a random rough interface with total internal reflection, *J. Opt. Soc. Am. A.*, 9(9), 424, 1992.
29. Sanchez-Gil, J.A. and Nieto-Vesperinas, M., Light scattering from random rough dielectric surfaces, *J. Opt. Soc. Am. A*, 8(8), 1270, 1991.
30. Nieto-Vesperinas, M., *Scattering and Diffraction in Physical Optics*, 2nd ed., World Scientific, New Jersey, CA, 2006.
31. Cocorullo, G., Della Corte, F.G., and Rendina, I., Temperature dependence of the thermo-optic coefficient in crystalline silicon between room temperature an 550 K at the wavelength of 1523 nm, *Appl. Phys. Lett.*, 74(22), 3338, 1999.
32. Jellison, E., Jr. and Burke, H.H., The temperature dependence of the refractive index of silicon at elevated temperatures at several laser wavelengths, *J. Appl. Phys.*, 60(2), 841, 1986.
33. Zhu, S., Yu, A.W., Hawley, D., and Roy, R., Frustrated total internal reflection: A demonstration and review, *Am. J. Phys.*, 54(7), 601, 1986.
34. Razali, N., Mohamed, R., Ehsan, A.A., Kuang, C.S., and Shaari, S., Thermo-optic coefficient of different photosensitive acrylate polymers for optical application, *J. Nonlinear Opt. Phys. Mater.*, 14(2), 195, 2005.
35. Bohm, D., *Quantum Theory*, Prentice-Hall, Englewood Cliffs, NJ, 1951, p. 240.
36. Eisberg, R.M. and Resnick, R., *Quantum Physics*, Wiley, New York, 1974.
37. Zhang, X.M., Cai, H., Lu, C., Chen, C.K., and Liu, A.Q., Design and experiment of 3-dimensional micro-optical system for MEMS tunable lasers, *19th IEEE Inter. Conf. MEMS (MEMS 2006)*, January 22–26, 2006, Istanbul, Turkey, pp. 830–833, paper MP45.
38. Iodice, M., Della Corte, F.G., Rendina, I., Sarro, P.M., and Bellucci, M., Transient analysis of a high-speed thermo-optic modulator integrated in an all-silicon waveguide, *Opt. Eng.*, 42(1), 169, 2003.
39. Li, J., Zhang, Q.X., Liu, A.Q., Goh, W.L., and Ahn, J., Technique for preventing stiction and notching effect on silicon-on-insulator microstructure, *J. Vacuum Sci. Technol. B*, 21(6), 2530, 2003.
40. Li, J., Liu, A.Q., Zhang, X.M., and Zhong, T., Light switching via thermo-optic effect of micromachined silicon prism, *Appl. Phys. Lett.*, 88(24), 243501, June 12, 2006.
41. Zhong, T., Zhang, X.M., Liu, A.Q., Li, J., Lu, C., and Tang, D.Y., Thermal-optic switch by total internal reflection of micromachined silicon prism, *IEEE J. Selected Top. Quantum Electron.*, 12(2), 1, 2007.

4

PhC Microresonators Dynamic Modulation Devices

Selin Hwee Gee Teo and Ai-Qun Liu

CONTENTS

4.1 Introduction 129
4.2 Single-Line Defect Waveguides with Embedded Microresonators 130
4.3 Photonic Crystal Line Intersections Optical Resonators 132
    4.3.1 Principles of Four Ports PhC Optical Intersections 132
    4.3.2 Analytical Formulations Using Coupled Mode Theory 136
4.4 Design of PhC Modulation Device 141
    4.4.1 Theory of Silicon PhC Optical Modulations 142
    4.4.2 Critical Challenges and Key Designs 144
    4.4.3 Photonic Crystal Device Effects of Optical Modulation 148
        4.4.3.1 Optically Generated Plasma 148
        4.4.3.2 Thermal Effects of Optical Excitation 150
4.5 Optical Measurements of Photonic Crystal Devices 151
    4.5.1 Photonic Crystal Test Structures, Calibrations, and Measurements 155
        4.5.1.1 Photonic Crystal Waveguide Propagation Loss Coefficients 157
        4.5.1.2 Silicon Waveguides—Coupling and Bending Losses 158
    4.5.2 Static Photonic Crystal Device Experiments 163
    4.5.3 Dynamic Modulation of Photonic Crystal Intersection 166
4.6 Summary 171
References 171

This chapter focuses on photonic integrated circuit (PIC) devices with a new class of material known as photonic bandgap (PBG) crystals for control of photons not achievable with legacy dielectric waveguide systems. Specifically, silicon is the material of choice being investigated for the PBG PIC devices designed to address the objectives of macro-to-micro optical system integration, ultracompact optical device footprint, high optical performance, realization of deep submicrometer length scale photonic crystal devices, and demonstration of PIC PBG functionalities.

4.1 Introduction

Based on the physical and mathematical derivations for different PBG structures, the various aspects of PIC PBG were studied and modeled. Simulations were carried out based on the plane wave method (PWM), the finite difference time domain (FDTD) method, and the coupled mode in time analyses. In addition to the dynamic frequency responses obtained through the transmission characteristic simulations, dispersion relations of band structures in PBGs that detailed the eigenmodes allowed or prohibited from existing within the PBG lattice may be obtained for purpose of device design. Together with gap maps generated for parametric analyses, design optimizations could be carried out to yield large absolute bandgap lattices in a high symmetry cubic Brillouin zone PBG.

In device simulations, PIC applications to designs of single-line embedded and orthogonally crossing optical intersection resonators are described and demonstrated experimentally. To enable such experiments to take place, the nontrivial fabrication skill required to realize PBG structures useful for the fiber optical communication wavelengths centered at 1550 nm must first be successfully addressed, as discussed in Chapter 11. Only then may the devices fabricated from silicon-on-insulator (SOI) substrates be tested experimentally, with effective macro- to microlevel coupling between the optical fibers (of more than 100 μm diameter) and the PBG PIC waveguides (of less than a micrometer width). Through the use of background level isolating designs and adiabatically tapered waveguide structures, experiments based on the proper setup of pump and probe lasers (consisting of high and low photon energy beams, respectively) may then be effected. Passive and dynamic optical measurement results could then be obtained for dispersion characterizations and optically/thermally induced switching in output levels by free carriers absorption and refractive index modulations. The chapter concludes with discussions on the specific requirements for localized application of impulses and absorption limits. These novel experimental demonstrations of such PBG PIC devices not only verified theoretical design predictions as proof of concept, but will also provide insights critical to potential applications in future designs of PIC devices.

4.2   Single-Line Defect Waveguides with Embedded Microresonators

Microcavity resonators refer to structures capable of confining light in small volumes by the action of resonant recirculation. Photonic microcavities have been identified as one of the most elusive resonators for design and implementation because design rules developed for its counterparts in the physical, sound, or microwave domains are not applicable in the design of optical microcavities. To investigate this, PhC cavities will be presented here in progressive designs of single-line resonators, and optical intersection configurations.

In the single-line PhC resonator configuration, the line defect waveguide is created by removing a line of rods (or reducing the diameter of a line of rods within a periodic lattice), coupled together with a resonator of a reduced/enlarged rod embedded within some PhC rods of the bulk PhC structure. The waveguiding mode of a reduced-diameter PhC-line-defect waveguide is exhibited in Figure 4.1, where there is an additional PhC guiding mode within the bandgap frequencies. For slab-type PhC with finite dimensions in the vertical plane, this mode diagram is further divided by a diagonal "light line" arising from the material cladding—in this case, air. PhC modes that are guided within the bandgap frequencies become resonant above the light line, such that they extend deeply into the surroundings, losing their lossless guiding properties.

For such configurations of single-line resonators, a single rod on each side of the resonator gives a 3×3 structure [1], whereas two rods on each side of the resonator gives a 5×5 structure, and so on, for the case of three rods on each side (see Figure 4.2). Hence, with increasing number of PhC rods on each side of the resonator, increasing reflectivity of the confining cavity is obtained (analogous to the mirrors of a confining Fabry–Pérot (F-P) cavity [2]).

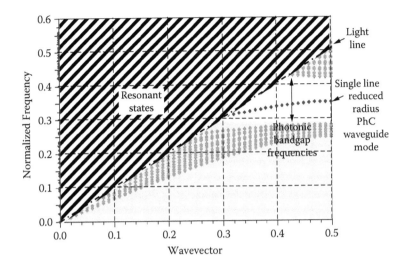

FIGURE 4.1
Diagram of photonic bandgap mode for a single-line-defect waveguide, with addition of the light line indicating the resonant states.

Hence, it is confirmed that for a higher number of rods placed on each side of the embedded resonator, a higher quality (Q) factor for the cavity mode results. Figure 4.3 gives the frequency response of the 1×1, 3×3, and 5×5 embedded PhC resonator with normalized transmission. From the spectrums, it can be seen that for progressively increased orders of the single resonator structures, the transmittance increases in Q-factors. This is accompanied by a corresponding reduction in the bandwidth, which is the range of transmitted wavelengths about the center resonance wavelength. Such increase in Q-factors for these increasing orders of single-line resonators, therefore, results in correspondingly reduced full widths at half maximum (FWHM) for the 1×1, 3×3, and 5×5 single-line resonator structures, respectively.

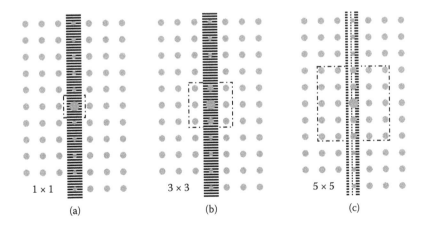

FIGURE 4.2
Schematic of single-line photonic crystal waveguides with a single embedded resonator: (a) 1×1, (b) 3×3, and (c) 5×5, where the shaded areas indicate high refractive index material set against a low index background such as air.

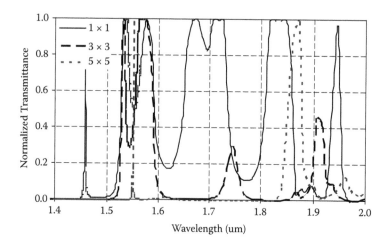

FIGURE 4.3
Transmission spectrums for progressively higher order single-line resonators.

With such frequency response results, the effect of changes to the center resonator rod sizes and refractive index may be predicted for their sensitivity, which corresponds directly with the Q-factors of the resonator structures.

4.3 Photonic Crystal Line Intersections Optical Resonators

This is similarly the case for the orthogonally crossing single-line defect waveguides. From the basic device structure of a square lattice rod-type periodic PhC, removal of a single-line of PhC rods introduces defects that create the single-line waveguide. By intersecting orthogonally two such waveguides, PhC optical intersections (Figure 4.4a) with fourfold symmetries [3] as illustrated in Figure 4.4b are formed.

These PhC optical intersections can have compact working areas (as small as $5 \times 5 \ \mu m^2$) and can be applied in optical integrated circuits for small-form, large-angle crossings [4,5]. This is an important component of photonic integrated circuits as they allow for interaction of many waveguides within a small footprint [1].

4.3.1 Principles of Four Ports PhC Optical Intersections

The advantage of small layout size for compact integrated circuit purpose, however, comes with its challenges in device design. Due to the subwavelength scale of such crossing waveguides, inputs undergo diffraction at the intersections [5], resulting in high cross talk and reflection levels lowering effective transmissions (as illustrated graphically in Figure 4.5a). This is explained by the steady-state FDTD simulation results of Figure 4.5b, in which the electric field propagating along the z axis with alternating maximum positive and negative amplitudes along the waveguides are shown in the lowest order, 1×1 form, without any center cavity rod. In this figure, the circles drawn in the schematic indicate the positions of the PhC silicon rods set against an air background. For optical intersections with the inclusion of an optimized center cavity rod, cross talk may be eliminated to obtain high transmission as shown in Figure 4.5c.

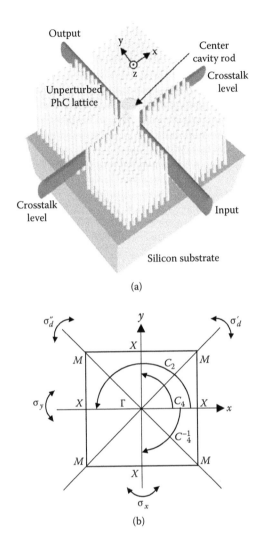

FIGURE 4.4
(a) Schematic representation of a photonic crystal optical intersection as an orthogonal crossing of two single-line-defect waveguides in a 2-D array of silicon rods with square lattice periodicity. (b) Illustration of high symmetry, fourfold operation in the Brillouin zone of the square lattice photonic crystal reciprocal space representation.

Here, such an optical intersection with a single cavity rod in the center can be referred to as a 1×1 intersection cavity, similar to that of the single-line resonator convention. If increasing layers of bulk PhC rods are introduced to each branch of the intersecting waveguides near this center cavity rod where rods were previously removed from lattice sites to form the line defect waveguide, progressively larger cavities of 1×1, 3×3, and 5×5 orders (as shown in Figure 4.6) are formed. For these progressively higher order cavities, there is increasing confinement effect at the center cavity rod (because such additional layers of PhC rods around the center cavity rods act like Bragg mirror layers similar to that found in a typical F-P resonator). This condition leads to correspondingly increasing cavity Q-factors, but smaller frequency bandwidths, as given by the impulse frequency responses of the FDTD simulation results (Figure 4.7) for progressively higher order microresonators intersections.

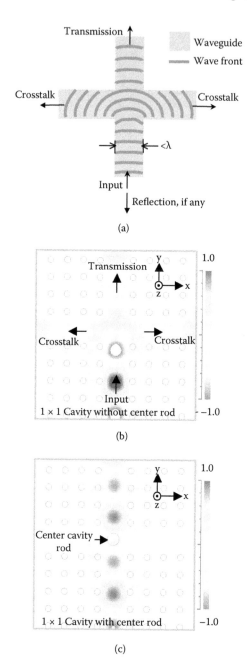

FIGURE 4.5
(a) Schematic of diffracting radiation, through subwavelength intersecting waveguides. (b) Steady state FDTD simulation of photonic crystal intersection without center cavity resonator rod. (c) Steady state FDTD simulation of photonic crystal intersection with center cavity resonator rod.

For these configurations of the PhC optical intersections, two degenerate modes in the frequency range of interest may be supported [6]. Figures 4.8a and 4.8b depict the electric field mode profiles of the in-phase and out-of-phase superposition of these degenerate modes for the 1 × 1 optical intersection. Similar to Figure 4.5, the field representations

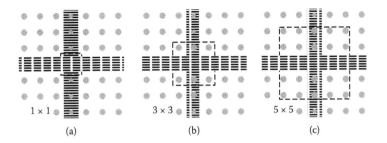

FIGURE 4.6
Photonic crystal optical intersection configurations with progressively higher-order configuration for correspondingly higher factors.

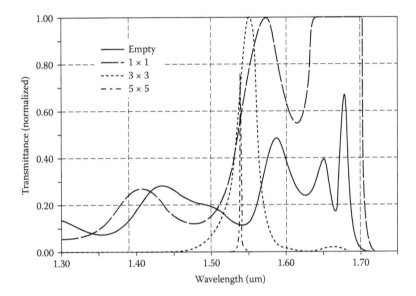

FIGURE 4.7
Frequency response of the different configurations of photonic crystal optical intersection resonators.

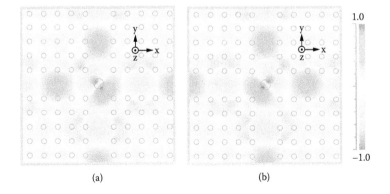

FIGURE 4.8
Profile plots of electric field modes for the in-phase and out-of-phase superposition of the degenerate modes in the 1×1 photonic crystal optical intersection.

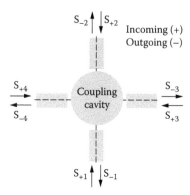

FIGURE 4.9
Illustration of a four-port photonic crystal intersection device consisting of four waveguides as ports connected to an arbitrary cavity.

of the mode profiles are given by the maximum positive and negative amplitudes of the propagating electric field in and out of plane along the z axis as labeled. From these mode plots, it is therefore easy to visualize how the cancellation of opposite phases at the cross-talk branches by the two degenerate modes enables the elimination of cross talk.

4.3.2   Analytical Formulations Using Coupled Mode Theory

To analyze such four-port PhC optical intersections, the coupled mode theory in time (CMT) [7] is used to qualitatively describe the optical properties of the 1×1 optical intersection. For an arbitrary PhC optical intersection with a center rod for coupling cavity (such as that shown in the "black box" diagram of Figure 4.9, where there is a center coupling cavity joined to four orthogonal side branches), there may exist the two degenerate modes as described earlier in the previous section, with amplitudes denoted as a and b, and resonance frequency ω_0 in the frequency of operation.

Qualitatively, the behavior of this system described by CMT [6,7] is formulated as follows: After intrinsic loss with decay rate $1/\tau_0$, and a waveguide branch transfer rate $1/\tau_e$ and input excitation s_{+1} with coefficient κ describing the coupling between the intersection cavity and the excitation radiation, the effective mode amplitude is given by

$$\frac{da}{dt} = j\omega_0 a - \frac{1}{\tau_0}a - \frac{1}{\tau_e}a + \kappa s_{+1} \tag{4.1}$$

For source at frequency ω, with $e^{j\omega t}$ as time dependence, at steady state:

$$a = \frac{\kappa s_{+1}}{\frac{1}{\tau_0} + \frac{1}{\tau_e} - j(\omega - \omega_0)} \tag{4.2}$$

To express κ in terms of τ_e, assuming no internal loss with no source and only one waveguide branch, the mode decay

$$\frac{d|a|^2}{dt} = -\frac{2}{\tau_e}|a|^2 = -|s_-|^2 \tag{4.3}$$

where s_- is the wave traveling away from the coupling cavity. For the time-reversed solution, it is converted into an incident wave, s_+, such that energy in the coupling cavity builds up rather than decays, with time dependence e^{2t/τ_e}. For "positive frequency" amplitudes of time reversed solution $\hat{a}$,

$$\frac{d\,|\hat{a}|^2}{dt} = \frac{2}{\tau_e}|\hat{a}|^2 \tag{4.4}$$

the time reversed solution is "driven" by incident wave $\hat{s}_+$ at frequency ω_0 and grows at the rate $\frac{1}{\tau_e}$, such that the frequency of the drive is $\omega = \omega_0 - \frac{1}{\tau_e}$. Therefore,

$$\hat{a} = \frac{\kappa \hat{s}_+}{\frac{2}{\tau_e}} \tag{4.5}$$

Further, because

$$|\hat{s}_+|^2 = \frac{2}{\tau_e}|a|^2 = \frac{2}{\tau_e}|\hat{a}|^2 \tag{4.6}$$

$$\kappa = \sqrt{\frac{2}{\tau_e}} \tag{4.7}$$

Because the phase of κ can be neglected due to the phase of a with respect to s_+, it can be defined arbitrarily,

$$\frac{da}{dt} = j\omega_0 a - a\left(\frac{1}{\tau_0} + \frac{1}{\tau_e}\right) + \sqrt{\frac{2}{\tau_e}}s_{+1} \tag{4.8}$$

For a linear system, s_- can be determined as proportional to s_+ and a, such that

$$s_- = c_s s_+ + c_a a \tag{4.9}$$

From Equation 4.3,

$$c_a = \sqrt{\frac{2}{\tau_e}} \tag{4.10}$$

where the phase factor of c_a is fixed at unity by choice of reference pane at which s_- is to be evaluated. Next, by energy conservation, the net power flowing into the coupling cavity has to be equal to the rate of energy buildup and the rate of energy dissipation

$$|s_+|^2 - |s_-|^2 = \frac{d}{dt}|a|^2 + 2\left(\frac{1}{\tau_0}\right)|a|^2 \tag{4.11}$$

From Equation 4.8,

$$\frac{d\,|a|^2}{dt} = -2\left(\frac{1}{\tau_0} + \frac{1}{\tau_e}\right)|a|^2 + \sqrt{\frac{2}{\tau_e}}(a^* s_+ + a s_+^*) \tag{4.12}$$

Comparing Equations 4.11 and 4.12,

$$|s_+|^2 - |s_-|^2 = -\frac{2}{\tau_e}|a|^2 + \sqrt{\frac{2}{\tau_e}}(a^* s_+ + a s_+^*) \tag{4.13}$$

Eliminating a from Equation 4.13 by Equations 4.9 and 4.10,

$$c_s = -1 \tag{4.14}$$

$$s_- = -s_+ + \sqrt{\frac{2}{\tau_e}}\, a \tag{4.15}$$

Therefore, for the coupling cavity with the fourfold symmetry waveguide coupling branches, for the two degenerate modes with amplitudes a and b, the amplitude of outgoing waves are given as:

$$s_{-1} = s_{+1} + \sqrt{\frac{2}{\tau_a}}\, a + \sqrt{\frac{2}{\tau_b}}\, b \tag{4.16a}$$

$$s_{-2} = \sqrt{\frac{2}{\tau_a}}\, a + \sqrt{\frac{2}{\tau_b}}\, b \tag{4.16b}$$

$$s_{-3} = \sqrt{\frac{2}{\tau_a}}\, a - \sqrt{\frac{2}{\tau_b}}\, b \tag{4.16c}$$

$$s_{-4} = \sqrt{\frac{2}{\tau_a}}\, a - \sqrt{\frac{2}{\tau_b}}\, b \tag{4.16d}$$

such that the transmittance T, cross talk X, and reflectance R levels can be expressed as

$$\frac{s_{-1}}{s_{+1}} \equiv R = -1 + \frac{\frac{2}{\tau_a}}{j(\omega - \omega_0) + \frac{4}{\tau_a}} + \frac{\frac{2}{\tau_b}}{j(\omega - \omega_0) + \frac{4}{\tau_b}} \tag{4.17a}$$

$$\frac{s_{-2}}{s_{+1}} \equiv T = \frac{\frac{2}{\tau_a}}{j(\omega - \omega_0) + \frac{4}{\tau_a}} + \frac{\frac{2}{\tau_b}}{j(\omega - \omega_0) + \frac{4}{\tau_b}} \tag{4.17b}$$

$$\frac{s_{-3}}{s_{+1}} \equiv X_L = \frac{\frac{2}{\tau_a}}{j(\omega - \omega_0) + \frac{4}{\tau_a}} - \frac{\frac{2}{\tau_b}}{j(\omega - \omega_0) + \frac{4}{\tau_b}} \tag{4.17c}$$

$$\frac{s_{-4}}{s_{+1}} \equiv X_R = \frac{\frac{2}{\tau_a}}{j(\omega - \omega_0) + \frac{4}{\tau_a}} - \frac{\frac{2}{\tau_b}}{j(\omega - \omega_0) + \frac{4}{\tau_b}} \tag{4.17d}$$

Here, it is noted that in the limit when the Q-factor becomes significant enough for increasing layers of rods surrounding the center cavity, this model tends to become a quantitative description of the system (as there is an approximate tenfold increase in Q-factor for each progressive increase from the 1×1 to 3×3 and from the 3×3 to 5×5 cavity configuration). Hence, for decay rates of both modes being equal, such that $1/\tau_a = 1/\tau_b$ cross-talk levels remain zero for all frequencies. When this optimized condition is altered, such that $1/\tau_a \neq 1/\tau$ cross talk inevitably increases from negligible to decrease the transmittance levels as can be seen from Figure 4.10.

When the PhC optical intersection operating point moves away from such optimized condition, cross-talk levels, however, increase, as no effective cancellation can occur (Figure 4.11).

From Figure 4.10, it can be seen that for degenerate modes with similar decay rates, the higher decay rates from the cavity results in correspondingly broader spectrums with

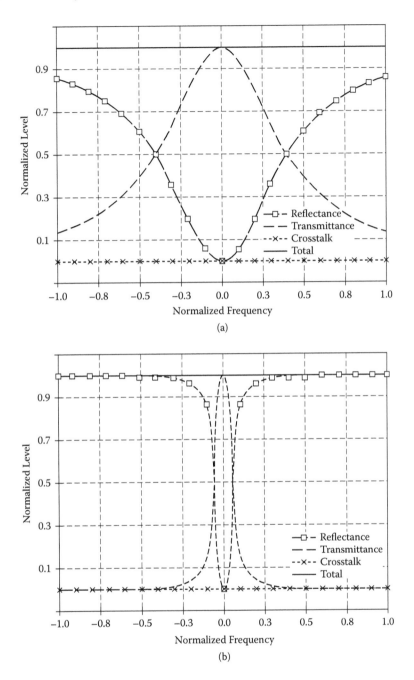

FIGURE 4.10
Coupled mode theoretical prediction of four-port optical intersection characteristics for degenerate modes with (a) high decay rates and (b) lower decay rates.

lower Q-factors (Figure 4.10a) whereas lower decay rates give transmission and reflection spectrums that are much sharper with higher Q-factors (Figure 4.11b). This scenario corresponds to the case for which the optimized PhC optical intersection had degenerate modes designed to have the same coupling efficiency with the connecting ports. On the

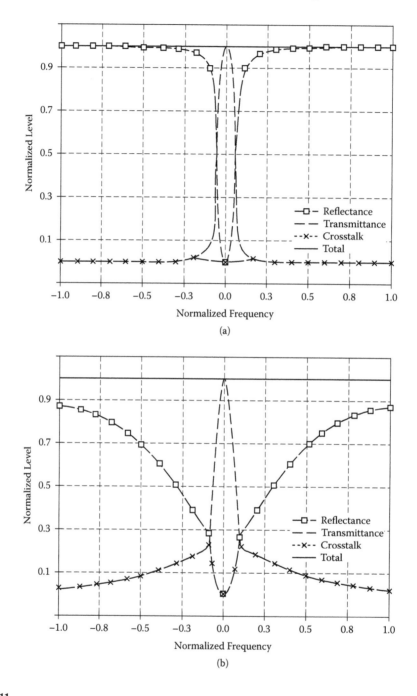

FIGURE 4.11
Coupled mode theoretical prediction of four-port optical intersection characteristics for modes (a) and (b) with different decay rates with (a) one order of magnitude difference, and (b) two orders of magnitude difference.

other hand, for modes which do not have the same coupling efficiency with the connecting ports, incomplete cancellation gives rise to cross talk, which increases from 3 to 23% for increase in difference of decay rates from one order of magnitude (Figure 4.11a) to two orders of magnitude (Figure 4.11b).

4.4 Design of PhC Modulation Device

Moving on from the discussion of the PhC optical intersecting waveguides as a small, compact-layout, photonic-integrated circuit device, it is further desirable that such devices in PIC be made responsive to dynamic effects for changing of states. For such development of dynamic functionality, which will also be useful for the purposes of fabrication deviations compensation and modulation of signals, etc., many methods have been proposed. Examples include the use of thermo-optic effects [17]; liquid crystal tuning [18]; mechanical manipulations [19]; material nonlinearity [8,9]; and optical absorption induced index tuning [11,12]. Of these, the slower modulation methods, such as the thermo-optical and mechanical tuning mechanisms, can be used for state changing applications, whereas the faster modulation methods, such as optically controlled mechanism of absorption and nonlinearity induced material index changes, can be used for real-time switching in dynamic network applications, etc. Yet, there are many challenges in the path towards the realization of tunable optical PhC device operations.

First, fabrication of PhC devices working at the optical communication wavelengths has always been much more challenging (see Chapter 10) than at longer wavelengths, because such PhCs have correspondingly much smaller critical dimensions (in the nanometer scale). Second, even with the myriad of modulation mechanisms available for tunable device performance, it remains a challenge to incorporate flexibility of properties tuning into PhC devices, as there is not only a need for sufficient sensitivity to the control mechanism, but also the requirement of spatial homogeneity (that needs to be induced) within the necessary device volume by the modulation mechanism. Especially in the ultrafast optical absorption mechanism for index change, the need for localized application of pump radiation with sufficient power density leads to problems in coupling and spatial averaging, constituting nontrivial experimental issues.

Therefore, only one method of choice—optically induced modulations—has been selected for a focused approach toward this subject matter. The key advantage of this approach lies in its inherent characteristics of ultrafast switching, low-power thresholds, immunity to electromagnetic interference, and its fundamental compatibility with PhC integration, permitting strategic implementation of optically induced signal modulations on appropriately designed PhC optical devices. The following sections establish the theories and techniques required to realize optical modulation for deep submicrometer scale, rod-type PhC devices, including the study of both static and dynamic optical device/material modulation principles.

First, in terms of the dynamic optical processes required, it is well known that the electrical properties of semiconductors are greatly affected by the presence of charge carriers into otherwise empty bands of energy levels. Although such carriers are typically thermally excited, a beam of laser radiation may also induce real populations of charge carriers temporarily onto excited states. In optical terms, the extent of excitation depends upon the balance between the rate of absorption of energy and the corresponding relaxation mechanisms.

Granted that there are also many different types of mechanisms involved with laser irradiation onto a semiconducting material—such as third harmonic generation, optical Kerr, Pockels, and Raman effects, etc.—these effects are not directly relevant to the experiments carried out in this work and would therefore be only points of reference. In this work where the material of choice was silicon, saturation conditions were found to be varied from as low as 1 W/cm^2 to as high as 1 GW/cm^2 [8] for different devices. Also, the range of relevant induced-carrier densities is between 10^{14} and 10^{22} carriers per cm^3.

4.4.1 Theory of Silicon PhC Optical Modulations

Any real, physical oscillating system will exhibit a nonlinear response when it is over-driven. In an optical system, a nonlinear response can occur when there is sufficiently intense illumination [9]. The nonlinearity is exhibited in the polarization ($\vec{P}$) of the material, which is often represented by a power series expansion of the total applied optical field ($\vec{E}$)

$$\vec{P} = \varepsilon_0 \chi^{(1)} \vec{E} + \varepsilon_0 \chi^{(2)} \vec{E}^2 + \varepsilon_0 \chi^{(3)} \vec{E}^3 + \cdots \tag{4.18}$$

where $\chi^{(1)}$ is the linear susceptibility representing the linear response and $\chi^{(2)}$, $\chi^{(3)}$ are the second and third order nonlinear susceptibilities, respectively. This macroscopic expression describes the expansion of components in the polarization of the material, which in turn, may be further categorized for effects, in terms of susceptibility and frequencies, for the passive and active processes, respectively [8]. The passive process is associated with the real parts of $\chi^{(1)}$, $\chi^{(2)}$, and $\chi^{(3)}$. As second order effects are always passive, only first- and third-order processes may be active, associated with the imaginary parts of $\chi^{(1)}$, and $\chi^{(3)}$.

For an indirect bandgap semiconductor like silicon, the minimum energy of the conduction band and the maximum energy of the valence band occur at different wave-vector (κ) values. In this case, optical absorption from above bandgap radiation results from the additional interaction of a long wave-vector phonon allowing momentum conservation to be satisfied. An electron excited by a photon in a virtual transition requires a phonon scattering off it to compensate for the energy mismatch. Like two-photon absorption, this is a two-particle interaction, and absorption is difficult to evaluate because all possible combinations of intermediate and final states must be summed up and the electron–phonon interaction changes considerably with κ-value.

Optical absorption effects may be used to modulate the device output levels through free charge carriers' plasma generation. When photons with sufficient energy are absorbed in a dielectric material, electrons in the valence band are promoted to the conduction band. The optically excited electron-hole pairs increase the free carriers' density within the material and, therefore, alter its optical properties as described by the Drude model [10]. Such behavior is, however, dependent on the intensity of the incident radiation. For silicon, optical absorption at low-incident power density levels scale linearly with intensity I as αI, where α is the linear absorption coefficient of silicon (which tends to zero for photon energies less than its bandgap energy E_g). At high incidence intensity, an additional nonlinear TPA effect becomes significant, scaling with optical intensity as βI, where β is the two-photon absorption coefficient of silicon, which vanishes for photon energies larger than half the silicon material's electronic bandgap energy.

In optical experiments, high-intensity pump radiation of photon energies greater than the silicon bandgap energy 1.1 eV, was used to excite the free charge carriers of electrons and holes in the optically pumped material. A low-intensity signal at telecommunication wavelengths may then be used as an optical probe input; the lattice length scale of the PhC structures was designed to be sensitive to this. In such a case, as the optical absorption effect is mainly linear, the Drude model can be used to describe well the changes in dielectric index for the silicon material (for induced charge carriers density not exceeding 10^{22}—corresponding approximately to 10% of the total valence-band population [11]). The change in real dielectric constant [12] may, therefore, be expressed as

$$\Delta \varepsilon = -\frac{Ne^2}{m^* \varepsilon_0 (\omega^2 + \tau_d^{-2})}, \tag{4.19}$$

where N is the free electron concentration, e is the electron charge, m^* is the optical reduced mass, ε_0 is the optical frequency dielectric constant, ω is the frequency, and τ_d is the Drude damping time.

Thus, the induced change in refractive index via change in dielectric constant is directly determined by the optically induced free charge carriers' density. Here, to enable effective modulation of material and device properties, the minimum activation dimensions need to be less than the carrier absorption length [13] for the excited free carrier plasma, which is given as

$$l_{abs} = \left(\frac{\omega}{\omega_p}\right)^2 \tau_d c \tag{4.20}$$

where c remains the speed of light, and the plasma frequency is given by

$$\omega_p = \sqrt{\frac{Ne^2}{m^* \varepsilon_0}} \tag{4.21}$$

The corresponding carrier density profile expanded is given by

$$N(z) = \frac{I(z)\tau_{pump}}{\hbar\omega_{pump}}\left[\alpha + \frac{1}{2}\beta I(z)\right] \tag{4.22}$$

where ω_{pump} is the pump laser frequency, τ_{pump} is the pump pulse duration; the factor of half before the TPA coefficient indicates the need for two photons to be absorbed before the generation of an electron-hole pair, and the intensity depth profile is given by

$$I(z) = \frac{I_0 e^{-\alpha z}}{1 + (\beta I_0/\alpha)(1 - e^{-\alpha z})} \tag{4.23}$$

where I_0 is the intensity at the interface and z is the displacement.

Figure 4.12 illustrates the exponentially inverse relationship between pump beam intensity and the carrier absorption length. Here, it can be seen that the absorption length is more than 8 μm for an incident pump power of 20 mJ/cm^2, corresponding to an induced carrier density of 5.36×10^{20} cm^{-3}.

The absorption length is critical for the experiments as the progressive decrease in absorption span affects the limitation and effectiveness of the method. Figure 4.13 shows the intensity depth profile for the different linear absorption and TPA absorption coefficients. In Figure 4.13a, only linear absorption mechanism is present and the intensity of the incident radiation onto the material decreases by 3 dB at less than a unit length. In Figure 4.13b, where the linear and TPA coefficients are equal to 0.5, slightly longer absorption depth is obtained. Finally, the best scenario for absorption penetration depth is obtained for the case whereby TPA is dominant as shown in Figure 4.13c.

To quantify the relationship between the absorption coefficients and the pumping homogeneity, critical for assessing the limitations and effectiveness of such optical modulation techniques, a homogeneity length as the span for the carrier density remaining within 10% of its surface value can be defined as

$$l_{hom} \cong \frac{0.1}{f} \times \left[\frac{1}{N(z)}\frac{dN(z)}{dz}\right]_{z=0} \tag{4.24}$$

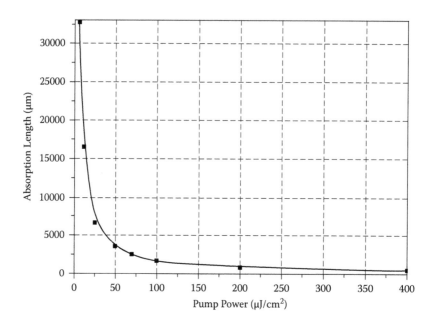

FIGURE 4.12
Absorption length as a function of the intensity of applied pump beam inducing free-carriers plasma.

where f is the filling factor of the silicon as compared to a bulk material, and $N(z)$ is the carriers' density as a function of the incident displacement. As f is always smaller than unity for PhC as compared to bulk material, the homogeneity length increases correspondingly. For instance, the l_{hom} of the PhC system is greatly increased by the low f of the rod-type structure. For an estimated mean free path (MFP) length of 63 unit cells, at the frequency of 5000 cm^{-1} absorption length of 116 unit cells, the homogeneity length spans 4.1 unit cells. Hence, the smaller the lattice-filling parameter, the greater the homogeneity length may be. Considering the extinction of the pump light due to the random scattering inside the PhC material, the homogeneity length of light for the PhC becomes

$$l_{hom} = 0.1\left(\frac{1}{l_{abs}} + \frac{1}{l_{mfp}}\right)^{-1} \tag{4.25}$$

where l_{mfp} is the MFP length of the scattering mechanism and l_{abs} is the absorption length. Figure 4.14 gives a surface plot of the l_{hom} for varied absorption and l_{mfp}, the latter arising from the scattering mechanisms within the PhC material.

From Figure 4.14, it follows that the more extended the l_{mfp} for the free carriers within the PhC material, and the longer the l_{abs}, the larger l_{hom}. These formulations, therefore, lay the foundation for the experiments and provide a guide for the limitation and effectiveness of the method.

4.4.2 Critical Challenges and Key Designs

The optical modulation mechanisms described by Equations 4.19 and 4.20 can be effectively implemented in many PhC devices, particularly those with small isolated cavity structures as the application of the high energy and intensity laser source may be localized for greater pump efficiency and lower thresholds in smaller switching volumes.

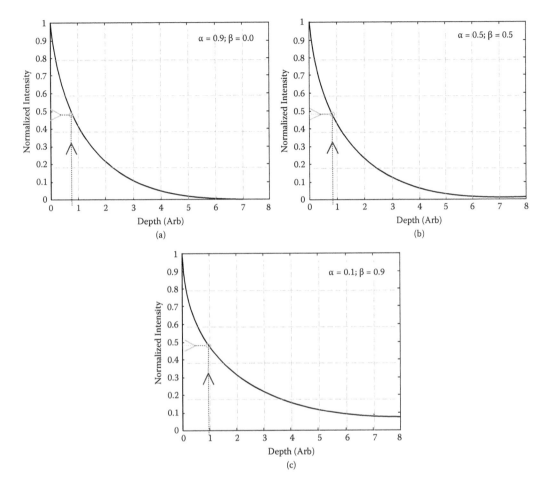

FIGURE 4.13
Intensity depth profile for (a) only linear absorption; (b) linear absorption and two-photon absorption coefficient equal to 0.5; and (c) dominantly two-photon absorption.

Figure 4.15 depicts that the probe beam at longer wavelengths may be applied through side-coupling to the PhC device, whereas the pump beam at shorter wavelength and higher photon energy may be beamed from the top of the device. By applying an optical pump beam at the center of the PhC intersection cavity's center resonator rod, localized application of optical pump radiation may be directed through a focusing lens onto the entire center resonator rod (as it has a diameter of only a few hundred nanometers) to induce effective refractive index modulations in the cavity rod. To demonstrate such PhC resonator modulations experimentally, many finer nuances need to be addressed. First, there is the requirement to develop deep submicrometer-sized PhC device fabrication processing techniques. Second, coupling of probe radiation into the subwavelength-sized optical intersections is critical in order to enable characterization of the PhC optical intersections. Finally, optical focusing and level control of high-power pump beam radiation precisely onto the center cavity rod of the optical intersection is also a nontrivial issue.

The issues for application of localized pump radiation onto the specific center resonator rod of the PhC optical intersection is mainly concerned with the diffraction limits imposed on the focusing of the pump beam, and also the temporal and spatial differentiation

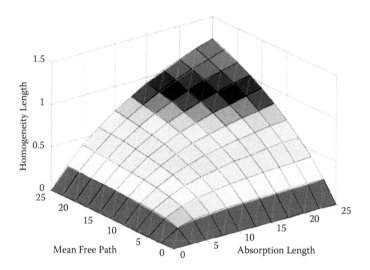

FIGURE 4.14
Homogeneity length of free carriers from the surface of incident radiation, plotted with respect to both absorption and scattering mechanisms.

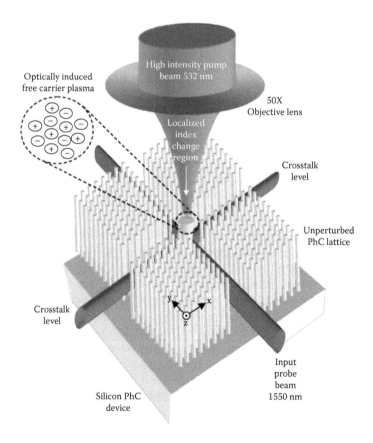

FIGURE 4.15
Illustration of high intensity pump radiation onto center of photonic crystal optical intersection, inducing localized material property changes in the center resonator rod.

between the pump and probe beam. Although the current experiments do not seek to distinguish between the spatial (i.e., to contain the probe beam within the pump beam) and temporal (i.e., to have determinate time delays between the pulses of the pump and probe beam) differences relative to the pump and continuous wave probe beam, the resolving power of the lens for localized application of pump beam onto the PhC resonator is critical.

The resolving power of any lens is fundamentally limited by diffraction. The aperture of a lens acts like a 2-D "single slit," such that light passing through a lens aperture inevitably interferes with itself to create a diffraction pattern known as an *Airy disk*. This empirical diffraction limit, also known as the *Rayleigh criterion*, is given by

$$\sin\theta = 1.22\frac{\lambda}{D_{lens}} \tag{4.26}$$

where θ is the angular resolution, λ is the wavelength of light, D_{lens} is the diameter of the lens, and 1.22 denotes the position of the first dark ring in the diffraction pattern in the Airy disk. This value coincides with the first positive zero of the Bessel function of the first kind, of order one, divided by π. For an ideal lens of focal length FL, Equation 4.26 yields a minimum spatial resolution as

$$\Delta l = 1.22\frac{FL \cdot \lambda}{D_{lens}} \tag{4.27}$$

which states the smallest spot radius that a focused laser beam can be collimated onto.

Hence, the shorter the wavelength of pump light used, the smaller the corresponding focused beam spot can be. This is simulated for the 100-fold objective lens used in the experiment with focal length of 2 mm, for wavelength range of 0 to 2 μm. As shown in Figure 4.16, the spatial resolution is linearly proportional to the wavelength range of

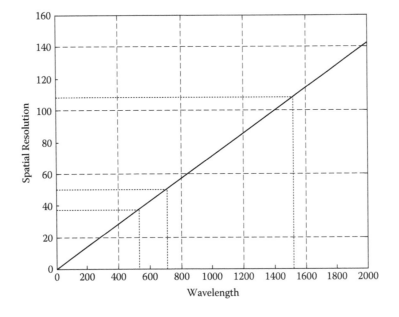

FIGURE 4.16
Spatial resolution for different optical wavelengths as limited by the diffraction limit of light.

interest. Here, it is also noted that for a telecommunication wavelength of 1550 nm, the minimum spot size that can be obtained is 108 nm. For a titanium:sapphire (Ti:Sa) laser output at 780 nm, the smallest spot size that can be obtained is reduced to 50 nm. Finally, for the neodymium:yittrium aluminum garnet (Nd:Yag) laser used in the experiments with 532 nm output, the smallest spot size that can be obtained is as low as 36 nm. This is much smaller than the size of the PhC resonator rod, and hence the diffraction limit of light in focusing does not limit the application of Nd:Yag pump beam for optical modulation in this work.

Yet, even though the application of pump radiation onto the PhC resonator rod is not diffraction limited, the focusing of the pump beam was difficult to achieve as the peak intensity of the pump beam should be targeted for the middle of the resonator rod height instead of at the surface of the device, which is easier to observe from the microscope. Hence, during the experiment, the vertical focus of the device needs to be adjusted for largest effect contrary to observations in the sharpness of the beam spot at device surface. Once the optical alignment of the probe beam was secured in place, it was critical to ensure that further alignment of the pump beam to the PhC device was carried out without altering the side-coupling of the optical fibers to the devices.

4.4.3 Photonic Crystal Device Effects of Optical Modulation

For optically controlled optical PhC response, the index modulations induced by the absorption of optical pump radiation in the center cavity rod results in the lowering of refractive index in the affected silicon material. To characterize such effects within the miniature size PhC devices, both HAR and slab-type configurations were investigated in both static and dynamic measurements. These include the variation of altered cavity conditions induced by the changes in resonator rod sizes, the application of continuous wave and pulsed radiation to produce the effects of carrier-plasma and thermo-optic operations. For an applied pump beam at wavelength 532 nm, from an Nd:Yag laser, the quantum energy of the photons is 2.3 eV, which is much greater than the silicon optical energy gap of 1.1 eV at room temperature of 295 K. As a result, free-carrier plasma can be generated by the laser beam which can in turn alter the optical properties and establish an optical nonlinearity of the material. Here, there is a resonant effect which primarily depends on the physics of the second step, i.e., the influence of the free carriers' on the optical properties.

Two mechanisms have been found in the past to be relevant for silicon. First, the free-carrier plasma gives rise directly to a change in the optical absorption and refraction which can be described by the classical Drude model. Based on this model, excellent agreement between prediction and experiment can be found in literature. Experimentally, this type of optical nonlinearity has been observed using short optical pulses from a Q-switch laser system to study phenomenon such as four-wave mixing, free-carrier gratings self-modulation, etc. [14]. Second, the thermo-optical properties also play an important role, especially when the measurements are performed by a continuous wave laser. In this case, the lattice of the semiconductor silicon is heated up by the recombining charge carriers. The resulting change of the refractive index then leads to optical bistability in silicon. In this section, the physical mechanisms for the optical plasma effect and the thermo-optic effects are detailed.

4.4.3.1 Optically Generated Plasma

Optical generation of stable plasmas is readily obtained in semiconductors, through cascaded linear processes in the formation-free electron-hole pairs that are generated by bandgap excitation with incident photons. The optically produced carriers augment the background electron-hole density and the free carriers' plasma remains electro-statically neutral. If the generation of an excess amount of plasma exceeds the rate of loss (by recombination or diffusion) on the timescale of interest, the plasma modifies the linear optical properties of the material. This effect is wall-modeled by the classical Drude model, where the refractive index of the material is given as

$$n = n_0 \sqrt{1 - \frac{\omega_p^2}{\omega^2}} \tag{4.28}$$

where ω is the angular frequency of the light, n_0 is the linear index in the absence of significant free-carrier density, and ω_p is the density-dependent plasma frequency given by Equation 4.21.

With increasing pump power, the carriers' density increases as

$$N = \frac{\alpha P_{pump}}{\hbar \omega} \tag{4.29}$$

P_{pump} is the optical pump power, α is the absorption coefficient, and $\hbar$ is the reduced Planck constant. This is as shown in Figure 4.17, where the densities of the induced free carriers increase proportionally with the power of the applied pump beam, based on the Drude formulations [13].

Hence, as the carrier density increases, the plasma frequency increase, which leads to a corresponding decrease in the material's refractive index as described by Equation 4.26. Figure 4.18 plots the effect of increasing pump power on the real and imaginary indices of the silicon PhC material.

Directly proportional to the density of the free charge carriers' plasma generated by the absorption of pump radiation onto the PhC material, the increasing optical pumping

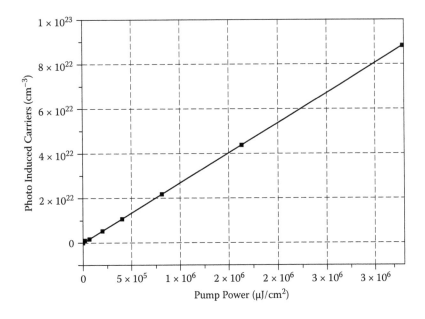

FIGURE 4.17
Photo induced carriers' density with 532 nm pump laser being directly proportional to the applied pump power, based on the Drude formulations.

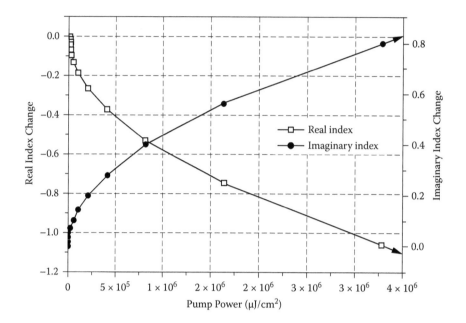

FIGURE 4.18
Induced free carriers effect on silicon refractive index by optical pumping.

power applied reduces the dielectric constant of the center cavity rod correspondingly, as shown in Figure 4.18. As can be seen, the increase in real index is also accompanied by a corresponding increase in the imaginary component, which indicates greater absorption of the material. Such trade-off, therefore, brings about a direct consequence onto the l_{abs} of the excited carrier plasma which needs to be carefully managed to prevent inhomogeneous device modulation.

The timescale for such processes is determined by the cavity charging time and the carrier relaxation time which, according to the rate equation, is approximately 80 picoseconds for silicon. For example, Figure 4.19 shows the transmission spectrums of a 3×3 PhC optical intersection when the refractive index of the center resonator rod is reduced by optical pumping. For the case whereby the index contrast of the PhC optical intersection compared to air was 2.65, the FWHM was 27 nm. This was reduced to 15 nm for PhC resonator rod index contrast reduced to 2.55, and to 13 nm for index contrast reduced to 2.45.

Such optical modulation becomes ever greater in sensitivity for optimized resonator rod dimensions. In the case where the thermo-optic effects become dominant, the reversed effects are to be observed, except for the additional phenomenon of resonant wavelength shifting arising from the effects of larger area modulations. The details of these differences will be described in the following section.

4.4.3.2 Thermal Effects of Optical Excitation

Linear absorption of light must result in energy deposition in the irradiated material. If the rate of energy deposition significantly exceeds its rate of removal, the thermo-optic effect takes place through the heat generated with the optically produced carriers. As the energy of a collection of atoms and molecules increases, their macroscopic optical properties will be altered. The physical interpretation of the thermo-optic effect arises from the induction of refractive index changes, which are proportional to the total energy deposited in the

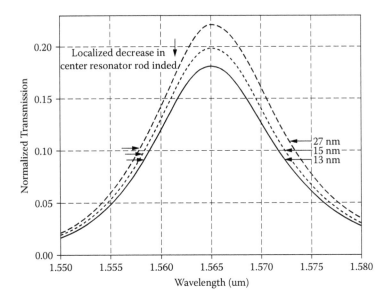

FIGURE 4.19
Simulation results of decreased full width at half maximum for photonic crystal resonator rod refractive index reduced by carriers' plasma effects.

medium, in the form of the integral of the absorbed intensity that the thermo-optic effect is linearly and positively proportional to temperature T for silicon.

Although the carriers' plasma effect is ultrafast, the timescale for the thermo-optic effects is much slower, typically eight order of magnitudes slower than that of the carriers' plasma effect, in the timescale of milliseconds. At the same time, due to the conduction effect of the thermo heating that is induced, not only will the center resonator rod undergo refractive index modulation, but also the surrounding rods. With increasing heating, the index contrast between the PhC resonator rod and the surrounding rods will increase to result in the reverse effects obtained in Figure 4.19. This is illustrated by the simulation result of Figure 4.20, whereby not only the center resonator rod index was modulated, but also the surrounding partially reflecting PhC rods. In this case, not only does the increase in refractive index contrast bring about a corresponding increase in bandwidth, but the center wavelength also shifts towards the longer wavelengths to result in a red shift.

4.5 Optical Measurements of Photonic Crystal Devices

This section presents the necessary experimental principles and techniques required to realize optical measurements for deep submicrometer length scale PhC devices. These include the study of measurement techniques, experiment design, and the methodology of measurement equipment set-ups. In the experiments with base set-up depicted schematically in Figure 4.21, the input signals (falling within the C- and L-band wavelengths in the optical communication spectrum) were obtained using a Lightwave LDX-3412 continuous wave laser source and an ANDO AQ4321 tunable laser system (TLS). For these two types of input sources, the output from the PhC optical intersection devices were then probed using a Newport 1835C optical meter and the TLS synchronized ANDO AQ6317 optical

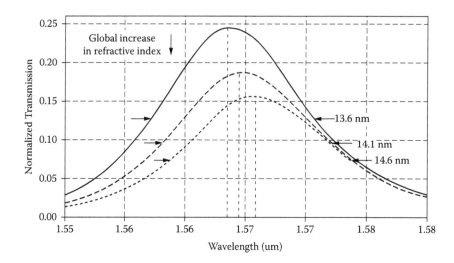

FIGURE 4.20
Simulated effect of thermo-optic modulation on the optical photonic crystal intersection with conduction resulting in the alteration of surrounding rods refractive index.

spectrum analyzer (OSA), respectively. In addition, broader band measurements were made, using a super-luminescent light emitting diode (SLED) DenseLight DL-BD9, which was similarly measured using an OSA.

To apply the input laser and probe the output signals from the very small PhC devices, optical fibers were used as the interconnecting medium for its excellent optical (attenuation of less than 0.2 dB/km) and mechanical properties. To perform side coupling of the optical fibers to the PhC optical intersection devices, Melles Griot NanoMax-TS alignment stages were used. For the coupling of the optical fibers with cladding diameter 125 μm, core diameter 8.2 μm, and mode field diameter of 10.4 μm, to the HAR PhC device (with device

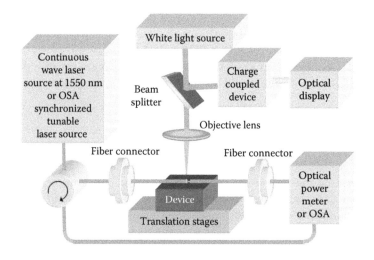

FIGURE 4.21
Schematic of experiment set-up, where the 2-D photonic crystal device was optically coupled to the input tunable laser source and output probing measurement apparatus with observations made from the top plane with a charge coupled device.

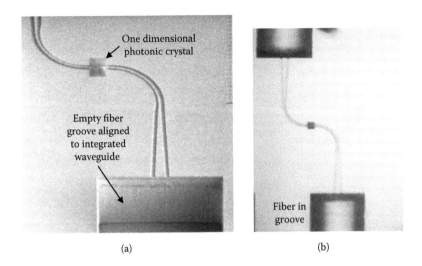

(a)   (b)

FIGURE 4.22
(a) SEM of integrated fiber grooves without fibers in grooves; (b) top-down microscope image of fibers aligned in grooves.

height 13 μm), alignment of the input and output optical fibers was achieved via physical guiding of the optical fibers along the monolithically integrated optical testing fiber grooves, which were aligned (in-plane) to the PhC device with high lithographic accuracy. Tilt of the fibers was minimized through alignment to the etched sidewalls, which was optimized to have etched angles ~90°. Figure 4.22 shows the coupling fiber grooves, with and without fibers in groove, together with the probe needles during experiment. Out-of-plane, microprobe needles with tips of 2 μm diameters were used to ensure alignment of fibers to the lower surface of the fiber grooves so that vertical alignment between the core of the optical fibers and the PhC device was achieved (see Figure 4.23).

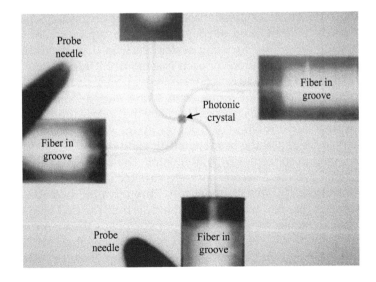

FIGURE 4.23
Optical microscope image of probe needles used to apply fibers in grooves.

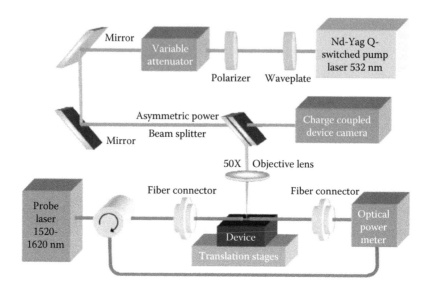

FIGURE 4.24
Schematic of experiment set-up, where the pump beam was focused with high intensity on the resonator from the top, while the weaker, longer wavelength probe beams were applied and detected laterally.

Figure 4.24 depicts the experiment set-up designed to enable dynamic optical modulation experiments as opposed to simpler static measurements, with coupling alignments, device search with imaging, and spatially localized focusing of high intensity pump beam onto the center of the optical intersections. Using such a set-up, both static and the dynamic properties of the PhC optical intersections may be probed. Specifically, in the static experiments, properties of PhC optical intersections with different center cavity rod sizes were characterized. Based on this first set of experiments, the center cavity rod modulations principles were tested for its dynamic properties through optical excitation for inducement of changes in refractive index via optically induced free charge carriers plasma effect.

For such experiments where localized material property change effected by high intensity pump laser needs to be induced in conjunction with application of the side coupled probe signal laser inputs, a laser focusing system such as that illustrated in Figure 4.15 allows for independent manipulations of each laser source. Here, the pump laser being a Q-switch Nd:YAG-based 532 nm direct-coupled system consisted of a laser head containing the optical pump and resonator components with a closed-loop chiller. At the output of the laser head, a half-wave plate was used to rotate the polarization of the linearly polarized pump beam such that a rotation of θ degree by the wave plate about the beam axis would rotate the polarization of the beam by 2θ. In addition to the wave plate, a polarizer was used in combination to form a first attenuator. The pump beam was then directed (by steering mirrors) through a second discrete variable attenuator and beam splitter to reach the top plane of the PhC device under test. Through a 100-fold aberration corrected Mitutoyo objective lens, the pump laser beam spot was then focused onto the center cavity rod of the PhC optical intersection, to a beam spot size of less than a micrometer.

The spatial application of such localized pumping was enabled through alignments made with the sample stage, and observation of the pump beam (adjusted to low-power level) through a charge-coupled device camera fed to a display card on a computer which outputs the feedback image. When the pump laser power was adjusted to a high level, the resulting pump intensity on the PhC optical intersection's center cavity rod became very high because the beam was focused on a very small spot size. Therefore, through optical absorption of the pump beam by the PhC intersection's center cavity rod, high free charge carriers' density result in the PhC material, which induces modulations in the device. These effects are measured with a low intensity probe laser (for which the PhC structures were designed to be sensitive to). As the linear absorption of the low-intensity probe beam at 1550 nm wavelength is considered negligible for the PhC material (as the photon energies is lower than the bandgap energy of silicon), modulations detected by the probe beam may be attributed entirely to the effect of the localized high intensity and high photon energy pump beam. At the same time, the cross talk levels measured were verified to be free from pump radiation through monitoring of readings, while toggling the application of pump laser power when the probe laser was switched off.

4.5.1 Photonic Crystal Test Structures, Calibrations, and Measurements

In addition to the experiment set-ups, an important factor for consideration in PhC demonstrations at the optical communication wavelength is the need to incorporate macro- to microcoupling testing structures. There are mainly two critical interfaces in optical coupling required for PhC testing: the first is the interface between the PhC device and the integrated waveguide, and the second is the coupling between the integrated waveguide and the macroscopic probing medium (e.g., an optical fiber or a set of collimating lens coupled to an optical meter). The focus here will be mainly on the interfaces between optical fibers, integrated silicon waveguides, and the PhC devices. First, integrated fiber grooves were fabricated, and Figure 4.25 gives the cross-sectional SEM image of the monolithic integration scheme involving both the PhC devices with the deeply etched optical fiber grooves.

Here, the optical fibers were designed to have their cores placed at the same level as the microscale PhC devices in both vertical and lateral directions. Careful control of etch rates—which differs greatly for varied etch conditions—are required to ensure vertical coupling; lateral alignments were ensured with a series of search and alignment marks

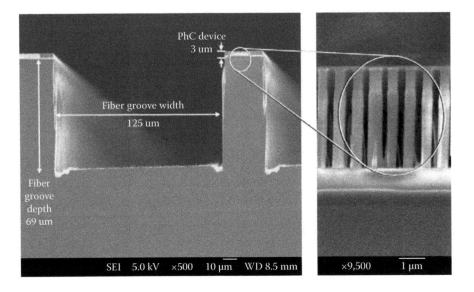

FIGURE 4.25
Cross-sectional SEM image of (a) deeply etched fiber grooves with the deep submicrometer scale photonic crystal devices; (b) blown-up view of photonic crystal device with respect to fiber grooves.

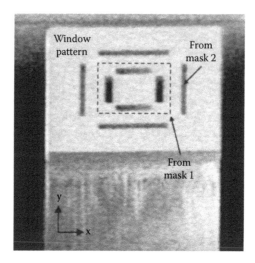

FIGURE 4.26
Cross-sectional SEM image of deeply etched fiber grooves with the deep submicrometer scale photonic crystal devices.

made around the periphery of the optical mask. Figure 4.26 shows the overlay marking that indicates how well the PhC devices were aligned to the optical fiber grooves. According to the measurements obtained, the average misalignment in the x and y directions were only 0.085 and 0.157 μm, respectively. Although such alignment accuracies would still need to be improved for multiple-layer-stacked 3-D PhC fabrications, nanometers scale accuracy is sufficient for good optical coupling between the silicon waveguides and optical fibers.

At the same time, beside the vertical and lateral alignment of the fiber mode field with the microscale PhC devices, appropriate coupling between the two well-aligned optical modes is also critical. For this purpose, the waveguide taper structures are necessary. As can be seen from Figure 4.27, a singular taper at the interface near the PhC waveguide

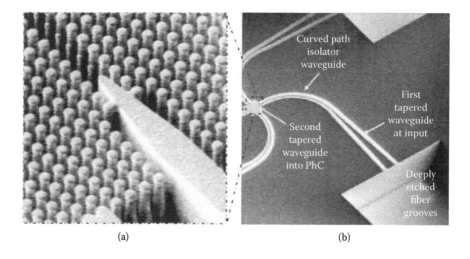

(a)   (b)

FIGURE 4.27
Micro- and macro-tapered silicon waveguides for coupling between (a) silicon waveguide and photonic crystal; and (b) fiber and silicon waveguide.

would not be sufficient to enable coupling between macroscopic fibers with modes of ~10 μm to submicrometer widths of PhC waveguides. Hence, two tapers are required: for slab type PhC devices, a bigger one near the fiber tip and a smaller one to connect to the PhC optical intersections, as shown in Figure 4.27b and 4.27a, respectively.

At the same time, on top of such taper waveguide structures, it was necessary to make use of curve waveguide structures to avoid the occurrence of straight through optical paths. In this way, optical leakages that occur, either under or over the PhC devices (through the substrate or through the upper cladding of air, respectively), can be eliminated. To quantify the wave propagating characteristics of such curved waveguides, many similar curved waveguides were connected in tandem and tested. Figure 4.28 exhibits the microscope image of two such test structures with two and four curved waveguides, respectively. The results of these test structures measurements will then be presented in the next section, based on principles similar to the cutback method employed for the quantification of the very small scale optical waveguide propagation.

Determination of the guiding and coupling interfacing components characteristics then allows for subsequent calibration and measurements of the PhC devices, in both HAR and slab-type configurations.

4.5.1.1 Photonic Crystal Waveguide Propagation Loss Coefficients

In the experiments, the HAR PhC devices were fabricated to obtain the high density rod-type PhC square lattice patterns of period 570 nm and rods radii 115 nm—with high resolution and OPCs along the line defect waveguides. Record HAR etching of PhC rods with ultrasmooth etch sidewall scallop depths of 12 nm were fabricated in the intersection configuration with different center cavity rod sizes, as shown in the SEM image of Figure 4.29.

At the same time, slab-type PhC of much thinner device silicon thickness was fabricated similarly. These two configurations are differentiated mainly by their aspect ratio and also

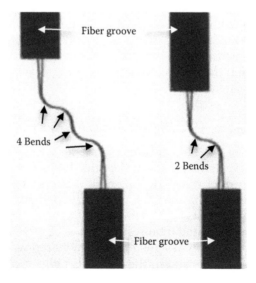

FIGURE 4.28
Optical microscope image of test structures with two tapered input waveguides and varied number of curved waveguides.

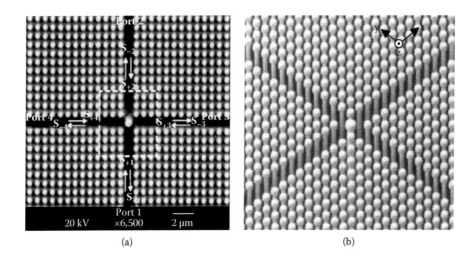

(a)   (b)

FIGURE 4.29
High aspect ratio photonic crystal rods SEM images: (a) top view with coupled mode analyses ports notations, (b) orthogonal view.

their out-of-plane guiding mechanism. Figure 4.30 shows the top and orthogonal view SEM images of the slab-type PhC with reduced radius line defect waveguide to confine the optical beam in the vertical direction. Fabricated in 8″ SOI wafers, of a 3-μm device thickness, a buried oxide thickness of 1 μm, and handle silicon layer of 725 μm, the bulk PhC array has a lattice constant of 517 nm, bulk rods with a diameter of 223 nm, and sub-100 nm reduced radius line defect rods with a diameter of 89 nm. These line defects not only confine optical radiation in-plane via PBG action, but also use index assistance for out-of-plane waveguiding.

For the fabricated PhC devices with the monolithically integrated fiber grooves, Figure 4.31 illustrates how the buried oxide layer acts as the lower cladding, whereas optical coupling was achieved via fibers side coupled to the devices through deeply etched fiber grooves. In the slab-type PhC, tapered silicon waveguides were used for mode size conversion instead of the PhC tapers used for the HAR PhC devices. In this way, optical alignment of the macrosize fibers and the sub-micrometer size PhC waveguides may be achieved in both lateral and vertical directions.

4.5.1.2 Silicon Waveguides—Coupling and Bending Losses

The spectrum response of the PhC optical intersections were obtained using the OSA synchronized tunable laser sweeping wavelengths 1520–1620 nm in steps of 0.05 nm. The transmission spectrum of the HAR PhC optical intersection with center cavity rod of radius 300 nm as measured by the OSA when synchronized to the tunable laser was obtained as shown in Figure 4.32.

Here, as expected from the simulation results, it was observed that the reflection from this PhC optical intersection with designed center cavity rod was lower than the case for which there was no center cavity rod in the PhC optical intersection. Next, characterization of a simple straight PhC waveguide property was carried out using the F-P resonance method. In Figure 4.33, the reflection spectrum was measured with changing wavelength in the telecommunication range.

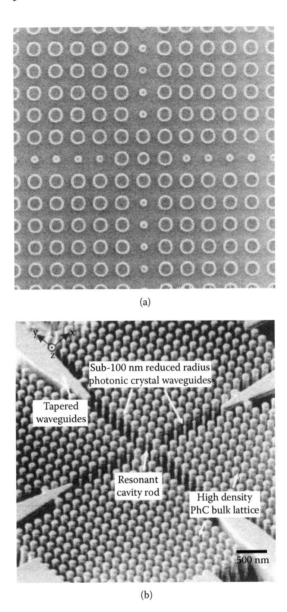

(a)

(b)

FIGURE 4.30
SEM of slab type photonic crystal intersection device with reduced radius rods waveguide in (a) top view, and (b) orthogonal view.

For this F-P resonance method, a constant power input light was used and the output end of the waveguide was butt-coupled directly by another cleaved single-mode fiber. From the spectrum obtained, the propagation loss coefficient is given by

$$\alpha = -\frac{1}{L}\ln\left(\frac{1}{R} \cdot \frac{\sqrt{I_{max}/I_{min}}-1}{I_{max}/I_{min}+1}\right) \tag{4.30}$$

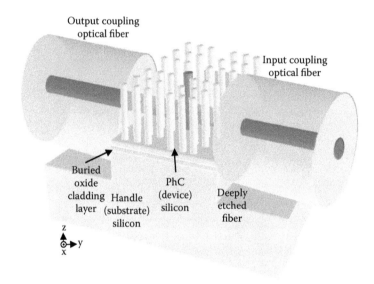

FIGURE 4.31
Schematic diagram of slab type photonic crystal (in rod configuration), implemented in silicon-on-insulator wafers and coupled with optical fibers.

where L is the waveguide length, R is the facet modal reflectivity, I_{max} and I_{min} are the peak and bottom intensities of each resonance, respectively. Compared to the conventional cutback method, the use of the I_{max}/I_{min} ratio, which is independent of coupling condition of input and output light, makes the evaluation of such high index waveguides more precise.

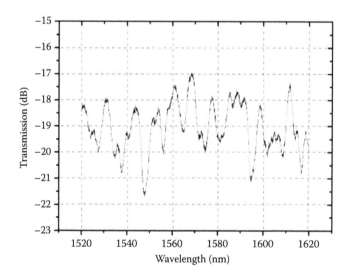

FIGURE 4.32
Transmission spectrum measurement of the photonic crystal optical intersection with center cavity rod radius of 300 nm, measured using an optical spectrum analyzer synchronized to a wavelength sweeping tunable laser source.

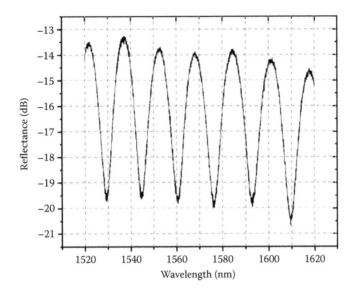

FIGURE 4.33
Reflection spectrum measurement of straight photonic crystal waveguide for Fabry–Pérot resonance method evaluation of propagation characteristics.

Here, the effective index is given by

$$n_{eff} = \frac{\lambda^2}{2L\Delta\lambda} \tag{4.31}$$

where $\Delta\lambda$ is the FSR, indicating the band width between neighboring resonance peaks. Based on such calculations derived from the measured peak/bottom intensities of each resonance, together with the FSR, the propagation loss coefficient of the HAR PhC waveguides may be estimated at 0.032 dB/μm.

Next, for the slab-type PhC structures, which allows for vertical confinement and therefore better waveguiding properties over longer propagation distances, the propagation loss coefficient may be obtained based on the cutback test structures. Here, instead of the destructive method of reducing the waveguide length after each measurement set, many waveguides with exactly the same input coupling and output probing conditions were tested for their propagation loss sustained from different waveguide lengths. Using similar input powers, PhC waveguides with different propagation distances were fabricated as shown in the SEM images of Figure 4.34.

The propagation loss coefficients obtained for single-line-defect, slab-type PhC waveguides with and without reduced radius waveguides were 0.46 dB/μm and 0.72 dB/μm, respectively. The higher rate of propagation loss in the devices without reduced radius waveguides may be attributed to the lack of vertical confinement, otherwise present in the index-assisted, reduced-radius PhC-line-defect waveguides. Here, the waveguide propagation loss seems greater than that of the HAR case because the HAR throughput was measured without isolation structures such as bends, etc. Hence, the straight through-path of the measurement resulted in greater transmission coefficients. At the same time, even though the slab structure has index guiding in the out-of-plane direction, coupling with TE-like modes form the symmetry-breaking substrate will result in such waveguides being leaky, too.

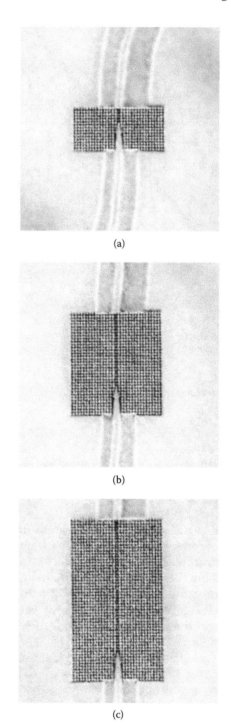

(a)

(b)

(c)

FIGURE 4.34
Photonic crystal line defect waveguides of varying length: (a) shortest at 15 a length, (b) mid-length at 35 a length, and (c) longest at 55 a length.

Following which, to obtain the losses incurred by each 90° waveguide, differences in transmission levels for varied numbers of right angle bends were obtained so that the unit loss was estimated to be ~3 dB. This loss may be attributed to both the small bending radius and multimode nature of the silicon waveguides, which promotes mode scrambling at the tight bends that result in losses through coupling to the radiative modes.

Here, it was noted that the coupling loss per port for the HAR was much lower than that of the slab type PhC devices. Although the total coupling loss was only 2.7 dB for the HAR PhC devices, the total coupling losses for the slab type PhC devices were ~50 dB. This was due to the incompatible optical mode sizes between the fiber and the much thinner silicon waveguides in the slab configuration. This loss is high; the cause of it was due to the lack of an appropriate mode converter between the very large optical fiber mode and the slab type silicon waveguide. This type of loss due to incompatible mode sizes had been shown to be effective circumvented with many methods; it had been shown experimentally that an adiabatic mode connector [15] may be used to achieve connection loss as low as 0.8 dB per port [16] for slab type PhC with dielectric claddings. Yet, as such adiabatic mode connector does not work well with PhC slabs with air claddings, further design and investigation is needed.

Therefore, the many characteristics of the testing structures, such as the propagation loss coefficients, coupling losses, and bending losses were measured using many dies, and the PhC devices may be measured and evaluated with these foundations.

4.5.2 Static Photonic Crystal Device Experiments

Having established the optical properties of the OTS assistance devices and the basis properties of the PhC waveguide propagation characteristics, the various functionalities of the 1-D and 2-D PhC optical intersection devices are then measured—first, in the static configuration. For both single-line resonators, and two orthogonally intersecting single-line-defect PhC waveguides (forming small form factor optical intersections), the cavity modes can be controlled by the size of the coupling cavity rod at the intersection, even for the case of the 1×1 resonator. To unveil the relationship between the size of the center cavity rod and the performance of the HAR PhC optical intersections, transmission measurements were repeated for different PhC optical intersection devices fabricated with different center cavity rod sizes (see Figure 4.35) at wavelength 1550 nm.

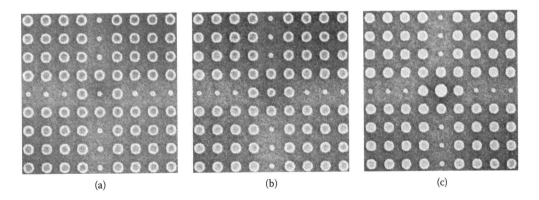

(a)   (b)   (c)

FIGURE 4.35
SEM images of photonic crystal optical intersections with progressively larger center resonator rod dimensions.

For HAR optical intersections with center tunneling cavity rods of varying radii (normalized to the lattice pitch of the PhC), the cross-talk and transmission level measurements were made as shown in Figure 4.36a and 4.36b, respectively. In the plots, the experimental results are indicated by filled symbols, whereas the FDTD simulation results are plotted in a continuous line. In the plots, the abscissa indicates the normalize radii of the center

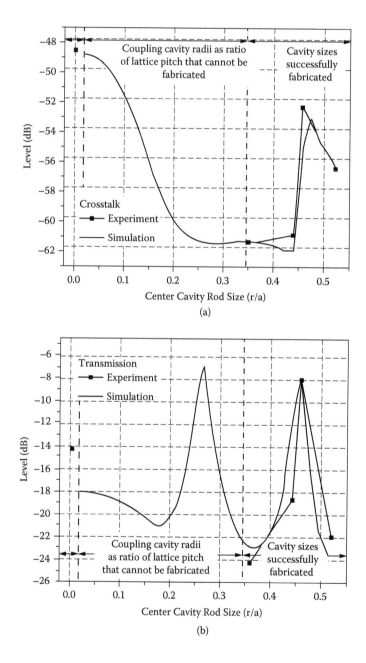

(a)

(b)

FIGURE 4.36

Optical experiment measurements (filled-symbols) and FDTD simulation results (smooth line) for (a) cross talk, and (b) transmittance levels, of the photonic crystal optical intersection devices with progressively increased center cavity rod size.

cavity rods (expressed as a ratio of the PhC lattice period), whereas the ordinate axis represents the measurement levels in dB.

Here, measurements of PhC optical intersections with center rod radius of less than 200 nm (i.e., normalized radius less than 0.35) cannot be realized due to limitations in fabrication resolution. Hence, the normalized radius of the tunneling cavity cannot be tested for small values ranging between 0 and 0.35 times the lattice period of 570 nm (i.e., between 0 to 190 nm due to limit in fabrication process resolution). For the other tunneling cavity sizes of PhC intersecting waveguide measurements, both the measured cross talk and transmission results corresponded well with the FDTD simulation results with good repeatability. From the experiments, the modulations of the transmission and cross-talk levels by the different cavity sizes were measured to be greater than 16 and 12 dB, respectively. Here, it is to be noted that as the optical path of the cross-talk measurement was not a straightforward one (unlike that of the transmission branch), the cross-talk measurements made can be attributed solely to the effect of PhC wave guiding and not from stray leakage radiations.

Further, for similar PhC intersections in the slab type configuration, where missing rod waveguides were replaced by reduced radii PhC rods, higher Q-factor resonators was tested with presence of vertical confinement mechanism. Here, Figure 4.37 exhibits the Lorentzian response of the single-line configuration 3×3 PhC optical resonators with one surrounding PhC bulk rod around the center cavity resonator.

For resonator sizes varied as 0, 84, and 136 nm, the expected blue shift of the optical responses for increase in resonator rod radius size from 0 to 84 nm was found to be 19 nm. At the same time, increases in transmission spectrum levels were also observed for optical intersections with radii 0 and 84 nm as compared to that of radii 136 nm. For the a range of center resonator rod dimensions and their corresponding effects, Figure 4.38 shows the transmitted levels of a 5×5 slab type PhC optical intersection at the telecommunication wavelength of 1550 nm. Here, the filled symbols represent the optical measurements, while the unfilled ones represent the 3-D finite deference time domain simulation results.

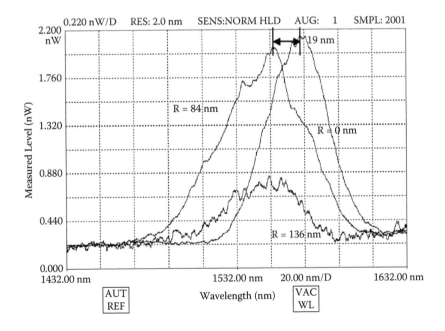

FIGURE 4.37
Measured optical spectrums of single-line resonators with varied center resonator rod sizes.

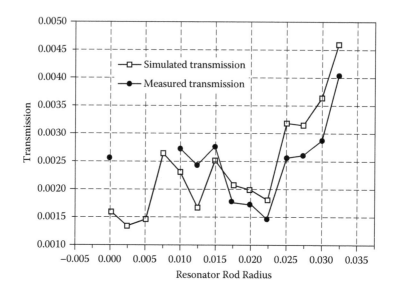

FIGURE 4.38

Photonic crystal optical intersection with two bulk rods on each side of the resonator and its simulated and measured transmittances for varied center resonator sizes.

Hence, both HAR and slab-type PhC structures have been experimentally measured to demonstrate the predicted characteristics of sensitivity towards the center cavity resonator rod sizes. At the same time, as expected, Lorentzian responses were observed for PhC intersections with order higher than the 1 × 1 configuration, as demonstrated in the slab-type PhC optical intersection measurements. Although the slab-type PhC suffered from high coupling losses due to the incompatible coupling between the input fibers and the slab-type waveguide mode, much higher quality optical intersections were successfully demonstrated due to the improved confinement in the out-of-plane direction. Hence, with these static measurement results, the next section will present the dynamic modulation aspects of the fabricated PhC devices.

4.5.3   Dynamic Modulation of Photonic Crystal Intersection

The unique structure of the orthogonally intersecting, line-defect, photonic crystal waveguides with optical characteristics dependent on the center resonator rod properties is useful for dynamic optical modulation purposes. This is due to the enhanced optical nonlinearity imposed by the PhC optical intersections. The experiments realized for these modulation principles and designs will be described in greater detail in the following sections.

Hence, the application of an optical pump beam onto the center resonator rod of the PhC intersection is capable of effectively modulating the optical responses of such devices. A demonstration of this prediction is shown for a center cavity rod size of 0.53 times the PhC lattice period for an HAR PhC optical intersection. Figure 4.39 shows the modulated response of the cross-talk levels in the HAR PhC optical intersection when the power of the pump beam was decreased progressively through increasing attenuation. Here, the abscissa axis gives the intensity of the pump laser, and the ordinate axis plots the normalized changes in cross-talk levels. The measured cross-talk levels are plotted as round symbols in Figure 4.39, whereas the FDTD simulation results are plotted as a solid line.

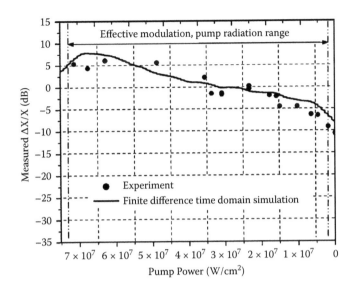

FIGURE 4.39
Normalized cross-talk level modulated by optical pumping.

For pump beam radiation of intensity less than 60 MW/cm², the measurements of the modulated signals were well repeatable. However, for pump beam radiation of more than 75 MW/cm², the PhC material began to shows signs of damage which, under SEM inspection, revealed melted material within the altered intersection. Such damage to the PhC device material being irreversible therefore changes the cavity property of the optical intersections permanently (an effect remaining even after removal of the optical pump beam). Hence, based on these experiments, an effective modulation range of normalized cross talk being 15 dB was, therefore, found for the HAR PhC optical intersection devices with applied pump radiation kept below 60 MW/cm² (where reversible modulations of the PhC optical intersections are possible). To explore the possibility of improving these results, higher Q-factor PhC optical intersections in the slab-type configuration were tested.

Similar to the procedures used for the HAR PhC optical intersections, the slab type PhC intersections were modulated dynamically with a pump laser applied with spatial localization from the top onto the center resonator rod. For a 3 × 3 slab-type PhC optical intersection, it can be seen that the Lorentzian spectrums were shifted in magnitude instantaneously with the on-and-off toggling of the applied continuous wave pump laser. Figure 4.40 gives the transmission spectrums of such a 3 × 3 slab-type PhC optical intersection, with and without application of optical pump power. From the spectrums, it can be seen that the modulated spectrum peak was increased by ~57%, and the FWHM was decreased from 40 to 36 nm. When the same device was melted down, the transmission signal was indeed lost as shown in the measured spectrum of Figure 4.40.

Here, the optical power required to achieve such modulations were much reduced in such slab-type PhC optical intersections of higher order configurations than in the HAR 1 × 1 configuration. As a result, the power applied was only 0.318 MW/cm² for the case of instantaneous modulations demonstrated in Figure 4.40. This is due to the higher Q-factors in such cavities as compared to the HAR case without partially reflecting PhC rods surrounding the center cavity resonator. Consequently, much lower power was observed for the meltdown thresholds (~1.06 MW/cm²), because higher order cavities are better capable of retaining radiation for a longer lifetime. Hence, the free carrier-plasma effect

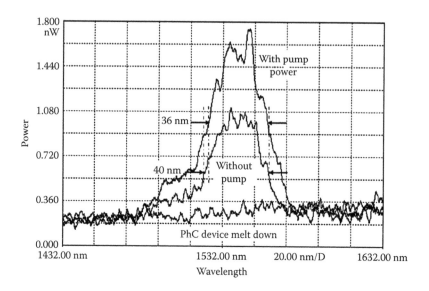

FIGURE 4.40

Experimental results of instantaneous photonic crystal output modulation by the on/off toggling of the high intensity applied continuous wave pump laser.

was observed; the timescales of such operations are limited mainly by the relaxation time of the carriers. When the application of optical pump radiation was used to heat up the lattice, through fast energy exchange between the optically excited electronic subsystem and the lattice, thermo-optic effect modulations were instead measured.

Thermo-optic switching operations may be derived using a similar experiment set-up as shown in Figure 4.24. For pump laser pulses with repetitive frequency of 50 kHz, the effect of increasing the thermo levels of the PhC intersection devices can be shown to yield both expected red shifts in the resonant spectrums (toward the longer wavelength) and also corresponding bandwidth increases, in the much slower thermo modulation effect. The thermo-optic effect arises from excessive circulating optical flux within the lattice of the PhC silicon material, which gets heated up by the recombining charge carriers. Typically, continuous wave laser beams used for optical pumping are more prone to the thermo-optic effects, due to the continuous bombardment of photons onto the material. However, as the continuous wave laser used in our experiments had much lower intensity than the Q-switched pulsed mode of the pulse-type laser output, we were able to induce thermo-optic effect modulations, only via the pulsed pumping mode. Figure 4.41 gives the thermo-optic modulation effect on the bandwidth of the 3×3 PhC slab-type optical intersection device as increasing optical pumping power was applied with onto the center cavity resonator rod. For incident pump power increased from 0 to 0.110 mW, the bandwidth of the PhC device was increased by up to 3 nm. Bandwidth modulations of such magnitudes indicate that the induced refractive index change was larger than 0.2, as was simulated in Figure 4.20. At the same time, the bandwidth widening was accompanied by a corresponding red-shift in the center wavelengths of the Lorentzian transmission spectrums as shown in Figure 4.42.

Figure 4.42 plots the resonant wavelength shifts of the 3×3 slab-type PhC optical intersection device under thermo modulation, as progressively increasing optical pumping power was applied. Such wavelength shift towards the longer wavelengths is also known as the red shifts. From the experiment, with applied optical power up to 0.11 mW, the center wavelengths was increased by 2.2 nm for increasing optical pump power applied.

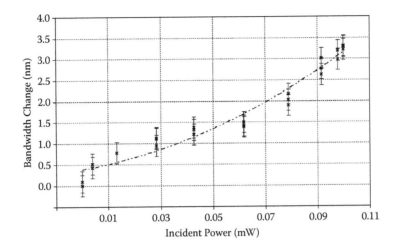

FIGURE 4.41
Progressive bandwidth change to a photonic crystal optical intersection resonator as it was thermo-optically modulated by progressively higher pump power.

From the measurement, it may be observed that fine shifts of 0.1 nm was observed for such modulation effects with applied pump power increased in steps of ~ 0.01 mW. Based on the FDTD simulation results presented in Section 5.3, center wavelength red-shifts of 2 nm would therefore correspond approximately to refractive index changes of nearly 0.2, which is corroborated by the induction made based on the bandwidth widening effect. Such index changes in turn correspond to large increases in temperature which therefore cause localized melting of the device under test, as shown in Figure 6.25.

From these results, the effects of the carriers' plasma and thermo-optic type modulation were very easily and clearly distinguished. In the carriers' plasma modulation effect, the changes made to the device spectrum returned instantaneously to its original position

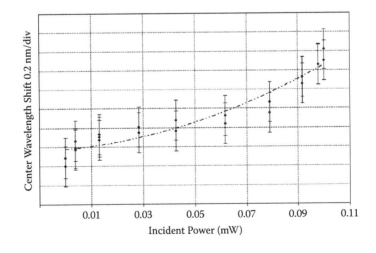

FIGURE 4.42
Corresponding center wavelength shifts for the 3 × 3 photonic crystal optical intersection resonator as it was thermo-optically modulated by progressively higher pump power.

once the pump laser beam was switched off. On the other hand, for the thermo-optic-induced effect, the modulated device spectrum took a long time to return to its original spectrum position. This may be attributed to the much slower rate of thermo cooling in the thermo-optic effect (required to return the device to room temperature after getting heated up by the pump laser) as compared to the fast recombination and relaxation processes in the former carriers' plasma process.

While the thermo-optic effect provides an alternative means of optically induced modulation for the PhC devices, its distinctively longer timescale makes it suitable mainly for state changing applications, and not optical switching operations, which require as short a switching time as possible. At the same time, as expected, the modulation of refractive index by the carriers' plasma effect was larger than that of the thermo-optic-induced refractive index change. The use of thermo-optic effect for modulation would also require

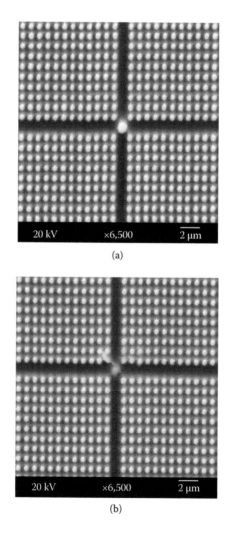

(a)

(b)

FIGURE 4.43
SEM of extreme effect in localized high intensity optical pumping: (a) photonic crystal intersection device prior to complete melt down, but slightly deformed after pumping, and (b) photonic crystal after excessive pumping, inducing cavity rod melting and collapse onto a nearby neighboring rod.

the careful application of laser irradiance onto the PhC device, as any occurrence of melting in the silicon PhC material is irreversible. Figure 4.43 shows not only how well the pump beam laser may be focused onto the center of the resonator rod to cause localized melting (Figure 4.43a), but also how the unbridled application of laser irradiance can lead to melting of the silicon material (Figure 4.43b) at the PhC resonator rod.

4.6 Summary

This chapter presented the principles for the theory and designs of PhC microresonator devices for small-size integrated photonic circuits, together with the principles for their dynamic modulations. Specifically, the design of embedded PhC microresonators, in single-line and orthogonal crossing waveguides form were first presented, followed by the analysis of a four-port PhC crossing in illustration of the coupled mode theory for prediction of the performance for the various designs of the PhC optical intersections with different degenerate intersection modes. Based on such theoretical formulations, predictions for the various cavity structures were made. All these theoretical and design works therefore lay the foundation for further exploration of the physical, architectural, and phenomenological effects of PhC microresonators suitable for dynamic modulations. Here, the indirect bandgap and small, nonlinear coefficient of silicon as the material of the PhC was taken into consideration for the selection of the modulation method by free carriers' inducement. By proper design of device and experiments, optical measurements of such submicrometers-scale PhC optical intersection circuits were demonstrated. Coupling between macroscopic optical source and detectors, to the microscopic scale PhC circuits were achieved through designed integrated alignment systems. The experimental results of the fabricated PhC devices have shown that the PhC devices is capable of achieving integrated circuit level performances such as low optical propagation loss coefficient, compact size, large tuning range, and high tuning accuracy. In the optical modulation experiments making use of excited carriers' plasma to affect the material optical properties within the sensitive resonator cavity volume, it was shown that light-tuned-light signal modulations of more than 12 dB can be implemented in the native optical form with spatial homogeneity based on localized free carriers' injection on small PhC devices with times limited by the relaxation times of the carriers. An additional benefit of such methods is that batch fabrication enables easy IC integration to other electronic of photonic circuit device and applications, which would not only reduce cost of implementation, but also broaden the application areas. Moreover, the miniaturization possible makes it advantageous to construct very small-size optical modulation circuits of the subwavelength dimension, which advances the state of the art for optical length-scale PhC devices capable of high-density integration.

References

1. Johnson, S.G. and Joannopoulos, J.D., *Photonic Crystals: The Road from Theory to Practice*, Kluwer Academic Publishers, Boston, 2002.
2. Yariv, A. and Yeh, P., *Optical Waves in Crystals: Propagation and Control of Laser Radiation*, John Wiley & Sons, New York, 1984.
3. Sakoda, K., *Optical Properties of Photonic Crystals*, Springer, Berlin, 2001.
4. Lin, H.B., Su, J.Y., Cheng, R.S., and Wang, W.S., Novel optical single-mode asymmetric Y-branches for variable power splitting, *IEEE Journal of Quantum Electronics*, 35, 1092, 1999.

5. Johnson, S.G., Manolatou, C., Fan, S., Villeneuve, P.R., Joannopoulos, J.D., and Haus, H.A., Elimination of cross talk in waveguide intersections, *Opt. Lett.*, 23, 1855, 1998.

6. Manolatou, C., Johnson, S.G., Fan, S., Villeneuve, P.R., Haus, H.A., and Joannopoulos, J.D., High-density integrated optics, *J. Lightwave Technol.*, 17, 1682, 1999.

7. Haus, H.A., *Waves and Fields in Optoelectronics*, Prentice-Hall, Upper Saddle River, NJ, 1984, p. 197.

8. Miller, A., Miller, D.A.B., and Smith, S.D., Dynamic nonlinear optical processes in semiconductors, *Adv. Phys.*, 30, 697, 1981.

9. Sheik-Bahae, M. and Hasselbeck, M.P., *Third Order Optical Nonlinearities*, Preprint of *OSA Handbook of Optics*, 4, 2000.

10. Leonard, S., Van Driel, W., Schilling, M., and Wehrspohn, J., Ultrafast band-edge tuning of a two-dimensional silicon photonic crystal via free-carrier injection, *Phys. Rev. B*, 66, 161102-1, 2002.

11. Sokolowski-Tinten, K. and Von Der Linde, D., Generation of dense electron-hole plasmas in silicon, *Phys. Rev. B*, 61, 2643, 2000.

12. Esser, A., Seibert, K., Kurz, H., Parsons, G.N., Wang, C., Davidson, B.N., Lucovsky, G., and Nemanich, R.J., Ultrafast recombination and trapping in amorphous silicon, *Phys. Rev. B*, 41, 2879, 1990.

13. Euser, T.G. and Vos, W.L., Spatial homogeneity of optically switched semiconductor photonic crystals and of bulk semiconductors, *J. Appl. Phys.*, 97, 043102-1, 2005.

14. Haug, H., *Optical Nonlinearities and Instabilities in Semiconductors*, Academic Press, San Diego, CA, 1988.

15. Shinya, A., Notomi, M., Kuramochi, E., Shoji, T., Watanabe, T., Tsuchizawa, T., Yamada, K., and Morita, H., Functional components in SOI photonic crystal slabs, *SPIE Proc.*, 5000, 104, 2003.

16. Shoji, T., Tsuchizawa, T., Watanabe, T., Yamada, K., and Morita, H., Low loss mode size converter from 0.3 μm square Si wire waveguides to single-mode fibres, *Electron. Lett.*, 38, 1669, 2002.

17. Tinker, M.T. and Lee, J.B., Thermo-optic photonic crystal light modulator, *Appl. Phys. Lett.*, 86, 221111, 2005.

18. Busch, K. and John, S., Liquid-crystal photonic-band-gap materials: the tunable electromagnetic vacuum, *Phys. Rev. Lett.*, 83, 967, 1999.

19. Park, W.J. and Lee, J.B., Mechanically tunable photonic crystal structure, *Appl. Phys. Lett.*, 85(21), 4845–4847, 2004.

5

MEMS Variable Optical Attenuators

Xuming Zhang, Hong Cai, and Ai Qun Liu

CONTENTS

5.1 Introduction 174
5.2 Configurations of MEMS VOAs 177
    5.2.1 Direction-Coupling-Type MEMS VOA 177
    5.2.2 Interference-Type MEMS VOA 178
    5.2.3 Diffraction-Type MEMS VOA 179
    5.2.4 Refraction-Type MEMS VOA 182
    5.2.5 Reflection Type of MEMS VOAs 183
5.3 Specifications of Different MEMS VOA Configurations 186
5.4 Optical Attenuation Model of Single-Shutter VOA 190
    5.4.1 General Model 191
    5.4.2 Far-Field Attenuation Model 193
    5.4.3 Near-Field Attenuation Model 196
5.5 Optical Model and Tuning Schemes of Dual-Shutter VOA 199
    5.5.1 Attenuation Model 199
    5.5.2 Analysis of Linear Tuning Schemes 201
    5.5.3 Analysis of Tuning Resolution 203
5.6 Analyses of Temperature, Wavelength, and Polarization Dependencies 206
    5.6.1 Temperature-Dependent Losses 206
    5.6.2 Wavelength-Dependent Losses 208
    5.6.3 Polarization-Dependent Losses 209
5.7 Experimental Studies of MEMS VOAs 211
    5.7.1 Surface-Micromachined Single-Shutter VOAs 211
        5.7.1.1 Design and Device Description 211
        5.7.1.2 Experimental Results 214
    5.7.2 Deep-Etched Dual-Shutter VOAs 215
        5.7.2.1 Device Description 215
        5.7.2.2 Single-Shutter Tuning Scheme 216
        5.7.2.3 Linear Tuning Scheme 217
        5.7.2.4 Wavelength- and Polarization-Dependent Losses 220
    5.7.3 Deep-Etched Elliptical Mirror VOAs 220
    5.7.4 Deep-Etched Parabolic Mirror VOAs 224
    5.7.5 Experimental Studies on TDL, WDL, and PDL 226
        5.7.5.1 Device Description and Measured Characteristics 226
        5.7.5.2 Temperature-Dependent Losses 227

        5.7.5.3   Wavelength-Dependent Losses 230
        5.7.5.4   Polarization-Dependent Losses..230
5.8   Summary...233
References 233

This chapter studies different types of micro-electro mechanical systems (MEMS) variable optical attenuators (VOAs), covering various topics such as classifications, specifications, working principles, optical modeling, and experimental studies. Compared with optical switches in the previous chapter that work in either fully open state (i.e., on state) or fully closed state (i.e., off state), VOAs provide many intermediate states, thus facilitating controllable output power. The optical design of VOAs in this chapter can also be used for micro-optical coupling systems in tunable lasers, which will be discussed in Chapters 6, 7, and 8.

In this chapter, an introduction on the applications and developmental history of MEMS VOAs will be first given in Section 5.1. Then, configurations of MEMS VOAs will be classified into different types and will be subclassified into different designs in Section 5.2. To compare the performance of different types of MEMS VOAs, achieved specifications of several types of developed VOAs will be listed and discussed in Section 5.3. The theoretical analyses and optical modeling of MEMS VOAs will be elaborated in next three sections: Section 5.4 will focus on the attenuation model of single-shutter VOAs, Section 5.5 on the tuning schemes of dual-shutter VOAs, and Section 5.6 on the temperature-, wavelength-, and polarization-dependent losses of VOAs. These sections are the core part of this chapter. Finally, the experimental demonstration of different types of VOAs will be presented in Section 5.7, including a surface-micromachined single-shutter VOA, a deep-etched asymmetric dual-shutter VOA, an elliptical mirror VOA, a parabolic mirror VOA, and some experimental studies on dependent losses. For the verification of their effectiveness, theoretical models will be used for data interpretation and will also be compared with experimental results.

5.1   Introduction

VOAs play a key role in the optical power management of many optical systems, especially in optical networks [1]. Fiber-optic systems operate over a large span of optical power level, from very strong signals (>1 W) emitted from high-power laser sources to very weak signals (several microwatts) after transmitting over long distances. VOAs are key components for adjusting optical powers to different levels. They were initially used to protect optical receivers from overhigh incident powers and simulate different span losses in short-span systems. However, the development of dense wavelength division multiplexed (DWDM) systems technology has opened up new application areas for VOAs [1,2].

The typical applications of VOAs in the fiber networks are shown in Figure 5.1. First, VOAs can be used to equalize the power levels of different wavelength sources as illustrated in Figure 5.1a. Light power from a laser source is sensitive to temperature and injection current, and also varies with time owing to laser aging. A VOA can be used to compensate the power variation of one laser source, and an array of VOAs can be used to equalize the power levels of different light sources (corresponding to different wavelength channels) before they are combined by a multiplexer (MUX) into an optical fiber for further

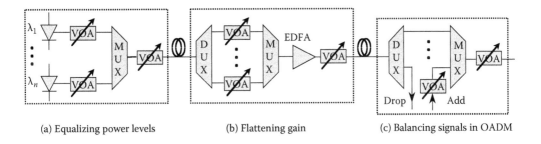

(a) Equalizing power levels   (b) Flattening gain   (c) Balancing signals in OADM

FIGURE 5.1
Typical applications of VOAs in fiber-optic communication systems: (a) equalizing power levels, (b) flattening gain, and (c) balancing signals in OADM.

transmission. In addition, a VOA can be used to limit the total power coupled into the optical fiber below the threshold level to avoid nonlinear effects in the fiber or for safety reasons. For example, to comply with the eye-safety standard, an upper limit of 17 dBm (e.g., 50 mW) is imposed on the total power that can be transmitted in an optical fiber.

VOAs are also able to get a flat gain for different wavelength channels in optical amplifiers, as shown in Figure 5.1b. In optical communication networks, optical amplifiers such as erbium-doped fiber amplifiers (EDFAs) are employed to amplify weak light signals. However, multiple wavelength channels arriving at a node may pass through different paths and experience different losses. In optical amplifiers, the gain of each channel depends on the power levels of other channels because of the cross-gain modulation. To get a flat gain, the multiple wavelength channels from a transmission fiber are first demultiplexed by a demultiplexer (DUX), and then, their powers are equalized by an array of VOAs before entering the optical amplifier. Although the gain varies with wavelength as well, such variation can be easily compensated by attenuators so as to obtain identical output powers over all amplified channels.

Moreover, VOAs are capable of balancing signals in optical add/drop multiplexers (OADMs), as shown in Figure 5.1c. In an OADM, the power levels of signals to be added cannot be too high, otherwise the existing channels will be overwhelmed; also, the dropped signals with high powers need to be attenuated before entering the local network. One attenuator is necessary for each add/drop channel. Besides, the total power is varied when one (or more channels) is added or dropped; therefore, an attenuator is needed to stabilize the power coupled into the transmission fiber.

Several technology choices have been investigated for VOAs. Conventional sliding-block optomechanical VOAs provide large attenuation range (>50 dB) and good linearity (0.1 dB), but they are bulky, costly, and slow (0.5–1.0 s), limiting their applications. Waveguide-based thermal-optic/electro-optic VOAs have a high-speed response (~1 ms), but encounter some difficulty in achieving large attenuation range with low insertion loss [3]. Other physical mechanisms have also been investigated, such as evanescent field coupling and photoacoustic interaction, but suffer from drawbacks similar to those of waveguide VOAs. In contrast, MEMS technology has attracted broad interest owing to its potential advantages of fast response, compact size, low power consumption, as well as superior optical and mechanical performance [1]. A specification comparison between MEMS and other technologies was elaborated in Reference 4. Although the comparison is for optical switching, most of the conclusions apply equally to VOAs.

The development of MEMS VOAs began in the 1990s because the invention of surface micromachining made it possible to fabricate complicated MEMS structures. MEMS

VOAs should have emerged much earlier because many studies on torsional and scanning micromirrors [5,6] could be easily adapted to VOA applications. However, this important application area was overlooked. The early MEMS VOAs were, in some sense, the by-products of high-speed light modulators [7,8]. Later, some other VOAs apparently borrowed ideas from MEMS optical switches, such as the 2 × 2 design [9], pop-up mirror design [10], stress-induced lift-up shutter design [11], etc. The first work was probably the deformable grating modulator demonstrated by Solgaard et al. [7], which employed surface-micro-machined silicon nitride grating lines to reflect the incident light and to interfere with the light reflected by the substrate. By adjusting the grating height relative to the substrate, the interference at a certain receiving angle can be constructive or destructive. As a result, the output power can be attenuated. A contrast of 16 dB and a speed of 1.8 MHz were obtained using only 3.2 V driving voltage [7]. Although the initial purpose was for fast light modulation, the same design has led to commercial success in MEMS VOAs.

The first MEMS device developed specifically for VOA application was presented by a group of researchers from Lucent Technologies [12,13]. The design directly came from the light modulator called mechanical antireflection switch (MARS) [8]. This work had a deformable membrane hung over a reflection substrate. The reflection state between two extreme ones (total reflection and total antireflection) can be obtained by adjusting the air gap of the membrane. It reported a 3 dB insertion loss, 31 dB attenuation under 35.2 V, and an amazingly fast response time of <3 μs. Since then, tremendous efforts have been made to develop new MEMS VOAs using various configurations, different mechanisms, and advanced processes. More details will be given in Section 5.2.

On the theoretical studies of MEMS VOAs, not much specialized work on MEMS attenuators has been done as the optical design is relatively simple, especially in VOAs that are based on direct fiber coupling, laser interference, and light reflection. Most of the optical design can be performed following the classic equations of geometrical and wave optics [14]. For VOAs that use the refraction of light, optical simulation can be done by commercial software such as ZEMAX, Code V, etc. In VOAs based on laser diffraction, the picture is more complicated because of the presence of simultaneous laser beam cutting-off, diffraction, and fiber coupling. Generally speaking, the attenuation can be numerically simulated on the basis of knife-edge diffraction of Gaussian beam using the Fresnel–Kirchhoff formula [15], as was suggested in [2,16]. The same idea was used to also calculate the symmetric-shutter design [17]. However, a closed-form analytical expression is always preferred because it reflects the physical relationship between the parameters. For this purpose, Zhang derived the formulas of light propagation, diffraction, and coupling in single-shutter VOAs and obtained simple expressions for far-field and near-field conditions [18–20]. Under both conditions, the relationship between transmission efficiency (i.e., the attenuation) and the shutter position can be expressed as an insertion loss multiplied by an error function. The same method was also extended to symmetric-shutter design that uses multiple shutters [21].

Further analysis for dual-shutter VOAs was also developed [22]. It inspired the implementation of special tuning functions such as linear attenuation relationship and ultrafine tuning. However, these studies are based on scalar diffraction theory, without considering the influence of the polarization state. A detailed study on wavelength-dependent loss (WDL), polarization-dependent loss (PDL), and temperature dependence was given in Reference 23. To suppress PDL, some recently proposed techniques include tailoring the shutter shape and coating material Reference 23. All these analyses give a deeper insight into physical mechanisms and provide guidelines for design and operation of MEMS VOAs.

5.2 Configurations of MEMS VOAs

Over the years, many designs have emerged for free-space MEMS VOAs, differing in various aspects such as optical axial arrangements, material choices, fabrication processes, fiber types, actuation methods, etc. In VOAs, the most important is the attenuation mechanism, that is, the way in which the attenuation is obtained. It determines the configuration and performance of VOAs. Depending on this feature, the developed MEMS VOAs can be classified into several types, including direct coupling, interference, refraction, diffraction, and reflection. This section will present these VOA configurations, describing their working principles, origins, and pros and cons as well.

5.2.1 Direction-Coupling-Type MEMS VOA

The simplest configuration is the direct-coupling-type, VOA as shown in Figure 5.2. It uses only two fibers without involving any other optical components. These two fibers are separated by a narrow gap (<10 μm), ensuring a low insertion loss in the initial state. According to the fiber-coupling theory, the attenuation can be obtained by spoiling the coupling between the two fibers by introducing misalignment, fiber separation, relative angle, or lateral position [14]. Many optomechanical VOAs make use of micrometer screws or motors to change fiber separation. However, optical fiber is very stiff and heavy compared to MEMS structures, making it extremely difficult to transport (or rotate) the whole piece of optical fiber. A practical way is to have one of the two fibers clamped as a cantilever beam while fixing the other fiber entirely, corresponding to the output fiber and input fiber in Figure 5.2. Then, attenuation can be obtained by applying a force, using an MEMS actuator, to the free end of the output fiber so as to cause lateral displacement. This idea was proposed by Haake et al. in 1998 [24]. A recent experiment demonstrated 48 dB attenuation range, 1 dB insertion loss, and <5 ms response speed [25]. The WDL measured 0.4 dB at the insertion loss, 1.4 dB at the 42 dB level, and 2.1 dB at 48 dB level. It increases with higher attenuation level, whereas the diffraction effect gradually plays an important role when the fiber is shifted more and more. In another work, similar performance was obtained [26]. As the fiber is really bulky and stiff compared to MEMS structures, an array of actuator elements (most commonly, thermal, magnetic, or piezoelectric actuators because they are more efficient in generating large force) are used for bending the fiber. As a result, the direct-coupling design has a comparatively large footprint and high power [24–26]. In the demonstrated work [26], the VOA is 100 mm^2 in size and consumes a power of ~6 W.

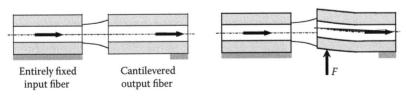

Entirely fixed   Cantilevered
input fiber   output fiber

Initial high coupling state   High attenuation state

FIGURE 5.2
Direct-coupling type of variable optical attenuators that adjust the attenuation level by introducing misalignment between the directly coupled fibers.

5.2.2 Interference-Type MEMS VOA

The interference type of VOA makes use of two-beam or multibeam interference to adjust the attenuation level. This type can have three different designs such as deformable grating design, deformable membrane design, and etalon design, as illustrated in Figure 5.3. The deformable grating design makes use of two sets of grating lines: one is movable, whereas the other sits on the substrate. The normal incident light is completely reflected when reflections by two grating line sets have a phase difference of $2m\pi$ (m is an integer), for example, when the gap is kept at λ as illustrated in Figure 5.3a. In contrast, reflection lights in the normal direction will be completely suppressed if the gap is changed to $3\lambda/4$, corresponding to a phase difference of $(2m + 1)\pi$. In this state, the optical power is diffracted to the other direction. This design was invented by Solgaard et al. [7]. In this work, the movable grating was made by depositing and then patterning a silicon nitride film (213 nm thick) suspended over the silicon substrate by a layer of silicon dioxide (also 213 nm thick). In the following step, the oxide layer was removed by hydrofluoric acid to release the movable grating. Finally, a 50 nm thick aluminum was sputtering over the whole area. The reflectivity of the movable grating was enhanced, and the substrate areas between the gaps of movable grating lines were also coated and acted as the fixed grating. A contrast of 16 dB and a speed of 1.8 MHz were obtained using only 3.2 V driving voltage [7]. However, the grating design suffers from WDL and PDL because the interference and diffraction are strongly affected by the wavelength and polarization. For example, the

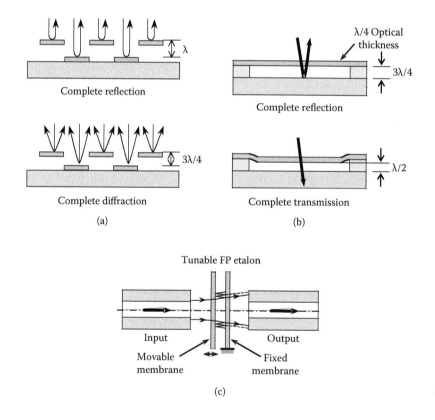

FIGURE 5.3
Interference type of variable optical attenuators that make use of constructive and destructive interference. (a) The deformable grating design, (b) deformable membrane design, and (c) etalon design.

WDL could reach as high as 6.7 dB for a 40 nm wavelength range at a 15 dB attenuation level. Some further work has been done to optimize the grating fill factor (i.e., width of grating lines relative to the grating period) and to replace the ribbon-shaped grating lines with round disks [27]. In this way, the WDL and PDL can be suppressed to very low levels. Typically, these are <0.2 dB.

The deformable membrane design has a transparent membrane suspended over a transparent substrate, as illustrated in Figure 5.3b. When reflections from the top surfaces of the membrane and substrate are in phase in the far field, the incident light is completely reflected. For example, when the membrane has an optical thickness of $\lambda/4$ (considering the refractive index) and the gap is $3\lambda/4$, the two reflections have a phase difference of 4π. In contrast, when the membrane is lowered down to have a gap of $\lambda/2$, the phase difference will be 3π and the two reflections will cancel each other, corresponding to a complete transmission state (or in other words, antireflection state). This design was developed by a group of researchers of Lucent Technologies [8,12,13]. The movable membrane is a single silicon nitride layer (194 nm thick, optical thickness $\lambda/4$) or a multilayer (consisting of a silicon nitride layer of 194 nm thickness sandwiched between two polysilicon layers with optical thicknesses $\lambda/2$ and $\lambda/4$, respectively). The multilayer membrane was found to have less wavelength dependence. It measured an insertion loss of 3 dB, attenuation range of 31 dB using 35.2 V, and an amazingly fast response time of <3 μs. Another similar work on the megahertz optical modulator was also presented two membranes using to form a Fabry–Pérot filter [28].

The etalon design uses two membranes directly to form a Fabry–Pérot filter, as shown in Figure 5.3c. One membrane is movable, whereas the other can be kept static. When the movable membrane is translated by an MEMS actuator, the cavity length will be changed, and thus, the transmitted optical power will be varied. A similar design was used for narrowband filters [29] and tunable laser external reflectors [30] but has not been demonstrated for MEMS VOAs. It can be easily implemented using the deep etching fabrication method.

The three designs are all based on the same principle of multibeam interferences if the grating diffraction is regarded as the interference of the reflections of many grating lines, but there are certain differences. For example, the deformable membrane is a whole piece, whereas the grating lines are sliced pieces of membranes. Consequently, the grating design has less air-damping problem (i.e., better mechanical response). The diffraction of grating also helps to diffract unwanted light out of the acceptance angle of the output fiber, in comparison to the only multibeam interference in the deformable membrane design. The etalon design is a natural variant of the deformable membrane design. The difference lies in that the deformable membrane design receives the reflection as the output, whereas the etalon type uses the transmission. This difference affects the positioning of the input and output fibers and thus the packaging form. In the deformable membrane design, the input and output fibers should be placed on the same side (called *retroaxial arrangement* to be discussed in the next section) and thus can be packaged in penlike packaging forms, whereas in the etalon type, these fibers are placed along the same line (called *coaxial arrangement*).

5.2.3 Diffraction-Type MEMS VOA

To attenuate the light, the straightforward method is to use shutters to block the light. It is commonly called the *shutter-type* VOA, but in this chapter it is classified as the diffraction-type VOA because the blocking of light by the shutter introduces diffraction. There are typically three designs, namely, single-shutter design, dual-shutter design, and multishutter design, depending on the number of shutters being used. In the single-shutter design, as shown in Figure 5.4a, the incident light beam exits from the input fiber and

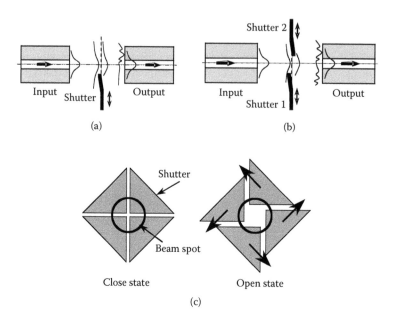

FIGURE 5.4
Diffraction type of variable optical attenuators that make use of movable shutters to block the light. (a) The single-shutter design, (b) dual-shutter design, and (c) multishutter design.

keeps diverging on its path to the shutter, and is then partially blocked by the shutter. The cut-off light is further diffracted to the output fiber before it is coupled inward. To suppress the back-reflection caused by the reflection on the shutter, in many MEMS VOAs the shutter is titled by a certain angle, typically 8° or larger. The first demonstration of the single-shutter design is the surface-micromachined VOA from Lucent [2,16]. The VOA was manually assembled after fabrication and release. It employs a vertical shutter to move up into the light path between two fibers. The shutter sits at one end of a seesawlike structure, whereas a parallel capacitor actuator is present at the other end. When a driving voltage is applied to the actuator to attract the moving plate down, the shutter will be lifted up and inserted into the light path. It reported >50 dB attenuation range at 1 dB insertion loss. Such design is inspiring, but the pivot of the seesaw has to maintain mechanical contact; thus, wear and tear, not to mention the mechanical instability in the presence of environmental vibration is a possible issue. An improved version was demonstrated by Zhang et al. using a drawbridge-like MEMS structure to pull up a shutter over the light path [18,31]. The structure is also fabricated by surface micromachining and is integrated by manual assembly. However, the movement of the shutter is produced by the bending of supporting beams, avoiding the mechanical contact and wear. However, from the standpoint of industrial production, surface-micromachined shutter VOAs are not very attractive, because careful assembly of fragile thin-film-based structures is involved, which is time consuming and low yielding. With the emergence of deep reactive ion etching (DRIE) technology, many research interests were shifted to deep-etched VOAs. A pioneering work was presented by Marxer et al. in 1999 [32]. The VOA device (including the actuators, fiber alignment grooves, shutters, and other auxiliary parts) was etched on a silicon-on-insulator (SOI) wafer in a single step. It reported 57 dB attenuation range, 1.5 dB insertion loss, and <5-ms response speed. The back-reflection (i.e., return loss) was well suppressed by tilting the shutter by 8° relative to the fiber facets. The benefit of this work is that all the VOA elements can be integrated on a single chip by simply pushing the fibers into the etched fiber grooves. The clamping and alignment of fibers are obtained automatically, without the need for a complicated assembly. In addition, MEMS structures are more robust and stable. Many other VOAs have also been reported using single shutters, differing in the actuator structures, actuation mechanisms, and waveguide designs, such as the pop-up shutter [10], movable surface-micromachined 3-D shutter [33], deep-etched rack-and-tooth mechanism for latching and digitalizing the attenuator [34], electrostatic torsional actuator [35], electromagnetic actuators, piezoelectric actuators [36], and so on.

VOAs using two shutters have been proposed recently, giving birth to the dual-shutter design, as illustrated in Figure 5.4b. The shutters can be moved symmetrically (at the same rate as in symmetric directions) or individually. On the basis of this movement feature, they can be categorized into two groups, the symmetric-shutter and asymmetric-shutter design. It is worth making such a categorization because these two groups have different specifications. The benefit of the symmetric-shutter design is that the aperture can be opened and closed more efficiently, but the drawback is a strong nonlinear relationship between the attenuation level and the shutter position. When the aperture is fully open, the symmetric moving in of the shutters causes little change in the attenuation level. However, when the aperture is nearly closed, a small movement of the shutters will cause a drastic increase in attenuation. There exists a singularity that the attenuation ideally goes up to infinity when the aperture is fully closed. Such nonlinearity is undesirable for the stability and control of VOAs in real-world applications. It also causes nonuniform tuning resolution. MEMS VOA using symmetrically moved dual shutters was demonstrated by Li [17]. The device was fabricated on an SOI wafer by a commercial foundry service SOIMUMPs. The shutters use the top surface of the silicon structural layer instead of the deep-etched sidewalls. It measured an insertion loss of 1.4 dB and an attenuation range of 29 dB using a driving voltage of 34 V (applied simultaneously to the two comb-drive actuators). A strong nonlinear relationship between the driving voltage and attenuation level was observed. To solve these problems, an asymmetric dual-shutter design was recently proposed [22]. In this design, the two shutters are moved independently, providing an additional degree of freedom for attenuation adjustment. In the operation, one shutter is for coarse tuning and the other for fine tuning; that is, one is used as the main shutter, whereas the other, as the auxiliary shutter, is moved back and forth to maintain a certain relationship between the attenuation level and the main shutter's control parameters (e.g., position, applied voltage, etc.). It was proved that ideally the tuning could be started from any available working point, linear to any controlling parameter, at any slope of linearity, and with any tuning resolution. In the experiment, the fabricated VOA device demonstrated linear tuning over a 20 dB range with respect to the driving voltage of one shutter, and it also realized simultaneous coarse tuning (2.5 dB/V) and fine tuning (0.1 dB/V) by the two shutters.

The multishutter design employs more shutters (4, 6, 8, or even more) to adjust the aperture, as shown in Figure 5.4c. This idea originates from the iris diaphragm of cameras that uses a series of overlapping metal blades to form a circle with a hole in the center whose diameter can be increased or decreased as desired. Therefore, the multishutter design is a form of symmetric-shutter design. This design was demonstrated by Syms in 2004 [21]. Figure 5.4c shows the operation of a four-shutter design. In the initial closed state, these four shutters almost contact one another, resulting in a minimum aperture. The narrow gap in between is to reduce the risk of sticking. Whenever the shutters are moved, the aperture is increased, resulting in the open state. The demonstrated device was fabricated on a bonded SOI wafer. All the four shutters are etched on the same layer of silicon (25 μm thick). The VOA obtains a 17 dB attenuation range using thermally expanded core fibers. It consumes 1.1 W power owing to the use of thermal actuators. Although the performance can be greatly improved, this work proves the viability of the multishutter design. The limited range is due to the presence of a narrow gap in the closed state. It can be overcome by overlapping the shutters if they can be fabricated on different layers using a multilayer process. The high power consumption can be alleviated by the use of other types of actuators such as electrostatic actuators instead of thermal actuators. When more shutters are used, there may be a problem in arranging the actuators owing to the limited space. A multilayer process can also solve this problem by fabricating each shutter and its actuator on one separate layer.

5.2.4 Refraction-Type MEMS VOA

The refraction type of VOA makes use of the refraction of light to adjust the attenuation level. Two typical designs are shown in Figure 5.5; one uses a single wedge, whereas the other employs two wedges. As the photoenergy of infrared light (e.g., $\lambda = 1.55$ μm) is lower than the bond energy of silicon (about 1.1 eV, corresponding to about 1.1 μm wavelength), the silicon is transparent to the infrared light. To adjust the attenuation level, the silicon wedge can be used in two ways. The first is to insert the wedge partially into the light path. The light passing through the wedge will be reflected by total internal reflection or deflected out of the wedge at an angle larger than the accepting angle of the output fiber, whereas the remaining light is transmitted and diffracted to the output fiber. The function of the wedge is similar to that of the shutter. It needs only one wedge to exercise its function and is thus called the *blocking-wedge design*, as shown in Figure 5.5a. This was investigated by Lee et al. [37,38]. It yielded 0.6 dB insertion loss with 43 dB attenuation range, but 0.5 dB WDL and 1.1 dB PDL at the 10 dB attenuation level. It saves the metal-coating step in the fabrication (like in the shutters), but the WDL and PDL are quite high because of the presence of diffraction. The other operational method, proposed by Medina et al. [39], to have the wedge fully cover the light path. The wedge is not uniformly thick across two sides (the profile of each side can be straight or curved); therefore, the movement of the wedge engages different amounts of position shift of the deflected light and eventually achieves different attenuation levels. To attenuate the move effectively, two wedges can be used [40]. This design is thus called the *deflection-wedge design*. The PDL was reduced to 0.5 dB, but the insertion loss could be increased owing to the Fresnel reflection at two or four Si–air interfaces. Each Si–air interface will reflect approximately 31% of incident light power.

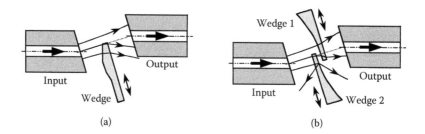

FIGURE 5.5
Refraction type of variable optical attenuators that make use of semitransparent silicon wedges to steer the light beam. (a) The blocking-wedge design, and (b) deflection-wedge design.

5.2.5 Reflection Type of MEMS VOAs

The reflection type of VOA makes use of mirrors to steer away the light beam with respect to the output fiber aperture [41–45]. Because the mirror reflection maintains the unity of the laser beam field, the reflection-type VOA is expected to have a low WDL and PDL. Compared with the aforementioned VOA types, the reflection type has attracted the most research and development interest. Many designs have been proposed, as shown in Figure 5.6. The simplest design is to have a flat mirror reflect the incident light back to its incoming direction, such as the back reflection. It is thus called the back-reflection design, as shown in Figure 5.6a. A focus lens is used to improve the optical coupling and thus reduce the insertion loss. The adjustment of attenuation can be obtained by either translating or rotating the mirror so as to spoil optical coupling [41]. This design was studied by Andersen et al. in 2000 [42]. The mirror is a gold-coated surface (size 500 μm × 500 μm) fabricated by surface micromachining. The tilting angle of the mirror can be controlled precisely from 0 to 3°. After packaging, the VOA device obtains an insertion loss of 1.0 dB and an attenuation range of 18 dB using 10 V. The PDL is only 0.06 dB at the attenuation level of 15 dB. Similar work was also presented by Isamoto et al., who fabricated the reflection mirror using the SOI wafer [43]. The mirror thickness is 30 μm and uses the polished top surface of the SOI wafer directly, yielding robust mechanical structure (resistance to shock up to 500 G) and superior reflection in comparison to the previous surface-micromachined VOAs [42]. It claimed an attenuation range of 40 dB using 5 V voltage (corresponding to a mirror angle of 0.3°), and a fast response of <5 ms. In general, the back-reflection design has the lowest PDL among all VOA designs because the incident angle is nearly normal to the mirror surface.

A similar design is the flat mirror design in which the input and output make an angle (typically 90°), as shown in Figure 5.6b. As the incident angle is 45°, the Fresnel reflection at the mirror surface has different efficiency for different polarization states [15]. Therefore, the flat mirror design has larger PDL than the back-reflection design. Another difference is the arrangement of the optical axis. The back-reflection design has two fibers leading from the same direction (i.e., retroaxial arrangement), whereas in the flat mirror design they are crossed (i.e., cross-axial arrangement). Such an axial arrangement will affect the packaging formats. Further studies have demonstrated flat mirror VOAs using translational comb drives [44,45] or rotary comb drives [46]. Typical specifications are >50 dB attenuation range and <3 ms response speed. The WDL was measured to be 0.19, 0.25, 0.61, and 0.87 dB for attenuation levels of 0, 3, 10, and 20 dB, respectively [46]. For PDL, it is 1.1 dB at 20 dB attenuation and 2.13 dB at >40 dB attenuation [23].

Two variant forms of the flat mirror design are shown in Figure 5.6c,d; one uses two flat mirrors to form a corner mirror to reflect the input back to its incoming direction [47–49], whereas the other uses two mirrors in parallel to shift the optical axis [45], similar to a periscope. They are thus called the *corner mirror design* and the *periscope design*, respectively. Because of the involvement of more mirrors and increase of fiber separation, the PDL, WDL, and insertion loss are higher than those of the back-reflection and flat mirror designs. Therefore, lensed fibers (or other methods of achieving focus and collimation) have to be used to maintain a reasonable level of insertion loss. The corner mirror design was investigated by Lim et al. using a rotational corner mirror and lensed fibers [47]. It reported a 0.5 dB insertion loss and 45 dB attenuation range. The PDL is 0.2 dB at 20 dB attenuation. Some further work was presented by Chen et al. and Lee et al. [48,49]. Similar specifications were obtained. The periscope design was proposed by Kim and Kim as another way to shift the optical axis of the output light [45].

Another interesting variant of the flat mirror design is the digital mirror array design, as shown in Figure 5.6e. In the flat mirror design, the reflector is a whole piece of mirror

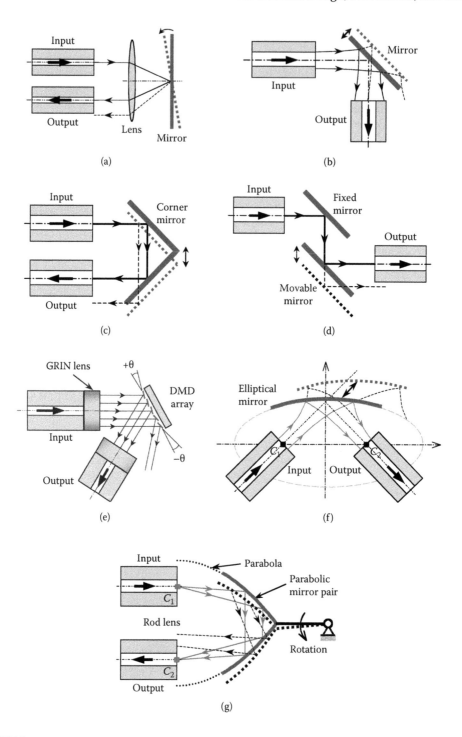

FIGURE 5.6
Reflection-type variable optical attenuators that make use of micromirrors to reflect the light beam are: (a) The back-reflection design, (b) flat mirror design, (c) corner mirror design, (d) periscope design, (e) digital mirror array design, (f) elliptical mirror design, and (g) parabolic mirror design.

surface, whereas in the digital mirror array, it is separated into many small pieces of micromirrors, each of which can be rotated independently. In the demonstrated work, a two-dimensional digital micromirror device (DMD) was employed as the reflector [50]. Each micromirror could stay in only one of the two angular states, $+\theta$ or $-\theta$. If the reflection from the micromirrors at one angular position was received as the output (for example, $+\theta$ in Figure 5.6e), the coupling efficiency could be controlled by discrete steps of $1/N$ (when N is the number of total micromirrors); that is, the coupling efficiency was digitized. In the reported work, 11 b resolution was obtained. However, the insertion loss is 8 dB and the PDL is 4.8 dB because the commercial DMD is designed from visible wavelength.

Recently, more interest has focused on using nonflat mirrors, such as the elliptical mirror design and parabolic design, as illustrated in Figure 5.6f,g. Two reasons are driving this development. One is the focusing effect and the other, more importantly, is the linear attenuation relationship. Various designs using mirrors, as discussed earlier, have no focusing effect on the incident light. Therefore, lensed fibers or bulky collimation lenses have to be used, increasing the cost and volume of MEMS VOAs. A straightforward idea is to curve up the mirror to provide the focusing effect. The mirror shape can be elliptical or parabolic. As the photopatterning and deep etching allow for arbitrary profiles, such mirror shapes can be implemented easily without the need for any excessive process steps. According to the geometric optics, any ray of light passing through one focus of the ellipse would pass through the other focus after a single bounce. Therefore, rays originating from one of the foci can be fully directed to the other. In an arrangement in which input and output fibers are positioned at the foci of the elliptical mirror as shown in Figure 5.6f, a low insertion loss can be obtained even with standard cleaved single-mode fibers, with the need for the lensed fibers or other collimating lenses are. More importantly, the attenuation (in dB) can be linearly related to the mirror displacement (in μm) if the mirror parameters are chosen properly. In the demonstrated elliptical mirror MEMS VOA [50], a good linearity is observed over the first 30 dB range of the total 44 dB attenuation range. With respect to the other specifications, the insertion loss is 1.0 dB, and the response speed is 220 μs. The PDL remains below 0.85 dB over the entire 44 dB attenuation range. It is 0.3 dB in the first 15 dB range and increases rapidly between the levels of 15 and 25 dB. At higher attenuation levels, the PDL settles at the level of 0.8 dB. Such linearity is a special merit of the elliptical mirror design; it is not found in any other VOA design except for the dual-shutter design, which uses a complicated combination of the movements of two shutters. In contrast, the elliptical mirror design realizes linearity in a simple and elegant way. Although the elliptical mirror provides the light-focusing effect in the horizontal plane, the light is kept diverging in the vertical plane. Therefore, the fiber separation should be minimized to obtain a reasonable insertion loss.

The attenuation linearity in the elliptical mirror design is not realized on purpose; it merely arouse as a surprise from the experiment. The deliberate realization of linearity was made using the parabolic mirror design recently [51]. The parabolic mirror design as shown in Figure 5.6g consists of a pair of parabolic mirrors arranged symmetrically; each mirror is a section of the parabola. The rays from the focus of a parabola can be collimated ideally after reflection, and vice versa. In the initial state, the fiber facets of the input and output are located at the two foci C_1 and C_2 of the parabolas, respectively. Therefore, the incoming rays from C_1 are collimated by the first mirror and are then converged to C_2, ensuring a good coupling.

Different attenuation levels can be obtained by spoiling the focus condition in three ways: horizontal translation of the mirror pair, vertical translation, or rotation about a remote pivot. Simulation shows that horizontal translation is not efficient; the attenuation is changed by only 3 dB when subjected to a translation of 50 μm. The vertical translation can go to high attenuation quickly, but it is nonlinear to the mirror displacement. In contrast, the rotation method is able to achieve high attenuation level without the need for a large rotation angle. Besides, a very good linear relationship can be obtained by proper choice of parameters. When the mirror is rotated, the incident rays are shifted from the focus and thus are not collimated. On the output side, the rays are further shifted and defocused. Such a combination of shift and defocus is essential to the linear relationship.

As the parabolic mirror provides optical coupling only in the horizontal plane, a rod lens (commonly, a section of optical fibers) can be introduced for vertical convergence, enabling 3-dimensional (3-D) optical coupling [53] to further reduce the insertion loss. In the cited work, the parabolic mirror pair was fabricated by DRIE on an SOI wafer (structural layer: 75 μm). The insertion loss is as low as 0.6 dB when the fibers are antireflection coated. When the rotation angle is increased from 0.5 to 2.6°, the attenuation rises up linearly to 62.6 dB. The linear approximation is $y = 29.66\ x - 14.53$, where y is the attenuation in dB and x is the angle in degree. For the other specifications, the WDL over 100 nm range is 0.3 dB at the 20 dB attenuation level, 0.7 dB at the 40 dB level, and 1.3 dB at the maximum attenuation level. The PDL is always <0.5 dB over the entire attenuation range. The PVOA has a temperature dependence of 0.01 dB/K at the 20 dB level.

5.3   Specifications of Different MEMS VOA Configurations

The specifications of the different MEMS VOA configurations are compared in Table 5.1. Different applications require different VOA specifications. Typically, commercial VOAs should meet the standards of Telcordia, which are the de facto industrial standards for VOAs. The basic standard is Telcordia GR-910, which is specific to VOAs [54]. Some further require meeting the standards of Telcordia GR-1209 and 1221, which define the performance as well as environmental and mechanical reliability. In academic research, it may not be necessary to follow the standards strictly, but they should be kept in mind. With regard to VOA specifications, nearly all applications require low insertion loss (<1 dB), large attenuation range (>20 dB or even >40 dB), fast response speed (<1 ms), low back reflection, low WDL, low PDL, and low temperature dependence. These specifications can be considered the basic requirements, as listed in Table 5.1. Most of the developed MEMS VOAs meet these requirements very well.

Real-world applications may also require VOAs to perform well in other respects, which can be regarded as the advanced requirements as listed in Table 5.1. For example, linear attenuation relationship and fine-tuning resolution are needed for the convenience of control and adjustment. In this chapter the terms *linear* and *nonlinear* refer to the relationship between the attenuation level (in dB, the unit is most commonly used in VOAs) and controlling parameters (such as the voltage, mirror displacement, rotation angle, etc.). In production, it is more economical to package VOA devices into standard optoelectronic-packaging formats such as the TO can, mini dual-in-line, or butterfly, all of which have only one exit hole for the fiber leads. In this way, VOAs can be given a small form factor and can be used conveniently as modules for higher-level devices or systems. Therefore, it is preferable to arrange the input and output fibers in parallel, and thus, they can be led out from the same exit hole.

From the point of view of the primary optical axis, the fiber axis should be folded back, similar to those in the VOAs with the corner mirror [47–49] and parabolic mirror [52] designs. Such an arrangement is thus called retroaxis. Two other axial arrangements are also widely used in MEMS VOAs. One is the coaxis, in which two fibers share the

TABLE 5.1

Comparison of the Specifications of Different Types of Free-Space MEMS VOAs

	Refraction Type			Diffraction Type (i.e., Shutter-Type)		Reflection Type (i.e., Mirror Type)			
	Direct-Coupling Type [26]	Interference Type [13]	Blocking-Wedge [37] Deflection-Wedge [40]	Symmetric or Single Shutter [16,17,21,31,32]	Asymmetric Shutter [22]	Back-Reflection and Flat Mirror [43]	Corner Mirror [48,49]	2-D Digital Mirror Array [50]	Elliptical and Parabolic Mirrors [51,52]
Basic Requirements									
Insertion loss (dB)	1.0 dB	3 dB	0.6 dB	<1.5 dB	1.2 dB	0.8 dB	0.5 dB [49]	8 dB	0.6–1.0 dB
Attenuation range (dB)	48 dB	28 dB	43 dB, 45 dB	~55 dB [32]	>60 dB	40 dB	45 dB [49]	37.8 dB	44 dB [76], 62 dB [52]
Response speed	<5 ms	<3 µs	4 ms	90 µs [17], 5 ms [32]	~40 µs	<5 ms	<5 ms [49]	15 µs	1 ms
Back reflection (dB)	~18 dB	—	>39 dB	37 dB [17]	—	—	50 dB [48]	—	—
WDL over 40 nm (dB) @ 20 dB	—	<1 dB	~0.5 dB	0.9 dB [31]	<1.2 dB	0.8 dB	0.3 dB [48]	—	0.3 dB
@ 40 dB	1.4 dB	—	—	—	—	—	—	—	0.7 dB
PDL (dB) @ 20 dB	—	0.06 dB	~1.1 dB, <0.5 dB, 1.5 dB	<0.5 dB [16]	~0.8 dB	0.2 dB	0.2 dB [49]	4.8 dB	0.1 dB
Temperature dependence	—	—	—	—	—	0.5 dB	—	—	0.01 dB/K [52]
λ range (nm)	—	40 nm	>100 nm	—	>100 nm	>100 nm	—	—	>100 nm
Advanced Requirements									
Tuning linearity	Nonlinear	Nonlinear	Nonlinear	Highly nonlinear	Linear	Nonlinear	Nonlinear	Nonlinear	Linear
Tuning resolution	—	—	—	—	0.01 dB	0.1 dB	—	Digital, 2^{-11}	—
Axis arrangement	Coaxial	Retroaxial	Coaxial	Coaxial	Coaxial	Retroaxial	Retroaxial	Crossaxial	Crossaxial [51] Retroaxial [52]
Scalability	No	Yes	Yes	Yes	Yes	Yes	Yes	No	Yes
Size of core part	~100 mm²	φ 300 µm	—	1.2 × 1.2 mm² [17]	4 × 3 mm²	~5 mm²	18 mm²	>400 mm²	<2 mm²
On-chip integration	Yes	No	Yes	Yes	Yes	No	Yes	No	Yes

(Continued)

TABLE 5.1 (CONTINUED)

Comparison of the Specifications of Different Types of Free-Space MEMS VOAs

	Refraction Type				Diffraction Type (i.e., Shutter-Type)		Reflection Type (i.e., Mirror Type)			
	Direct-Coupling Type [26]	Interference Type [13]	Blocking-Wedge [37]	Deflection-Wedge [40]	Symmetric or Single Shutter [16,17,21,31,32]	Asymmetric Shutter [22]	Back-Reflection and Flat Mirror [43]	Corner Mirror [48,49]	2-D Digital Mirror Array [50]	Elliptical and Parabolic Mirrors [51,52]
Actuation methods	Thermal	Electrostatic parallel plate	Electrostatic comb drive	Electrostatic comb drive	Electrostatic	Electrostatic comb drive	Electrostatic	Electrostatic comb drive	Electrostatic	Electrostatic rotary comb drive
Power consumption	~6 W	100 mW	—	—	2 μW [16]	—	10 mW	—	—	<1 mW
Driving voltage or driving current	10 V, 0.6 A	35.2 V	15 V	30 V	30–40 V	10.5 V	5 V	21 V	28 V	25 V
Maximum optical power	—	>100 mW	—	—	1.1 W [21]	—	300 mW	—	—	—
Fabrication method	DRIE	Surface micromach	DRIE	DRIE	Surface micromach or DRIE	DRIE	DRIE	DRIE	Surface micromach	DRIE

same line of optical axis (also called in-line VOAs), similar to those in many diffraction-type VOAs [2,10,16–20,31–34] and refraction-type VOAs [37–40]. The other is the crossaxis (also called off-axis), in which the axes of two fibers make an angle (typically, 90°), as in the flat mirror [44–46] and elliptical mirror designs [51]. In many cases, a number of VOAs should be used, for example, to equalize the power levels of different DWDM channels. For this purpose, the single unit of MEMS VOAs should be easily scaled up to form an array. Therefore, the scalability of MEMS VOAs could be an important issue. In MEMS VOAs, many factors affect the scalability. For example, it will be more cost-effective if the core part of the MEMS VOA has a small footprint because many of them can be fabricated within a limited area. In addition, it will be more convenient if MEMS VOAs allow for on-chip integration; that is, the assembly of optical fibers and the fabrication of control circuit can be performed on the same chip of the MEMS structures. For large-array or field uses, power consumption becomes an important issue. The actuation method is the decisive factor. Among the common methods, thermal and magnetic actuators typically have relatively high power consumption (~1 W) owing to the presence of both voltage and current and, thus, joule heating; in contrast, electrostatic actuators such as comb drives and parallel-plate capacitor types have much lower power consumption, typically, <1 mW. For the convenience of drive, MEMS VOAs should have the lowest possible maximum voltage and current. Voltage and current are also determined by the actuation method. Thermal actuators require low voltage (<5 V) but large current (>1 mA); the electrostatic ones are opposite, requiring large voltage (typically 10–30 V) but low current (<10 μA); whereas piezoelectric actuators require extremely high voltage (100–500 V). For the convenience of integration with the IC control circuit, the fabrication method of MEMS VOAs is also important. The silicon-based processes such as the surface micromachining and DRIE are complementary metal oxide semiconductor (CMOS) compatible; that is, the IC circuit can be fabricated near to MEMS structures in a separate process. In contrast, a German acronym for "Lithographic Galvano formumg Abformung (LIGA)" and electroplating are less CMOS compatible. Fortunately, most of the developed MEMS VOAs directly use the silicon-based processes, or at least similar results can be replicated using the silicon-based processes. All these aspects can be regarded as the advanced requirements of MEMS VOAs. There are many other additional requirements for VOA products, such as maximum handling optical power, repeatability, reliability, temperature range, etc. However, these additional requirements are mostly not dependent on the choice of MEMS part but on the packaging and other auxiliary parts (e.g., cooling system, control). Therefore, they will not be discussed in this chapter.

In comparing the different types of MEMS VOAs as listed in Table 5.1, the direct-coupling type excels in the PDL because the fiber mode is circular. The cited data in Table 5.1 are based on normally cleaved fibers [25,26]. The back reflection can be reduced to a very low level by the use of antireflection-coated fibers or angled fibers. However, the requirement of large force, and thus the choice of thermal actuator, causes the highest power consumption (~6 W) among all the MEMS VOA types. The interference-type VOA has the fastest response speed (~3 μs) because of the use of narrow but stiff grating lines (thus, low air dumping and high resonant frequency). However, the nature of interference and diffraction causes considerable WDL and PDL. Fortunately, this problem has been overcome by replacing the grating lines with circular plates and by tailoring the fill factor, leading to commercial success. In the refraction-type VOA, the blocking-wedge design uses the wedge in a way similar to the shutter and thus has to deal with the WDL and PDL, whereas the deflection-wedge design preludes this by letting the incident light refract out of the acceptance angle of the output fiber [39,40]. However, as discussed previously, the insertion loss may be a challenge because of the involvement of at least two air–silicon interfaces unless antireflection coating is applied (but the fabrication will be complicated).

The diffraction-type VOA enjoys high attenuation range because the shutters can easily block out the entire light path [16,21,22,31], and a dynamic range as high as 70 dB is obtainable. As mentioned previously, a severe drawback is the high WDL and PDL caused by the knife-edge diffraction, especially at higher attenuation levels. Tailoring the shutter shape and coating the surface material with dielectric material could alleviate this problem [23]. Among the designs of the diffraction type, the single-shutter and symmetric-shutter designs have high nonlinearity between the attenuation level and the shutter position (or applied voltage). When the shutters are moved at a constant rate from the fully open to the nearly closed state, the attenuation increases slowly at the beginning but shoots up at the high attenuation level. It makes attenuation control difficult, causes low tuning resolution, and also introduces instability in the presence of environmental vibration. On the contrary, the dual-shutter design is able to provide linear attenuation and fine-tunes resolution by moving one shutter for coarse tuning and the other for fine tuning [22]. For the reflection-type VOA, the WDL and PDL are suppressed, as the full-field distribution of incident laser beam is maintained after the reflection [41–49]. The PDL can be further reduced in the back-reflection design when the input and output fibers are arranged nearly normal to the mirror plane (i.e., retroaxial) [41–43] as compared to the 45° incident arrangement (i.e., cross-axial) [44–46].

The corner mirror design exhibits similar performance. The 2-D micromirror array design provides a special benefit of discrete level of efficiency (11 bit) [50]. However, attenuation in the dB scale is more convenient for applications. The stepwise change of transmission in the linear scale is not stepwise in the dB scale, which considerably undermines the significance of digitalization. This design can be constructed using a commercial module of the mirror array, but the high cost and bulky size can be a hindrance. Besides, the module is not optimized for the communication wavelength in regard to the materials of the exit window and mirror reflection, which causes large insertion loss [50]. The designs of back-reflection, flat mirror, corner mirror, and 2-D micromirror array have large fiber separation and thus have to use optical coupling lens or lensed fibers; moreover, they do not support linear attenuation. In contrast, the elliptical mirror design [52] provides a natural coupling between the normally cleaved fibers and also realizes linear attenuation, but the arrangement is cross-axial. The parabolic mirror design [51] employs the better retroaxial arrangement while maintaining the features of coupling and linearity in the elliptical mirror design.

The direction comparison of the values of specifications is not fair and meaningful because different types of VOAs are in different developmental stages, some have succeeded in commercialization, some have just proved the concepts, and many are caught in between. However, the specification comparison reflects the uniqueness of each type of VOAs, which depends on their attenuation mechanisms.

5.4   Optical Attenuation Model of Single-Shutter VOA

This section will focus on theoretical analysis of single-shutter VOAs. First, a general model will be developed in the form of a series of integrals. It is not analytically integrable, but is well suited to numerical simulation. Then, the analytical models will be derived for the far-field and near-field conditions.

Three assumptions are used throughout the derivation: (1) Scalar optical assumption: that is, the scalar optical theory is valid unless vector theory is involved explicitly, for example, in discussion of the PDL; (2) Gaussian beam assumption: that is, the light transmitted in a weakly guided single-model fiber (SMF) can be approximated by a Gaussian distribution; (3) half-plane thin-shutter assumption: that is, the shutter thickness is negligible but thick enough to block all the light in the area it covers. The shutter has smooth and straight edges, it is infinitely long in the ξ direction, and moves along the η axis from $-\infty$. Two additional assumptions are used in deriving the far-field and near-field models, respectively: (4) far-field assumption (or far-field condition): that is, the output fiber is far from the shutter plane, more specifically, $z_2 \gg z_R$ (to be discussed later); (5) near-field assumption (or near-field condition): that is, the output fiber is very close to the shutter plane, more specifically, $z_2 \ll z_R$ (also to be discussed later).

5.4.1 General Model

The diagram illustrating the attenuation model of a free-space single-shutter VOA is shown in Figure 5.7. The conventional method uses light propagation and coupling theories in a straightforward manner, and tracks what happens in the VOA step by step. This method is called the general model in this chapter. Under the Gaussian beam assumption, the light from the input SMF is a Gaussian beam with its waist at the fiber-end facet. In the single-shutter VOA, the Gaussian beam from the plane Ω is first transmitted to the shutter plane and is partially blocked by the shutter. It is then diffracted to the facet plane Ω' of the output SMF and coupled into the output, as illustrated in Figure 5.7. Ω and Ω' stand for the positions of the beam waists of the input fiber and the output fiber, respectively. They are right at the end facets of the fibers. In Figure 5.7, (s, t, z), (ξ, η, z), and (x, y, z) stand for the coordinate systems in the input fiber facet, plane after the shutter, and output fiber facet, respectively. To simplify the notation, the plane right before the shutter is termed the *shutter plane* and that at the output fiber facet, the *facet plane*.

The fundamental mode, $U_0(t, s)$, of the light beam in SMFs can be approximated by

$$U_0(t,s) = \sqrt{\frac{2}{\pi w_0^2}} \exp\left\{-\frac{t^2+s^2}{w_0^2}\right\} \tag{5.1}$$

where w_0 is the waist radius of Gaussian beam. The term $\sqrt{2/\pi w_0^2}$ is used to normalize the light energy to 1 when it is integrated over the whole (t, s) plane.

After passing through a distance z_1, the light beam reaches the shutter plane, and the field distribution $U(\xi, \eta)$ can be expressed by the Fresnel–Kirchhoff formula [15]

$$U(\xi,\eta) = \frac{1}{2j\lambda} \iint_\Omega U_0(t,s) \frac{\exp\{jkr_{01}\}}{r_{01}}(1+\cos\theta_{01})\,dt\,ds \tag{5.2}$$

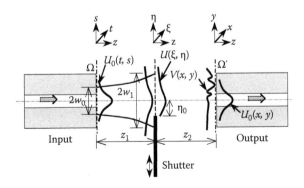

FIGURE 5.7
Diagram of attenuation model for the free-space single-shutter VOA.

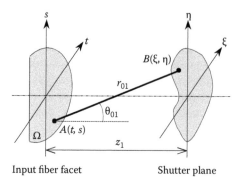

FIGURE 5.8
Diagram of light beam diffraction from the input fiber to the shutter plane.

where λ stands for the wavelength and k is the wavenumber defined by $k = 2\pi/\lambda$. The integration is over the whole Ω plane. The variables r_{01} and θ_{01} represent, respectively, the distance and angle between the point $A(t, s)$ at the input fiber end facet and the point $B(\xi, \eta)$ in the shutter plane, as shown in Figure 5.8. They are given by

$$r_{01} = \sqrt{(x - \xi)^2 + (y - \eta)^2 + z_1^2} \tag{5.3a}$$

$$\theta_{01} = \arccos(z_1/r_{01}) \tag{5.3b}$$

The light beam is then diffracted to the facet plane. The field distribution $V(x, y)$ can be expressed again using the Fresnel–Kirchhoff diffraction formula [15]

$$V(x,y) = \frac{1}{2j\lambda} \iint_{\Sigma} U(\xi,\eta) \frac{\exp\{jkr_{12}\}}{r_{12}} (1 + \cos\theta_{12}) d\xi d\eta \tag{5.4}$$

where z_2 is the separation between the shutter and the output fiber facet. The variables r_{12} and θ_{12} have similar meanings to r_{01} and θ_{01}; the difference is that r_{12} and θ_{12} are for the point projection from the shutter plane to the facet plane. Σ is the integral area that is not blocked by the shutter. Note that the integration is not over the whole shutter plane, or in other words, $U_0(\xi, \eta) = 0$ when $\eta < \eta_0$; here η_0 is the position of the shutter edge, as shown in Figure 5.7. The integration of $U_0(\xi, \eta)$ is over the region $-\infty < \xi < +\infty$ and $\eta_0 \leq \xi < +\infty$.

The coupling efficiency E_c represents how much the field distribution $V(x, y)$ is coupled to the fundamental mode $U_0(x, y)$ of the output fiber. It can be obtained by the mode-overlap integral [15].

$$E_c = \left| \int \int_{-\infty}^{+\infty} V(x,y) U_0^*(x,y) dx dy \right|^2 \bigg/ \left(\int \int_{-\infty}^{+\infty} |V(x,y)|^2 dx dy \cdot \int \int_{-\infty}^{+\infty} |U_0(x,y)|^2 dx dy \right) \tag{5.5}$$

where $U_0^*(x, y)$ is the conjugate of $U_0(x, y)$, which can be expressed by

$$U_0(x,y) = \sqrt{\frac{2}{\pi w_0^2}} \exp\left\{ -\frac{x^2 + y^2}{w_0^2} \right\} \tag{5.6}$$

$U_0(x, y)$ is the same as Equation 5.1 except that the coordinate system is changed to (x, y) rather than (t, s) because the output fiber is identical to the input fiber. The physical meaning of the mode-overlap integral is that $V(x, y)$ can be expanded on the bases of a set of orthogonal fiber modes, and Equation 5.5 expresses the fractional power on the base of $U_0(x, y)$.

The overall transmission efficiency T from the input to the output is given by

$$T = E_c \int \int_{+\infty}^{-\infty} |V(x,y)|^2\, dx dy \bigg/ \int \int_{+\infty}^{-\infty} |U_0(\xi, \eta)|^2\, d\xi d\eta = \left| \int \int_{-\infty}^{+\infty} V(x,y) U_0^*(x,y) dx dy \right|^2 \tag{5.7}$$

The condition that $U_0(x, y)$ and $U_0(\xi, \eta)$ have unit energy is used. Finally, the attenuation of the VOA is given by

$$L_a = -10 \log(T) \tag{5.8}$$

where log is the logarithm on base of 10.

The general model expressed in Equations 5.1 to 5.8 uses the accurate diffraction formula; thus, it is valid for various optical attenuators. In addition, it can be used to study the field pattern in any position between the two fibers. Figure 5.9 illustrates the calculated diffracted patterns after the shutter plane and facet planes. Figure 5.9a shows the amplitude and phase distributions after the shutter plane when $z_1 = 10\ \mu m$ and $\eta_0 = 0\ \mu m$ (i.e., half of the beam is blocked). Within the $15\ \mu m \times 15\ \mu m$ observation window, the amplitude pattern undergoes a sudden change across the shutter edge (due to cutoff by the shutter), and most of the energy is concentrated in a small region. The phase also experiences great changes in the window area (from 20 to $-150°$). However, it does not vary significantly in the high-amplitude region. Figure 5.9b,c indicate the diffracted patterns in the facet plane when $z_2 = 10\ \mu m$ and $z_2 = 500\ \mu m$, respectively. Values of z_2 are intentionally selected to represent the near-field and far-field conditions.

In Figure 5.9b, the amplitude has smooth but irregular contours and is still confined in a small region. The phase exhibits rapid change in the direction perpendicular to the shutter edge but slow change in the other direction. This is due to the knife-edge diffraction. In Figure 5.9c, the amplitude and phase distribute smoothly. The amplitude contours appear to be elliptical, whereas the phase tends to be circular. One interesting point is the change of the pattern centers. In Figure 5.9a, both the amplitude and phase centers are at $\eta = 0\ \mu m$. In Figure 5.9b, they shift nearly identically to about $\eta = 2\ \mu m$. However, in Figure 5.9c, they are separated. The amplitude center moves to $\eta = -2.5\ \mu m$, whereas the phase center changes to $\eta = 4\ \mu m$. These observations suggest that the near-field case is very different from the far-field one, and therefore, different treatments may have to be adopted in the simplification and analysis. Other parameters in this diffraction pattern study are $\lambda = 1.55\ \mu m$ and $w_0 = 5.2\ \mu m$ (Corning® SMF-28™ single-mode fiber).

The general method has to employ time-consuming numerical integrals because Equations 5.2, 5.4, and 5.7 are commonly not integrable. Besides, the general model cannot show the analytical relationships between the parameters. In the design and optimization of the attenuator, any change of the parameters results in another round of numerical calculation, making it very tedious. Some simplifications are necessary to obtain traceable analytical expressions.

5.4.2 Far-Field Attenuation Model

Under the far-field assumption, $z_1 \gg w_0$ and $z_2 \gg w_0$ for common single-mode fibers. Therefore, the propagation from input fiber to shutter plane and diffraction from shutter plane to the facet plane can both be treated as far-field diffractions. As a result, the attenuation model becomes a far-field model for an optical attenuator.

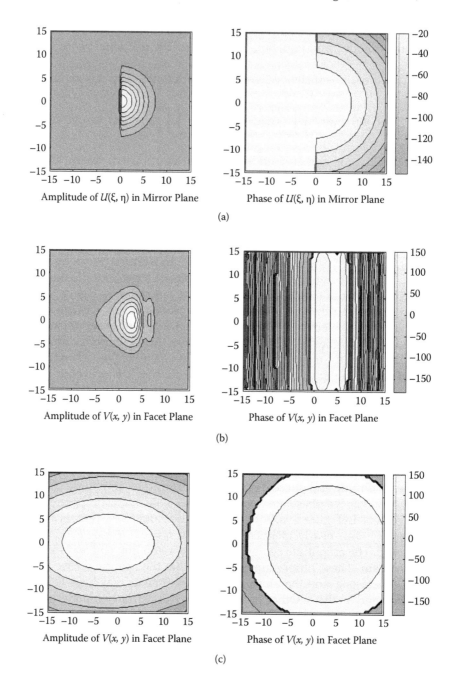

FIGURE 5.9
Amplitude and phase patterns after the shutter plane and at the facet planes when half of the input light is blocked ($\eta_0 = 0$ μm). (a) Cutoff light field after the shutter plane ($z_1 = 10$ μm); (b) near-field patterns in the facet plane when $z_2 = 10$ μm; and (c) far-field patterns in the facet plane when $z_2 = 500$ μm.

To calculate the closed-form expression of $U(\xi, \eta)$, the variables r_{01} and θ_{01} can be approximated by $r_{01} = z_1 + [(x - \xi)^2 + (y - \eta)^2]/2z_1$ and $\theta_{01} \approx 1$. Therefore, Equation 5.2 can be integrated as

$$U(\xi, \eta) = \frac{\sqrt{2\pi}w_0}{j\lambda(z_1 - jz_R)} \exp\left\{ jk\left[z_1 + \frac{\xi^2 + \eta^2}{2z_1\left(1 + z_R^2/z_1^2\right)} \right] \right\} \exp\left\{ -\frac{\xi^2 + \eta^2}{w_1^2} \right\}$$

$$= \frac{\sqrt{2\pi}w_0}{\lambda(z_R + jz_1)} \exp\{jkz_1\} \cdot \exp\left\{ -\left[\frac{1}{w_1^2} - \frac{j\pi z_1}{\lambda\left(z_1^2 + z_R^2\right)} \right](\xi^2 + \eta^2) \right\} \tag{5.9}$$

$$= \sqrt{\frac{2}{\pi w_1^2}} \exp\{jkz_1 - \phi_t\} \cdot \exp\{-a_t(\xi^2 + \eta^2)\}$$

where z_R and w_1 represent the Rayleigh range and waist radius of the Gaussian beam in the shutter plane, respectively, as defined by

$$z_R = \pi w_0^2/\lambda \tag{5.10a}$$

$$w_1 = w_0\sqrt{1 + z_1^2/z_R^2} \tag{5.10b}$$

and the temporary variables a_t and ϕ_t are given by

$$a_t = \frac{1}{w_1^2} - \frac{j\pi z_1}{\lambda\left(z_1^2 + z_R^2\right)} \tag{5.10c}$$

$$\phi_t = \arctan(z_1/z_R) \tag{5.10d}$$

Ignoring the constant-phase term in Equation (5.9):

$$U(\xi, \eta) = \sqrt{\frac{2}{\pi w_1^2}} \exp\{-a_t(\xi^2 + \eta^2)\} \tag{5.11}$$

This is the field distribution right before the shutter plane, expressed in a simple and beautiful form. It has the form of the standard Gaussian distribution. The only difference is that the variable a_t is complex rather than real. After the shutter plane, $U(\xi, \eta) = 0$ if $\eta < \eta_0$ due to the cutoff of the shutter.

To calculate the overall transmission efficiency in Equation 5.7, the variables r_{12} and θ_{12} can be approximated in a similar way. Then, by swapping integration orders (i.e., first integrate over the (x, y) plane and then over the ξ direction), T can be finally simplified after a long derivation as

$$T = \frac{T_0}{4}\left| 1 - \text{Erf}\left(\sqrt{p - iq} \, \frac{\eta_0}{\sqrt{w_1 w_2}} \right) \right|^2 \tag{5.12}$$

where p and q are unitless variables defined by $p = \frac{w_1}{w_2} + \frac{w_2}{w_1}$ and $q = \frac{w_1 z_2}{w_2 z_R} + \frac{w_2 z_1}{w_1 z_R}$, respectively; T_0 represents the initial loss given by $T_0 = \frac{4}{p^2 + q^2} = \frac{1}{1 + (z_1 + z_2)^2/4z_R^2}$; $w_2 = w_0\sqrt{1 + z_2^2/z_R^2}$ are the waist radius after the Gaussian beam is propagated by a distance of z_2, respectively; and *Erf* is the complex-variable error function defined by $\text{Erf}(x) = \frac{2}{\sqrt{\pi}}\int_0^x e^{-t^2} dt$ (here, t is complex). When the shutters are fully open (i.e., $\eta_1 = -\infty$), the error function converges to −1, and yields $T|_{\eta=-\infty}=T_0$. Therefore, T_0 is the insertion loss. The form of T_0 is exactly the same as the widely used formula of insertion loss in Reference 14 (by taking $n = 1$ and $w_T = w_R$ in Equation 34 of Reference 14). It validates the effectiveness of Equation 5.12. Compared with the previous theoretical study of single-shutter VOAs in References 18–20, Equation 5.12 simplifies the insertion loss term and also arrives at a simple form.

The key features in the mathematical operation of this far-field model are swapping the integral order and expanding r_{01} and r_{12} under the far-field condition. As a result, the integrals become integrable. It reveals the direct relationship between the shutter position and attenuation. The attenuation can be estimated without engaging time-consuming numerical integrals. It is noted that when z_1 and z_2 are fixed while the shutter moves, the first term in Equation 5.12 contributes to the attenuation like a constant bias (i.e., insertion loss), whereas the second term defines the shape of the curve—such that the function of transmission efficiency is determined by the error function (i.e., proportional to it).

5.4.3 Near-Field Attenuation Model

In most of the MEMS VOAs the near-field assumption is valid. Under this assumption, it is reasonable to assume that the beam pattern in the facet plane does not vary significantly from that in the shutter plane. This consideration enables the diagram of the attenuation model to be simplified, as shown in Figure 5.10. The physical concept is that the shutter is shifted to the facet plane. The light from the input fiber first propagates to the combined shutter plane and facet plane; it is then blocked by the shutter and coupled into the output. In contrast, in the far-field model, the light first propagates to the shutter plane is immediately blocked by the shutter, and then propagates to the facet plane before being coupled into the output. Viewed from the standpoint of calculation order, the difference is that the light in the far-field model is first blocked and then diffracted, whereas in the near-field model the order is reversed. From Equation 5.12, the beam pattern V_{nf} in the facet plane can be simplified as

$$V_{nf}(x,y)=\begin{cases}\sqrt{\dfrac{2}{\pi w_2^2}}\exp\left\{-\left(\dfrac{1}{w_{12}^2}-\dfrac{j\pi(z_1+z_2)}{\lambda\left[z_R^2+(z_1+z_2)^2\right]}\right)\cdot(x^2+y^2)\right\} & y\geq\eta_0\\0 & y<\eta_0\end{cases}$$

(5.13)

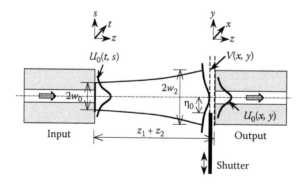

FIGURE 5.10
Attenuation model of the near-field condition.

where w_{12} is the waist radius given by $w_{12} = w_0 \sqrt{1 + (z_1 + z_2)^2 / z_R^2}$. Then, the transmission efficiency T_{nf} can be derived to be

$$T_{nf} = \frac{T_0}{4} \left| 1 - \mathrm{Erf}\left(\sqrt{f - jg} \; \eta_0 \right) \right|^2 \tag{5.14}$$

where f and g are variables representing $f = \frac{1}{w_0^2} + \frac{1}{w_{12}^2}$ and $g = \frac{\pi(z_1 + z_2)}{\lambda [z_R^2 + (z_1 + z_2)^2]}$. Here, by derivation, the constant term T_0 has exactly the same form as that of Equation 5.12.

Equation 5.14 directly links the attenuation to the shutter position under the near-field condition. Similar to that in the far-field model, when the fiber separation (i.e., $z_1 + z_2$) is fixed and the shutter moves, the first term of Equation 5.14 contributes a fixed attenuation, whereas the second term given by the extended error function defines the shape of the curve. The key feature of this model is making the integrals integrable by replacing the knife-edge diffraction with the Gaussian-beam diffraction between the shutter plane and the facet plane. From Equations 5.12 and 5.14, it is interesting to observe that both the far-field and near-field models have the same expression of insertion loss T_0. The only difference between the models is the term inside the error function.

Under a special condition, when $z_1 \ll z_R$ and $z_2 \ll z_R$, it is reasonable to let $w_2 \approx w_0$. As a result, $f \approx 2/w_0^2$ and $g \approx \frac{\pi(z_1 + z_2)}{\lambda z_R^2} \approx \frac{z_1 + z_2}{2 z_R} \cdot \frac{2}{w_0^2} \ll f$. Therefore, Equation 5.14 can be further simplified to be

$$T_{nf} = \frac{1}{4} [1 - \mathrm{Erf}(\sqrt{2}\eta_0 / w_0)]^2 \tag{5.15}$$

Here, the term inside the error function is real rather than complex as in Equations 5.12 and 5.14.

The diagrams of the attenuation models in Figures 5.7 and 5.10 assume implicitly that the shutter moves perpendicularly to the optical axis between the two fibers. However, the shutter can also be configured to tilt at a certain angle as shown Figure 5.11. In this way, the back reflection can be greatly reduced if the angle is properly chosen (8° or 12°) [32]. The near-field model can be used directly because the fiber separation does not change with the shutter position. In the far-field case, the far-field model can be used by varying z_1 and z_2 with the position of the shutter as expressed by

$$z_1 = z_{10} - \eta_0 \tan \varphi \tag{5.16a}$$

$$z_2 = z_{20} + \eta_0 \tan \varphi \tag{5.16b}$$

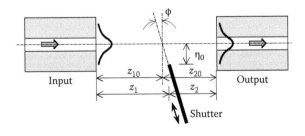

FIGURE 5.11
VOA using a tilting shutter.

where ϕ stands for the tilting angle, z_{10} and z_{20} denote the distances from the shutter to the input and the output fibers, respectively, when the shutter edge sits on the center of optical axis. It is noted that η_0 is negative when the shutter is below the optical axis.

Then, the attenuation models are compared with the general model in the cases of $z_1 = z_2 = 10$ μm and $z_1 = z_2 = 500$ μm, as shown in Figure 5.12a,b, respectively. The input and output fibers are identical; both use Corning SMF-28 single mode fiber ($w_0 = 5.2$ μm and $z_R = 54.8$ μm). So, it satisfies the near-field condition for $z_1 = z_2 = 10$ μm and the far-field condition for $z_1 = z_2 = 500$ μm. In the near-field case as shown in Figure 5.12a, the attenuation varies from about 0.14 dB to 95 dB when the shutter moves from −10 to 11 μm. The three models overlap one another. Discrepancy A shows the difference between the far-field and general models, and Discrepancy B shows the difference between the near-field and general models. The near-field model is nearly identical to the general model, especially when the shutter position is less than 5 μm. The maximum difference is −0.28 dB (−0.3%) when the shutter position is at 11 μm. In contrast, the far-field model predicts higher

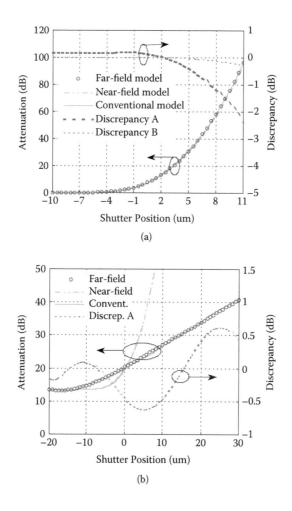

FIGURE 5.12

Comparison of the attenuation models. (a) $z_1 = z_2 = 10$ μm (near-field condition); (b) $z_1 = z_2 = 500$ μm (far-field condition). Discrepancy A represents the difference of the far-field model from the general model, and Discrepancy B shows the difference of the near-field model from the general model.

attenuation in the low attenuation level (+0.14 dB deviation, correspondingly +100% error) and a considerably lower value in the high attenuation range (maximum deviation –2.3 dB, correspondingly –2.5% error). Therefore, the near-field model is valid under the near-field condition.

In the far-field case, as shown in Figure 5.12b, the near-field model is obviously different from the general model, whereas the far-field model deviates only slightly from the conventional model. The attenuation given by the far-field model varies from 13.5 to 41.4 dB when the shutter moves from –20 to 30 μm. Discrepancy A represents the difference between the far-field and general models. The deviation ranges within –0.62 dB (–2%) to +0.61 dB (+2%). Therefore, the far-field model is suitable for the far-field condition.

It is also observed in Figure 5.12 that the attenuator has small insertion loss and high attenuation range if the fibers are aligned very close to one another (near-field configuration), but the attenuation curve is very nonlinear. On the contrary, the attenuator in the far-field configuration has linear attenuation curve, but high insertion loss and small attenuation range. A good attenuator design should trade off between them. Alternatively, lensed fiber or other focus/collimation components can be introduced to suppress the insertion loss, but the simulation becomes more complicated.

5.5 Optical Model and Tuning Schemes of Dual-Shutter VOA

In asymmetric dual-shutter VOAs, the inclusion of an additional shutter provides two degrees of freedom (DOF) for the choice of attenuation relationship in comparison to the previously considered one-DOF single-shutter design VOAs. By coordinating the movement of the two shutters, a linear relationship can be realized between attenuation and the shutter position or other control parameters (e.g., control voltage). Employing the analytical model in the preceding text, this section will first derive the analytical attenuation model, after which different tuning schemes will be analyzed.

5.5.1 Attenuation Model

The configuration of the asymmetric dual-shutter VOA is shown in Figure 5.13. Two independently movable shutters are inserted into the gap between the input and output fibers.

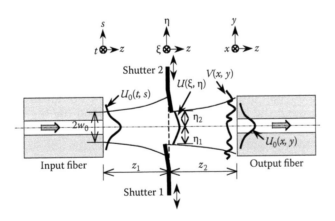

FIGURE 5.13
Schematic diagram of the dual-shutter VOA for attenuation modeling.

The tips of the shutters are deflected by a certain amount (typically, 8 or 12°) to allow them to engage with one another and also to reduce the back reflection [32]. The facets of input and output fibers are defined as (t, s) and (x, y), respectively, whereas the plane of the shutters is (ξ, η). In this model, all these planes are parallel and share a common z-axis. The analytical model can be obtained by following the same procedures for deriving the far-field and near-field models in the previous section. Here, the far-field condition is adopted for the attenuation model. According to Equation 5.12, the coupling efficiency can be expressed relative to the positions of two shutters as given by

$$T(\eta_1,\eta_2) = \frac{T_0}{4}\left|\mathrm{Erf}\left(\sqrt{p-iq}\,\frac{\eta_2}{\sqrt{w_1 w_2}}\right) - \mathrm{Erf}\left(\sqrt{p-iq}\,\frac{\eta_1}{\sqrt{w_1 w_2}}\right)\right|^2 \qquad (5.17)$$

where η_1 and η_2 indicate the shutter positions (see Figure 5.13), and p, q, w_1, and w_2 have the same definitions as those in the far-field model for single-shutter VOAs. In the derivation of Equation 5.17, it is assumed that the two shutters are on the same (ξ, η) plane, and they are infinitely long in the ξ direction. However, shutter 1 is in the lower part of the η direction, whereas shutter 2 in the higher part.

Finally, the attenuation in dB, by definition, is determined by

$$A(\eta_1,\eta_2) = -10\log[T(\eta_1,\eta_2)] \qquad (5.18)$$

The calculated attenuation (in dB) with respect to the positions of shutters 1 and 2 is contoured in Figure 5.14a, which exists only in a triangular region above the diagonal line $\eta_1 = \eta_2$. This is because the light path will be fully blocked, and theoretically, the attenuation tends to infinity if $\eta_1 \geq \eta_2$ as represented by the shaded region in Figure 5.14a. In the triangular region of $\eta_1 < \eta_2$, the attenuation is kept at the low level for $\eta_1 < -2\ \mu m$ and $\eta_2 > +2\ \mu m$; that is, the shutters are almost opened to let most of the incident light pass through. In such low-level regions, the gradient is small, which allows for fine attenuation tuning. At the corners of the triangular region, the attenuation is increased rapidly, and thus causes a high gradient. Such a contour provides a roadmap for the choice of attenuation tuning scheme.

Here, the term *attenuation tuning scheme* refers to coordinating the movement of the two shutters to obtain a certain tuning relationship between the attenuation and shutter 1's parameters (e.g., position, voltage, etc.). Two common tuning schemes can be readily presented in the contour, as shown in Figure 5.14a. Scheme I is for single-shutter tuning; that is, the attenuation is tuned by only shutter 1, whereas shutter 2 is always kept open. This actually represents the tuning relationship in many developed single-shutter VOAs [16,31,32,36] as given by Equation 5.12. The corresponding attenuation level with respect to shutter 1 is plotted in Figure 5.14b by the curve I, which is quite flat at the beginning but then becomes more and more steep. Such undesirable nonlinearity is because tuning scheme I passes through the low-gradient region and then the high-gradient region, as seen in Figure 5.14a. Scheme II is for symmetric-shutter tuning, in which the two shutters are positioned and moved symmetrically so as to control the attenuation level by adjusting only the aperture size, which is used in some studies [17,21]. It can be observed in Figure 5.14a that scheme II is mostly in the low-gradient region, but experiences a drastic gradient change when it moves close to the diagonal line. As can be seen in Figure 5.14b, the corresponding curve rises slowly but shoots up to $+\infty$ within a small range when $\eta_1 = \eta_2 \to 0$. Such nonlinearity leads to in low tuning resolution and control instability. In Figure 5.14b, the shaded region is inaccessible as it is limited by the single-shutter tuning.

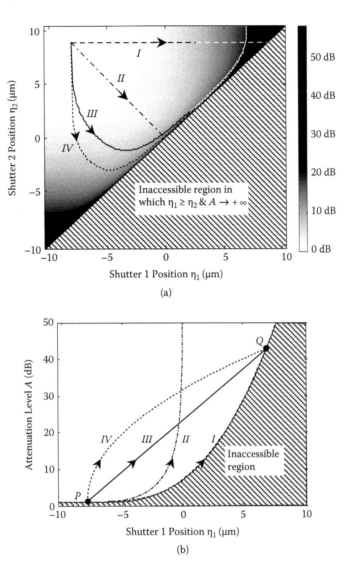

FIGURE 5.14
Calculated relationships between the attenuation level, the shutter 1 position, and the shutter 2 position. (a) Contour of the attenuation level A with respect to the positions of the two shutters η_1 and η_2; and (b) the loci of the two shutters' positions in different nonlinear and linear tuning schemes. Scheme I is for single shutter; that is, only shutter 1 is movable, whereas shutter 2 is kept open; scheme II is for symmetrical shutter design; that is, the two shutters are moved by the same amount; scheme III is for linear tuning, in which attenuation is linear with respect to shutter 1's displacement; and scheme IV is for linear tuning in which the attenuation is linear to the applied voltage of shutter 1.

In the calculation, the VOA has $w_0 = 5.2\ \mu m$, $\lambda = 1550$ nm, and $z_1 = z_2 = 28.5\ \mu m$. The parameters are the same as those of the VOA to be used in the experiment [22].

5.5.2 Analysis of Linear Tuning Schemes

In previous analyses, a straight locus of η_1 versus η_2 in the contour of Figure 5.14a led to a nonlinear attenuation tuning relationship in Figure 5.14b. To obtain a linear tuning, the

locus should be curved. The schemes III and IV represent two types of linear tuning relationships, in which the attenuation is linear to shutter 1's position and the driving voltage,
respectively. The reason for choosing these two linear schemes as examples is that in analysis it is easy to link the attenuation to the shutter 1 position, as given in Equation 5.3, whereas
most of real-world applications the driving voltage is the directly controllable parameter
(in most cases, electrostatic comb drives are used as the actuators). To compare the tuning
schemes, the same initial and final states, P and Q, from scheme I are chosen, as shown in
Figure 5.14b. For the state P, $\eta_1 = -8.0\ \mu m$ and $A = 1.0$ dB, whereas for Q, $\eta_1 = 7.0\ \mu m$ and $A =$
42.8 dB. Scheme III is a straight line in Figure 5.14b as given by $A(\eta_1, \eta_2) = A_0 + k_\eta(\eta_1 - \eta_{10})$;
here, the start point $A_0 = 1.0$ dB, slope $k_\eta = 2.8$ dB/μm, and shutter 1's initial position $\eta_{10} =$
$-8.0\ \mu m$. Scheme IV is given by $A(V_1, V_2) = A_0 + k_V V_1$; here, $k_V = 4.8$ dB/V, and V_1 and V_2
are the driving voltages to the actuators that control the positions of the shutters 1 and 2,
respectively. The driving voltage is related to shutter 1 position by the parabolic expression
$\eta_1(V_1) = \eta_{10} + \alpha_0 V_1^2$; here, $\alpha_0 = 0.2\ \mu m/V^2$. This parameter is just an example of the actuation
relationship of the electrostatic comb-drive actuators. It is also from the measured result to
be used in the experimental studies in the next section [22]. Because of the parabolic relationship between η_1 and V_1, scheme IV is a parabolic curve in Figure 5.14b. The loci of schemes
III and IV are both complicated curves; η_1 keeps increasing from $-8\ \mu m$ to $7\ \mu m$, whereas η_2
moves back and forth so as to maintain a linear increase of the attenuation level.

It is interesting to note in Figure 5.14a that both the loci for schemes III and IV have some
straight segments inclined at nearly 45°. This implies that when attenuation increases, both
shutters are practically moving at almost the same pace toward one direction, whereas the
aperture size remains constant. To make this clearer, the aperture size $\Delta\eta = \eta_2 - \eta_2$ is plotted in Figure 5.15 for both schemes III and IV. With the increase of the attenuation level,
aperture sizes of III and IV are reduced, but they are kept flat over the range of roughly
25–35 dB and are then further increased. The segment of constant aperture size is very
useful because linear attenuation can be achieved by merely moving aperture, rather than
reducing aperture size. This unique feature of the dual-shutter design may help eliminate

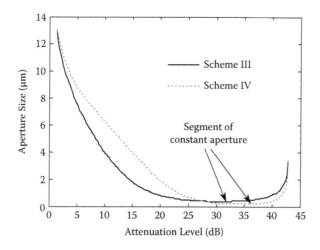

FIGURE 5.15
The aperture size as a function of attenuation level for linear tuning schemes III and IV. Segments of constant
aperture size can be observed, which is very useful for relaxing the position control at the high attenuation
level. Minimum aperture size can also be read from the curves, which are 0.35 and 0.21 μm for schemes III and
IV, respectively.

the need for extreme narrow apertures, and therefore, complicated control systems, even when high attenuation is attempted. To demonstrate this advantage, let us consider achieving 1 dB resolution at 35 dB level. The linear tuning schemes III and IV need both shutters to move at steps of 0.36 μm, whereas the single-shutter scheme I needs 0.18 μm and the symmetric-shutter scheme needs 0.01 μm. The accuracy of shutter movement is thus relaxed by 2–30 times just by using the linear tuning schemes. Another important feature is that the aperture size is not required to be smaller and smaller to obtain higher attenuation; instead, it has a minimum value. It is 0.35 μm for scheme III and 0.21 μm for scheme IV. This feature in another way relaxes the requirement of shutter position control.

Some diffracted beam patterns of scheme IV are shown in Figure 5.16 for three sets of shutter positions, that is, (a) $\eta_1 = -7.0$ μm, $\eta_2 = -1.0$ μm, and correspondingly $A = 11.7$ dB; (b) $\eta_1 = -3.5$ μm, $\eta_2 = -2.2$ μm, and $A = 22.8$ dB; and (c) $\eta_1 = 0$, $\eta_2 = 0.3$ μm, and $A = 31.4$ dB. Other parameters are $z_1 = z_2 = 28.5$ μm. For each case, the intensity field distribution after the shutter plane (right after the light beam is chopped by the two shutters) is shown together with the corresponding intensity and phase distributions in the output plane on the output fiber facet. It can be observed from Figure 5.16 that the discontinuous beam chopped by the shutters is restored to a continuous beam pattern after being diffracted to the output plane. The intensity varies quite fast in the x direction, but the phase distribution is merely flat. With reduction of the aperture size, the field distribution is actually extended in the y direction due to the thin-slit diffraction.

The schemes I–IV serve just as examples of the many ways to choose tuning schemes. As the dual-shutter design has two DOFs to choose the shutters' positions, it is feasible to make the attenuation linear to the other control parameters, which may have a complicated relationship with the shutter movement generated by other actuators, such as thermal [26,48], electromagnetic, and piezoelectric actuators [36]. In addition, the linear tuning can start from any point at any slope, only if the start and end points fall in the accessible region in Figure 5.14b. This brings in flexibility to set the operating conditions of dual-shutter VOAs.

5.5.3 Analysis of Tuning Resolution

Because of the asymmetry of the two shutters, it is possible to obtain coarse tuning and fine tuning in the same device simultaneously; for instance, moving shutter 1 close to the beam center would obviously give low tuning resolution, whereas keeping shutter 2 away from the center would give high tuning resolution. Here, the main objective is to find one pair of shutter positions (and also the aperture position) so that attenuation changes due to two shutters give an appropriate tuning contrast such as 1:25. Meanwhile, these two resolution values are neither too low nor too high (i.e., slow tuning) to be impractical. The tuning resolution of the two shutters can be described by

$$R_i = \left| \frac{\partial A}{\partial \eta_i} \right| = \frac{4.3}{T} \left| \frac{\partial T}{\partial \eta_i} \right|, \quad (i = 1, 2) \tag{5.19}$$

and the tuning contrast, which measures the ratio of two resolutions, can be defined as

$$C = \frac{R_1}{R_2} = \left| \frac{\partial T}{\partial \eta_1} \middle/ \frac{\partial T}{\partial \eta_2} \right| \tag{5.20}$$

In both definitions the variable A is in units of dB, whereas the variables η_1 and η_2 are in units of μm. R_1 and C as functions of shutter positions are computed and plotted in

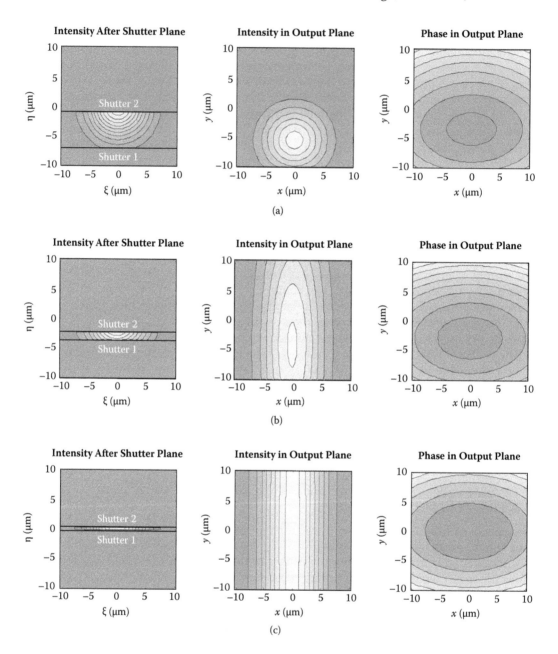

FIGURE 5.16
Diffraction beam patterns for the linear tuning scheme. The intensity distribution after the shutter plane and the intensity and phase distribution in the output fiber facet plane are presented for different shutter positions. (a) $\eta_1 = -7.0$ μm, $\eta_2 = -1.0$ μm, and correspondingly $A = 11.7$ dB; (b) $\eta_1 = -3.5$ μm, $\eta_2 = -2.2$ μm, and $A = 22.8$ dB; and (c) $\eta_1 = 0$, $\eta_2 = 0.3$ μm, and $A = 31.4$ dB.

Figure 5.17 for the same parameters as in Figure 5.14. R_2 is similar to R_1 and is thus not shown. Here, half of the plane is inaccessible because the shutter overlap ($\eta_1 \geq \eta_2$) is ignored, as mentioned earlier. For better visual effect, R_1 and C are plotted in the logarithm scale. Owing to the exchangeability of η_1 and η_2, the value of C in Figure 5.17b is replaced by $1/C$

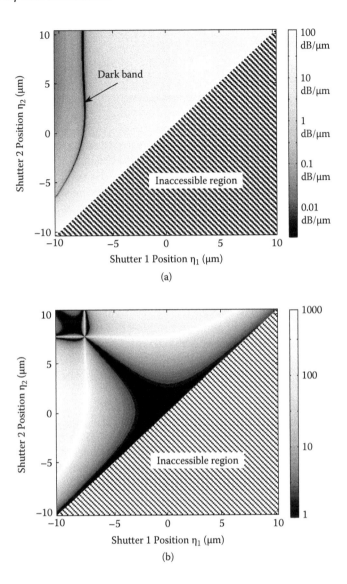

FIGURE 5.17
Contours of the tuning resolution. (a) Tuning resolution of shutter 1 and (b) resolution contrast of the two shutters.

if $C < 1$. The dark band in Figure 5.17a corresponds to the local maxima of the calculated attenuations and thus has little practical significance. As can be expected from Figure 5.14, the resolution can go up to 0.1 dB/μm if the shutters are in the low-gradient region in Figure 5.14a; that is, high resolution at the low attenuation level. Moreover, one can see in Figure 5.17b that high tuning contrast (e.g., 10–100) can be achieved if both shutters lie in the same side, which corresponds to the high-gradient regions in Figure 5.14a. That is to say, an off-axis aperture appears to be critical for satisfactory performance in both the resolution and tuning contrast.

In practice, sometimes it is more convenient to use the resolution's dependence on the driving voltage. In this regard, the actuator characteristic $\eta_1(V_1)$ has to be taken into account. If the actuator is an electrostatic comb drive for which $\eta_1(V_1)$ has a quadratic form as discussed earlier, the contrast C_V with respect to driving voltage is given by

$$C_V = \frac{dA/dV_1}{dA/dV_2} = \frac{V_1}{V_2}\frac{dA/d\eta_1}{dA/d\eta_2} = C\frac{V_1}{V_2} \qquad (5.21)$$

If V_1/V_2 has typical values of 1–5 and the required value of C_V is assumed to be 10–30, the desired contrast value C is estimated to be only 2–6. As can be seen from Figure 5.17b, this requirement can be met easily by the dual-shutter design. In this regard, the position control of these shutters is further relaxed. Although the actual feasibility and effectiveness of the VOA device are also strongly dependent on the reliability and repeatability of actuators, Figure 5.17b provides an important roadmap for the selection of working points at which a satisfactory tuning contrast is available without the need for high-resolution control of shutter positions.

5.6 Analyses of Temperature, Wavelength, and Polarization Dependencies

Almost all MEMS VOAs have attenuation values that vary with temperature, wavelength, and polarization. It is an important issue in practice, because the environmental temperature could vary significantly, the light wavelength may span a large range (e.g., in DWDM), and the polarization state is unpredictable and changing constantly. This section will try to pinpoint the physical mechanisms involved and estimate their contributions in different types of VOAs.

5.6.1 Temperature-Dependent Losses

When MEMS VOAs are heated up or cooled down, various factors could contribute to the TDL, such as thermal expansion of all the components, thermal softening of the silicon material, change of optical properties of fiber and metal-coated mirror, thermal stress, hot airflow, etc. The TDL is a combination of all these factors.

Generally, the variation of attenuation can be simply regarded as the change of fiber coupling. As the optical properties of optical fibers are altered very slightly in the presence of a certain temperature change (typically <100 K), their influence is negligible. The remaining factors would affect the optical coupling either by misaligning the fibers or by shifting the shutter (or mirror) relative to its nominal position, or both. For example, the thermal expansion of UV epoxy impairs the fiber alignment because of the mismatch of thermal expansion coefficients of the epoxy, substrate, and fibers. However, this misalignment is strongly dependent on the details of fiber fixation and epoxy application, and is thus difficult to quantify.

Different types of VOAs have different fiber positions, and thus, the fiber misalignment contributes differently. As discussed earlier, there are three fiber arrangements: coaxis, retroaxis, and cross-axis [52]. Coaxis and retroaxis are similarly influenced by thermal misalignment, so they are represented by only one example, the single-shutter VOA (SVOA), in the foregoing discussion. For the cross-axis, the flat mirror VOA (FVOA) and elliptical mirror VOA (EVOA) are chosen as examples. The choice of EVOA is to investigate the influence of light focusing, because SVOA and FVOA have no focusing effect if normally cleaved fibers are used. These examples are just for the convenience of discussion. The results can be applied equally to other types of VOAs.

In the SVOA, the input and output fibers are positioned coaxially; thus, the influence of thermal expansion might be counterbalanced if the epoxy were applied symmetrically to the input and output fibers. When the SVOA works at low attenuation level, the shutter is almost fully open. At this stage, the SVOA has a relatively large tolerance to fiber positioning, which also helps suppress the TDL. For example, an angular misalignment of 0.1° is estimated to introduce a negligible increase of loss (only 0.002 dB). In the FVOA and EVOA, the input and output fibers are perpendicular to each other, and thus the counterbalance does not exist. As the EVOA focuses light from the input to output, it has less tolerance to the misalignment than the FVOA. For example, an angular misalignment of 0.1° is estimated to cause an increase of loss by 1.0 dB, for the EVOA; but for the FVOA, it is only 0.2 dB.

In addition to misaligning the fibers, temperature change also affects VOA attenuation by slightly shifting the mirror, which is determined by the thermal stability of the actuator. For fair comparison, it is assumed that the SVOA, FVOA, and EVOA all use the same electrostatic comb drive etched on the silicon wafer, which is the most widely used type of actuator. In the comb drive, it uses two pairs of combs (one is fixed, whereas the other is movable) to move the mirror in or out, and employs a folded beam to suspend the moving structures. Very good design and analysis can be found in Reference 56. In most cases, the influence of residue stress and thermal stress can be left out as the folded beam is symmetric and commonly not coated with gold. In a comb-drive actuator, the displacement x is given by [56].

$$x = \frac{F}{k} = \frac{\varepsilon_0 N L^3}{2 E g b^3} V^2 \tag{5.22}$$

where F is the actuation force, k the stiffness of the folded beam, ε_0 the permittivity of air, N the number of fingers, E the Young's modulus, g the finger gap, V the driving voltage, and L and b the length and width of the folded beams, respectively. In MEMS VOAs, assume $N = 205$, $g = 2.5 \ \mu\text{m}$, $b = 2.5 \ \mu\text{m}$, and $L = 900 \ \mu\text{m}$ (taken from Reference 23). The temperature dependence of the mirror displacement can be expressed as

$$\frac{dx}{dT} = -(\alpha_0 + \beta_0)x_0 \tag{5.23}$$

where α_0 is the thermal expansion coefficient of silicon material (2.63×10^{-6} K^{-1} at room temperature) and β_0 is the temperature reduction coefficient of the Young's modulus as defined by $\beta_0 = \frac{1}{E}\frac{dE}{dT}$. When the temperature is not very high and not very low, Young's modulus of the single crystal silicon follows the Wachtman's model [57] as given by

$$E(T) = E_0 - BT \exp\{-T_0/T\} \tag{5.24}$$

where the constants are determined by experiment as $E_0 = 167.52$ GPa, $B = 14.41$ MPa/K, and $T_0 = 251$ K [58]. With these data, it can be easily derived that at room temperature $\beta_0 = -6.9 \times 10^{-5}$ K^{-1}. The negative value shows that the folded beam is softened with the increase of temperature (i.e., thermal softening). As the absolute value of β_0 is much larger than that of α_0, the thermal expansion of the structure can be neglected. From Equation 5.23, when $dx/dT > 0$, the actual mirror displacement would be larger than the nominal value. With a certain temperature change, for example, 55 K, the mirror displacement is estimated to increase by about 0.41%. In an experimental study, the comb drive produces a displacement of 12.74 μm when a driving voltage of 10 V is applied at room temperature [23]. When the VOA is heated up by 55 K, the displacement is increased by 0.06 μm; that is, 0.46%. It matches the estimated value of 0.41% well. As a larger displacement causes a higher attenuation, the thermal softening explains the observed phenomenon that a higher temperature induces larger attenuation [43,59]. From Equation 5.23, the mirror shift is proportional to its current position; therefore, TDL increases with current attenuation level. It explains the experimental observation that the TDL for the same temperature change increases with the increase of attenuation level.

The hot airflow can also contribute to the TDL. If VOAs are heated up in an open environment, it is observed that the mirror position would fluctuate owing to the hot airflow, regardless of how much voltage is applied to the comb drive. In an experimental study, the mirror shift is measured every 0.5 s when the device is heated up to 100°C, and no voltage is applied. The value of the shift is between −0.05 μm and 0.17 μm, and on average, 0.08 μm [23]. Such hot airflow could cause the attenuation to change by >10 dB and could make the comb drive unstable, especially at the high attenuation level and under high voltage for SVOAs and symmetric-design VOAs. It is also observed that this fluctuation does not happen either at room temperature or when VOAs are cooled down. To avoid the hot airflow, MEMS VOAs can be covered by a glass slide during the experiment. Such hot air fluctuation will not be a serious issue in real-world VOA products because most of them are encapsulated (or sealed) inside a packaging box. There is no open airflow.

5.6.2 Wavelength-Dependent Losses

Several physical mechanisms could cause the WDL, such as the wavelength dependence of the fiber property, diffraction of chopped Gaussian beam, diffraction or scattering at mirror edges, defects of mirrors (surface roughness and scalloping), and interference of forward stray lights (fiber-end reflection, diffracted or scattered light, etc.).

It is well known that the size of the Gaussian beam in fibers varies with the wavelength. For the SMF-28® fiber, it can be expressed as $w(\lambda) = 5.203 + 3.150\ \Delta\lambda + 1.627\ \Delta\lambda^2 + \cdots$, taking wavelength dependence of the refractive indices of the fiber core and cladding into consideration. The variable λ is the wavelength, $w(\lambda)$ is the beam waist radius, and $\Delta\lambda$ is the wavelength change relative to 1.55 μm; all are in units of micrometers. The diffraction is always strongly wavelength dependent, especially in the SVOA that has the input Gaussian beam partially chopped. On the basis of the attenuation model in Equation 5.12, the attenuation as a function of the wavelength and shutter position is calculated and contoured in Figure 5.18. At low attenuation levels, isolines are quite flat with the change

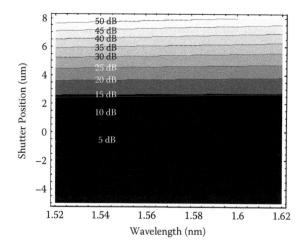

FIGURE 5.18
Simulated contour of the attenuation with the change of wavelength and shutter position in the single-shutter VOA.

of wavelength. However, as the attenuation increases, isolines become more and more inclined (i.e., the WDL increases). At the same shutter position, longer wavelength leads to lower attenuation. In the FVOA and EVOA, mirror edges may also contribute to WDL by diffraction and scattering as the mirror has limited depth. In MEMS VOAs, root-mean-square (RMS) roughness is typically < 40 nm and scalloping is <75 nm, and thus, the mirror surface causes Rayleigh scattering, which is proportional to $1/\lambda^4$. A part of the unwanted stray light could be transmitted forward and interfere with the main light. Such interference may dominate the WDL at a high attenuation level because the stray light becomes comparable to the main light in terms of optical power. The PDL may also contribute to the WDL in the experiment as it is difficult to maintain the polarization state during the measurement of WDL due to the presence of fiber connection and other components in the test setup. However, this contribution is left out, as the PDL will be discussed later.

The measured WDL is a combination of all these mechanisms. In test data, SVOAs commonly have a quasi-periodic change of attenuation with respect to wavelength change [2,19,20,22,23,31], indicating the presence of light interference (especially between the two fiber ends); sometimes it could become dominant [19]. For example, in the WDL curve in Reference [23], the period is 51.2 nm, corresponding to a cavity length of 23 μm. This cavity might be formed by the two ends of the input and output fibers. Whereas in the FVOA and SVOA, these curves have no obvious periodicity; thus, the interference is insignificant.

When the SVOA goes to a high attenuation level, the shutter blocks most of the light and its edge is far over the fiber axis. Therefore, the scattering could be comparable to the diffraction in delivering optical power from the shutter edge to the output fiber end. The WDL trend will be more complicated. With regard to FVOA and EVOA, the WDL should exhibit a slow increase with higher attenuation level. This is because the Gaussian beam is almost fully reflected (limited by the mirror height), and thus, the diffraction is not significant.

5.6.3 Polarization-Dependent Losses

The PDL comes from the high conductivity of the metal-coating material applied on the shutters and mirrors, typically gold or aluminum. Here, gold is chosen as an example for numerical demonstration. The different VOA configurations may be analyzed as follows. In the SVOA, the shutter edge excites different electromagnetic fields in response to the E-polarization (electric vector is parallel to the edge) and H-polarization (magnetic vector is parallel to the edge). Here, a simple case of the plane wave diffraction is presented, as shown in Figure 5.19. The detailed study can be found in Reference 15. It is assumed that gold has infinite conductivity, and that the shutter is infinitely long (along the x direction) and infinitely thin half-plane located in the plane of $z = 0$ with its edge at $y = 0$. In Figure 5.19, the incident light is shined from the top of the z axis. According to rigorous diffraction theory, the electric field excited by the H-polarized incidence have components $\vec{E}_x^H$ and $\vec{E}_y^H$ in the x and y directions, as shown in Figure 5.19a,b, whereas in E-polarization, the y component $\vec{E}_y^E$, is always equal to 0 ($\vec{E}_x^E$ is shown in Figure 5.19c). The maximum amplitudes of $\vec{E}_x^H, \vec{E}_y^H$ and $\vec{E}_x^E$ are 1, 0.43, and 0.97, respectively. It can be seen from Figure 5.19a,c that $\vec{E}_x^H$ and $\vec{E}_x^E$ have not much light diffracted to the shadow region, whereas $\vec{E}_y^H$ has. The difference of the total diffracted electrical fields in response to the two polarization states is shown in Figure 5.19d. There is obvious difference in the shadow region. When the SVOA works at a high attenuation level, the output fiber sits in the shadow; therefore, the PDL would become obvious. The reason is that $\vec{E}_y^H \neq 0$, whereas $\vec{E}_y^E = 0$. Note that in the case of H-polarization, the linearly polarized incidence would become elliptically polarized. This depolarization may not be acceptable in some applications, for instance, coherent communications.

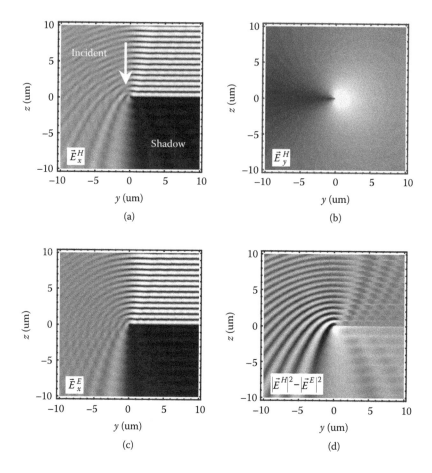

FIGURE 5.19

Diffracted electrical fields of H- and E-polarized incident light. (a) The x-component $\vec{E}_x^H$ of the diffraction of H-polarization, the maximum value is normalized to be 1; (b) the y-component $\vec{E}_y^H$ of the diffraction of H-polarization, the maximum value is 0.43; (c) the x-component $\vec{E}_x^E$ of the diffraction of E-polarization, the maximum value is 0.97, and the y-component $\vec{E}_y^E = 0$; and (d) the difference between the total electric fields of H- and E-polarizations, the maximum value is 0.40.

In the FVOA and EVOA, the light is reflected by the gold mirror as a Fresnel reflection [15], and thus, the PDL would vary with the incident angle, as shown in Figure 5.20. The simulation shows that the PDL is 0 at the incident angles of 0 and 90° (corresponding to normal incidence and grazing incidence, respectively), but reaches its maximum of 0.5 dB at 84°. In the FVOA, the Gaussian beam is incident at 45 ± 5.4° (considering the beam divergence) the average PDL in Figure 5.20 is only 0.07 dB. In the EVOA, the slope of the ellipse varies from 43.6 to 54°, and the average PDL is about 0.09 dB. As the attenuation becomes larger, the PDL is associated with the diffraction and thus a smooth increase occurs. To illustrate the effect of different materials, the curve of aluminum is also drawn in Figure 5.20 (n_{gold} = 0.559 + j 9.81, and $n_{aluminum}$ = 1.44 + j 16). Although aluminum has a maximum PDL that is 0.77 dB larger than the gold, no significant difference is seen near 45°. This implies that the coating of gold or aluminum does not affect the PDL of MEMS VOAs.

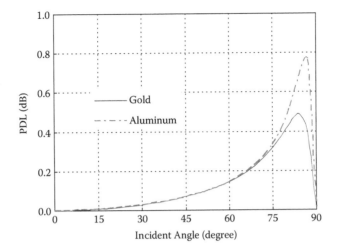

FIGURE 5.20
PDL induced by the Fresnel reflection of metal-coated surface.

5.7 Experimental Studies of MEMS VOAs

This section will present experimental studies on different types of VOAs and the influence of VOAs configuration on the TDL, WDL, and PDL.

5.7.1 Surface-Micromachined Single-Shutter VOAs

5.7.1.1 Design and Device Description

The schematic structure of a developed MEMS VOA is shown in Figure 5.21a. This idea was demonstrated in 2001 [31]. Two single-mode fibers are aligned very close to each other, one as the input fiber and the other as the output. A shutter is inserted into the light path between the two fibers to block a portion of the light. To obtain variable attenuation, a proprietary drawbridge actuator [60] is employed. The shutter located at the end of an L-shaped plate is mounted vertically on a mounting plate by a side holder. The mounting plate is hung over the substrate by a triangular drawbridge structure.

Figure 5.22 illustrates the triangular frame, which is composed of a holding plate and a drawing beam that are pinned to the substrate by microhinges. The holding plate has a T-shaped hole, whereas the drawing beam has a T-shaped head. By latching the T-shaped head into the T-shaped hole using manual assembly, the holding plate and the drawing beams form a firm triangular frame to tilt the holding plate at a certain angle. The tilting angle is determined by the lengths and positions of the holding plate and drawing beam, which in turn determines the working distance of the shutter. This drawbridge structure has an advantage over other actuators because it is easy to obtain large working distance up to several hundred microns. Several thin-bending beams are used to connect the mounting plate to the holding plate so as to reduce the stiffness. The shutter stands at the high position (as shown in Figure 5.21a) when no voltage is applied between the mounting plate and the electrode (on the substrate). And the shutter moves down while a potential is applied, as shown in Figure 5.21b. The deformation is mainly due to the

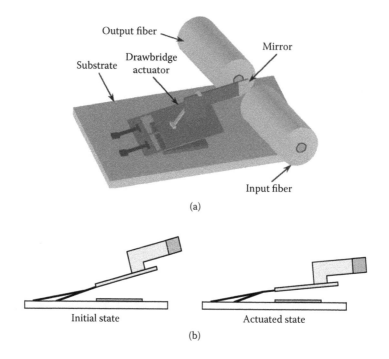

FIGURE 5.21
Schematic of the single-shutter MEMS VOA using a drawbridge actuator. (a) 3-dimensional structure and (b) the operational states of the drawbridge actuator.

bending beams, whereas the other parts of the drawbridge actuator are kept straight. The dimension of the bending beams determines the stiffness of the drawbridge actuator, and in turn, the response speed and the maximum driving voltage.

The MEMS VOA is fabricated using silicon surface micromachining technology, followed by manual assembly of the drawbridge actuators and vertical shutters. Finally, the

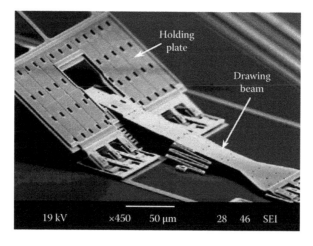

FIGURE 5.22
SEM of a triangular frame of the drawbridge actuator.

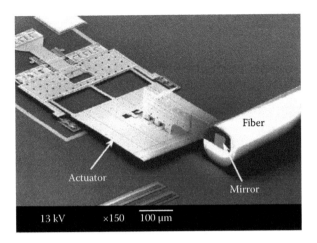

FIGURE 5.23
SEM micrograph of the MEMS VOA.

fibers are attached using the UV-curable epoxy. The fiber guides fabricated on the substrate help the alignment. An assembled VOA is shown in Figure 5.23. For clarity, only one fiber is shown packaged in Figure 5.23 (a workable VOA actually has two fibers). The distances between the fiber terminals and the gold mirror (used as the shutter) are both about 10 μm. The design parameters of the VOA are listed in Table 5.2. The flat shutter having a size of 40 × 40 μm is formed by coating a 0.5-μm-thick gold layer on a 1.5-μm-thick polysilicon layer. The gold layer helps to fully block the light because the polysilicon layer is nearly transparent to infrared light. The lower edge of the shutter is designed to be 66 μm high over the substrate, which allows most of the light to pass through when no voltage is applied. The center of the shutter is at a distance of 646 μm to its rotation center. The mounting plate has an initial tilting angle of 4.0° after assembly, which corresponds to a working distance of 45 μm for the shutter. Note that the shutter can be controlled to move stably to only a small part of the 45 μm range because of the snap-down effect. For a butt-coupling MEMS VOA, a controllable displacement range of about 15 μm is enough for the shutter. Three bending beams are employed to reduce the stiffness; each has a size of 104 × 5 μm and is formed by a polysilicon layer that is 1.5 μm thick. The mounting plate has a dimension of 325 × 356 μm and is formed by a polysilicon layer of 2 μm thick. The electrode is 279 × 364 μm formed by a 0.5-μm-thick polysilicon layer deposited on the substrate.

TABLE 5.2

Main Design Parameters

Item	Parameter
Mirror (i.e., shutter)	40 μm × 40 μm × 0.5 μm
Mirror center to the rotation center	646 μm
Initial height of the mirror lower edge	66 μm
Initial tilting angle of the plate after assembly	4.0°
Bending beams	3 × 104 μm × 5 μm × 1.5 μm
Mounting plate	325 μm × 356 μm × 2 μm
Electrode	279 μm × 364 μm × 0.5 μm

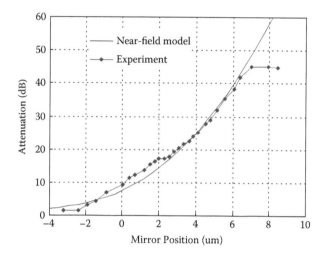

FIGURE 5.24
Comparison of the experiment with the near-field model.

Only the overlapping area of the mounting plate and electrode is effective for electrostatic actuation. The mounting plate is longer than the electrode to ensure they are well isolated even when the outer edge of the mounting plate touches the substrate.

5.7.1.2 Experimental Results

The measurement data are shown in Figure 5.24. For comparison, the values predicted by the near-field model according to Equation 5.14 are also plotted in Figure 5.24. The VOA has an insertion loss of 1.5 dB at 1.55 μm wavelength when no voltage is applied. This insertion loss is mainly introduced by three sources: 1.0 dB from the angular/lateral misalignment between the input and output fibers that are fixed to the substrate by the adhesive; 0.4 dB from the Fresnel reflection of air–glass interfaces in fiber ends; and 0.1 dB from the fiber separation. To make the simulation comparable to the experiment, an insertion loss of 1.4 dB is added to the simulation data. When the shutter is actuated to move in the light path, the attenuation continuously increases to its maximum value of 45 dB when the shutter is moved to about –7 μm. Further movement of the shutter does not produce higher attenuation as predicted by the near-field model because the signal power becomes too weak and reaches the detector noise floor of the power meter used. In Figure 5.24, the simulation result using the near-field model is very close to the experimental result, showing the effectiveness of the near-field model.

The wavelength dependence of the VOA is shown in Figure 5.25. In the measurement, the VOA is first set at a certain attenuation level, and then a tunable laser source (ANDO AQ4321) is used to sweep the input wavelength from 1520 to 1620 nm. Figure 5.25 illustrates the attenuation fluctuations when the VOA is set at 20 and 40 dB, respectively. Both curves are irregular, but have the same trend. The 40-dB curve is not as smooth as the 20 dB curve, probably owing to the noise. In Figure 5.25, the peak-to-peak variation is measured to be about 0.20 dB at 20 dB level and 0.38 dB at 40 dB level. The VOA also has the variations of 0.17 and 0.35 dB at the attenuation levels of 10 and 30 dB, respectively (not shown in Figure 5.25). The wavelength dependence variation tends to increase with the increase of attenuation level.

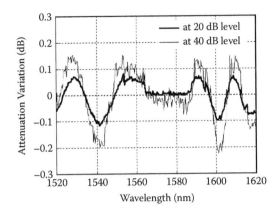

FIGURE 5.25
Wavelength dependence of the VOA at different attenuation levels.

5.7.2 Deep-Etched Dual-Shutter VOAs

5.7.2.1 Device Description

An asymmetric dual-shutter VOA has been designed and fabricated by a two-mask process. Overview of the assembled device as well as a close-up of the shutter part is shown in Figure 5.26. This work was demonstrated recently by Zhao [22]. In the actuator design, each shutter is driven by a pair of bidirectional comb drives so as to allow it to move both forward and backward for easy adjustment of the initial working condition. The mechanical structures are designed for a translation of 22 μm in both directions with a maximum applied voltage of 10.5 V. The tips of the shutters are deflected by 8°. When they begin to engage each other, the tips are still separated by 4 μm in the z direction to avoid being stuck together.

The device is fabricated on an 8 in. SOI wafer with a 75-μm-device layer and a 2-μm-buried oxide layer. In the first step, a plasma-vapor-deposited aluminum layer is patterned and etched to form the electrodes, after which the structures, including the comb drives, shutters, and fiber trenches, are patterned and etched by DRIE in one step. Movable structures are then released by wet etching. Finally, the shutter part is coated with aluminum to block the light. The fiber grooves are designed such that the spacing between the assembled input and output fibers is 57 μm, and the shutter plane is equidistant to both fibers facets. The 57 μm space is occupied by some fiber stoppers to avoid breaking the shutters unintentionally during the assembly (see Figure 5.26b).

Single-mode fibers (outer diameter 125 μm and beam waist radius 5.2 μm) are assembled onto the fabricated chip, and then the whole device is packaged into a butterfly box. As the depth of the fiber grooves is 77 μm, the fiber core is 14.5 μm below the top surface of the device after the assembly. According to the foregoing calculation, the stray light passing through the top of the shutters is negligible. A critical step of the assembly is the fixation of fibers by UV epoxy. Some protection trenches are designed, as can be seen in Figure 5.26a. Those trenches near the fibers are used to prevent the epoxy from spreading into comb-drive actuators in case too much epoxy is applied accidentally, whereas those next to the shutters are used to stop the crawling of the epoxy along the fiber surface to the fiber end facets to spoil the shutters. Such trench design significantly reduces the assembly difficulty and improves the yield to nearly 100%.

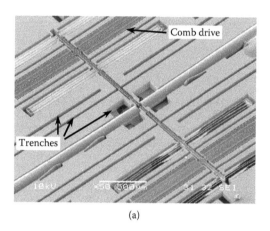

(a)

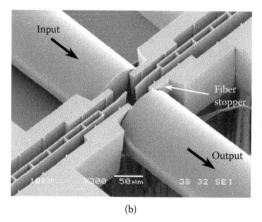

(b)

FIGURE 5.26
Fabricated dual-shutter VOA device. (a) Overview of the assembled device, and (b) close-up of the shutter part.

To have a direct comparison between theory and experiment, the performance of comb drives is first examined. Because four comb drives are involved for bidirectional movement of the two shutters, their electrostatic coupling could cause unwanted movement or vibration of the structure, resulting in jittering, leaping, and mutual attraction of shutters, which is a serious issue. These phenomena, however, can be suppressed by grounding all suspended electrodes, and therefore, the accumulated charge can be removed. As a result, the static displacement of comb drives is found to be well fitted by the quadratic relation $\eta = \alpha_0 V^2$, and exhibits satisfying reliability and repeatability. As stated earlier, the ratio α_0 is 0.200 ± 0.002 μm/V^2 for all the four comb drives. With this movement quality, the actuation behavior of comb drives can be used with confidence for the verification of the attenuation models in Sections 5.4 and 5.5. The shutter 1 is initially rested at $\eta_1 = -8$ μm and is moved upward for higher attenuation, whereas shutter 2 is initially at $\eta_2 = +8$ μm and is moved downward for higher attenuation. In an experiment, it is more convenient to monitor the driving voltage than the actual shutter position.

5.7.2.2 Single-Shutter Tuning Scheme

Single-shutter tuning (i.e., scheme I in Figure 5.14) is measured to examine the validity of the far-field attenuation model in Section 5.4. It is realized by withdrawing one shutter far

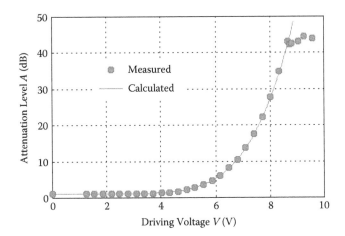

FIGURE 5.27
Attenuation curve of the single-shutter tuning scheme.

away from the light beam and keeping it stationary. As shown in Figure 5.27, the experimental results agree well with the calculation up to the attenuation level of 43.9 dB. Over this tuning range, measured attenuation is slightly higher than the prediction by an almost constant difference of 0.5 dB. This discrepancy may result from some additional losses not included in the calculation model. In particular, the insertion loss of the device is 1.2 dB, a bit higher than the predicted value of 1.0 dB. When the attenuation reaches 43.9 dB, however, the measured attenuation begins to deviate from the prediction and is clamped at 44.5 dB. This phenomenon strongly implies that a constant amount of stray light, which comprises approximately 0.01% of the input, can always be coupled into the output fiber and is unaffected by the moving shutter. It is noted that this saturated attenuation value does not represent the largest tuning range achievable by the device. The highest attenuation rate is attained in the symmetric-moving scheme, which is realized by applying the same voltage on both shutters. It can go up to >60 dB. However, very large attenuation range is not the primary target of this study and is thus not further pursued.

5.7.2.3 Linear Tuning Scheme

The linear tuning performance (i.e., scheme IV in Figure 5.14) of the device is also tested. For convenience of explanation, shutter 1 is chosen as the main shutter, whose driving voltage is to vary linearly with the attenuation, whereas shutter 2 is as the auxiliary shutter. For ease of demonstration, the attenuation range is chosen from 1.2 to 20.7 dB, which corresponds to a driving voltage from 0.0 to 7.4 V. Generally speaking, the shutter movement strategy is implemented by using shutter 1 to provide the baseline attenuation level (following Figure 5.27) and then by moving shutter 2 to increase the attenuation to the desired level. However, in this experiment, two methods can be used to realize the linear tuning function because three variables are involved (i.e., V_1, V_2, and A). The first method is something similar to trial and error. For a targeted linear attenuation–voltage relationship, shutter 1 is positioned according to the calculation, whereas shutter 2 is moved back and forth till the desired attenuation is obtained. The resultant voltage of shutter 2 is then recorded and compared to values from the calculation model. In other words, this method yields V_1 and A to test V_2. The second method employs a look-up table of the required voltages calculated from the models. The voltages are then directly applied to the comb drives.

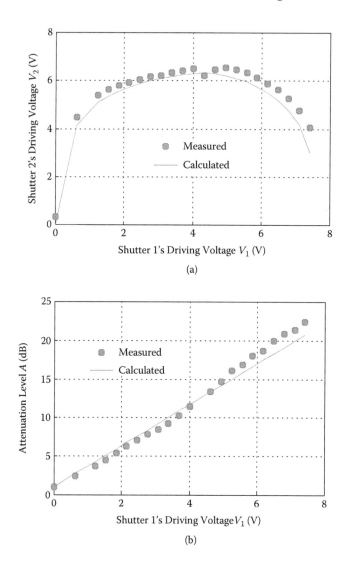

FIGURE 5.28
Demonstration of linear tuning schemes. (a) Comparison of the measured and calculated values of V_2 when the values of V_1 and A are given, (b) comparison of the measured and calculated values of A when the values of V_1 and V_2 are given according to a look-up table.

The obtained attenuation is then compared with a linear one. That is, it gives V_1 and V_2 to test A. The obtained data for these two methods are shown in Figure 5.28a,b, respectively. One can see from Figure 5.28a that the measured voltages of shutter 2 are larger than the calculated value by less than 1 V. It may be due to the fact that light reflection through the shutter edges should be taken into account [61], which can be balanced by a slight extra-movement of shutter 2. Further experiment shows that this deviation could become more obvious (2–3 V), which tends to support this argument.

The measured attenuation curve obtained by the look-up table method is shown in Figure 5.28b. It is seen that the attenuation level develops with satisfactory linearity up to the voltage of 4.6 V, after which the attenuation curve shifts up almost uniformly by about 1.5 dB. In the good match region, the attenuation is slightly lower than the required values by less than 0.5 dB, which is obviously due to the same reason: light reflection by the shutter edges that accounts for the discrepancy in Figure 5.28a. For the region above 4.6 V, however, it is noticed that there is a turning point in measurement from which shutter 2 begins to move backward, as seen in Figure 5.28a. Therefore, the residue stress of actuators could be responsible for the discrepancy.

Because the measurement is conducted progressively from low to high voltage, the whole structure becomes sluggish when it begins moving back at the point 4.6 V; and therefore, the measured attenuation is slightly higher than the required attenuation, which is exactly the case in Figure 5.28b. The linear tuning performance can therefore be improved by suppressing residue stress through better actuator design. Anyway, the discrepancies in Figure 5.28a,b caused by the light reflection of shutter edges and the residue stress do not affect the actual applications of linear VOAs because they can be easily corrected by a look-up table from the experimental data.

The measured resolution performance of the device obtained by moving two shutters is illustrated in Figure 5.29. The attenuation level is chosen to be 11.1 dB, and the shutter positions are $\eta_1 = +1.0$ μm (moving from the initial position of -8 μm) and $\eta_2 = +6.2$ μm (from $+8$ μm), corresponding to the driving voltage of 6.6 and 2.5 V, respectively. Hence, while shutter 1 may be adjusted to provide attenuation fine-tuning, shuttle 2 may be used for coarse attenuation tuning. At this working point, the applied voltages are changed over a small range (<1.1 V), and the attenuation shift is examined. In Figure 5.29, two different resolutions of 2.5 and 0.1 dB/V are realized simultaneously. This represents a tuning resolution contrast of 25 times, which appears to be suitable for practical use. It is seen that the attenuation can be easily tuned by steps less than 0.01 dB (i.e., ultrafine tuning).

In Figure 5.29, the measured local attenuation curve (and therefore, the resolution) again agrees well with the calculated values within the small region considered. As the required working point for a given resolution contrast differs from the attenuation level, the calculation model can be used to choose the appropriate shutter positions. It is worth noting that the sensitivity of attenuation value with respect to the two shutters' movement is strongly dependent on their positions, that is, the selected working point. When the coarse-tuning

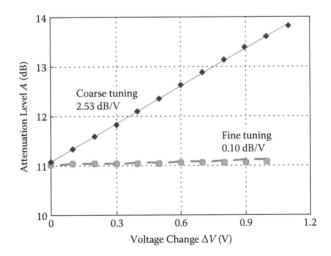

FIGURE 5.29
Demonstration of both coarse tuning (2.53 dB/V) and fine tuning (0.10 dB/V) in one device. Shutter 1 is under 6.6 V for coarse tuning, whereas shutter 2 is under 2.5 V for fine tuning. The starting attenuation is 11.1 dB.

shutter is moved by a large step, however, the attenuation varies steeply and the original working point is lost. A new working point has to be set.

5.7.2.4 Wavelength- and Polarization-Dependent Losses

Some other properties of the dual-shutter VOA are also characterized, such as the WDL and PDL. The WDL and PDL cannot be measured in a regular way because two tuning variables are involved. Their absolute values are strongly dependent on the positions of both shutters. Although the same level of attenuation can be realized by many combinations of two shutter positions (i.e., different aperture position and corresponding aperture size), the WDL could be very different. To investigate this aperture effect, the WDL at a nominal 25 dB level is measured as shown in Figure 5.30. In measurement, the size and position of aperture are varied but the attenuation at 1550 nm is kept unchanged. From Figure 5.30, it is interesting to see that the attenuation oscillates over the whole wavelength range when the aperture is moved away from the center. For some particular wavelength such as 1580 nm, the attenuation is independent of the aperture position (the independence at 1550 nm is because it is chosen as the reference). In contrast, at several wavelengths such as 1570 and 1600 nm, the attenuation changes sharply from a local maximum to a local minimum with the shift of the aperture. The reason for this phenomenon is still not clear. In regard to the influence of the aperture position, it can be seen from Figure 5.30 that the fluctuation is alleviated when the aperture is close to the beam center, leading to reduced wavelength dependence. The lowest PDL of this device is 0.8 dB at the 25 dB level over the wavelength range of 1520–1620 nm. Polarization dependence of the attenuation is measured with a deterministic method and also turns out to be a function of shutter positions even at the same attenuation level. However, its actual dependence on aperture position exhibits no special behavior. The attenuation variation on polarization is measured to be 1.0 dB at the 25 dB level.

5.7.3 Deep-Etched Elliptical Mirror VOAs

An elliptical mirror MEMS VOA was recently developed by Cai et al. [51]. In the design of the VOA, the parameters were chosen for low insertion loss and easy assembly. The mirror

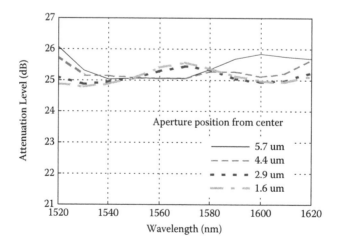

FIGURE 5.30
Influence of the aperture position on the wavelength-dependent loss. The nominal attenuation level is kept constant at 25 dB for the wavelength of 1550 nm regardless of the aperture position.

is a part of an ellipse that has 150 μm and 106 μm for its major and minor axes, respectively. The light path between the two foci is therefore 150 μm. The length of the major axis is $\sqrt{2}$ times that of the minor axis. This is to ensure that the input and output fibers are orthogonally related, which facilitates the assembly and positioning of fibers. To obtain symmetry of the overall VOA arrangement to the minor axis, the mirror center is chosen to be the vertex of the minor axis. Besides, this part of the ellipse has a small curvature. In the initial state, most of the light power is reflected by the central part of the mirror, which is close to a flat mirror. The orthogonal positioning of fibers allows a relatively large tolerance for the fiber alignment. A misalignment of 1 μm in the short axis direction introduces about 0.2 dB loss. In contrast, if the mirror center is at the vertex of the long axis (having large curvature), the same misalignment causes 0.7 dB loss. The mirror has a length of 125 μm, which is large enough to handle the light beam (spot size is about 15 μm in diameter on the mirror). To obtain variable attenuation, an electrostatic comb drive is used to displace the mirror along the axis of the input fiber.

A packaged VOA is shown in Figure 5.31a. MEMS structures, including the elliptical mirror and the electrostatic comb-drive actuator, are fabricated on an SOI wafer using the DRIE process. The structures are released by backside etching. The mirror surface is coated with a layer of gold (0.2 μm thick) using the shadow mask, the reflectivity is more

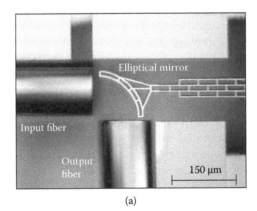

(a)

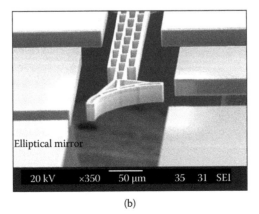

(b)

FIGURE 5.31
Micrographs of the VOA using an elliptical mirror. (a) A packaged VOA with fibers attached, (b) close-up of the elliptical micromirror.

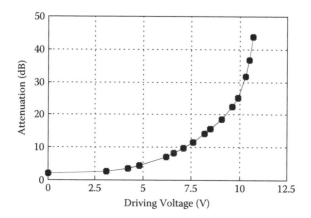

FIGURE 5.32
Optical attenuation versus driving voltage.

than 95%. A close-up of the elliptical mirror is shown in Figure 5.31b. The fiber grooves are wide enough for active alignment. The input and output fibers are first positioned roughly at the marks labeled near the elliptical mirror, then finely adjusted by monitoring the insertion loss, and finally fixed to the substrate by the UV-curable epoxy.

The attenuation versus the driving voltage and the mirror displacement are plotted in Figures 5.32 and 5.33, respectively. At the initial position, there is an insertion loss of 1.0 dB. When the applied voltage is increased, the attenuation increases accordingly and reaches its maximum (44 dB) at 10.7 V. In this maximum state, the elliptical mirror is displaced by 20 μm along the input fiber axis. As a result, the center of the focused beam is shifted by about 21.5 μm from the center of the output fiber. The beam shape is quasi-elliptical; the major and minor axes are estimated to be 33 and 14 μm, respectively. It is observed from Figure 5.33 that the attenuation varies almost linearly with the mirror displacement up to an attenuation level of 30 dB. Compared with the strong nonlinearity in most of the

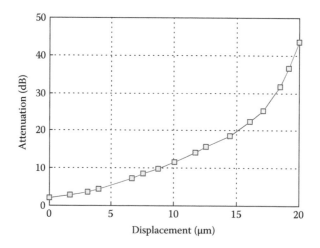

FIGURE 5.33
Optical attenuation versus mirror displacement, which is linear over the first 30 dB range.

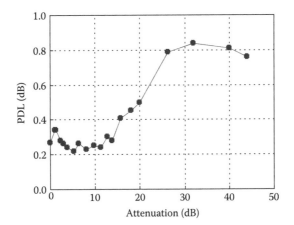

FIGURE 5.34
Polarization-dependent losses of the elliptical mirror VOA at different attenuation levels.

other VOAs such as the single-shutter VOAs [2,16,18,31,32], symmetric multishutter VOAs [17,21], and direct-coupling VOAs [24–26], linearity is a special merit of this VOA. This feature facilitates the application of VOA control in optical networks. In the dynamic test, the VOA measures a response time of 220 μs when a step-driving voltage (from 0 to 10 V) is applied.

The PDL is measured at different attenuation levels (at a wavelength of 1550 nm) as shown in Figure 5.34. It can be seen that it remains below 0.85 dB over the entire 44 dB attenuation range. It is 0.3 dB in the first 15 dB range and experiences fast increases between the levels of 15 and 25 dB. At a higher attenuation level, the PDL settles at the level of 0.8 dB. The relationship between the WDL and the attenuation is shown in Figure 5.35. The wavelength range is from 1520 to 1620 nm (covering the C-band and L-band). The WDL remains below 1.3 dB when the attenuation is less than 20 dB. Then it is increased rapidly to 4 dB at the range of 20–30 dB. Over the rest of the attenuation range, it rises up to 4.7 dB. The value of WDL is comparable to the typical data of reflection-type VOAs at low attenuation level, but

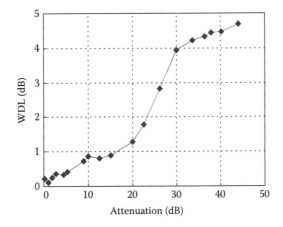

FIGURE 5.35
Wavelength-dependent losses of the elliptical mirror VOA at different attenuation levels (from 1520 to 1620 nm).

TABLE 5.3

Measured Specifications of the Fabricated VOA

Specification	Value
Insertion loss	1.0 dB
Maximum attenuation	44 dB @ 10.7 V
PDL	0.5 dB @ 20 dB
WDL	1.3 dB (1520 ~ 1620 nm) @ 20 dB
Linearity (attenuation level versus mirror displacement)	Linear attenuation over 30 dB
Response time	220 μs

a bit larger than those at a high level (e.g., 0.7 dB at 25 dB level) [41–49]. This could be due to the scattering of the etched mirror surface (higher roughness at larger depth position) and the diffraction in the mirror edge (limited mirror depth). Improvement of fabrication would help reduce the WDL.

The performance of the VOA is summarized in Table 5.3. Compared with a recent work involving a reflection VOA that uses lensed fibers [49], this VOA has a comparable performance in terms of attenuation range and the PDL. It also has better performance in driving voltage, linearity, and response time. This VOA uses only normal single-mode fibers. The relatively large value of the insertion loss is mainly due to the lack of focusing in the vertical direction resulting from deep etching. It can be reduced by shortening the fiber separation.

5.7.4 Deep-Etched Parabolic Mirror VOAs

An integrated parabolic mirror VOA (PVOA) was presented recently [52]. It consists of a pair of parabolic mirrors arranged symmetrically; each mirror is a section of the parabola. According to the simulation, the optimized design value is $p = 250$ μm (the parabola $y^2 = 2px$) and the rotation arm length is $\rho = 330$ μm from the apex of the mirror pair. A rotary comb-drive actuator is employed to rotate the parabolic mirror pair. The MEMS mirror pair is fabricated by the DRIE technique on an 8 in. SOI wafer (structural layer 75 μm). To release the movable parts, dry release method and backside open are both implemented, but they do not result in any significant difference in device performance. The etched quality is determined by measuring a reference flat sidewall. In the depth direction, the sidewall surface is smooth in the top 40 μm region (RMS roughness 20–30 nm) and gradually deteriorates toward the bottom. The fidelity of the fabricated mirror profile is measured by taking SEMs of the top and bottom views of the mirror pair after breaking it out. The image process is used to fit the profiles to the ideal parabola. The standard deviation is 0.5% at the top edge and 4% at the bottom edge. As the top region of the sidewall is the main reflection surface in the initial state, the bottom region has a negligible effect on the insertion loss.

The parabolic mirrors have an open angle of 60° with respect to the fibers, large enough to handle the laser beam in the horizontal plane. A section of single-mode fiber (ϕ125 μm) is used as the rod lens. The separation from input to output is 1100 μm. The insertion loss would be >13 dB without the rod lens. The rotary actuator for driving the mirror pair follows the same design of the previous work [62]. The rotation pivot is provided by a double-clamped beam as discussed in Reference 62. It has 40 pairs of comb fingers (width 2 μm and separation 2.5 μm). An ideal virtual pivot is provided by a double-clamped beam (length 150 μm, width 3 μm). The actuator measures a rotation angle of 3° under 25 V (the measured actuation curve $\theta = 4.8 \times 10^{-3} V^2$, V in volt and θ in degree). For convenience of testing, a

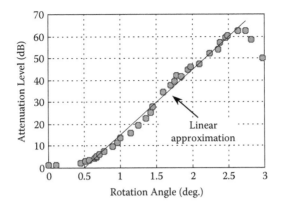

FIGURE 5.36
The measured attenuation versus the mirror rotation angle.

strong actuator is selected. For real-world applications, the voltage can be easily reduced to <3 V by using more comb fingers or a longer pivot beam (i.e., lower rotational stiffness).

The attenuation relationship is shown in Figure 5.36, in which the attenuation level and rotation angle are in units of dB and degree, respectively. For the rotation angle <0.5°, the attenuation does not change much. After that, it rises up to as high as 62.6 dB when the rotation angle is increased to 2.6°. An attenuation range of 62 dB is obtained. Further rotation causes a reduction of attenuation. A linear relationship can be seen over the entire range ($y = 29.66 x - 14.53$, y is the attenuation in dB, and x is the angle in degree).

The WDL is measured over 1520–1620 nm. It is 0.3 dB at the 20 dB attenuation level, 0.7 dB at the 40 dB level, and goes up to 1.3 dB at the maximum attenuation level. The PDL is measured by a PDL multimeter. Over the whole attenuation range, it is always <0.5 dB. It is only about 0.1 dB at the 20 dB attenuation level, and 0.3 dB at the 40 dB level. For easy comparison, the specifications are listed in Table 5.4. The PVOA has a temperature dependence of 0.01 dB/K at the 20 dB level.

It is worth noting that the properties of linear relationship and large range of the PVOA come from the combination of the retroaxial parabolic mirror configuration and the rotary actuator [46,49]. Compared with some recent studies [46,48], this PVOA has similar insertion loss, such as PDL and WDL, but enjoys larger attenuation range, and particularly, a linear relationship over the entire attenuation range. It is noted that the obtained linear attenuation is relative to the mirror rotation, not to the driving voltage of the rotary comb drive that is more desirable for control. It is possible to obtain the linear relationship directly between the attenuation and the driving voltage by properly designing the parabolic mirror profile and the rotation arm length.

TABLE 5.4

Specifications of the MEMS VOA with the Parabolic Mirror Design

Specification	Value
Insertion loss	0.6 dB
Attenuation range	62 dB
PDL	0.1 dB @ 20 dB, 0.3 dB @ 40 dB
WDL (over 100 nm)	0.3 dB @ 20 dB, 0.7 dB @ 40 dB
Linearity	Good linearity over the whole range

5.7.5   Experimental Studies on TDL, WDL, and PDL

5.7.5.1   *Device Description and Measured Characteristics*

For studying the influence of the mirror configuration on the dependent losses, three types of VOAs are fabricated by the DRIE on SOI wafers, followed by the integration of normally cleaved single-mode optical fibers using UV-curable epoxy. The three VOAs are SVOA, FVOA, and EVOA. This work was presented recently in Reference [23]. The structure layer has a depth of 80 μm. Figure 5.37 shows the close-ups of the devices; all of them share the same structure (including a bidirectional comb-drive actuator, electrical pads,

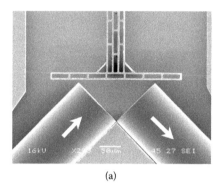

(a)

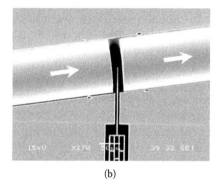

(b)

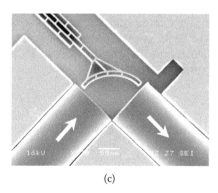

(c)

FIGURE 5.37
Scanning electron micrographs of packaged MEMS VOAs. (a) Flat mirror VOA, (b) shutter-type VOA, and (c) elliptical mirror VOA.

and wirings) except for the mirrors and the corresponding fiber positioning. To avoid the influence of residue stress on the temperature dependence study, a shadow mask is used to coat only the mirror part, the pads, and wires with a thin gold layer (0.2 μm thick). The FVOA uses the sidewall of a frame as the mirror, as shown in Figure 5.37a; this is to reduce the bending and thus ensure high flatness in the presence of gold coating and residual stress. The fibers are aligned 45° relative to the mirror surface, with an initial light path length of about 150 μm. In the SVOA, the mirror (more accurately, the shutter) is formed by a silicon slab (4 μm thick), as shown in Figure 5.37b. The coaxial fibers have a separation of about 20 μm. The optical axis is tilted by 8° relative to the shutter, which is a common design to reduce the return loss [32]. The EVOA in Figure 5.37c has the same mirror design as that in Section 5.7.3 [51], but the actuator is different. The mirror is seated on the inner surface of a frame structure and has an elliptical shape, with the input and output fibers positioned at the two foci of the ellipse, respectively, as shown in Figure 5.37c. The two fibers are separated also by 150 μm. As the ellipse provides an ideal mutual coupling between the foci in the horizontal plane, the insertion loss of EVOA is ideally half that of FVOA. Before fixation by the UV epoxy, the fibers are roughly aligned by the trenches and markers, and then finely adjusted by monitoring the coupling power. Minimal insertion losses are 0.85 dB, 4.6 dB, and 2.5 dB for SVOA, FVOA, and EVOA, respectively. After UV curing, the losses are increased to 1.2, 5.0, and 3.2 dB, respectively. The SVOA measures the lowest insertion loss, whereas the FVOA has the highest. Large insertion losses of the FVOA and EVOA are due to the long fiber separation (150 μm), because they both suffer divergence in the vertical direction (i.e., out of the plane). Their insertion losses can be reduced to less than 1 dB by introducing an optical fiber as the rod lens to form a 3-dimensional coupling system to converge the light [53]. However, this section does not introduce this design because the primary target is to study the dependent losses.

5.7.5.2 *Temperature-Dependent Losses*

The measured and simulated relationships between the mirror position and attenuation are shown in Figure 5.38a. The input has a wavelength of 1550 nm and a fixed polarization state. Although the attenuation range and linearity are not among the prime interests of this chapter, they affect the TDL significantly and thus are examined in detail. It can be seen from Figure 5.38a that the SVOA experiences a slow increase of attenuation level before the shutter position reaches 5 μm (relative to the fiber axis). After that, the attenuation shoots up and achieves a maximum attenuation of 46.6 dB (dynamic range about 45 dB). The simulation curve for SVOA is calculated on the basis of the theoretical model in Reference 20. For the FVOA, the attenuation is altered gradually over most of the range, but shoots up at the final stage. A mirror displacement of 19.4 μm yields an attenuation of 42.5 dB (dynamic range about 37 dB). The simulation of FVOA is from the fiber-coupling model in Reference 14. For the EVOA, the attenuation level increases steadily with mirror displacement, and reaches 45.4 dB (dynamic range about 42 dB) with a mirror translation of 16.4 μm. It can be seen from Figure 5.38a that very good linearity is maintained over the entire attenuation range. This is a special merit of the EVOA. The simulation result of EVOA using the ZEMAX software also suggests a better linearity than the SVOA and FVOA. Although the three simulated curves do not closely match the corresponding experimental data, perhaps owing to oversimplification, they describe the general trend. It is verified that the dynamic range of the SVOA is limited by the finite depth of the shutter (the upper edge is only about 17.5 μm over the fiber axis). A similar experiment is carried out by inserting a razor blade into two fibers, which obtains a dynamic range of more than

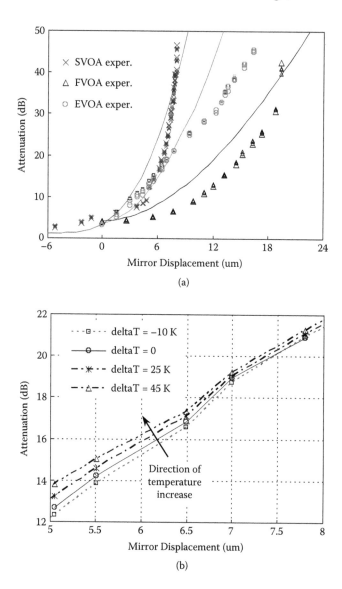

FIGURE 5.38
Relationship between the attenuation and mirror displacement. (a) Measured data and simulated curves for MEMS VOAs, and (b) a small region of attenuation data of the EOA at different temperatures.

70 dB. With FVOA and EVOA, the dynamic ranges might be limited by the diffraction at the edges of mirrors and the scattering on the rough surface.

The attenuation varies with the temperature even if the mirror position is kept constant. The measured points in Figure 5.38a are collections of data at different temperatures (i.e., $\Delta T = -10$ K, 0, 25 K, and 45 K relative to the room temperature $22 \pm 0.2°C$) by sending current through a TEC cooler. A magnified view is shown in Figure 5.38b for the EVOA. It can be observed that attenuation tends to increase with temperature, and this relationship is found in all the three packaged VOAs. Figure 5.39 presents the relationship between the TDL and the attenuation level at different temperatures. The straight lines represent the linear approximation. Here, the TDL is defined as the attenuation value of

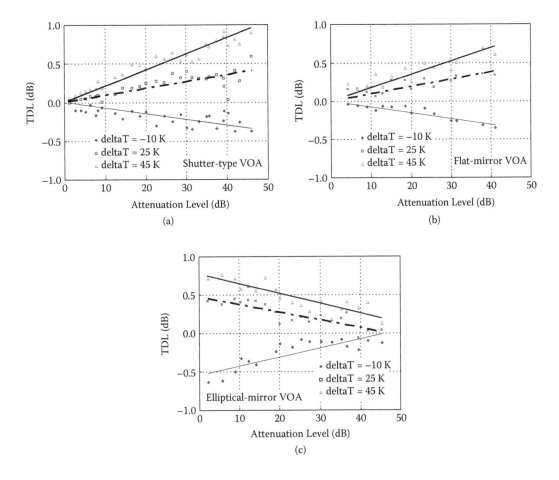

FIGURE 5.39
Temperature-dependent losses of MEMS VOAs at different temperatures. The straight lines represent the linear approximations. (a) Shutter-type VOA, (b) flat-mirror VOA, and (c) elliptical-mirror VOA.

current temperature relative to the value of room temperature. From Figure 5.39a, the TDL of SVOA is small at the low attenuation level, and the absolute value increases with higher attenuation level. The data points are quite divergent; however, linear relationships can be observed, as represented by trend lines. At the attenuation of 20 dB, the TDL values are −0.12, 0.23, and 0.45 dB for the temperature conditions of −10 K, 25 K, and 45 K, respectively. At the maximum attenuation, they are −0.38, 0.59, and 0.90 dB, respectively. In other words, when the temperature is increased by 55 K, the value of attenuation is changed by 0.67 dB at the 20 dB level and by 1.28 dB at about the maximum attenuation level.

The FVOA exhibits a similar TDL change as the SVOA. However, the TDL is not small at the low attenuation level, but the slope is less steep in comparison to that of the SVOA. For example, at the 20 dB attenuation level, the TDLs are −0.16, 0.15, and 0.26 dB for the three temperature conditions, respectively; whereas at the maximum attenuation level, they are −0.35, 0.33, and 0.60 dB, respectively. Correspondingly, the 55 K temperature change induces an attenuation variation of 0.42 dB at the 20 dB level, and 0.95 dB at the maximum attenuation level.

The EVOA exhibits a different temperature dependence in comparison to the SVOA and FVOA, as shown in Figure 5.39c. The TDL is large at the low attenuation level but

TABLE 5.5

Comparison of the Specifications of the Three Types of Packaged MEMS VOAs

Specifications	SVOA	FVOA	EVOA
Insertion loss	1.2 dB	5.0 dB	3.2 dB
Dynamic range	45 dB	37 dB	42 dB
Linearity	Highly nonlinear	Nonlinear	Linear
TDL for $\Delta T = 55$ K			
At 20 dB attenuation level	0.67 dB	0.42 dB	0.61 dB
Maximum TDL	1.28 dB	0.95 dB	1.36 dB
WDL over 100 nm			
At 20 dB attenuation level	0.51 dB	0.54 dB	0.41 dB
Maximum WDL	2.10 dB	0.84 dB	0.65 dB
PDL			
At 20 dB attenuation level	0.55 dB	1.10 dB	0.73 dB
Maximum PDL	4.8 dB	2.13 dB	2.56 dB

gradually reduces with increase of attenuation level. This trend is the same for all the three temperature conditions. At the attenuation level of about 20 dB, the TDLs measure −0.14, 0.12, and 0.47 dB, respectively. The attenuation variation corresponding to the 55 K temperature change is 0.61 dB, which is comparable to that of the SVOA but larger than that of the FVOA. At the maximum attenuation level, the TDLs are reduced to −0.12, 0.04, and −0.13 dB, respectively, resulting in an attenuation of 0.25 dB for 55 K temperature change. This value is smaller than those of the SVOA and FVOA. The maximum attenuation TDL for a 55 K temperature change reaches 1.36 dB at the low attenuation level. For comparison, the specifications of the packaged MEMS VOAs are summarized in Table 5.5.

5.7.5.3   *Wavelength-Dependent Losses*

The attenuation varies with the wavelength of the input light even if the mirror position is fixed. Figure 5.40a shows an example of the relative attenuation change over the wavelength range of 1520–1620 nm when the VOAs are set at an attenuation level of 37 dB, which roughly corresponds to the largest WDLs for all the VOAs. For the SVOA, the curve shows a quasi-periodic change with a peak-to-peak difference of 2.1 dB. For the FVOA and EVOA, the curves reach their minima at 1580 nm and rise at other wavelengths. Peak-to-peak differences are reduced to 0.84 and 0.69 dB, respectively. The WDL over the full attenuation range is shown in Figure 5.40b. At low attenuation level, the three types of VOAs have a small WDL between 0.1 and 0.4 dB. At an attenuation of 20 dB, the WDLs are increased to 0.51, 0.54, and 0.41 dB for the SVOA, FVOA, and EVOA, respectively. These values of WDLs are comparable to the typical data of 0.2–1.2 dB in previous studies [45,47,51], and are also listed in Table 5.5. In the following range, SVOA experiences a rapid increase and then a drop of the WDL, whereas the FVOA and EVOA exhibit a smooth increase. It can be observed that the SVOA has the largest WDL, whereas the FVOA and EVOA have similar values.

5.7.5.4   *Polarization-Dependent Losses*

The value of PDL is commonly obtained by measuring the power of the state of polarization (SOP), which is generated using a polarization controller. The difference of the power extrema is regarded as the PDL. As the PDL is quite small and could be easily

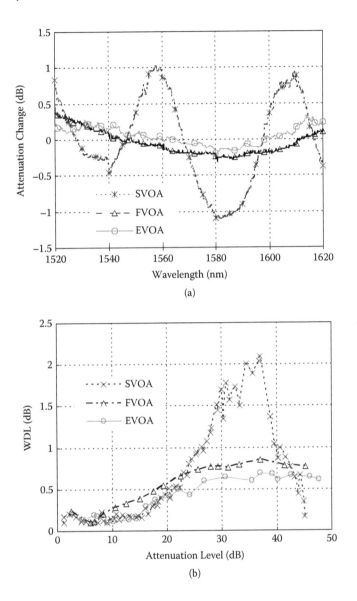

FIGURE 5.40
Wavelength-dependent losses of MEMS VOAs. (a) Variation of the attenuation over a wavelength range of 100 nm when the attenuation is about 37 dB, and (b) change of the wavelength-dependent loss with attenuation level.

overwhelmed by the noise and fluctuation of the photodetector readings, the all-state measurement tends to give values larger than the real data. To overcome this problem, a deterministic method is adopted, as shown in Figure 5.41a. The light from a laser source is isolated and then passes a polarization controller. An inline polarometer is used to monitor the SOP before the light enters the device under test (DUT). Four well-defined polarization states (i.e., *x* polarized, *y* polarized, +45°, and right-hand circular polarizations) are generated; the corresponding output powers are measured by an optical power meter. The PDL can be calculated by the Mueller matrix method. In this method, the exact SOP of the input beam right at position of the shutter/mirrors is still unknown because the fiber from

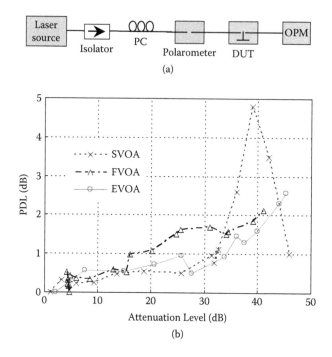

FIGURE 5.41

Polarization-dependent losses of MEMS VOAs. (a) Setup for deterministic measurement of polarization-dependent losses. PC, DUT, and OPM stand for polarization controller, device under test, and optical power meter, respectively; and (b) comparison of polarization-dependent losses.

the polarometer to DUT and connectors in between may introduce additional change to the SOP. Therefore, the input fiber has to be regarded as part of the MEMS VOA device, and the PDL represents the whole packaged VOA device.

The measured PDLs of the three VOAs as a function of the attenuation level are shown in Figure 5.41b. The SVOA has a small PDL before the attenuation level of 30 dB, and then shoots up to 4.8 dB at the attenuation of about 38 dB, followed by a rapid drop to 1.0 dB. In contrast, the FVOA and EVOA experience a gradual increase in the PDL over the entire attenuation range. At the attenuation of 20 dB, the PDLs are 0.55, 1.10, and 0.73 dB for the SVOA, FVOA, and EVOA, respectively. The maximum PDLs over the entire attenuation range are 4.80, 2.13, and 2.56 dB, respectively. These values are also comparable to the typical data of 0.4–2.9 dB in previous works [37,45,49,51].

On the basis of the previous experimental results and theoretical analyses, it can be seen that the effects of physical mechanisms are also dependent on the VOA configuration. With respect to the TDL, the fiber misalignment due to thermal expansion of UV epoxy gives rise to the TDL at low attenuation level, whereas the mirror shift due to thermal softening of silicon material determines the increase of TDL when the temperature and attenuation level go higher, as predicted in Section 5.6. The VOA configuration affects both the trend and amount of the TDL change. In regard to the WDL, the diffraction of chopped Gaussian beam and the Rayleigh scattering of the rough surface/edge affect all VOAs, but the SVOA is more seriously affected, and the interference of the fiber-end reflection and stray lights further reduces SVOA performance further. On the topic of the PDL, the high conductivity of gold is the cause; it plays its role in the SVOA through the knife-edge diffraction and in the FVOA and EVOA through Fresnel reflection.

Some rules for VOA design can be inferred from measurement results and theoretical analyses. To minimize the TDL, temperature stabilization should be introduced to MEMS VOAs. If the stabilized temperature is higher than the ambient temperature, the hot airflow may cause a serious displacement fluctuation of the comb drive. To tackle this problem, encapsulation should be implemented in MEMS VOAs. To minimize the WDL, chopping of the Gaussian beam should be avoided. A preferred method is to obtain attenuation by steering away the Gaussian beam while maintaining its integrity, as in reflection-type VOAs. However, the mirror should be large enough to avoid trimming the laser beam. To minimize the PDL, it is recommended that dielectric materials be used rather than gold for light absorption (in shutter-type VOAs) and reflection (for reflection-type VOAs). This would also solve the depolarization problem. As subwavelength defects of the mirror affect the WDL and PDL through Rayleigh scattering, the mirror surface should be very smooth. All these considerations could suppress the dependences more or less, but cannot eliminate them. To further improve the VOA specifications, closed-loop control may have to be used to monitor the output power and to make adjustment in real time.

5.8 Summary

In conclusion, this chapter has investigated different aspects of MEMS VOAs, including configurations, classifications, applications, designs, optical analyses, and experimental studies. Empowered by advanced fabrication technologies and integration capabilities, different types of VOAs have been demonstrated based on the fiber direct coupling, laser interference, diffraction, refraction, and reflection; each has several different designs. Most of them meet the basic requirements of typical applications in regard to insertion loss, dynamic range, and response time while excelling in various aspects of advanced requirements, and well exploiting and demonstrating the advantages of MEMS technology over conventional optomechanical and waveguide-based VOAs. Several VOAs have been successfully commercialized, and many others are on the way. Recent development has led to higher levels of performance. The recently proposed asymmetric dual-shutter VOAs and parabolic mirror VOAs facilitate linear attenuation relationship and ultrafine tuning schemes. Theoretical analyses of the attenuation models, tuning schemes, and dependent losses enable deeper understanding of the underlying physical mechanisms, and guide the further improvement of VOA designs and specifications. The experimental studies, though mostly academic, well demonstrate the vast scope for innovating new VOA designs. With full-fledged research and development effort, MEMS VOAs could be one of the killer applications of MEMS.

References

1. Wu, M. C., Solgaard, O., and Ford, J. E., Optical MEMS for lightwave communications, *J. Lightwave Technol.*, 24, 4433, 2006.
2. Giles, C. R., Aksyuk, V., Barber, B., Ruel, R., Stulz, L., and Bishop, D., A silicon MEMS optical switch attenuator and its use in lightwave subsystems, *IEEE J. Sel. Topics Quantum Electron.*, 5, 18, 1999.
3. Lee, S.-S., Jin, Y.-S., Son, Y.-S., and Yoo, T.-K., Polymetric tunable optical attenuator with an optical monitoring tap for WDM transmission network, *IEEE Photon. Technol. Lett.*, 11, 590, 1999.

4. Ma, X. and Kuo, G. S., Optical switching technology comparison: optical MEMS versus other technologies, *IEEE Opt. Commun.*, 41, S16, 2003.

5. Petersen, K. E., Micromechanical light modulator array fabricated on silicon, *App. Phys. Lett.*, 31, 521, 1977.

6. Hornbeck, L. J., Deformable-mirror spatial light modulators, In *Proc. Spatial Light Modulators and Applications III, SPIE Critical Reviews (SPIE 1150)*, San Diego, California, 1989, 86.

7. Solgaard, O., Sandejas, F. S. A., and Bloom, D. M., Deformable grating optical modulator, *Opt. Lett.*, 17, 688, 1992.

8. Goossen, K. W., Walker, J. A., and Arney, S. C., Silicon modulator based on mechanically-active anti-reflection layer with 1 Mbit/s capability for fiber in the loop applications, *IEEE Photon. Technol. Lett.*, 6, 1119, 1994.

9. Marxer, C., Thio, C., Grétillat, M.-A., de Rooij, N. F., Bättig, R., Anthamatten, O., Valk, B., and Vogel, P., Vertical mirrors fabricated by deep reactive ion etching for fiber-optic switching applications, *IEEE J. Microelectromech. Syst.*, 6, 277, 1997.

10. Lee, C., Lin, Y.-S., Lai, Y.-J., Tasi, M. H., Chen, C., and Wu, C.-Y., 3-V driven pop-up micromirror for reflecting light toward out-of-plane direction for VOA applications, *IEEE Photon. Technol. Lett.*, 16, 1044, 2004.

11. Chen, R. T., Nguyen, H., and Wu, M. C., A high-speed low-voltage stress-induced micromachined 2×2 optical switch, *IEEE Photon. Technol. Lett.*, 11, 1396, 1999.

12. Ford, J. E., Walker, J. A., Greywall, D. S., and Goossen, K. W., Micromechanical fiber-optic attenuator with 3 μs response, *J. Lightwave Technol.*, 16, 1663, 1998.

13. Ford, J. E. and Walker, J. A., Dynamic spectral power equalization using micro-opto-mechanics, *IEEE Photon. Technol. Lett.*, 10, 1440, 1998.

14. Yuan, S. and Riza, N. A., General formula for coupling-loss characterization of single-mode fiber collimators by use of gradient-index rod lenses, *Appl. Opt.*, 38, 3214, 1999.

15. Born, M. and Wolf, E., *Principles of Optics*, 6th ed., Pergamon University Press, Oxford, 1999.

16. Barber, B., Giles, C. R., Askyuk, V., Ruel, R., Stulz, L., and Bishop, D., A fiber connectorized MEMS variable optical attenuator, *IEEE Photon. Technol. Lett.*, 10, 1262, 1998.

17. Li, L. and Uttamchandani, D., Design and evaluation of a MEMS optical chopper for fibre optic applications, *IEE Proc. Sci. Meas. Technol.*, 151, 77, 2004.

18. Liu, A. Q., Zhang, X. M., Lu, C., Wang, F., Lu, C., and Liu, Z. S., Optical and mechanical models for a variable optical attenuator using a micromirror drawbridge, *J. Micromech. Microeng.*, 13, 400, 2003.

19. Zhang, X. M., Liu, A. Q., and Lu, C., New near-field and far-field attenuation models for free-space variable optical attenuators, *J. Lightwave Technol.*, 21, 3417, 2003.

20. Zhang, X. M., Liu, A. Q., Lu, C., Wang, F., and Liu, Z. S., Polysilicon micromachined fiber-optical attenuator for DWDM applications, *Sens. Actuators A*, 108, 28, 2003.

21. Syms, R. R. A., Zou, H., Stagg, J., and Veladi, H., Sliding-blade MEMS iris and variable optical attenuator, *J. Micromech. Microeng.*, 14, 1700, 2004.

22. Zhao, Q. W., Zhang, X. M., Liu, A. Q., Cai, H., Zhang, J., and Lu, C., Theoretical and experimental studies of MEMS dual-shutter VOA for linear attenuation relationship and ultra-fine tuning, *J. Lightwave Technol.*, (in press).

23. Song, W. Z., Zhang, X. M., and Liu, A. Q., Theoretical and experimental studies of polarization dependence loss of shutter-type MEMS VOAs, In *Proc. Asia-Pacific Conference of Transducers and Micro-Nano Technology (APCOT 2006)*, Singapore, 2006, paper 95-OMN-A0119.

24. Haake, J., Wood, R., and Duhler, V., In-package active fiber optic microaligner, In *Proceedings of SPIE*, 3276, 207, 1998.

25. Unamuno, A. and Uttamchandani, D., MEMS variable optical attenuator with vernier latching mechanism, *IEEE Photon. Technol. Lett.*, 18, 88, 2006.

26. Syms, R. R. A., Zou, H., Yao, J., Uttamchandani, D., and Stagg, J., Scalable electrothermal MEMS actuator for optical fibre alignment, *J. Micromech. Microeng.*, 14, 1633, 2004.

27. Godil, A., Diffractive MEMS technology offers a new platform for optical networks, *Laser Focus World*, 38, 181, 2002.

28. Marxer, C., Grétillat, M. A., Jaecklin, V. P., Baettig, R., Anthamatten, O., Vogel, P., and de Rooij, N. F., Megahertz opto-mechanical modulator, *Sens. Actuators A*, 52, 46, 1996.

29. Yun, S.-S. and Lee, J.-H., A micromachined in-plane tunable optical filter using the thermo-optic effect of crystalline silicon, *J. Micromech. Microeng.*, 13, 721, 2003.

30. Heikkinen, V., Aikio, J. K., Alojoki, T., Hiltunen, J., Mattila, A.-J., Ollila, J., and Karioja, P., Single-mode tuning of a 1540-nm diode laser using a Fabry-Pérot interferometer, *IEEE Photon. Technol. Lett.*, 16, 1164, 2004.

31. Zhang, X. M., Liu, A. Q., Lu, C., and Tang, D. Y., MEMS variable optical attenuator using low driving voltage for DWDM systems, *Electron. Lett.*, 38, 382, 2002.

32. Marxer, C., Griss, P., and de Rooij, N. F., A variable optical attenuator based on silicon micro-mechanics, *IEEE Photon. Technol. Lett.*, 11, 233, 1999.

33. Li, L., Zawadzka, J., and Uttamchandani, D., Integrated self-assembling and holding technique applied to a 3-D MEMS variable optical attenuator, *IEEE J. Microelectromech. Syst.*, 13, 83, 2004.

34. Syms, R. R. A., Zou, H., Stagg, J., and Moore, D. F., Multistate latching MEMS variable optical attenuator, *IEEE Photon. Technol. Lett.*, 16, 191, 2004.

35. Hashimoto, E., Uenishi, Y., Honma, K., and Nagaoka, S., Micro-optical gate for fiber optic communication, in *Proc. Int. Conf. Solid-State Sens. Actuators (Transducers '97)*, Chicago, 1997, 331.

36. Debéda, H., Freyhold, T. V., Mohr, J., Wallrabe, U., and Wengelink, J., Development of miniaturized piezoelectric actuators for optical applications realized using LIGA technology, *IEEE J. Microelectromech. Syst.*, 8, 258, 1999.

37. Lee, J. H., Kim, Y. Y., Yun, S. S., Kwon, H., Hong, Y. S., Lee, J. H., and Jung, S. C., Design and characteristics of a micromachined variable optical attenuator with a silicon optical wedge, *Opt. Commun.*, 221, 323, 2003.

38. Lee, J. H., Yun, S.-S., Kim, Y. Y., and Jo, K.-W., Optical characteristics of a refractive optical attenuator with respect to the wedge angles of a silicon optical leaker, *Appl. Opt.*, 43, 877, 2004.

39. Medina, M., Schreiler, D., Kin, D., Glushko, B., Kryov, S., and Ben-Gad, E., U.S. Patent Application 2004/0136680 A1, 2004.

40. Glushko, B., Krylov, S., Medina, M., and Kin, D., Insertion type MEMS VOA with two transparent shutters, In *Proc. Asia-Pacific Conference of Transducers and Micro-Nano Technology (APCOT 2006)*, Singapore, 2006, Paper 95-OMN-A0594.

41. Riza, N. A. and Sumriddetchkajorn, S., Fault-tolerant variable fiber-optic attenuator using three-dimensional beam spoiling, *Opt. Commun.*, 185, 103, 2000.

42. Andersen, B. M., Fairchild, S., and Thorsten, N., MEMS variable optical attenuator for DWDM optical amplifiers, In *Proc. Optical Fiber Commun. Conf. (OFC 2000)*, Baltimore, MD, 2000, 2, 260.

43. Isamoto, K., Kato, K., Morosawa, A., Chong, C., Fujita, H., and Toshiyoshi, H., A 5-V operated MEMS variable optical attenuator by SOI bulk micromachining, *IEEE J. Sel. Topics Quantum Electron.*, 10, 570, 2004.

44. Chen, C., Lee, C., and Lai, Y.-J., Novel VOA using in-plane reflective micromirror and off-axis light attenuation, *IEEE Opt. Commun.*, 41, S16, 2003.

45. Kim, C.-H. and Kim, Y.-K., MEMS variable optical attenuator using a translation motion of 45° tilted vertical mirror, *J. Micromech. Microeng.*, 15, 1466, 2004.

46. Yeh, J. A., Jiang, S.-S., and Lee, C., MOEMS variable optical attenuators using rotary comb drive actuators, *IEEE Photon. Technol. Lett.*, 18, 1170, 2006.

47. Lim, T.-S., Ji, C.-H., Oh, C.-H., Kwon, H., Yee, Y., and Bu, J.U., Electrostatic MEMS variable optical attenuator with rotating folded micromirror, *IEEE J. Sel. Topics Quantum Electron.*, 10, 558, 2004.

48. Chen, C., Lee, C., and Yeh, J. A., Retro-reflection type MOEMS VOA, *IEEE Photon. Technol. Lett.*, 16, 2290, 2004.

49. Lee, C. and Yeh, J. A., Development of electrothermal actuation based planar variable optical attenuators (VOAs), *J. Phys.: Conf. Series*, 34, 1026, 2006.

50. Sumriddetchkajorn, S. and Riza, N. A., Fault-tolerant three-port fiber-optic attenuator using small tilt micromirror device, *Opt. Commun.*, 205, 77, 2002.

51. Zhang, X. M., Liu, A. Q., Cai, H., Yu, A. B., and Lu, C., Retro-axial VOA using parabolic mirror pair, *IEEE Photon. Technol. Lett.*, 19, 692, 2007.

52. Cai, H., Zhang, X. M., Lu, C., Liu, A. Q., and Khoo, E. H., Linear MEMS variable optical attenuator using reflective elliptical mirror, *IEEE Photon. Technol. Lett.*, 17, 402, 2005.

53. Zhang, X. M., Cai, H., Lu, C., Chen, C. K., and Liu, A. Q., Design and experiment of 3-dimensional micro-optical system for MEMS tunable lasers, In *Proc. 19th IEEE Int. Conf. Micro Electro Mechan. Syst. (MEMS 2006)*, Istanbul, Turkey, 2006, 830.

54. *GR-910-CORE generic requirement for fiber optic attenuators*, Telcordia, December 1998.

55. Takahashi, M., Variable light attenuator of improved air-gap type with extremely low return light, In *Proc. IEEE Instrum. Meas. Technol. Conf. (IMTC '94)*, Hamamatsu, Japan, 1994, 947.

56. Legtenberg, R., Groeneveld, A. W., and Elwenspoek, M., Comb-drive actuators for large displacements, *J. Micromech. Microeng.*, 6, 320, 1996.

57. Wachtman, J. B., Jr., Tefft, W. E., Jr., Lam, D. G., and Apstein, C. S., Exponential temperature dependence of young's modulus for several oxides, *Phys. Rev.*, 122, 1754, 1961.

58. Varshni, V. P., Temperature dependence of the elastic constants, *Phys. Rev. B*, 2, 3952, 1970.

59. Morimoto, M., Morimoto, K., Sato, K., and Iizuke, S., Development of a variable optical attenuator (VOA) using MEMS technology, *Furukawa Rev.*, 23, 26, 2003.

60. Liu, A. Q., Murukeshan, V. M., Zhang, X. M., and Lu, C., U.S. Patent 6,788,843, 2004.

61. Maaty, H., Bashir, A., Saadany, B., and Khalil, D., Modelling and characterization of a VOA with different shutter thickness, In *Proc. 3rd Workshop Photon. Applic. Egyptian Engin. Faculties and Institutes*, Giza, Egypt, 2002, 117.

62. Zhang, X. M., Liu, A. Q., Lu, C., and Tang, D. Y., A real pivot structure for MEMS tunable lasers, *IEEE J. Microelectromech. Syst.*, 16, 269, 2007.

6

MEMS Discretely Tunable Lasers

Xuming Zhang, Hong Cai, and Ai-Qun Liu

CONTENTS

6.1 Introduction 238
6.2 Design of Micro-Optical-Coupling Systems of External Cavity 242
    6.2.1 Curved-Mirror Design and the Three-Dimensional
        Optical-Coupling System...242
    6.2.2 Optical Analysis of Curved-Mirror Configuration........................244
        6.2.2.1 Ray Transfer Matrix of Curved Mirror.............................245
        6.2.2.2 Coordinate System Conversion on Laser Facet 246
        6.2.2.3 Feedback-Coupling Efficiency Using the
               Mode-Coupling Method..247
6.3 Theory of Continuous Wavelength Tuning..250
    6.3.1 Variation of Laser Gain and Phase with Mirror Translation...........251
    6.3.2 Variation of Wavelength with Mirror Translation 252
    6.3.3 Comparison of the Extended Feedback Model with the Conventional
        Weak Feedback Model 253
    6.3.4 Wavelength-Tuning Range and Stability of Mirror Configuration 256
6.4 Theory of Discrete Wavelength Tuning...258
    6.4.1 Laser Modes and Frequency Shift..259
    6.4.2 Transition between Continuous and Discrete Tuning 260
    6.4.3 Conditions and Hopping Range of Discrete-Dominant Tuning 262
6.5 Characterization of MEMS Discretely Tunable Lasers..............................262
    6.5.1 Device Description 263
    6.5.2 Wavelength Tunability..264
    6.5.3 Mode Stability 266
6.6 Discussions on the Various Cavity and Integration Scheme Configurations...........266
    6.6.1 Short Cavity versus Long Cavity..267
    6.6.2 Hybrid Integration versus Single-Chip Integration 268
6.7 Summary..268
References 269

This chapter covers theoretical studies and experimental verifications of discretely tunable external cavity lasers using microelectromechanical systems (MEMS) technology. This chapter and the next two will each focus on a particular type of MEMS laser. The discretely tunable laser is considered first because of its relatively simple configuration.

In this chapter, MEMS discretely tunable lasers will be first reviewed in Section 6.1. Then micro-optical-coupling systems will be discussed in Section 6.2, which enable a full range of cavity length, from extremely short (<10 μm), very short (<100 μm), and short (<1 mm) to long, without the need for conventional bulky optical components (lenses, reflectors, etc.). In Sections 6.3 and 6.4, respectively, theoretical studies of continuous and discrete tuning behaviors will be elaborated. This is to provide theoretical guidance for the design of discrete-tuning-dominant lasers and an interpretation of the observed phenomena. Next, an experimental demonstration and characterization of an MEMS discretely tunable laser will be presented in Section 6.5. In Section 6.6, some guidelines will be given for the choice of cavity length and integration approaches.

6.1 Introduction

With the rapid development of MEMS technology, miniaturization of external cavity tunable lasers has recently attracted increasing research and development interest, and some have even led to commercial success [1–3]. MEMS technology makes use of photolithography to fabricate micromechanical structures that facilitate complicated movement with fast response, high accuracy, and enhanced mechanical stability. MEMS has a natural synergy with external cavity tunable lasers in various respects. For example, a semiconductor chip is about several hundred microns long, close to the overall size of a MEMS device (typically 1 mm, including all the functional parts such as microactuators and suspensions). The wavelength of most tunable lasers is about 0.4–2 μm, on the same scale as the MEMS fine feature. Some types of lasers need a displacement of only about half a wavelength to tune over the whole range [1,2]; such displacement can be easily obtained using microactuators. In addition, high precision and stable movement of the MEMS structures enable fine-tuning of the wavelength with high repeatability. Moreover, micromechanical structures make it feasible to form very short external cavities (<100 μm), which cannot be realized using bulky optical components. Most applications such as optical communications and biomedical studies require tunable lasers with a wide tuning range (>40 nm), continuous or stepwise (i.e., discrete) wavelength change, high speed (<1 ms), high accuracy/repeatability, high reliability/stability, etc. Some further require light weight, easy integration, and low cost. The introduction of MEMS to tunable lasers brings in significant improvement in most of these specifications. The small size of MEMS causes the lasers to be lightweight and portable, and makes it easy to form large arrays with enhanced mechanical-related properties (e.g., tuning speed, tuning resolution/accuracy, and mechanical reliability/stability) in comparison to conventional lasers. The photolithography of the MEMS process benefits lasers in accurate positioning/alignment, batch fabrication, easy integration with IC control/monitor circuits, and potentially low cost. More important, the ability to construct short cavities makes it a promising technology to obtain wide tuning range. In a sense, MEMS technology is revitalizing tunable lasers.

The MEMS external cavity tunable laser could be used broadly to describe all lasers that make use of MEMS structures for wavelength change. For example, micromachined movable mirrors have been widely used in the vertical-cavity side-emitting lasers (VCSELs) [4]. Typically, the top and bottom mirrors are two oppositely doped distributed Bragg reflectors (DBRs) with a cavity layer (consisting of an active layer and an air gap) in between. The bottom mirror sits on the substrate, whereas the top mirror is suspended by a soft beam, which can be a cantilever beam or a membrane. The air gap can be adjusted by applying an electrostatic potential to attract down the top mirror. As a result, it changes the cavity length and eventually tunes the oscillation wavelength to as much as 30–40 nm [5]. The rapid progress has changed MEMS VCSELs from a concept to a commercial success [6,7]. Because of very short air gaps, MEMS VCSELs have no space for optical components and do not need different optical configurations. From this point of view, design concerns and the actual tuning mechanism of MEMS VCSELs are quite different from those of edge-emitting tunable lasers. As the MEMS VCSEL has been extensively reviewed [5,8], this chapter will not cover the topic. MEMS rotational mirrors have also been employed to select a wavelength from an array of distributed-feedback (DFB) lasers [9]. However, because the wavelength of each laser element is kept unchanged during the selection, the DFB is not a tunable laser. This chapter focuses only on edge-emitting lasers that make use of MEMS technology to construct external cavities.

Before MEMS, conventional macroscopic tunable lasers had significantly developed since their advent. Details can be found in numerous review articles and textbooks [10–13], which provide a substantial technological database for MEMS lasers. With the progress of MEMS fabrication capability from wet etching and surface micromachining to deep reactive ion etching (DRIE) [14], MEMS tunable lasers have evolved from simple to complicated, from noncontinuous tuning to continuous tuning, from small to large tuning range, and from hybrid integration to single-chip integration [1–3]. Various configurations have been demonstrated [16–23], differing in many aspects of the external cavity such as external reflectors (i.e., mirrors, etalons, gratings, etc.), optical-coupling systems, laser chips (normally cleaved or coated), and integration methods (single-chip or hybrid). Some are simple replications but miniaturized versions, of conventional lasers [3], whereas the others present something new to a certain extent [1,2,15–17,19–22]. In theory, although MEMS tunable lasers follow the same principles as their conventional bulky counterparts, laser theories and designs should be reexamined, because some parameters cannot be simply scaled down. For example, MEMS technology is able to construct very short external cavities (e.g., <10 μm) [1,2,15–17,22–26], much smaller than the long cavity in bulky tunable lasers (typically >10 cm). In such short cavities, the wavelength dependence of the refractive index of the laser gain medium is no longer negligible [20,27].

The terms related to wavelength tuning should be clarified before further detailed discussions of the specifications and behaviors of tunable lasers. The change of wavelength in the tunable lasers can be realized with two approaches: one is continuous shift and the other is mode hopping. The former represents the gradual wavelength change if any one of the modes of the laser cavity is monitored. It occurs in either the single-longitudinal-mode lasers or multiple-longitudinal-mode lasers when subjected to an external feedback, but generally in single-mode lasers, continuous shift over a large range is the primary target while mode hopping is suppressed. The latter means peak mode jumps among the cavity modes, leading to abrupt wavelength changes. It happens mostly in multimode lasers. Many works have not mentioned the difference in claiming the wavelength tuning range [2,3,15–23]. For the same amount of continuous shift and mode hopping, the technical difficulties are quite different. For an accurate description of the complicated tuning behaviors of MEMS tunable lasers, the terms *continuous shift* and *mode hopping* are used to distinguish between two tuning approaches; correspondingly, *shift range* and *hopping range* refer to tuning ranges in each of the approaches. The term *wavelength tuning* remains unchanged, describing the alteration of wavelength by either of (or both) the shift and hopping approaches; the *wavelength tuning* range refers to the overall range of wavelength change.

The tunability of the many developed MEMS tunable lasers is mainly determined by the properties of the external cavity, such as the type of external reflectors, cavity length, feedback strength (i.e., coupling efficiency of the external cavity), and micro-optical systems. In tunable lasers, the reflectors determine the general behavior of laser tenability.

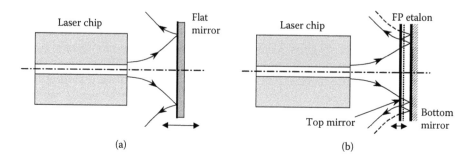

FIGURE 6.1
Typical configurations of the developed MEMS tunable lasers. (a) The mirror configuration that use a flat mirror as the external reflectors in the external cavities flat mirror and (b) the Fabry–Pérot configuration that use an etalon.

Typical reflectors include mirrors and etalons (grating is also a common type but mainly for continuous wavelength tuning). The mirror is a nondispersive element and presents no filtering effect to the laser beam; if there is no internal wavelength selection mechanism in the laser cavity, the output commonly has multiple longitudinal modes, and thus mode hopping and competition can easily happen during wavelength tuning. In the Fabry–Pérot (FP) configuration [22–26], the etalon can select a peak mode while suppressing the others; single longitudinal mode is feasible, but mode hopping may appear. Once the reflector type is given, the cavity length and feedback strength will determine more specifically the performance of the ECTL, for example, tuning range, resolution, and tuning behavior (discretely, quasi-continuously or continuously, linearly or nonlinearly, stably or chaotically, etc.).

The common configurations of the MEMS lasers developed are illustrated in Figure 6.1, which make use of a flat mirror or an FP filter as the external reflector (named as the flat-mirror configuration and FP configuration, respectively). In the flat-mirror configuration, the laser emission coming out of the laser diode is propagated to the mirror and is then reflected and coupled back into the laser internal cavity. The flat-mirror configuration has the simplest configuration, without any focusing component in the external cavity. As the laser beam emitted from the end facet has a small spot size and a large divergence angle, a very short external cavity length should be chosen, so as to maintain an acceptable feedback efficiency. Typically, the Gaussian beam emitted has an elliptical shape with beam waist radii of about 1.0–2.5 μm in the horizontal direction and 0.5–1.8 μm in the vertical direction, whereas the half-divergence angle is commonly 15°–40°. MEMS lasers developed with the flat-mirror configuration always have very short cavity lengths (<20 μm, typically 5–10 μm).

Estimations of the coupling efficiency η_{ext} (in terms of optical power, or intensity) with the change of external cavity length L are plotted in Figure 6.2, with two cases in the literature as examples. In one case (curves A and B), the laser parameters are from Reference 2, in which the wavelength is 1538 nm, and the beam radii, 1.75 μm and 1.53 μm, respectively. The other case (curves C and D) is from Reference [22]; the laser parameters are 980 nm wavelength, and beam radii 2.5 μm and 0.6 μm, respectively. It is assumed in the estimation that the external reflector has 100% reflectivity and perfect alignment with the laser chip. The calculation is based on the mode-coupling method developed in Reference 28, which takes into account the divergence and phase change of the Gaussian beam during propagation. It can be seen in curves A and C that the coupling efficiency decreases rapidly with greater cavity length. At a cavity length of 5 μm, the coupling efficiencies

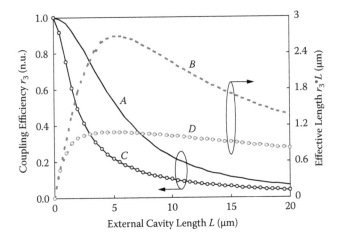

FIGURE 6.2
Variations of the coupling efficiency and effective external cavity length with a change of real external cavity length. Curves labeled A and B are estimated according to laser parameters in Reference 2, whereas C and D are based on those in Reference 31 (n.u. stands for normalized unit).

have already dropped to 53 and 21%, respectively, for the two cases; at 10 μm, they are 22 and 10%, whereas at 20 μm, they are as low as 7 and 4%. Curve A is always higher than C, implying better coupling for laser beams that have larger beam size (relative to wavelength). This explains why MEMS lasers developed in the flat-mirror configuration always have very short cavity lengths (<20 μm, typically 5–10 μm). Curves B and D in Figure 6.2 represent the product $\eta_{ext} * L$, which determines the tuning range. This will be elaborated in the next section.

As MEMS has some difficulty in constructing a high-quality micro-optical system in the external cavity, especially when the structures are fabricated by the deep-etching process, the flat-mirror configuration was historically the primary choice of many early MEMS lasers. For example, such a configuration was used to detect the resonance of a cantilever beam in 1993 [1]. In that work, a freely suspended cantilever microbeam is sandwiched between two FP laser diodes, all etched on the same GaAs substrate. One laser diode (LD2) emits intense pulsed laser to shine on the sidewall of the cantilever. As a result, the induced thermal stress forces the cantilever to vibrate. The maximum vibration amplitude can be obtained when the driving frequency coincides with the resonant frequency of the cantilever. The other laser diode (LD1) emits weak but continuous laser light. It makes use of the other side of the cantilever (other than that of the external mirror) to form an external cavity, with a separation of only 3 μm. The cantilever vibration alters the cavity length; consequently, the output power of LD1 is varied along with the wavelength. In this way, the power change represents vibration amplitude directly. Although wavelength tuning was not the focus of the work, it had all the trappings of a real MEMS tunable laser such as the use of external cavity and the change of wavelength. It was also proposed that this idea be applied to optical disk reading [15]. In 1995 and 1996, the same research group presented revised versions of wavelength-tuning structures [16,17]. One utilized an anisotropically etched silicon cantilever beam as the mirror (no metal coating), 10 μm away from a cleaved laser diode facet [16]. A hopping range of about 3 nm was obtained by discrete intervals of 0.3 nm. Nonlinear movement of the cantilever was produced by a parallel capacitor actuator. In the other work [17], many improvements were introduced compared with the previous one. The mirror was fabricated by nickel plating, which had higher reflectance. The external cavity was shortened to 5 μm for higher coupling efficiency. Antireflection (AR) coating was applied to the laser facet that faced the nickel mirror, leading to better spectral purity. A comb-drive actuator was used to translate the nickel mirror, providing linear and precision displacement and thus accurate wavelength tuning. As a consequence of the improvement, it reported a wavelength change of 20 nm by combination of mode hopping and continuous shift (~1 nm range at 0.01 nm accuracy). A surface-micromachined mirror was also proposed for tunable lasers [29]. A prototype was demonstrated in 2001 [2], in which an in-plane polysilicon mirror was assembled to the vertical direction and was then integrated with a laser chip, obtaining an external cavity length of only 10 μm. A wavelength change of 16 nm was obtained by an alternation of continuous shift and mode hopping.

The FP filter configuration employs an FP etalon (as illustrated in Figure 6.1b) for wavelength selection, which improves the spectral purity of the laser output, such as single longitudinal mode and high side-mode-suppression ratio (SMSR). The concept was proposed and realized by a series of works [24–26]; extensive analyses and experimental studies can be found in Reference 19. As shown by the schematic diagram in Figure 6.1b [23], the etalon consists of a movable top mirror and a fixed bottom mirror, which is formed by many alternative quarter-wave layers of polysilicon and silicon dioxide. The reflectance of the etalon is 97%. By applying voltage between the mirrors, the top mirror is moved toward the bottom one, and thus, the spectral transmission band is tuned to a shorter wavelength as given by the relationship $m\lambda = 2l$ (m is the interference mode number and should be an integer, $m = 3$ in this work; λ is the central wavelength of FP pass band, and l is the effective etalon separation between the two mirrors). The central wavelength is 1540 nm, and the external cavity length is chosen to be 75 μm. By carefully adjusting the cavity length using a piezoelectric stage to move the entire etalon relative to the laser chip and by controlling the etalon separation using electrostatic force, both multimode and single-mode tunings were demonstrated with a range of 13 nm and improved spectral quality (narrower line width and higher SMSR). This configuration has three FP cavities: the etalon, the internal laser cavity, and the external cavity. Limited by space for the laser chip, the movable top mirror of the etalon is located between the laser facet and the fixed bottom mirror because the top and bottom mirrors are deposited on a substrate. With the movement of the top mirror, the external cavity length is changed simultaneously with the etalon length. A careful choice of the initial condition and good match of the cavities are critical for stable wavelength tuning [22]. Otherwise, the tuning can be complicated and unstable.

6.2 Design of Micro-Optical-Coupling Systems of External Cavity

This section investigates micro-optical systems to obtain high coupling efficiency and greater cavity length at the same time. Particular attention will be paid to the optical modeling for estimation of the coupling efficiency and alignment tolerance.

6.2.1 Curved-Mirror Design and the Three-Dimensional Optical-Coupling System

In the simple configurations that use only the flat mirror or the FP etalon (see Figure 6.1), no optical system is presented in the external cavity. However, the length of the external cavity has to be <20 μm. Otherwise, the coupling efficiency will be too low to produce meaningful wavelength tuning, as stated previously. In many cases, it is necessary to extend the external cavity length because longer external cavities have some technical benefits, such as less line width, larger tuning range, and more space for insertion of other components.

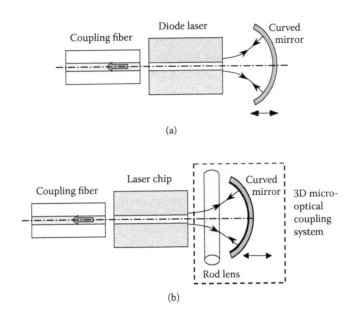

FIGURE 6.3
Configurations of MEMS tunable lasers that use optical-coupling systems in the external cavity: (a) curved-mirror design and (b) three-dimensional optical-coupling system.

Two designs of the optical-coupling system have been recently demonstrated, as shown in Figure 6.3; one employs a curved mirror as the external reflector [18,30] (named as curved-mirror configuration), whereas the other uses a three-dimensional (3-D) coupling system [21,31] (named as 3-D system configuration). The former extends the external cavity up to 100 μm, whereas the latter is suitable for longer cavities. Combined with the flat-mirror configuration, these configurations provide a full range of cavity length, from a few micrometers to several millimeters.

To improve the coupling efficiency, a simple solution is to curve up the mirror surface to converge the reflected laser beam as illustrated in Figure 6.3a. The concave surface helps to focus the diverging laser light and therefore improves the feedback efficiency. As the concave surface remains a challenge for the deep-etching process, a curved shape might be suitable. The curved mirror converges the laser light in one direction while leaving it diverging in the other direction. Even so, the coupling efficiency can be significantly improved. The beauty of this design is that it makes use of the MEMS process to pattern the mirror profile, without the need for any further process step or external components.

For even longer external cavities >100 μm, the convergence of the laser beam should be applied in both the horizontal and vertical directions, i.e., in 3-D space. A direct approach is to fabricate 3-D focus lenses by micromachining and then to integrate with the MEMS lasers. Tremendous efforts have been put in on developing Fresnel zone plates [32] or thermal reflowed photoresist microlenses [33,34]. However, it remains a challenge to fabricate 3-D microlenses directly on the sidewall of a deep-etched structure. Most of the current methods fabricate the microlenses on the surface of the substrate or holder and then lift up and fix them to the vertical direction [32,33]. An alternative approach to forming a 3-D focus is to use two cylindrical lenses, one in the horizontal direction and the other in the vertical direction. In a recent work [35], the cylindrical lenses are fabricated by SU-8 with a height of 200 μm. The lenses are taken from the substrate, rotated 90°, and then pressed into the hosts (trenches in the SU-8 layer) in the desired position. Although it reported good results of collimation and focusing, it is difficult to incorporate it into MEMS tunable lasers because of assembly difficulty and accuracy.

Another solution is to introduce a rod lens (i.e., cylindrical lens) into the curved-mirror scheme, as shown in Figure 6.3b. The rod lens provides collimation and focus of the laser beam in the vertical direction and thus ensures high coupling efficiency even at greater cavity length. The 3-D coupling design was presented in a recent work [21]. The curved mirror was deeply etched on an SOI wafer, and then the laser chip and the rod lens were assembled and integrated. The curved mirror has a curvature radius of 220 μm. In the optimized alignment, the 3-D micro-optical system obtains a power coupling efficiency of 46.5%, much higher than the values of about 1% for the flat-mirror configuration and 13% for the curved-mirror configuration.

One of the interesting parts of the 3-D system configuration is that a piece of single-mode optical fiber can be used as the rod lens. It offers many advantages as the fiber is easily available and convenient for handling, aligning, and packaging. The diameter of the fiber can be made smaller, into required dimensions, by HF etching (etching rate 3.5 μm/min for Corning SMF-28 in 49% HF solution). The fiber core has negligible influence on the focus/collimation effect because the refractive index difference is only 0.36%, and the core is only 8.2 μm in diameter compared to 125 μm of the fiber cladding [36]. By carefully designing the curved mirror and the fiber rod lens, the external cavity can be extended to a greater distance. In this way, the MEMS tunable laser may have a full range of cavity length, from extremely short (<10 μm), very short (<100 μm), and short (<1 mm) to long cavity, without a need for conventional bulky optical components (lenses, reflectors, etc.).

6.2.2 Optical Analysis of Curved-Mirror Configuration

To estimate optical coupling efficiency and the influence of position misalignment, optical modeling has to be done. Here, the curved mirror configuration is used as an example for the detailed analysis and is compared to the flat mirror configuration.

For a simple comparison between the two configurations, it is assumed that they have perfect mirrors with no angular/lateral misalignment between the laser and the mirrors. The coupling efficiencies of the flat mirror and the curved mirror can be estimated on the basis of insertion loss analysis of the Gaussian beam [28], as expressed by

$$R_{3\text{flat}} = T_2 \left(1 + \frac{L^2\lambda^2}{\pi^2 w_x^4} \right)^{-\frac{1}{2}} \left(1 + \frac{L^2\lambda^2}{\pi^2 w_y^4} \right)^{-\frac{1}{2}} \tag{6.1a}$$

$$R_{3\text{curved}} = T_2 \left(1 + \frac{L^2\lambda^2}{\pi^2 w_y^4} \right)^{-\frac{1}{2}} \tag{6.1b}$$

where $R_{3\text{flat}}$ and $R_{3\text{curved}}$ are the effective coupling efficiencies in the flat mirror configuration and the curved mirror configuration, respectively. λ is the wavelength, T_2 is the transmittance of the laser facet, and w_x and w_y are, respectively, the laser-beam waists in the horizontal and vertical directions. As an example, with an external cavity length of 66 μm, the flat mirror leads to a feedback-coupling efficiency of about 2%, whereas the curved mirror gives about 10%. Other parameters used in the example are as follows: the laser spot size is 1.75 μm and 1.53 μm, respectively, in an elliptical shape; the laser wavelength is 1.55 μm; and the transmittance of the laser facet is 70%. Therefore, the curved-mirror configuration is preferred because its higher coupling efficiency places a lower requirement for alignment accuracy in the laser assembly as compared to the flat mirror configuration.

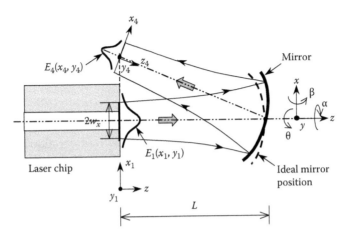

FIGURE 6.4
Analytical model for the coupling efficiency of mirror in the presence of misalignment.

To compare the two laser configurations in the presence of position misalignment, a detailed coupling analysis is developed on the basis of the Collins integral [37] and the mode-coupling theory [38]. The former calculates the transformation of the laser field emitted by the mirror (either curved mirror or flat mirror), whereas the latter works out the percentage of the transformed laser that can be coupled back into the laser internal cavity. The analytical model of the mirror coupling is shown in Figure 6.4. The mirror can be either a curved mirror or a flat mirror. The curved mirror has a curvature in the x_1-z plane, but is straight in the y_1 direction. The calculation has three steps: (1) ray transfer matrix, (2) coordinate system conversion, and (3) mode coupling. Following this sequence, the coupling efficiency of the laser configuration is estimated as detailed in the following text.

6.2.2.1 Ray Transfer Matrix of Curved Mirror

To calculate coupling efficiency using the Collins integral, the ray matrix of the curved mirror is first derived. Suppose the initial light field $E_1(x_1, y_1)$ emitted from the laser chip can be expressed as

$$E_1(x_1, y_1) = \sqrt{\frac{2}{\pi w_x w_y}} \exp\left\{-\frac{x_1^2}{w_x^2} - \frac{y_1^2}{w_y^2}\right\} \tag{6.2}$$

where the coordinate system (x_1, y_1) is on the end facet of the laser as shown in Figure 6.4. The term $\sqrt{\frac{2}{\pi w_x w_y}}$ is used to normalize the input power to be 1.

According to the Collins integral [37], the transformed laser field $E_1(x_4, y_4)$ is given by

$$E_4(x_4, y_4) = \frac{-j\exp\{jkL\}}{\lambda\sqrt{\det(\bar{B})}} \iint E_1(x_1, y_1) \times \exp\left\{j\frac{\pi}{\lambda}\left[\vec{r}_1'(\bar{A}\bar{B}^{-1})\vec{r}_1 + \vec{r}_4'(\bar{D}\bar{B}^{-1})\vec{r}_4 - 2\vec{r}_1'\bar{B}^{-1}\vec{r}_4\right]\right\} dx_1 dy_1 \tag{6.3}$$

where k is the wave number defined by $k = \frac{2\pi}{\lambda}$, L is the distance from the laser end facet to the mirror center, and the plane (x_4, y_4) is perpendicular to the central line of the reflected light, its origin being in the (x_1, y_1) plane (see Figure 6.4). For the matrices and vectors,

$\vec{r}_1 = \begin{pmatrix} x_1 \\ y_1 \end{pmatrix}$, $\vec{r}_4 = \begin{pmatrix} x_4 \\ y_4 \end{pmatrix}$, $\vec{B}^{-1} = \text{inv}(\vec{B})$, and $\det(\vec{B})$ is the determinant. $\vec{A}$, $\vec{B}$, and $\vec{D}$ are 2×2 matrices defined by

$$\begin{pmatrix} \vec{A} & \vec{B} \\ \vec{C} & \vec{D} \end{pmatrix} = \vec{M} \tag{6.4}$$

Here, $\vec{M}$ is a 4×4 matrix that stands for the overall ray transfer matrix for the laser field transformations, including the propagation from the facet to the mirror, rotation of the mirror, tilting of the mirror, reflection/focus of the mirror, and the propagation from the mirror to the (x_4, y_4) plane. It can be expressed as

$$\vec{M} = \vec{M}_{fs}(L_0) \cdot [\vec{R}^{-1}(\alpha) \cdot \vec{T}(\theta) \cdot \vec{R}(\alpha)] \cdot \vec{M}_{fs}\left(\frac{L_0}{\cos 2\theta}\right) \tag{6.5}$$

where $\vec{M}_{fs}(L)$ is the ray transfer matrix for the free-space propagation of a distance L, and $\vec{R}(\alpha)$ for the rotation about the z axis by an angle α (see Figure 6.4), as given by

$$\vec{M}_{fs}(L) = \begin{pmatrix} 1 & 0 & L & 0 \\ 0 & 1 & 0 & L \\ 0 & 0 & 1 & 0 \\ 0 & 0 & 0 & 1 \end{pmatrix} \tag{6.6a}$$

$$\vec{R}(\alpha) = \begin{pmatrix} \cos\alpha & \sin\alpha & 0 & 0 \\ -\sin\alpha & \cos\alpha & 0 & 0 \\ 0 & 0 & \cos\alpha & \sin\alpha \\ 0 & 0 & -\sin\alpha & \cos\alpha \end{pmatrix} \tag{6.6b}$$

$\vec{T}(\theta)$ is the matrix representing both mirror reflection/focus and mirror tilting by angle θ about the y axis. For the curved mirror and the flat mirror, $\vec{T}(\theta)$ can be expressed as

$$\vec{T}_{curved}(\theta) = \begin{pmatrix} 1 & 0 & 0 & 0 \\ 0 & 1 & 0 & 0 \\ \dfrac{-2\cos\theta}{\rho} & 0 & 1 & 0 \\ 0 & 0 & 0 & 1 \end{pmatrix} \tag{6.6c}$$

$$\vec{T}_{flat}(\theta) = \begin{pmatrix} 1 & 0 & 1 & 0 \\ 0 & 1 & 0 & 0 \\ 0 & 0 & 1 & 0 \\ 0 & 0 & 0 & 1 \end{pmatrix} \tag{6.6d}$$

6.2.2.2 Coordinate System Conversion on Laser Facet

With Equations 6.2 through 6.6d, the laser field in the (x_4, y_4) plane can be obtained. However, to couple the laser field $E_4(x_4, y_4)$ into the laser internal cavity, the coordinate system (x_4, y_4, z_4) should be converted to the coordinate system (x_1, y_1, z) according to the following rules as illustrated in Figure 6.5:

$$x_4 = (x_1 - L\sin 2\theta)\cos 2\theta$$
$$y_4 = (y_1 - L\sin 2\beta)\cos 2\beta$$

(6.7a)

The projection of (x_4, y_4, z_4) into (x_1, y_1, z) would result in a distance change as given by

$$\Delta z_x = (y_1 - L\sin 2\beta)\sin 2\beta$$
$$\Delta z_y = (y_1 - L\sin 2\theta)\sin 2\theta$$

(6.7b)

Therefore, when $E_4(x_4, y_4)$ in the (x_4, y_4, z_4) system is converted to $E_4(x_1, y_1)$ in the system (x_1, y_1, z), a phase change should be compensated as follows:

$$\Delta\phi = k(\Delta z_x + \Delta z_y)$$

(6.7c)

6.2.2.3 Feedback-Coupling Efficiency Using the Mode-Coupling Method

According to the mode-coupling theory [38], the coupling efficiency E_0 (in optical power) of $E_4(x_1, y_1)$ being coupled into the laser internal cavity can be obtained by the mode-overlap integral as given by

$$E_0 = \frac{\left|\int\int_{-\infty}^{+\infty} E_4(x_1,y_1)E_1^*(x_1,y_1)dx_1dy_1\right|^2}{\int\int_{+\infty}^{-\infty}\left|E_4(x_1,y_1)\right|^2 dx_1dy_1 \cdot \int\int_{+\infty}^{-\infty}\left|E_1^*(x_1,y_1)\right|^2 dx_1dy_1}$$

$$= \frac{\left|\int\int_{-\infty}^{+\infty} E_4(x_1,y_1)E_1^*(x_1,y_1)dx_1dy_1\right|^2}{\int\int_{+\infty}^{-\infty}\left|E_4(x_1,y_1)\right|^2 dx_1dy_1}$$

(6.8)

where $E_1^*(x_1,y_1)$ is the conjugate of $E_1(x_1,y_1)$, which has unit power.

On the basis of the preceding analytical model, the feedback-coupling efficiencies of the curved mirror and the flat mirror are compared in the presence of different types of misalignments. In the simulation, the curved mirror has a curvature of 66 μm, the wavelength is 1.55 μm, the reflectivity of the mirror surface is assumed to be 100%, and the laser beam is taken as elliptical Gaussian in shape with major and minor axes of 1.75 μm and 1.53 μm, respectively. As shown in the simulation results of Figure 6.6a, the curved mirror has a feedback-coupling efficiency ranging from 4 to 9% when the mirror is moved back and forth within ±4 μm relative to its original position. In contrast, the flat mirror has a range of only 0.5 to 0.6%. When the laser wavelength sweeps from 1.25 to 1.85 μm, the curved mirror produces a reduction of coupling efficiency from about 12 to 7%, whereas the flat mirror has a range of only 1 to 0.5%. This proves that the curved mirror has a much higher coupling efficiency.

When there are certain angular misalignments α, β, and θ, which correspond to the roll, yaw, and pitch of the curved mirror as defined in Figure 6.4, the curved mirror has different performance compared with the flat mirror. As shown in Figure 6.7a, when it is purely

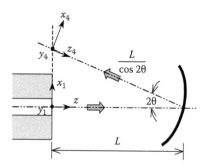

FIGURE 6.5
Analytical model for coordinate system conversion and phase compensation.

α rotation (i.e., $\beta = 0$), the coupling efficiency of the curved mirror varies between 8 and 9% when the value of α is as much as $\pm 90°$. With an increase of β, the peak coupling efficiency is reduced, and the efficiency drops rapidly with a larger α value. However, even at $\beta = 2°$, the efficiency is still higher than 7% when $-3° < \alpha < +3°$. It shows that the curved mirror can

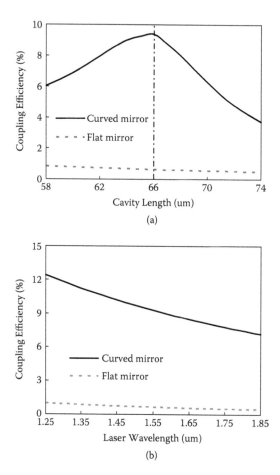

FIGURE 6.6
Comparison of the feedback-coupling efficiencies of the two configurations that use the curved mirror and the flat mirror as the external reflector: (a) changing the cavity length and (b) changing the wavelength.

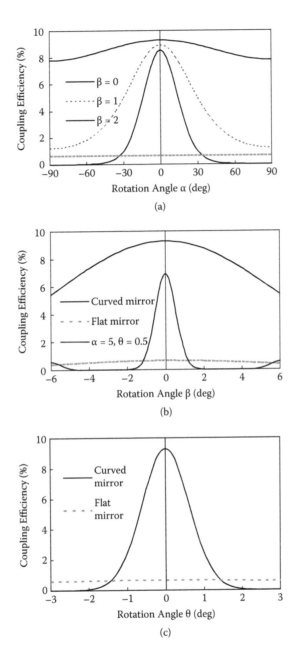

FIGURE 6.7
Comparison of coupling efficiencies at different angular misalignments: (a) rotation angle α, (b) rotation angle β, and (c) rotation angle θ.

sustain quite large angular misalignments in the α direction. In contrast, the flat mirror provides an efficiency of as low as 0.7%, but it remains nearly constant with an increase of α. In the curve of the coupling efficiency versus β as shown in Figure 6.7b, the coupling efficiency falls from 9 to 5% when the rotation angle β increases from 0 to 6°. It can tolerate a certain angular misalignment in β. However, when there is any other type of angular misalignment, for example, $\alpha = 5°$ and $\theta = 0.5°$, the toleration for β is reduced drastically. The coupling efficiency would fall from its peak of 6% to nearly 0 within ±2° of angle β

rotation. In contrast, the flat mirror keeps going smoothly at a low lever of 0.6% with the change of rotation angle β. The influence of the rotation angle θ is shown in Figure 6.7c. The curved mirror can provide an efficiency ranging from 9 to 6% when the value of θ is increased from 0 to 0.5°. The efficiency will go down to almost 0 when θ is about ±2°. In contrast, the flat mirror has nearly constant efficiency at 0.6%.

In summary, the curved mirror is about 10 times more efficient than the flat mirror in coupling the feedback laser light when the misalignment is not very serious. The flat mirror has a much smoother relationship with misalignment; however, its efficiency is too low. For the curved mirror, it can tolerate a quite high α and moderate β, but it is very sensitive to θ.

6.3   Theory of Continuous Wavelength Tuning

This section focuses on studying the continuous wavelength shift of the mirror configuration (for any types of mirrors, such as flat mirror, curved mirror, etc.). The main objective is to derive a theoretical model that is suitable for MEMS tunable lasers, which may not have very weak feedback strength and cannot be handled by the available weak feedback models [13,39]. For this purpose, an extended-feedback model will be developed to describe the gain variation and wavelength shift for a broad range of feedback strengths [40]. The range and stability of wavelength tuning will also be addressed.

Before the start of the analysis, some clarifications should be made. The terms *frequency* and *wavelength* will both be used on different occasions in this chapter, but they will be regarded as the same because they have, strictly, a one-to-one relation. Commonly, it is more convenient to use the term frequency in the analysis and wavelength in experiments. In addition, the wavelength tuning theory is developed on the basis of the following conditions and assumptions:

- FP laser condition: The bare laser chip for the external cavity laser is a normal multilongitudinal mode FP laser.
- Nondispersive mirror condition: The external reflector is a mirror whose reflectivity is wavelength independent, unlike the gratings or etalon used in the other types of tunable lasers.
- Short external cavity length condition: The length of the external cavity is less than the effective length of the internal cavity.
- Short mirror translation condition: The mirror translation is relatively small compared with the initial external cavity length.

Homogeneous assumption: The mode with minimum threshold gain oscillates, whereas the others are suppressed. When the laser is operated slightly over the threshold (for example, 28% over the threshold current), the laser output can be single mode owing to the homogeneous broadening, even though it is an FP laser, which usually has multiple modes.

In this chapter, the terms *weak feedback* and *medium feedback* are used to represent cases of different feedback strengths, whereas the terms *extended feedback model* and *weak feedback model* represent two different models. The weak feedback model is valid only in the case of weak feedback, whereas the extended feedback model covers both weak and medium feedbacks. The condition for the weak feedback model is $\xi \ll 1$, whereas that for the

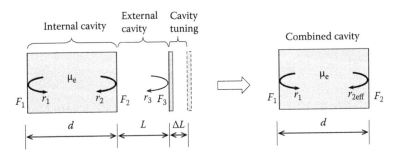

FIGURE 6.8
Analytical model for the wavelength tuning of the external cavity laser.

extended feedback model is $r_2 r_3 \ll 1$. The definitions and more details will be given in the following text.

6.3.1 Variation of Laser Gain and Phase with Mirror Translation

A schematic diagram of a tunable laser is shown in Figure 6.8. It consists of an internal cavity representing a laser chip and a passive external cavity formed by a mirror. According to the FP laser condition and the nondispersive mirror condition, the laser chip provides an FP cavity and an optical gain, and the mirror is nondispersive and can be translated to change the external cavity length.

The extended feedback model is based on the three-mirror model [13,39,40] as shown in Figure 6.8. F_1, F_2, and F_3 denote the two facets of the FP laser and the external mirror; r_1 and r_2 are real variables representing the amplitude reflectances of F_1 and F_2, respectively, whereas r_3 denotes the effective amplitude reflectance of the external mirror F_3. Here, r_3 is not directly the reflectance of the mirror as coupling losses due to mirror misalignment and divergence of the laser beam should be taken into account. The notation μ_e is the effective refractive index of the laser internal cavity, and d and L stand for the lengths of the laser internal cavity and the external cavity, respectively. The reflection from the external mirror can be treated by combining it with the reflection of the laser end facet F_2 as shown in Figure 6.8, yielding an effective reflectance r_{2eff} given by [41]

$$r_{2eff} = r_2 + \frac{1-r_2^2}{r_2} \sum_{k=1}^{\infty} [r_2 r_3 \exp(j2\pi\tau_L v)]^k$$

$$= \frac{r_2 + r_3 \exp\{j2\pi\tau_L v\}}{1 + r_2 r_3 \exp\{j2\pi\tau_L v\}} \tag{6.9}$$

$$= R_2 \exp(j\phi_L)$$

where v is the frequency of the diode laser, τ_L is the round-trip delay of the external cavity given by $\tau_L = 2L/c_0$, and c_0 denotes the light velocity in vacuum. R_2 and ϕ_L represent the amplitude and phase of the effective reflectance r_{2eff}, respectively.

Assuming $r_2 r_3 \ll 1$, the multiple reflections in the external cavity are neglected. Then R_2 and ϕ_L can be expressed as

$$R_2 = r_2 T \tag{6.10}$$

$$\phi_L = \arcsin[\xi \sin(2\pi\tau_L v)/T] \tag{6.11}$$

and

$$\xi = r_3\left(1 - r_2^2\right)/r_2 \tag{6.12}$$

$$T = \sqrt{1 + \xi^2 + 2\xi \cos(2\pi\tau_L v)} \tag{6.13}$$

where ξ is the relative reflectance of the external mirror, and T is a variable having a period of $1/\tau_L$.

For an initial electrical field E_0 inside the internal cavity, after traveling a roundtrip through the internal cavity and the external cavity, it would become

$$E_1 = E_0 r_1 r_{2eff} \exp[-j2\pi\tau_d v + (g - \gamma)d] \tag{6.14}$$

where E_1 is the amplified field, $\tau_d = 2\mu_e d/c_0$ is the roundtrip delay of the internal cavity, and g and γ represent the modal gain and modal loss of the internal cavity, respectively. To obtain a stationary laser oscillation, the amplitude and phase conditions should be met [13], yielding

$$r_1 R_2 \exp[(g - \gamma)d] = 1 \tag{6.15}$$

$$\phi = \phi_L + 2\pi\tau_d v = 2\pi m, \quad (m = \text{integer}) \tag{6.16}$$

where ϕ denotes the roundtrip phase of the laser beam after traveling through the internal cavity and external cavity.

Based on the amplitude condition (see Equation 6.15), the threshold difference of the modal gain Δg due to the optical feedback is given by

$$\Delta g = -\ln(T)/d \tag{6.17}$$

where ln represents the natural logarithm.

From Equation 6.16, the total roundtrip phase change $\Delta\phi$ due to the frequency change is given by

$$\Delta\phi = \phi_L + \frac{4\pi d}{c_0}\Delta(\mu_e v) \tag{6.18}$$

6.3.2 Variation of Wavelength with Mirror Translation

The effective refractive index μ_e is changed with the frequency variation because optical feedback alters the carrier density ρ in the internal cavity. Adopting the derivation procedure in Chapter 9 of Reference 13, the change $\Delta\mu_e$ can be expressed by

$$\Delta\mu_e = \frac{\partial\mu_e}{\partial\rho}(\rho - \rho_0) + \frac{\partial\mu_e}{\partial v}(v - v_0) \tag{6.19}$$

where ρ_0 is the threshold carrier density. The carrier density variation induces the change of the material gain Δg_{mat} as given by [13]

$$\frac{\partial\mu_e}{\partial\rho}(\rho - \rho_0) = \alpha\,\frac{\partial\mu_e''}{\partial\rho}(\rho - \rho_0) = -\alpha\,\frac{\partial g}{\partial\rho}\frac{c_0}{4\pi v}(\rho - \rho_0) = -\frac{\alpha c_0}{4\pi v}\Delta g_{mat} \tag{6.20}$$

where μ_e'' denotes the imaginary part of the complex refractive index of the internal cavity, α is the line-width enhancement factor defined by $\alpha = \partial \mu_e / \partial \mu_e''$, and μ_e'' is related to g by $\mu_e'' = \frac{c_0}{4\pi v} g$ (from Equations 2.41 and 2.42 of Reference 13).

It should be noted that in Equations 6.15 and 6.17, the gain is the modal gain, whereas in Equation 6.20 it is the material gain. The relationship is given by

$$\Delta g = \Gamma \Delta g_{mat} \tag{6.21}$$

where Γ is the optical confinement factor. This difference significantly alters the value because Γ is generally quite small (e.g., 0.1), but it has long been neglected in previous studies that dealt with similar cases [13,39,42].

It can be derived from Equation 6.19 and 6.21 that

$$\Delta(\mu_e v) = v\Delta \mu_e + \mu_e(v - v_0) = -\frac{\alpha c_0}{4\pi \Gamma}\Delta g + \bar{\mu}_e(v - v_0) \tag{6.22}$$

where $\bar{\mu}_e$ is the group effective refractive index in the internal cavity given by Reference 13:

$$\bar{\mu}_e = \mu_e + v\frac{d\mu_e}{dv} \tag{6.23}$$

Substituting Equation 6.17 and 6.22 into Equation 6.18:

$$\Delta \phi = 2\pi\, t_d(v - v_0) + \phi_L + \phi_g \tag{6.24}$$

where $t_d = 2\bar{\mu}_e d/c_0$ is the group roundtrip delay in the internal cavity, and ϕ_g denotes the roundtrip phase change due to the gain variation, given by

$$\phi_g = \alpha' \ln(T) \tag{6.25}$$

where $\alpha' = \alpha/\Gamma$.

To achieve stationary oscillation, the phase condition Equation 6.18 requires $\Delta \phi = 0$, yielding

$$
\begin{aligned}
v - v_0 &= -\frac{\phi_g\big|_{v=v_0} + \phi_L\big|_{v=v_0}}{2\pi t_d + \frac{\partial \phi_g}{\partial v}\big|_{v=v_0} + \frac{\partial \phi_L}{\partial v}\big|_{v=v_0}} \\
&= -\frac{T_0^2}{2\pi}\cdot\frac{\alpha' \ln(T_0) + \arcsin[\xi \sin(2\pi\tau_L v_0)/T_0]}{t_d T_0^2 + \tau_L \xi[\xi + \cos(2\pi\tau_L v_0) - \alpha' \sin(2\pi\tau_L v_0)]}
\end{aligned}
\tag{6.26}
$$

where $T_0 = T|_{v=v_0}$ represents the value of T at $v = v_0$. This is one of the key analytical results achieved in this chapter, and is unique in comparison to previous analyses [2,13,39,41].

6.3.3 Comparison of the Extended Feedback Model with the Conventional Weak Feedback Model

It is noted that Equations 6.17, 6.24, and 6.26 are derived under the assumption that $r_2\, r_3 \ll 1$, which is rather different from the weak optical feedback condition $\xi \ll 1$, which was widely used [2,39,41]. However, they are also available for deducing the weak optical feedback condition. Assuming $\xi \ll 1$, Equations 6.17, 6.24, and 6.26 become

$$\Delta g = -\xi/d \ \cos(2\pi\tau_L v) \tag{6.27}$$

$$\Delta\phi = 2\pi t_d(v - v_0) + \xi\sqrt{1 + \alpha'^2} \ \sin[2\pi\tau_L v + \arctan(\alpha')] \tag{6.28}$$

$$v - v_0 = -\frac{1}{2\pi\tau_L} \frac{C\sin[2\pi\tau_L v_0 + \arctan(\alpha')]}{1 + C\cos[2\pi\tau_L v_0 + \arctan(\alpha')]} \tag{6.29}$$

where $C = (\tau_L/t_d)\xi \sqrt{1 + \alpha'^2}$ represents the external feedback strength. Equations 6.27–6.29 are exactly the same as the equations given in References [13] and [39] when α' is replaced by α (as mentioned previously, the optical confinement factor was not taken into account in their analyses).

In the tunable lasers using micromachined mirrors, r_3 may not be so weak (e.g., $r_3 = 0.15$), because of the short external cavity. If the end facet F_2 is coated with dielectric antireflection films, r_2 can be reduced to a lower value (e.g., $r_2 = 0.2$). In this example, $r_2 r_3 = 0.03 \ll 1$, whereas $\xi = 0.72$ does not meet the condition of $\xi \ll 1$. In another example from the literature, it can be seen in Figure 23 of Reference [22] that $r_3 = 0.3$–0.4 when the external cavity is 5–10 μm. If the laser facets are normally cleaved, then $\xi \approx 0.45$, which is no longer a weak feedback. With the same parameters, $r_2 r_3 \approx 0.18$, the condition of $r_2 r_3 \ll 1$ is met roughly. Here, a simple classification is given: the conditions of $\xi \leq 0.1$ and $0.1 < \xi < 0.8$ are regarded as weak feedback and medium feedback, respectively, whereas the other represents strong feedback. When $\xi \gg 1$, it requires that $r_3 \gg r_2$, and the internal cavity has negligible influence on wavelength selection. Wavelength tuning is merely continuous, and is out of the scope of this chapter. The values of 0.1 and 0.8 are chosen somewhat arbitrarily. Physically, ξ represents the amplitude ratio of the light reflected by facet F_2 relative to the light transmitted to the external cavity and then reentering the internal cavity (only one round). When $\xi \leq 0.1$, the light energy fed back from the external cavity is much smaller than that fed back by facet F_2 of the internal cavity. It is a weak feedback. When $\xi > 0.8$, coherence collapse may easily occur [39], which invalidates the analysis. In micromachined tunable lasers that work in the extended feedback region, Equations 6.17, 6.24, and 6.26 are more suitable.

To compare the difference between the extended feedback model and the weak feedback model, the weak feedback model is extended to the extended feedback region. The threshold gain difference induced by the optical feedback varies as the length of external cavity changes. The predictions by both models are compared in Figure 6.9 (assuming $\xi = 0.72$). Both predictions have a period of $\lambda/2$, as implied by Equations 6.24 and 6.27. However, the extended feedback model gives out a larger gain increment, $\Delta g_{max} = -\ln(1 - \xi)/d$, and a smaller gain reduction, $\Delta g_{min} = -\ln(1 + \xi)/d$, compared with those ($\Delta g_{max} = \xi/d$) and ($\Delta g_{max} = -\xi/d$) predicted by the weak feedback model. The discrepancy between these two models increases with larger ξ. In addition, the extended feedback model predicts broad valleys in the curve of threshold gain difference. When the external mirror is initially adjusted to the valley regions, the threshold gain changes very little with the change of external cavity length. Thus, the output power of the dominating mode is stable. However, the small slopes of the gain valleys are harmful for side-mode suppression.

Wavelength variations predicted by Equations 6.26 and 6.29 are compared in Figure 6.10. The weak feedback model shows a cosinelike relationship between wavelength tuning and external cavity length, whereas the extended feedback model has sharp valleys and smooth peaks. This implies that, in the micromachined tunable laser, the wavelength can

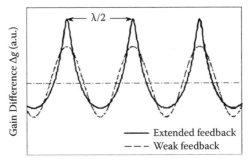

FIGURE 6.9
Comparison of the extended feedback model and the weak feedback model in predicting gain variation with change of external cavity length ($\xi = 0.72$; a.u. represents arbitrary unit).

be tuned slowly with a movement of the external mirror if the light is initially adjusted to one of the smooth peaks, but must be tuned rapidly if adjusted to a sharp valley. In the extended feedback model, the achievable maximum and minimum wavelengths are smaller than those predicted by the weak model. However, the extended feedback model gives a larger tuning range (i.e., the achievable maximum wavelength minus the minimum) than the weak feedback model. Based on Equations 6.26 and 6.29, the wavelength-tuning range given by the two models is compared in Figure 6.11. When the value of ξ is small, the two models match very well. However, with increase of ξ, the extended feedback model tends to give a larger tuning range than the weak feedback model.

In tunable lasers that use micromachined mirrors to form external cavities, the short distances between the mirrors and the laser facets are found to induce significant feedbacks as shown in Figure 6.2. Under this circumstance, the extended feedback model should be adopted instead of the weak feedback model so as to avoid a large discrepancy in analyzing tuning characteristics.

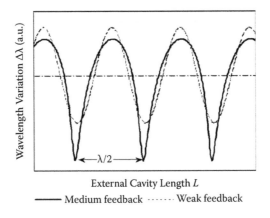

FIGURE 6.10
Comparison of wavelength variation predicted by the extended feedback and weak feedback models ($\xi = 0.72$; a.u. represents arbitrary unit).

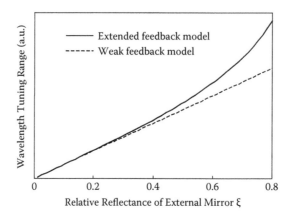

FIGURE 6.11
Comparison of wavelength-tuning ranges predicted by the extended and weak feedback models (a.u. represents arbitrary unit).

6.3.4 Wavelength-Tuning Range and Stability of Mirror Configuration

The preceding extended feedback model describes the frequency shift characteristics (see Equation 6.26). In many applications, the range and stability of wavelength tuning are of great importance. To have a simple but clear image of how the shift range is determined by the parameters, the analysis here goes back to the weak feedback model because of its simplicity.

Using Equation 6.29, such a wavelength shift $\Delta\lambda$ under the weak feedback strength can be expressed as [13,39]

$$\Delta\lambda = \frac{\lambda_0^2}{2\pi L} \cdot \frac{C \sin[2\pi\tau_L v_0 + \arctan(\alpha')]}{1 + C \cos[2\pi\tau_L v_0 + \arctan(\alpha')]} \qquad (6.30)$$

where λ_0 is the central wavelength. Equation 6.30 describes the basic relationship between continuous shift and the change of external cavity length. It should be noted that the derivation of Equation 6.30 is based on an explicit assumption $\xi \ll 1$ (called the weak feedback condition) and two implicit assumptions: one is that the laser is in single longitudinal mode (called the single-mode condition) and the other is $\tau_L \Delta v \ll 1$ (called the small tuning condition), where $\Delta v = c_0 \Delta\lambda/\lambda_0^2$ is the shift of central frequency. These three conditions define the applicability of the analysis. In MEMS lasers, they should be reexamined. The small tuning condition is equivalent to $\Delta\lambda \ll \lambda_0^2/2L$; that is, the tuning range is limited by the free spectral range (FSR) of the external cavity. It implies that a short cavity allows a larger tuning range. Under the single-mode condition, laser tuning is continuous with the increase of external cavity, at a period of half a wavelength. However, when this condition is violated, the tuning behavior becomes complicated. This explains why experiments employ a combination of continuous tuning and mode hopping when multiple longitudinal lasers are used [2,16,17,22].

Using Equation 6.30, the shift range $\Delta\lambda_r$ of a tunable laser can be derived as

$$\Delta\lambda_r = \frac{\lambda_0^2}{2\pi} \cdot \frac{1}{\sqrt{\frac{1}{A_0^2 r_3^2} - L^2}} \qquad (6.31)$$

where $A_0 = (1 - r_2^2)\sqrt{1 + \alpha^2}/r_2\mu_0 d_0$ is a variable determined by the parameters of internal laser cavity; its value increases with lower r_2 and shorter internal cavity length. Equation 6.31

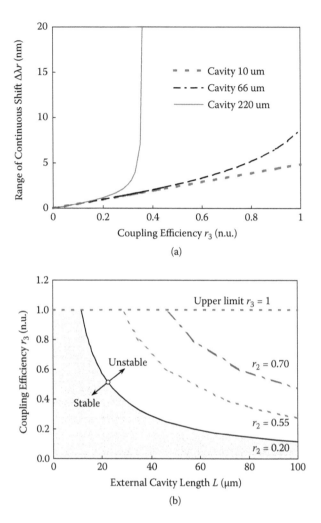

FIGURE 6.12
Influence of coupling efficiency and external cavity length on tuning behavior: (a) shift range for different cavity lengths and (b) stable region of laser operation at different reflectances (n.u. stands for normalized unit).

directly links the shift range to the external mirror reflectance and external cavity length. The relationship is illustrated in Figure 6.12a [21,31]. For a given cavity length, higher coupling efficiency results in larger shift range. In Figure 6.12a, the shift range for a cavity of 66 μm is increased gradually to 8.7 μm when r_3 rises from 0 to 1. However, at a given coupling efficiency, a larger external cavity length produces more continuous shift. For example, if coupling efficiency is fixed at $r_3 = 0.8$, the shift range is estimated to be 5.2 nm for an external cavity length of 10 μm, in comparison to a shift of 3.9 nm for a 66 μm cavity.

Stability (or in the other extreme, instability) of wavelength tuning is another important concern for tunable lasers. It can be seen from Figure 6.12a that, for a cavity length of 220 μm, the wavelength tuning shoots up at $r_3 \approx 0.36$, which corresponds to the stability problem (or coherence collapse) of tuning lasers. It was observed experimentally that the laser runs through five different regimes successively when external coupling increases from very weak to very strong in comparison to the laser facet reflectance [43]. Coherence collapse [44] occurs in the fourth regime (i.e., not very strong feedback) with the appearance of drastic broadening of line width and sudden loss of coherence of the light beam. The condition of stable operation can be easily derived from Equation 6.31, as given by

$$r_3 L < 1/A_0 \qquad (6.32)$$

plus one more limit $r_3 \leq 1$. Examples of the stable operation regions are shown in Figure 6.12b for different levels of laser facet reflectance. The shaded area is the stable region for the condition $r_2 = 0.20$. The parameters are taken as $\alpha = 3$, $\Gamma_0 = 0.1$, $\mu_0 = 3.482$, and $d_0 = 300$ μm. At $r_2 = 0.20$, it has $1/A_0 = 11.5$ μm. The area below $r_3 = 1$ and $r_3 L = 11.5$ defines the stable region. Out of this region, the laser operation becomes unstable. In other words, at a given level of r_3, the external cavity length cannot be too long. For example, in the case of perfect coupling (i.e., $r_3 = 1$), the external cavity length should be shorter than 11.5 μm. When r_2 is increased to 0.55 and then 0.70, the stable region is widened as illustrated in Figure 6.12b. That is to say, longer L is allowed with an increase of r_2 if r_3 is kept constant. This is because the increase of r_2 causes a decrease of laser facet transmittance and thus reduces the laser light that can be coupled back into the laser internal cavity.

The preceding analysis provides another way to understand the operation of external cavity lasers. For example, in an external cavity where r_3 changes with L, the product $r_3 \cdot L$ is more directly related to the shift range than the individual variables r_3 and L. Therefore $r_3 \cdot L$ is named as the effective length of the external cavity. It was demonstrated by numerical simulation in Reference 22 that the mirror configuration runs through three successive tuning regions with the increase of external cavity length: the reduced tuning region, in which the cavity length is close to 0 and has small wavelength tuning; the nonlinear tuning region, which has the largest wavelength tuning; and the linear tuning region, which has wavelength tuning limited by the FSR of the external cavity. The observation of the three regions was explained by treating the external cavity as an introduced loss curve to move over the gain curve [22]. The concept of effective length can provide a more straightforward explanation. From curves B and D of Figure 6.2, the effective length has roughly three different parts: it is always small at the beginning, and then increases gradually to the peak, followed by a decrease at larger L. As a longer effective length results in larger wavelength-tuning range, the three parts of the effective length curve directly correspond to the reduced, the nonlinear, and the linear tuning regions, respectively, as observed in Reference 22. The concept of effective length also provides a guideline for the design of mirror configuration: the external cavity length should be chosen near the peak of effective length if the primary target is to obtain a large tuning range.

6.4 Theory of Discrete Wavelength Tuning

This section represents the core analysis of this chapter. It aims at finding the conditions for discrete wavelength tuning on the basis of the analysis of continuous tuning in the previous section. It will provide a guideline for laser design and experimental demonstration discussed in the next section.

It is noted that the usefulness of the extended feedback model relies on its versatility in handling both weak and medium feedback strengths, which are the normal working conditions of short-cavity MEMS external cavity lasers. However, in dealing with purely discrete tuning, the weak feedback model is adopted for two reasons. One is its simplicity; the other is that, as is later proved, purely discrete tuning requires the condition of weak feedback.

As the mirror is a nondispersive component, the laser output may contain one or several of its cavity modes (i.e., single longitudinal mode or multiple longitudinal mode). The extended feedback model developed previously is used to predict the continuous wavelength tuning of any of the laser modes. It does not take into account mode hopping. However, it is commonly observed in experiments that, during tuning, the laser would experience a small range of continuous tuning and then suddenly jump to the adjacent mode by mode hopping (i.e., discrete tuning). In this section, this continuous-then-discrete tuning phenomenon will be explained theoretically by monitoring the tuning behavior of one laser mode, and then the conditions for a discrete-tuning-dominant wavelength tuning will be derived.

6.4.1 Laser Modes and Frequency Shift

In a multimode laser, based on Equation 6.14, the gain and the frequency at the threshold for the m-th mode can be expressed as

$$2\pi\tau_d v_m + \phi_L = 2\pi m \quad (m \text{ is an integer}) \tag{6.33}$$

$$g_{th} = \gamma - \ln(r_1 R_2)/d \tag{6.34}$$

where v_m is the m-th longitudinal mode of the laser cavity. Equations 6.33 and 6.34 govern the laser output in such a way that the former defines the possible output frequencies (i.e., laser modes), whereas the latter determines the gain of each mode. According to the homogeneous assumption, the mode having the lowest gain is selected for oscillation.

The amount of continuous tuning is the frequency shift of the laser modes relative to their original positions when there is no external cavity. According to Equation 6.20, the change of refractive index is given by

$$\mu_e - \mu_{e0} = \frac{c_0 \alpha'}{4\pi v_m}(g_{th} - g_{th0}) \tag{6.35}$$

where μ_{e0} and g_{th0} are the effective refractive index and the gain in the absence of external cavity, respectively.

Substituting Equation 6.35 into Equation 6.34 yields

$$v_m = \frac{1}{2\pi}[2\pi m + \alpha' \ln(R_2/r_2) - \phi_L] \cdot \Delta v_s \tag{6.36}$$

$$v_m - v_{m0} = \frac{1}{2\pi}[\alpha' \ln(R_2/r_2) - \phi_L] \cdot \Delta v_s \tag{6.37}$$

where $\Delta v_s = c_0/(2\mu_{e0}d)$ is the original mode spacing of the laser (s for spacing), and $v_{m0} = mc_0/(2\mu_{e0}d)$ represents the m-th laser mode in the absence of external cavity. Equation 6.36 describes the possible frequencies of the laser output, and Equation 6.37 expresses the relative frequency shift of any laser mode relative to its original position. It is noted that the frequency shift is continuous with the mirror translation.

Substituting R_2 and ϕ_L in Equations 6.10 and 6.11 into Equation 6.37 would yield an accurate but complicated expression of the laser frequency shift. For a more straightforward physical interpretation, the expression is simplified under the condition of weak feedback; i.e., $\xi \ll 1$. In this case, $r_2 r_3 = \xi \cdot [r_2^2/(1 - r_2^2)] \ll 1$ if r_2 does not approach 1 (i.e., not high-reflection-coated). Therefore, $R_2 \approx r_2[1 + \xi \cos(2\pi\tau_L v_m)]$ and $\phi_L \approx \xi \sin(2\pi\tau_L v_m)$. Using these relationships, Equation 6.37 can be further simplified as

$$v_m - v_{m0} = \frac{\xi\sqrt{1 + \alpha'^2}}{2\pi}\cos\left(2\pi\tau_L v_m + \frac{\pi}{2} - \arctan(\alpha')\right) \cdot \Delta v_s \tag{6.38}$$

Although v_m appears on both sides of Equation 6.38, its variation relative to the original mode spacing is limited to a small range as expressed by

$$-\frac{\xi\sqrt{1+\alpha'^2}}{2\pi} \leq \frac{v_m - v_{m0}}{\Delta v_s} \leq \frac{\xi\sqrt{1+\alpha'^2}}{2\pi} \tag{6.39}$$

In the case of the weak feedback $\xi \ll 1$ and the short external cavity length $L_0 < \mu_e d$, the continuous frequency shift is small, and

$$2\pi\tau_L(v_m - v_{m0}) \leq \frac{\xi L_0 \sqrt{1+\alpha'^2}}{\mu_e d} \ll 1 \tag{6.40}$$

Therefore,

$$v_m - v_{m0} \approx \frac{\xi\sqrt{1+\alpha'^2}}{2\pi} \cos\left(2\pi\tau_L v_{m0} + \frac{\pi}{2} - \arctan(\alpha')\right) \cdot \Delta v_s \tag{6.41}$$

The continuous shifts of the m_1-th, $(m_1 - 1)$-th, and $(m_1 - 2)$-th laser modes are shown in Figure 6.13a (m_1 is an integer). They are changed periodically at intervals of about $\lambda_0/2$ while maintaining a fixed phase difference of about $2\pi L_0/(\mu_{e0} d)$ (corresponding to a difference of $L_0 \lambda_0/(2\mu_{e0} d)$ in the L axis). L_0 and λ_0 are, respectively, the initial external cavity length and the laser output wavelength without external feedback. The thick solid lines represent the frequencies that have minimum threshold gain (to be discussed later in this chapter).

6.4.2 Transition between Continuous and Discrete Tuning

It is observed in the experiments that the wavelength varies continuously within a certain range, jumps to the adjacent mode (i.e., discrete tuning), and then varies continuously and jumps again [2,17]. The transition between continuous and discrete tuning appeared in the whole tuning wavelength range. This transition can be attributed to the mode selection mechanism in which only the mode with the lowest gain can be oscillated (i.e., homogeneous assumption).

Under the weak feedback condition, the variation of gain can be expressed as

$$g_{th} - g_{th0} = -\ln(R_2/r_2)/d \approx -\xi \cos(2\pi\tau_L v_m) \tag{6.42}$$

where $g_{th0} = \gamma - \ln(r_1 r_2)/d$ is the gain in the absence of external cavity. The threshold gain varies periodically with mirror translation as shown in Figure 6.13b. Similar to continuous shift, the threshold gain has a period of $\lambda_0/2$ and a fixed phase difference of $2\pi L_0/(\mu_{e0} d)$ between adjacent laser modes.

To investigate the transition between continuous and discrete tuning as a function of mirror translation, here is an example, assuming, when $L = L_1$ and $m = m_1$, that

$$\tau_{L_1} v_{m_1} = p \quad (p \text{ is an integer}) \tag{6.43}$$

According to Equation 6.42, the m_1-th mode has the lowest gain and therefore becomes the dominant oscillating mode. If the mirror is displaced slightly so that $L = L_1 + \Delta L$ (assuming ΔL is not large enough to cause mode hopping), the phase terms of the m_1-th and the $(m_1 - 1)$-th mode become

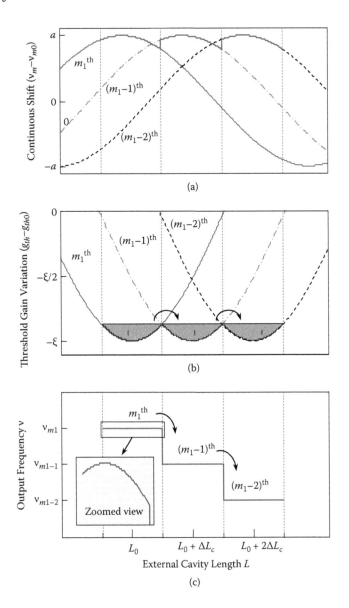

FIGURE 6.13
Frequency tuning with a change of external cavity length: (a) continuous frequency shift, (b) variation of threshold gain, and (c) output frequency after taking into account the transition between continuous tuning and discrete tuning; a stands for $\xi\sqrt{1+\alpha'^2}\Delta v_s/(2\pi)$.

$$\tau_{L_1+\Delta L}v_{m_1} = p + 2\Delta L/c_0 \tag{6.44a}$$

$$\tau_{L_1+\Delta L}v_{m_1-1} = p + 2\Delta L/c_0 - 2(L_1 + \Delta L)\Delta v/c_0 \tag{6.44b}$$

where Δv is the current mode spacing.

To keep the gain of the m_1-th mode lower than that of the $(m_1 - 1)$-th mode, it should satisfy the condition that

$$\Delta L < L_1/(2m_1 - 1) \tag{6.45}$$

In general,

$$|\Delta L| < \Delta L_c/2 \tag{6.46}$$

where $\Delta L_c = L_0\lambda_0/(2\mu_e d)$ represents the critical range within which the laser mode does not change (c for critical).

The thick solid lines in Figure 6.13b indicate the regions within which the minimum threshold gain is obtained. It can be seen that one laser mode, for instance, the m_1-th mode, has the minimum threshold gain when mirror translation is between $L_0 - \Delta L_c/2$ and $L_0 + \Delta L_c/2$. Accordingly, the output frequency is tuned continuously, following the thick solid lines in Figure 6.13a, also shown in the zoomed view in Figure 6.13c. With further translation of the mirror, the $(m_1 - 1)$-th mode obtains the minimum threshold gain and gets oscillated. As a result, the output frequency changes abruptly, then continuously, and abruptly again as shown in Figure 6.13c. The alternation of continuous and discrete tuning makes the laser sweep the whole tuning range.

6.4.3 Conditions and Hopping Range of Discrete-Dominant Tuning

A discrete-dominant tuning laser should have the drifts of mode position and mode spacing well suppressed; that is,

$$\xi\sqrt{1+\alpha'^2}/2\pi \ll 1 \tag{6.47}$$

which requires that $\xi \ll 1$, i.e., the weak feedback condition. From the preceding derivation, it also requires that the short external cavity $L_0 < \mu_{e0}d$. Therefore, the conditions for discrete-dominant tuning are $\xi \ll 1$ and $L_0 < \mu_{e0}d$.

To determine the maximum hopping steps, the external cavity can be approximately treated as an FP filter superimposed on the longitudinal modes of the internal laser cavity. According to Equation 6.42, the frequency hopping range v_r, the wavelength hopping range λ_h, and the maximum hopping steps M are given by

$$v_r \le c_0/2L_0 \tag{6.48a}$$

$$\Delta\lambda_h \le \lambda_0^2/2L_0 \tag{6.48b}$$

$$M_r \le \mu_{e0}d/2L_0 \tag{6.48c}$$

This implies that a short external cavity length has a large tuning range. In conventional tunable lasers, the external cavity length commonly ranges from centimeters to meters. The discretely tunable range is too small to have any practical use. In contrast, MEMS lasers have external cavity length ranging from several microns to hundreds of microns. They support a discrete tuning range as large as tens of nanometers. This is one of the fundamental advantages of MEMS lasers compared to conventional lasers.

6.5 Characterization of MEMS Discretely Tunable Lasers

This section will focus on the characterization of real MEMS discretely tunable lasers. Special attention will be paid to the experimental verification of the analyses developed in previous sections.

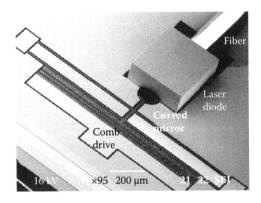

FIGURE 6.14
Overview of the MEMS discretely tunable laser.

6.5.1 Device Description

The preceding analysis provides guidance for the design of MEMS tunable lasers. A proto-type is shown in Figure 6.14, which was demonstrated in a recent work [18]. It is produced by integrating an FP diode laser with a MEMS mirror and a single-mode optical fiber. The fiber is for butt-coupling. After fabrication, the laser chip and optical fiber are integrated and pack-aged. The laser has an overall dimension of $1.5 \times 1 \times 0.6$ mm (not including the full length of the fiber). In the assembly, the laser in contact with the top surface is placed facedown in a trench in the silicon wafer (see Figure 6.14). The bottom of the trench is connected to the wafer surface; both are gold-coated. A layer of indium foil about 45 μm thick is sandwiched between the laser and the trench bottom. This foil serves two purposes. It bonds the laser to the trench, and therefore provides better ohmic contact and heat transfer. It also lifts up the optical axis of laser emission for fiber output coupling. In the experiment, two probes are employed to apply the injection current. The positive touches the wafer surface (i.e., it is connected to the contact surface of the laser), whereas the negative contacts the other laser surface.

A curved mirror is used as the external reflector, as shown in Figure 6.15. As stated previ-ously, the curved shape serves to improve the feedback-coupling efficiency. The mirror has

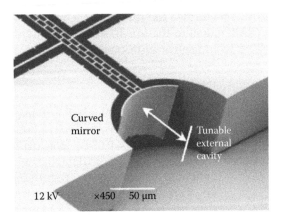

FIGURE 6.15
Close-up of the curved mirror fabricated on an SOI wafer by DRIE. The silicon structure layer is 75 μm thick. The mirror is actuated by an electrostatic comb drive.

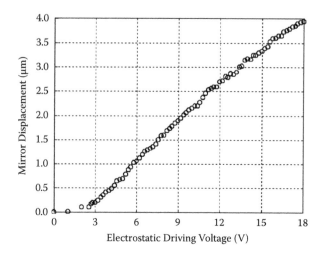

FIGURE 6.16
Static actuation relationship of the comb-drive actuator. Out of the initial dead zone, it exhibits an unexpected linear relationship between mirror displacement and driving voltage.

a curvature radius of 66 μm and is coated with a layer of gold (0.2 μm thick) to improve its reflectivity. The position of the mirror is controlled by applying an electrostatic voltage to a comb-drive actuator. The actuation relationship is shown in Figure 6.16. The mirror displacement increases almost linearly with higher driving voltage and reaches 3 μm at 30 V (displacement $\approx 0.11 \times$ voltage). The optical fiber is placed in a U-shaped groove etched on the silicon layer down to the bottom. The end of the fiber is very close to the emitting window of the laser diode, which ensures simple butt-coupling with an efficiency good enough for the characterization. With regard to butt-coupling, the fiber has a beam diameter of 10.4 μm, which is different from that of the laser output (beam size 1.75 $\times$ 1.53 μm in an elliptical shape). With respect to the optical coupling between the laser and the curved mirror, the optical axis is about 30 μm lower than the upper edge of the curved mirror. At the position of the curved mirror, the laser beam's dimensions are about 74 $\times$ 84 μm (in diameters). The curved mirror is large enough (>120 μm) to reflect the laser beam in the horizontal direction but has to cut off the laser beam in the vertical direction and thus introduces a loss of 3.4 dB. The surface roughness of the mirror due to etching and gold-coating is about 40 nm (RMS value) in the top 40 μm of the mirror. In the bottom part, it gets worse. Such roughness would cause additional losses by scattering.

6.5.2 Wavelength Tunability

The laser chip is made of multiple-quantum-well InGaAsP/InP materials and has a dimension of 210 $\times$ 300 $\times$ 100 μm. The laser facets are not coated (i.e., $r_2 = 0.56$). The effective reflectance of the curved mirror is evaluated to be $r_3 = 0.03$ by measuring the threshold current variation while moving the curved mirror. Using this data, $L_0 = 66$ μm, $\mu_{e0}d = 735$ μm, and $\xi = 0.037 \ll 1$. Therefore, the tunable laser satisfies the requirements for discrete wavelength tuning.

The wavelength-tuning property of the MEMS tunable laser is shown in Figure 6.17. The injection current is kept at 25.4 mA (about 28% over the threshold current 19.8 mA). At this level, the tunable laser can be operated in continuous wave (CW) without additional heat sink. Moreover, it is able to maintain single longitudinal mode (because of

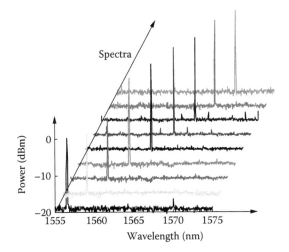

FIGURE 6.17
Laser spectra at different output wavelengths.

homogeneous broadening) when tuned to different wavelengths. The spectra of different output states are shown in Figure 6.17 (for convenience, "wavelength" is used instead of "frequency" in describing experimental data). The output power is about 1 mW, and varies with wavelength tuning (variation <6 dB). The suppression ratio of the side mode is >13 dB. Figure 6.18 shows both the measurement and the simulation of wavelength tuning with regard to mirror displacement. In the experiment, the laser spectrum is measured at different driving voltages, and then the voltage is converted to mirror displacement using the relationship shown in Figure 6.16. The wavelength is initially 1570.04 nm and remains constant at very small mirror displacements. Further movement of the curved mirror makes it suddenly drop to 1556.56 nm (the shortest wavelength). After that, the wavelength appears at the positions of the laser modes, and increases in constant steps of 1.69 nm. A tuning range of 13.5 nm is obtained by tuning the laser to nine modes. Obtaining such a tuning range requires the effective reflectance of the curved mirror to be $r_3 = 0.057$ according to Equations 6.12 and 6.29 (correspondingly, $C = 0.92$ and $\xi = 0.17$, assuming $\Gamma = 0.1$ and

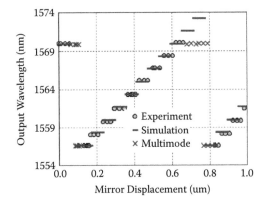

FIGURE 6.18
Measured and calculated wavelength-tuning relationships.

$\alpha = 6$). The value of 0.057 is in good agreement with the predicted coupling efficiency of 5 to 9% given in Figures 6.6 and 6.7. Simulation predicts the measured mode positions with marginal discrepancy, as shown in Figure 6.17. According to Equation 6.48c, eleven modes should be observed. In the positions of the two missing modes, the laser output becomes multimode as shown in Figure 6.17. This is due to the unevenness of the laser gain curve. The weak modulation of the gain threshold is not strong enough to keep only one mode oscillating while suppressing the others, especially when the laser needs to transit from the maximum output wavelength to the minimum. Measurement shows that further displacement of the curved mirror results in periodical wavelength change, which matches the theoretical prediction. The mirror translation corresponding to one period is about half the initial wavelength. The sudden drop and stepwise increase of wavelength are evidence of discrete wavelength tuning. Continuous wavelength change is well suppressed, to be less than 0.040 nm (estimated to be 0.043 nm by Equation 6.26 or 6.41).

6.5.3 Mode Stability

It is also observed that the output at any wavelength is stably single mode when the mirror displaces within a certain region (named as the stable region, about 0.05–0.07 μm of mirror displacement). Two adjacent modes can be in oscillation at the same time once the mirror moves out of this region, and thus mode competition occurs. Different output spectra during mode competition are shown in Figure 6.19a–e. Among them, one mode becomes increasingly weaker, whereas the other keeps increasing. However, they are not in time sequence. Instead, the relative power varies randomly, and therefore the laser output becomes unstable. The length of the unstable region is about 0.01 μm. This can be explained by the areas of minimum threshold gain in Figure 6.13b. Within the central part of any area, the threshold gain of the current mode is apparently lower than that of the adjacent one. Thus, the current mode is stable. However, at the edges of the areas, the gain difference between two adjacent modes becomes very small. Random effects such as fluctuations of temperature, current, and gain play their roles.

6.6 Discussions on the Various Cavity and Integration Scheme Configurations

In addition to the many benefits of MEMS technology (such as batch fabrication, small size, fast speed, etc.), the discretely tunable laser has some special advantages. For example, the laser has high mechanical stability and wavelength repeatability because the laser mode does not change when the external mirror moves within stable regions. A wide tuning range can be obtained by the use of a short external cavity. The laser configuration is simple and does not need any integrated optical lenses or gratings. Movement of the external cavity is simply a translation, in contrast to the complicated mechanical movement in continuously tuning lasers [11,45]. The external reflector needs only to displace a distance less than one wavelength to sweep the whole tuning range giving high-sensitivity output tuning. Another advantage is that all the longitudinal modes are automatically aligned to the ITU grids (if the initial output is deliberately adjusted to an ITU wavelength and the mode spacing is selected to match the ITU grid interval). So that no further wavelength conversions are necessary. Drawbacks include power fluctuation during tuning, low power in single-mode operation, and difficulty in obtaining a very wide tuning range. To obtain the single mode, the injection current should not be too high (typically less than two times the threshold current). Thus, the output power cannot be high

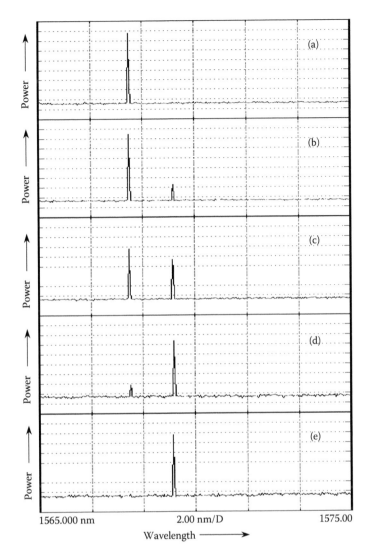

FIGURE 6.19
Mode competition in the unstable region during tuning.

(typically ~1 mW). As wavelength tuning is induced by weak feedback, the unevenness of the gain curve would fundamentally limit the tuning range. Outside a certain wavelength range, the gain of the laser modes would drop to such a low level that a weak modulation of the threshold gain cannot make them oscillate.

The design of MEMS tunable lasers should compromise between various parameters once the target is given. To prototype lasers, however, different integration approaches have been practiced. A brief discussion is presented here on the choice of cavity length and integration approach.

6.6.1 Short Cavity versus Long Cavity

The choice of cavity length is dependent on the goal of tunable lasers and parameters of the laser chip. If continuous tuning is the primary target, the laser facet that forms the external cavity should be AR coated. In this case, the wavelength-tuning range is determined by $\Delta\lambda_r/\lambda_0 = \Delta L/L_0$, where ΔL and L_0 are the cavity length change and initial cavity length, respectively. It can be seen that a small L_0 leads to a large $\Delta\lambda_r$; that is, a short external cavity is preferable for a large tuning range. If discrete tuning is the focus, a short cavity is also preferable according to Equation 6.40a through 6.40c. Only under one condition, that the reflectivity of the laser facet in the external cavity be larger than or at least comparable to the external coupling efficiency, is long cavity important in obtaining a large continuous shift as depicted by Equation 6.31.

6.6.2 Hybrid Integration versus Single-Chip Integration

The core part of an MEMS external cavity tunable laser typically consists of at least three separated components: a gain medium for optical amplification, an external reflector (mirror or grating) plus a microactuator for wavelength tuning, and an optical-coupling system in between. Two approaches have been involved in integrating all these components to form a stand-alone device: hybrid integration and single-chip integration. The former uses the MEMS structure to replace bulky tuning mechanics but keeps the other parts unchanged, whereas the latter aims at fabricating most of the tunable laser components (such as mirrors, grating, lenses, etc.) using the MEMS process and then integrating the whole laser system onto a single chip. Hybrid integration is often the first choice among many options, especially in commercial products; it allows a smooth shift from conventional tunable lasers to the MEMS ones and thus carries less risk. A successful example of hybrid integration was the MEMS continuously tunable lasers that were commercialized [3,45]. Such hybridization improves tuning speed and mechanical stability well, but the size cannot be reduced because the bulky optical components remain. Single-chip integration makes full use of MEMS technology and incorporates many advantages of MEMS into tunable lasers. Microfabricated reflectors and their micromechanical actuators are very light, yielding better tuning properties; the integration further reduces the overall dimension; on-chip fabrication of optical components makes optical packaging/assembly unnecessary, resulting in high yield and low cost; moreover, it facilitates more functional subsystems that consist of several MEMS components such as optical add/drop multiplexers and injection-locked lasers [46,47]. Therefore, single-chip integration represents the trend of further development. Its limitation is that the optical quality of the microfabricated components is still not comparable to that of their conventional bulky counterparts. A lot of work remains to be done before single-chip integration moves to commercial production.

6.7 Summary

This chapter focuses on optical designs, laser tuning theories, and experimental verification of discretely tunable lasers. In the optical design part, different optical coupling systems have been presented for different lengths of external cavity, from a few micrometers to several millimeters. In the theoretical part, the theory is derived for continuous shift and discrete hopping, more suitable for MEMS tunable lasers than the conventional models. In the experimental study, the characteristics of an MEMS discretely tunable laser are investigated. Discrete tuning is measured in constant steps of 1.69 nm over nine wavelength positions. The reasonable agreement of the theoretical prediction with experimental results verifies the effectiveness of the tuning theories developed in this chapter. The tunable laser demonstrates one of the important advantages of MEMS lasers, that is, the capability of constructing a very short cavity. The short cavity paves the way for a large tuning range, in contrast to the negligibly small tuning range in the conventional long-cavity tunable lasers. It can be seen that the inclusion of MEMS in tunable lasers not only downscales bulky lasers because of many engineering merits such as high speed, small size, batch fabrication, etc., but also opens up space for utilizing new physical methods.

References

1. Ukita, H., Uenishi, Y., and Tanaka, H., A photomicrodynamic system with a mechanical resonator monolithically integrated with laser diodes on gallium arsenide, *Science*, 260, 786, 1993.
2. Liu, A. Q., Zhang, X. M., Murukeshan, V. M., and Lam, Y. L., A novel device level micromachined tunable laser using polysilicon 3D mirror, *IEEE Photon. Technol. Lett.*, 13, 427, 2001.
3. Berger, J. D. and Anthon, D., Tunable MEMS devices for optical networks, *Opt. Photon. News*, 14, 42, 2003.
4. Chen, Q., Cole, G. D., Björlin, E. S., Kimura, T., Wu, S., Wang, C. S., MacDonald, N. C., and Bowers, J. E., First demonstration of a MEMS tunable vertical-cavity SOA, *IEEE Photon. Technol. Lett.*, 16, 1438, 2004.
5. Chang-Hasnain, C. J., Tunable VCSEL, *IEEE J. Sel. Topics Quantum Electron.*, 6, 978, 2000.
6. Vakhshoori, D., Tayebati, P., Lu, C.-C., Azimi, M., Wang, P., Zhou, J.-H., and Canoglu, E., 2mW CW singlemode operation of a tunable 1550 nm vertical cavity surface emitting laser with 50nm tuning range, *Electron. Lett.*, 35, 900, 1999.
7. Vail, E. C., Li, G. S., Yuen, W., and Chang-Hasnain, C. J., High performance micromechanical tunable vertical cavity surface emitting lasers, *Electron. Lett.*, 32, 1888, 1996.
8. Pezeshki, B., New approaches to laser tuning, *Opt. Photon. News*, 12, 5, 34–38, 2001.
9. Pezeshki, B., Vail, E., Kubicky, J., Yoffe, G., Zou, S., Heanue, J., Epp, P., Rishton, S., Ton, D., Faraji, B., Emanuel, M., Hong, X., Sherback, M., Agrawal, V., Chipman, C., and Razazan, T., 20-mw widely tunable laser module using DFB array and MEMS selection, *IEEE Photon. Technol. Lett.*, 14, 1457, 2002.
10. Coldren, L. A., Fish, G. A., Akulova, Y., Barton, J. S., Johansson, L., and Coldren, C. W., Tunable semiconductor lasers: a tutorial, *J. Lightwave Technol.*, 22, 193, 2004.
11. Littman, M. G., Single-mode operation of grazing-incidence pulsed dye laser, *Opt. Lett.*, 3, 138, 1978.
12. Liu, K. and Littman, M. G., Novel geometry for single-mode scanning of tunable lasers, *Opt. Lett.*, 6, 117, 1981.
13. Petermann, K., *Laser Diode Modulation and Noise*, Kluwer Academic, London, 1988.
14. Li, J., Zhang, Q. X., Liu, A. Q., Goh, W. L., and Ahn, J., Technique for preventing stiction and notching effect on silicon-on-insulator microstructure, *J. Vac. Sci. Technol. B*, 21, 2530, 2003.
15. Ukita, H., Uenishi, Y., and Katagiri, Y., Applications of an extremely short strong-feedback configuration of an external-cavity laser diode system fabricated with GaAs-based integration technology, *Appl. Opt.*, 33, 5557, 1994.
16. Uenishi, Y., Tsugai, M., and Mehregany, M., Hybrid-integrated laser-diode micro-external mirror fabricated by (110) silicon micromachining, *Electron. Lett.*, 31, 965, 1995.
17. Uenishi, Y., Honma, K., and Nagaoka, S., Tunable laser diode using a nickel micromachined external mirror, *Electron. Lett.*, 32, 1207, 1996.
18. Zhang, X. M., Liu, A. Q., Tang, D. Y., and Lu, C., Discrete wavelength tunable laser using microelectromechanical systems technology, *Appl. Phys. Lett.*, 84, 329, 2004.
19. Liu, A. Q., Zhang, X. M., Tang, D. Y., and Lu, C., Tunable laser using micromachined grating with continuous wavelength tuning, *Appl. Phys. Lett.*, 85, 3684, 2004.
20. Zhang, X. M., Liu, A. Q., Lu, C., and Tang, D. Y., Continuous wavelength tuning in micromachined Littrow external-cavity lasers, *IEEE J. Quantum Electron.*, 41, 187, 2005.

21. Zhang, X. M., Cai, H., Lu, C., Chen, C. K., and Liu, A. Q., Design and experiment of 3-dimensional micro-optical system for MEMS tunable lasers, In *Proc. 19th IEEE Int. Conf. Micro Electro Mech. Syst. (MEMS 2006)*, Istanbul, Turkey, 2006, 830.

22. Aikio, J. K., *Extremely Short External Cavity (ESEC) Laser Devices*, VTT publication 529, ESPOO 2004, http://www.vtt.fi.

23. Heikkinen, V., Aikio, J. K., Alojoki, T., Hiltunen, J., Mattila, A.-J., Ollila, J., and Karioja, P., Single-mode tuning of a 1540-nm diode laser using a Fabry–Pérot interferometer, *IEEE Photon. Technol. Lett.*, 16, 1164, 2004.

24. Aikio, J. K. and Karioja, P., Wavelength tuning of a laser diode by using a micromechanical Fabry–Pérot interferometer, *IEEE Photon. Technol. Lett.*, 11, 1220, 1999.

25. Sidorin, Y., Blomberg, M., and Karioja, P., Demonstration of a tunable hybrid laser diode using an electrostatically tunable silicon micromachined Fabry–Pérot interferometer device, *IEEE Photon. Technol. Lett.*, 11, 18, 1999.

26. Sidorin, Y., Karioja, P., and Blomberg, M., Novel tunable laser diode arrangement with a micromachined silicon filter: feasibility, *Opt. Commun.*, 164, 121, 1999.

27. Broberg, B. and Lindgren, S., Refractive index of $In_{1-x}Ga_xAs_yP_{1-y}$ layers and InP in the transparent wavelength region, *J. Appl. Phys.*, 55, 3376, 1984.

28. Yuan, S. and Riza, N. A., General formula for coupling-loss characterization of single-mode fiber collimators by use of gradient-index rod lenses, *Appl. Opt.*, 38, 3214, 1999.

29. Kiang, M.-H., Solgaard, O., Muller, R. S., and Lau, K. Y., Silicon-micromachined micromirrors with integrated high-precision actuators for external-cavity semiconductor lasers, *IEEE Photon. Technol. Lett.*, 8, 95, 1996.

30. Zhang, X. M., *Theory and Experiment of MEMS Tunable Lasers*, Ph.D. thesis, School of Electrical and Electronic Engineering, Nanyang Technological University, Singapore, 2006.

31. Liu, A. Q., Zhang, X. M., Lu, C., and Tang, D. Y., Review of MEMS external-cavity tunable lasers, *J. Micromech. Microeng.*, 17, R1, 2007.

32. Lee, S. S., Lin, L. Y., Pister, K. S. J., Wu, M. C., Lee, H. C., and Grodzinski, P., Passively aligned hybrid integration of 8×1 micromachined micro-Fresnel lens arrays and 8×1 vertical-cavity surface-emitting laser arrays for free-space optical interconnect, *IEEE Photon. Technol. Lett.*, 7, 1031, 1995.

33. King, C. R., Lin, L. Y., and Wu, M. C., Out-of-plane refractive microlens fabricated by surface micromachining, *IEEE Photon. Technol. Lett.*, 8, 1349, 1996.

34. Wu, M.-H. and Whitesides, G. M., Fabrication of two-dimensional arrays of microlenses and their applications in photolithography, *J. Micromech. Microeng.*, 12, 747, 2002.

35. Hsieh, J., Weng, C.-J., Lin, H.-H., Yin, H.-L., Hu, Y. C., Chou, H.-Y., Lai, C.-F., and Fang, W., The study on SU-8 micro cylindrical lens for laser induced fluorescence application In *Proc. IEEE/ LEOS Int. Conf. Opt. MEMS (Optical MEMS 2003)*, Hawaii, 2003, 65.

36. Maeda, M., Ikushima, I., Nagano, K., Tanaka, M., Nakashima, H., and Itoh, R., Hybrid laser-to-fiber coupler with a cylindrical lens, *Appl. Opt.*, 16, 1966, 1977.

37. Hodgson, N. and Weber, H., *Optical Resonators: Fundamentals, Advanced Concepts and Applications*, Springer, New York, 1997.

38. Kogelnik, H., Coupling and conversion coefficients for optical modes, In *Proc. Symposium Quasi-Optics*, New York, 1964, 333.

39. Wang, W. M., Gratten, K. T. V., Palmer, A. W., and Boyle, W. J. O., Self-mixing interference inside a single-mode diode laser for optical sensing application, *J. Lightwave Technol.*, 12, 1577, 1994.

40. Liu, A. Q., Zhang, X. M., Murukeshan, V. M., Lu, C., and Cheng, T. H., Micromachined wavelength tunable laser with an extended feedback model, *IEEE J. Sel. Topics Quantum Electron.*, 8, 73, 2002.

41. Sidorin, Y. and Howe, D., Laser-diode wavelength tuning based on butt coupling into an optical fiber, *Opt. Lett.*, 22, 802, 1997.

42. Kane, D. M. and Willis, A. P., External-cavity diode lasers with different devices and collimating optics, *Appl. Opt.*, 34, 4316, 1995.

43. Tkach, R. W., Kataja, K. J., Alajoki, T., Karioja, P., and Howe, D. G., Regimes of feedback effects in 1.5 μm distributed feedback lasers, *J. Lightwave Technol.*, LT-4, 1655, 1986.

44. Pan, M. W., Shi, B.-P., and Gray, G. R., Semiconductor laser dynamics subject to strong optical feedback, *Opt. Lett.*, 22, 166, 1997.

45. Anthon, D., Berger, J. D., Drake, J., Grade, J., Hrinya, S., Ilkov, F., Jerman, H., King, D., Lee, H., Tselikov, A., and Yasumura, K., External cavity diode lasers tuned with silicon MEMS, In *Proc. Opt. Fiber Commun. Conf. Exhibit (OFC2002)*, Anaheim, California, 2002, 97.

46. Li, J., Zhang, X. M., Liu, A. Q., Lu, C., and Hao, J. Z., A monolithically integrated photonic MEMS subsystem for optical network applications, *Opt. Commun.*, 249, 579, 2005.

47. Liu, A. Q., Zhang, X. M., Cai, H., Tang, D. Y., and Lu, C., Miniaturized injection-locked laser using microelectromechanical systems technology, *Appl. Phys. Lett.*, 87, paper 101101, 2005.

7

MEMS Continuously Tunable Lasers

Xuming Zhang and Ai-Qun Liu

CONTENTS

7.1 Introduction 274
7.2 Design of Laser Configurations 277
    7.2.1 Littrow Configuration 278
    7.2.2 Littman Configuration 279
    7.2.3 Nonstandard Grating Configuration 279
7.3 Design of Virtual and Real Pivots Using MEMS Mechanisms 282
    7.3.1 MEMS Virtual Pivot Designs 283
    7.3.2 MEMS Real-Pivot Designs 285
7.4 Analysis and Comparison of Real and Virtual Pivots 287
    7.4.1 Real Pivot Using a Double-Clamped Beam 287
    7.4.2 Virtual Pivot Using a Cantilever Beam 288
    7.4.3 Comparison of Real Pivot and Virtual Pivot 291
7.5 Theory of Continuous Wavelength Tuning 293
    7.5.1 Ideal Pivot Position 293
    7.5.2 Optimization of Pivot Position for Maximum Tuning Range 296
    7.5.3 Variation of Gain Medium Refractive Index with Grating Rotation 297
    7.5.4 Influence of Effective Reflectivity of the External Reflector 298
    7.5.5 Wavelength Dependence on Gain-Medium Refractive Index 299
    7.5.6 Comparison of Optimized Design and Conventional Design 300
7.6 Experimental Studies and Discussions 301
    7.6.1 Device Description 301
    7.6.2 Characteristics of Blazed Grating and Microlens 304
    7.6.3 Characteristics of Laser Gain Chip 306
    7.6.4 Wavelength Tunability 309
    7.6.5 Comparison and Discussions 312
7.7 Summary 313
References 313

This chapter focuses on the designs, analyses, and experimental issues of microelectrome-chanical systems (MEMS) tunable lasers for continuous wavelength tuning. In comparison to the discretely tunable lasers in the previous chapter, this type of laser is preferable for many applications, such as optical communications, biomedical studies, laboratory equipment, etc. The laser in this chapter will also serve as the key component for developing more complicated and more functional laser systems, to be discussed in Chapter 8.

As conventional continuously tunable lasers have been extensively studied, this chapter concentrates on new features enabled by MEMS technology, such as the new laser configurations, new working mechanisms, and new design issues. First, the typical laser configurations will be presented in Section 7.2. Then, different designs of virtual and real pivots will be discussed in Section 7.3, followed by the mechanical analysis and comparison of the real and virtual pivots in Section 7.4. Next, theoretical studies on the optimization of a fixed pivot will be detailed in Section 7.5. Finally, experimental implementation and characterization of the laser device will be presented in Section 7.6.

7.1 Introduction

Tunable lasers are of great importance in many applications such as optical communications, biophysics, and environmental engineering [1,2]. Among all the technological options, the external cavity tunable laser (ECTL) has led itself to the mainstream because of its advantages of simple configuration, wide tuning range, spectrum purity, high power, and continuous wavelength tuning [3]. Conventional ECTLs are in either the Littrow or Littman configuration [4–8]. The design of the Littrow laser can be traced back to the 1960s [4], soon after the emergence of the ECTL [5]. At that time, tunable dye lasers occupied the center stage in laser research, driven by their exciting applications for material spectroscopy, which required high spectrum purity, such as single mode, narrow line width, continuous tuning, etc. To overcome the problems of wavelength tuning discontinuity and multiple longitudinal modes associated with flat mirrors, gratings were introduced to the tunable lasers as a dispersive external reflector. Bradley et al. used a blazed grating arranged in the Littrow configuration in a dye laser (this laser configuration is thus named the *Littrow configuration*) and obtained 10 nm wavelength tuning by rotating the grating [4]. Another Littrow-pulsed high-power dye laser was also reported to tune a large range, from 437.5 to 642.0 nm, with a line width of only 0.3 nm [6]. The Littman configuration was proposed and analyzed by Littman and his colleagues in a series of studies in the late 1970s [7]. The initial purpose was to narrow the line width of dye lasers and obtain a single longitudinal mode. Later, the configuration was extended to the wavelength tuning of dye lasers [7]. The same configuration was also demonstrated to tune semiconductor lasers [8].

In both Littrow and Littman configurations, wavelength tuning is realized by the simultaneous rotation and translation of an external reflector [2,3], which remains a challenge for the conventional bulky mechanics to achieve fast tuning speed and high reliability. The development of MEMS technology has opened up a new opportunity to downscale a bulky tunable laser to a miniaturized device, while improving the tuning speed and mechanical reliability at lower fabrication cost [1,2]. Most of the early MEMS tunable lasers used mirrors [9–17] and Fabry–Pérot filters [18–20] as the external reflectors and were all not continuously tunable; the literature review is given in Chapter 5, so here we do not repeat it here. The first continuously tunable laser was implemented in a Littman configuration in 2001 and was later commercialized [10]. In that experiment, a silicon mirror was mounted on a specially designed comb-drive actuator [21], whose rotation by ±1.4° produces a continuous wavelength tuning over 40 nm. Another Littman laser was produced by constructing a virtual pivot using MEMS flexures [22]; a tuning range of 65 nm was reported with an angular rotation of 3.6°. In both lasers, the MEMS part was only used to rotate a mirror, and the other parts (optical lenses, gratings, etc.) were still conventional bulky components. All of them were integrated in a hybrid manner. This increased the tuning speed but did not reduce the overall dimensions. Recently, Cai et al. demonstrated a single-chip integrated Littman laser to obtain two output wavelengths at the same time the wavelength difference is being tuned [23]. In this study, all the basic components for a tunable laser, except for the laser chip and the coupling fiber, are made by the MEMS process and are integrated on the same MEMS chip. It makes full use of the MEMS fabrication capability and leads to clear advantages in high compactness, better mechanical stability, and savings in assembly.

Compared with the Littman configuration, the Littrow configuration has high coupling efficiency and simple tuning structure, and thus has attracted much interest from researchers. For example, Lohmann and Syms produced an integrated hybrid Littrow laser by using a deep-etched grating [24]. A ball lens is used for laser collimation, and a micromanipulator for grating movement. A wavelength tuning range of 120 nm was claimed. Their main purpose was to verify the optical quality of micromachined gratings rather than obtain a tunable laser, and thus, the compact integration was not of interest. The same group later contrived different micromechanical tuning flexures to provide remote pivots for blazed gratings, and implemented the flexures into several MEMS lasers [24,25]. These lasers have, in some sense, single-chip integration because the actuators of the gratings are now made by the MEMS process. However, the lenses are still external components. The first single-chip integrated laser was developed by Liu and Zhang, in which the microlens, blazed grating, and tuning mechanical structures were fabricated on the same chip [26–28]. After integration of the MEMS parts with the gain chip and optical fiber, the device had dimensions of only 2.0 × 1.5 × 0.6 mm, and could be continuously tuned over a range of 30.3 nm. Although the obtained output power (typically 1 mW) is quite weak in comparison to their conventional counterparts—mainly because of the low diffraction efficiency of the MEMS gratings and poor thermal conductivity of the silicon substrate—the developed lasers have demonstrated the merits of MEMS such as fast tuning speed, small size, and single-chip integration. Liu and Zhang also presented a series of studies on different aspects of tunable lasers such as theoretical analyses, new laser configurations, new reflector designs, and new coupling systems [26,27,29–32]. For example, the use of nonstandard grating to construct the external cavity gave rise to an additional option of laser configuration [29] in comparison to the standard Littrow and Littman configurations: the introduction of the three-dimensional (3-D) optical coupling system to the external cavity ensures high coupling efficiency [30] (refer to Section 6.2 for more details on the 3-D coupling system). The use of current injection to tuning wavelength in the MEMS coupled-cavity lasers [31] circumvents the limit of tuning speed by the micromechanical movement in all the previous MEMS tunable lasers. The tuning specifications have also been improved significantly: for instance, the tuning range, from <10 to >50 nm [28,29,32]; the output power, from ~0.1 to >1.5 mW [26,30]; and the sweeping speed, from ~1 ms to ~1 ns [27,31]. Recently, another laser was constructed using a deep-etched rectangular grating (not blazed), which was hung by a torsional spring on the back-opened silicon chip [33]. The wavelength tuning was obtained by rotation of the grating about its central line. However, the cavity length was not changed during the rotation of the grating. As a result, the tuning was not continuous, and hence, it is not discussed any further in this chapter.

The rotation center (i.e., the pivot) plays an important role in both the Littrow and Littman configurations. The wavelength tuning is obtained by rotating an external reflector, which is a grating for the Littrow and a mirror for the Littman. For the convenience of control, a fixed pivot is used in most conventional tunable lasers. As a result, the choice of pivot position determines tuning behavior and wavelength range. However, theoretical analysis of the pivot position was not undertaken until McNicholl provided a straightforward explanation [34]. In his analysis, the Littrow laser was abstracted into a mirror and a grating. The mirror and the grating represent the two ends of the overall laser cavity.

The length between the mirror and the grating in the optical axis is equal to the original laser's internal cavity length plus the external cavity length. If the rotation center of the grating is right at the intersection of the surface planes of the mirror and grating, the round-trip phase accrual is a constant that is independent of grating angle and wavelength. In this way, the returned wave is able to maintain a constant phase relative to the incident wave throughout the wavelength tuning, and thus, the wavelength of the laser cavity mode can always match the diffraction wavelength of the grating. In other words, the wavelength tuning is always continuous. Although McNicholl's study was based on the geometric optics and scalar diffraction theory, it was later proved correct by several other analyses [35,36] and many laser implementations. However, McNicholl's study neglected the cavity length change caused by the dispersive property of the laser medium and optical lens, which is acceptable in the bulky optomechanical lasers because the laser cavity length is much longer than the dispersion-induced change. A further study was done by Trutna et al., who presented a Doppler shift mode to explain the wavelength tuning and to predict the optimum pivot position [37]. Theoretical analysis for the MEMS Littrow lasers was presented by Zhang in Reference 26. This work reexamined the previous studies and found that in the Littrow lasers: (1) a mode-hop-free wavelength tuning requires a moving pivot; (2) more realistically, a fixed pivot has only a limited tuning range; (3) the refractive index change of the gain medium, which was neglected in previous studies [34–37], significantly affects the obtainable tuning range; and (4) the chromatic aberration of the microlens also affects the tuning range.

To solve these problems, two methods have been proposed. One is to optimize the pivot position [26], and the other is to employ an adjustable pivot [32]. The former is for a fixed pivot. Optimization takes into account the refractive index change and aberration of the microlens during the tuning. Theoretical analysis suggested significant improvement of the wavelength tuning range. In one experimental study, it was demonstrated that the wavelength range could be increased from 5.9 to 30.3 nm by optimizing the pivot position [26]. In the adjustable pivot solution, the pivot position is movable and controllable, or in other words, the rotation and translation of the blazed grating can be controlled separately and independently. In this way, the tuning range can reach an arbitrary high level. This idea was proposed and demonstrated in Reference 32. It employs two independent MEMS actuators to move the two ends of the grating at different rates. The average displacement yields the shift of the grating position and, thus, the change of the external cavity length, whereas the displacement difference of the two grating ends produces a rotation of the grating and, thus, the diffraction angle. In this way, the cavity length change and the grating angle can be controlled independently to maintain the wavelength tracking over an arbitrary wavelength range. From the pivot's point of view, it is no longer a fixed pivot; instead, its position is freely adjustable. Thus, it can be called the *adjustable pivot design*. It is a breakthrough in laser design as it can eliminate the mode-hopping problem that has long been associated with conventional tunable lasers. It also demonstrates the uniqueness of MEMS when applied to tunable lasers.

Compared with the conventional ECTLs, different concerns are associated with the MEMS tunable lasers [26]. For example, the components in the MEMS structure are pre-aligned with high accuracy (limited by the resolution of photolithography) and cannot be readjusted. Therefore, in the laser design the position should be accurate. Refractive index change may not, therefore, be a serious problem in conventional mechanical ECTLs because fine adjustment in the laser calibration stage would be able to eliminate any position error. However, in the MEMS tunable lasers, no further adjustment is possible. Another concern is that MEMS tunable lasers have a short cavity length (typically ~1 mm) in comparison with the long cavity in conventional lasers (typically ~1 cm to 1 m). For this reason, the cavity change due to the refractive index change of the gain medium is negligible in conventional lasers but significant in MEMS tunable lasers.

7.2 Design of Laser Configurations

This section will discuss different configurations that have been used in the developed MEMS continuously tunable lasers, especially the new features enabled by MEMS technology.

To tune the wavelength continuously while maintaining a single longitudinal mode, various laser designs have been proposed [1]. According to the arrangement of the gratings, the developed MEMS lasers can be simply classified into three configurations: (1) the Littrow configuration, (2) the Littman configuration, and (3) the nonstandard grating configuration, as shown in Figure 7.1a–c. The Littrow and Littman configurations are simply the miniaturized version of the conventional lasers [10,22–24,26–28], in which the gratings have straight baselines and require the incident beam well collimated. In these two configurations, lenses have to be used for laser collimation and focusing.

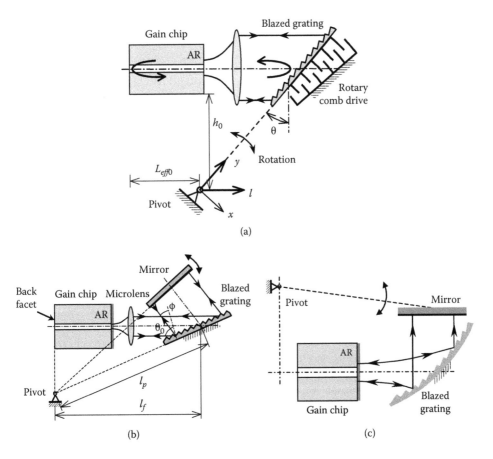

FIGURE 7.1
Typical configurations of the developed MEMS continuously tunable lasers as classified by the arrangement of external reflectors: (a) Littrow configuration, (b) Littman configuration, and (c) nonstandard grating configuration, a blazed grating having variable profile and position of teeth to diffract and collimate the laser beam simultaneously.

As the common materials of the MEMS structure layer have high optical refractive index, the use of lenses causes quite high Fresnel reflection losses in the interfaces with the air. For example, single crystalline silicon has refractive index $n = 3.482$ at 1550 nm, and each Si–air interface loses optical power by $R = (n-1)^2/(n+1)^2 = 30.6\%$. After a round trip in the external cavity, only 23.1% of laser power is left. To tackle this problem, it is desirable to use a reflector rather than a lens to collimate the laser beam. Moving the idea a step forward, it is possible to combine the reflector and grating to form a nonstandard grating that has a curved baseline and serves the purpose of collimation and diffraction simultaneously, as shown in Figure 7.1c [29]. As the MEMS process employs photolithography for patterning, nonstandard gratings can be easily fabricated without the need for an extra process step. This is one of the special merits that MEMS imparts to tunable lasers.

In the design of all these configurations, one common factor is that the wavelength-tracking condition should be met. In any of these configurations, there exist two factors that determine the oscillation wavelength at the same time. One is the Fabry–Pérot resonance of the total laser cavity (including both the internal and external cavities), and the other is the grating diffraction. To tune the wavelength continuously over a large range, the wavelengths given by these two factors should match initially and continue to match during the tuning. This is wavelength tracking, without which mode hopping may occur, and continuous tuning will fail. From this point of view, the different configurations are just to obtain wavelength tracking by different mechanical arrangements.

7.2.1 Littrow Configuration

In the Littrow configuration, the laser is collimated to and diffracted by the blazed grating at the same angle, as shown in Figure 7.1a. It is a single-pass diffraction; that is, the light is diffracted only once in the external cavity. Wavelength tuning is realized by rotating the grating about a remote pivot [24,26,27,34–37]. Wavelength tracking can be expressed to the first order as [26,34]

$$N_0\lambda = 2(L_{eff\,0} + h_0 \tan\theta) \tag{7.1}$$

$$m_0\lambda = 2p_0 \sin\theta \tag{7.2}$$

where λ represents the oscillation wavelength, N_0 the laser mode number, m_0 the grating diffraction order, p_0 the grating period, θ the diffraction and also the rotation angles of the pivot arm about the pivot point (the clockwise direction is set positive), h_0 the vertical distance from the pivot to the light path, and $L_{eff\,0}$ the initial effective total cavity length (including the internal cavity, the external cavity, and the lens thickness).

Similar to the conventional lasers, MEMS Littrow lasers suffer from the fundamental limitation that the wavelength-tuning range cannot be very large (typically <30 nm) if the pivot for the grating rotation is fixed [26,34–37]. This is due to the mismatching of wavelength selection mechanisms in the case of a fixed pivot. According to Equation 7.1, the wavelength selected by the laser cavity resonance is changed with the rotational range by a tangential curve. In contrast, it can be seen from Equation 7.2 that the diffracted wavelength selected by the grating is changed by a sine curve. That is to say, one is a tangential curve, and the other is sine; thus, the wavelength mismatch would accumulate with the increase of rotation angle (corresponding to the increase of wavelength tuning) and eventually break the wavelength tracking. This limitation is not due to MEMS but due to the geometric arrangement of the gain chip relative to the grating. The solution is to optimize the pivot position, or to use an adjustable pivot, as discussed in Section 7.1. Detailed explanations and analyses of these two solutions will be given in Sections 7.3 and 7.5.

7.2.2 Littman Configuration

The Littman configuration has one more mirror in comparison to the Littrow configuration, as shown in Figure 7.1b [10,22,23]. The Littman laser comprises a gain medium (e.g., dye cell), a blazed grating, and a tuning mirror [7]. The laser from the gain medium is first projected to the blazed grating, and then steered to the tuning mirror by diffraction; it is reflected back to the grating, and finally diffracted again back into the gain medium. This is double-pass diffraction i.e., the laser beam is diffracted twice in a round trip of the external cavity. Unlike the rotational grating in the Littrow configuration, the grating in the Littman configuration is fixed, and the mirror is rotated about a remote pivot for wavelength tuning. The zero-th order of the grating diffraction can be used as the output. Wavelength tuning is achieved by rotating the grating about a pivot. To track the grating diffraction wavelength over the entire tuning range, the pivot position should be chosen in such a way that the parameters satisfy the following conditions [7]:

$$N_0 \lambda = 2(l_f + l_p \sin \phi) \tag{7.3}$$

$$m_0 \lambda = p_0 (\sin \theta_0 + \sin \phi) \tag{7.4}$$

where θ_0 is the incidence angle, ϕ is the diffraction angle, l_f is the distance from the back facet of the laser chip to the grating surface, $l_p \sin \phi$ is the distance from the grating to the tuning mirror, and l_p is the distance from the fixed remote pivot to the grating center. Here, l_f is not the physical distance, instead it is the total effective optical path length.

By comparing Equation 7.3 with Equation 7.4, it is observed that the two equations can be always identical if

$$l_f = l_p \sin \theta_0 \tag{7.5}$$

This requirement can be easily satisfied if the pivot is chosen at the intersection of the grating surface and the laser back facet, as shown in Figure 7.1b. Note that the position of the laser back facet in this discussion is not present, but should be extended from the grating center by l_f. In this way, the geometric arrangement of this configuration does not impose any constraint on the tuning range [7]. Here, the Fabry–Pérot cavity length is changed by a sine curve (no more tangential curves) with the rotation angle, exactly the same way as with the sine relationship in the blazed grating. That is, exact wavelength tracking is achievable for all accessible wavelengths for designs meeting the condition of Eq. 7.5. Therefore, a large tuning range can be easily obtained with a fixed pivot, making it very attractive for industrial products.

Similar to the Littrow theory, this analysis also is based on geometric optics and scalar diffraction theory. When the size of the cavity becomes small, some other effects such as the Gaussian expansion of the laser beam and dispersion of the gain medium should be taken into account. The Littman configuration requires grazing incidence on the grating, which helps to cover more grating teeth at a given laser beam size and thus obtains narrower line width, but this results in low diffraction efficiency, especially when the grating is fabricated by micromachining. It can be seen from Figure 7.1b that the remote pivot always cuts across the gain chip and the light path because the pivot and the tuning mirror are on opposite sides of the light path. It hinders the construction of a real pivot, and thus, a complicated mechanical structure may have to be used to realize a virtual pivot.

7.2.3 Nonstandard Grating Configuration

The nonstandard grating configuration in Figure 7.1c represents a new arrangement enabled by MEMS technology. In a recent work, use of a variable grating profile obtained simultaneously the collimation and diffraction of a laser beam [29]. This grating is different from the standard blazed gratings that have straight or concave profiles, and so it is called a nonstandard grating. Similar to the standard Littman configuration, it also consists of an antireflection (AR)-coated gain chip, a fixed blazed grating, and a rotational micromirror. The laser light emitted from the gain chip shines directly onto the blazed grating; the selected wavelength is diffracted to the micromirror and is then reflected and diffracted back into the gain chip. In comparison with the standard Littman configuration shown in Figure 7.1b, the nonstandard grating configuration does not need a collimation lens. Instead, it employs a nonstandard blazed grating, which has a curved shape, and the grating teeth have a varying profile. This is to realize the optical coupling between the gain chip and the micromirror without the need for any optical lenses, while maintaining the wavelength selection ability of the blazed grating. This type of variable profile is not practical for conventional ruled gratings because of fabrication difficulties. However, because MEMS fabrication makes use of photolithography, it is easy to pattern a structure with an arbitrary 2-D shape. This is particularly useful for fabricating the nonstandard grating.

To determine the position and profile of the grating teeth, we first consider a simple case where the laser light emitted from the laser diode is collimated in the direction perpendicular to the optical axis of the laser diode after it is diffracted by the grating, as shown in Figure 7.2. Then we can find a solution to the problem of collimating/diffracting the light in any direction. The analysis that follows is based on the assumptions that (1) the light emitted from the laser diode can be regarded as a point light source, and (2) the geometric optics is valid.

In Figure 7.2, the emitting point $C(0, 0)$ of the laser diode is taken as the origin of the coordinate system (ξ, η). As the equiphase planes (representing the possible positions for the mirror) have fixed phase difference, it is reasonable to choose the optical axis of the laser diode as the primary reference plane. The light path length L_n for the light beam passing through the n-th tooth of the grating is given by

$$L_n = CA + AB = \xi_n / \cos(-\theta_n) + \xi_n \tan(-\theta_n) \tag{7.6}$$

where $n = ..., -2, -1, -, 1, 2, ...,$ and (ξ_n, η_n) is the position of the center of the n-th grating tooth; θ_n is the open angle (it is negative when n is a negative integer).

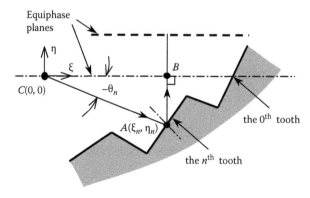

FIGURE 7.2
Determination of the position and profile of the grating teeth in the nonstandard grating configuration.

As required by the blazed diffraction, the light path should have fixed difference related to the wavelength, as given by

$$L_n = L_0 + n \cdot \frac{m\lambda}{2} \tag{7.7}$$

where m is the diffraction order, λ is the wavelength, and L_0 is the light path length for the zero-th grating tooth (right on the optical axis of the laser diode).

From Equations 7.6 and 7.7, the position of the n-th grating tooth can be obtained as follows:

$$\xi_n = \frac{\cos\theta_n(2L_0 + nm\lambda)}{2 - 2\sin\theta_n} \tag{7.8a}$$

$$\eta_n = \frac{\sin\theta_n(2L_0 + nm\lambda)}{2 - 2\sin\theta_n} \tag{7.8b}$$

For practical use, one more condition should be applied to the grating period. Here, we assume that the grating has an equal period when projected to any equiphase plane; that is,

$$\xi_n = L_0 + n\Delta\xi \tag{7.9}$$

where $\Delta\xi$ is the projected grating period in the equiphase plane. With this condition, the open angle of the n-th tooth can be expressed as

$$\theta_n = \arcsin\left(\frac{\xi_n^2 - L_n^2}{\xi_n^2 + L_n^2}\right) \tag{7.10}$$

Once Equations 7.8a and 7.8b are satisfied, the m-th order of the laser light at the wavelength λ will be diffracted to the perpendicular direction. If a mirror is put in any of the equiphase planes, the laser will be oscillated at the wavelength λ. Translation of the mirror does not influence the filtering property of the blazed grating (it may affect the external cavity length). This is a special merit of the new design and makes the alignment easy. In contrast, the standard Littman configuration has a strict requirement for the mirror position and does not allow the mirror to translate [7]. This particular design does not need a lens for collimation, and thus circumvents the losses induced by the micromachined lenses or the extra assembly of external lenses. In addition, it avoids the use of grazing incidence to the grating, which causes low diffraction efficiency in the Littman configuration. The incident angle for different teeth varies within a small range near 45° (the incident angle for the zero-th tooth).

Once the light is diffracted in the perpendicular direction, it is easy to design the grating profile to diffract it in any direction. As can be observed from Figure 7.2, the diffraction relationship is kept invariant when all the structures are rotated about the origin C. Therefore, it only needs to rotate all the grating teeth about C by the required angle so as to steer the diffracted light to any other angle.

The basic idea of the nonstandard grating is that by gradually varying the position and angle of each grating tooth, the optical paths that pass through adjacent grating teeth have a fixed phase difference of $m2\pi$ (where m is an integer denoting the order of diffraction) when they reach a certain plane (where the mirror sits). The rotation of the mirror introduces an extra phase difference and thus tunes the wavelength. Similar ideas can be found in the concave gratings and replicated holographic gratings [38] that sit the grating teeth on spherical or curved surface so as to focus and diffract the laser beam simultaneously.

Some recent work shared this idea in patterning the planar waveguide to form lenses and gratings for tunable lasers, though the tuning is not based on MEMS but on the refractive index change by injected electrical current [39]. In the work that first proposed this nonstandard grating [29], the tuning structure is fabricated by deep etching on a silicon-on-insulator (SOI) wafer, and is then integrated with the gain chip and optical fiber onto a single chip, with a footprint of only 3 × 2.5 mm. The grating teeth have varying position and profile, but share the same period of 3.32 μm if projected to the equiphase plane (i.e., the mirror surface). It reported a tuning range of 9.2 nm. Although the obtained tuning range is quite limited, it demonstrates the potential of MEMS in enabling new laser tuning configurations, which is not practical for conventional lasers because of the difficulty in fabricating gratings.

7.3 Design of Virtual and Real Pivots Using MEMS Mechanisms

This section will start with the requirements that pivots have to satisfy when they are used in MEMS tunable lasers. Then, different designs of virtual and real pivots will be presented. The new pivot designs enabled by MEMS technology will also be discussed.

Any continuous tunable external cavity laser would need a pivot for rotating the grating. These pivots can be simply categorized into real and virtual pivots, or fixed (i.e., unadjustable) and adjustable pivots, depending on the angle characterizing them. The real pivot has a pivot physically located on the mechanical beam, whereas the virtual pivot has the pivot positioned at a remote position (out of the beam). The virtual pivot can realize a pivot at a distance with a limited beam length and thus avoid conflicting space requirements for multiple components. It is especially useful for the Littman configuration, which commonly has the pivot arm (i.e., the line from the pivot to the grating) cutting through the gain chip or the external cavity [7,10,22,29]. However, the position of the virtual pivot is determined by the combination of several parameters, so that it is sensitive to condition changes such as fabrication error, temperature change, and residual stress [22]. For example, a small deviation of fabricated beam width or height could shift the pivot position significantly and actually make the whole design fail. In contrast, the real pivot has high tolerance and is more robust because the pivot position is determined by the presence of the beam itself. Strictly speaking, only the pivot realized using a bearing structure can be called a *real pivot*. It is commonly used in conventional bulky laser systems. However, the use of the term *real pivot* here will not cause any confusion as an in-plane bearing is not easily available for MEMS devices.

A fixed pivot has a stationary rotational center regardless of the working condition; it is rigid and robust enough to sustain the disturbance and uncertainty (such as fabrication error, external force, temperature change, etc.) to a certain extent. It is useful for all the laser configurations. In contrast, an adjustable pivot can shift the pivot position in a controllable manner. It is useful to maintain wavelength tracking for the Littrow lasers, to allow for very large tuning ranges. The study that follows is not intended to give an extensive analysis of all the pivots. It merely focuses on the MEMS mechanisms for making the virtual and real pivots, and at the same time investigates the pivot shift under loading and its influence on wavelength tunability. Further, the discussion is limited to pivot designs that provide in-plane rotation by the deformation of micromechanical beams, which are relevant to MEMS lasers. Other structures that produce bearings and out-of-plane rotation are not included.

Before a detailed discussion on pivot structures, the requirements of tunable lasers to the pivot should be examined. First, a large rotation angle is not absolutely necessary for

MEMS tunable lasers; instead, a small angle of about 4° is quite enough. As an estimation, in the Littrow configuration, the blazed grating should have $m_0\lambda = 2p_0 \sin\theta$, and thus $\Delta\theta = \Delta\lambda \tan\theta/\lambda$. (see Equation 7.1). Following the parameters $m_0 = 3$, $\lambda = 1.51$ μm, $p_0 = 3.2$ μm, and $\theta = 45°$, as given in Reference 26, a rotation angle of $\Delta\theta = 4°$ would correspond to the wavelength tuning $\Delta\lambda \approx 108$ nm. Such a tuning range is quite rare for a single piece of tunable laser chip. Second, in the tuning structure that employs a fixed pivot (either real or virtual pivot) [22,25–27,40–42], the shift of pivot position during the tuning should be minimized. Although in most cases the pivot is designed to be fixed, it could deviate from its optimized nominal position in the real devices owing to load, fabrication error, environmental change (e.g., temperature, humidity), etc. Such shifts can be decomposed into two directions, l and y, as shown in Figure 7.1a. The former is along the optical axis of the external cavity, whereas the latter is along the grating surface. The pivot shift in the l direction causes a variation of cavity length and may deteriorate the wavelength tracking. As a result, the continuous tuning wavelength range would be reduced. The pivot shift in the y direction does not change the cavity length, but it influences the operation of the microactuators. Commonly, a rotary comb drive is utilized to generate drive force because of its compactness, robustness, high rotation accuracy, and environment/temperature insensitivity [21,22,40–42]. The movable comb of a rotary comb drive should rotate concentrically with the fixed comb, as shown in Figure 7.1a. The shift in the y direction causes the movable fingers to have unequal gaps relative to the fixed fingers on both sides, and the imbalanced electrostatic force would further shift the movable comb in the radial direction until it suddenly touches the fixed comb (called a *pull-in problem*). As a result, the achievable rotation angle is restricted, which in turn limits the continuous tuning range. Because mode hopping would occur (i.e., loss of wavelength tracking) when the external cavity length is deviated by a half-wavelength, as a rule of thumb, the pivot shift in the l direction should be less than $\lambda/100$ (i.e., about 15 nm). Similarly, the pivot shift in the y direction should be smaller than 1/50 of the initial finger gaps. It is 40 nm if the finger gap is 2 μm. It is noted that the y and l directions may not be perpendicular to each other, as illustrated in Figure 7.1a. In the adjustable virtual pivot, the opposite is true; here, the pivot shift is desirable and controllable, an important requirement being that the accuracy and speed of position control should be high enough.

A proper design of the pivot should take into account other factors such as the integration method and pivot type. There are two solutions to integrate the MEMS tunable lasers, as discussed in Chapter 5. One is the hybrid integration method, which makes use of MEMS only for actuation, while keeping the conventional bulky components for other functions such as beam collimation and wavelength selection. The final device would have a large size and would involve manual assembly [10,25]. The other method is single-chip integration; this makes full use of micromachining to fabricate on the same silicon wafer as many components (actuators, lenses, gratings, mirrors, etc.) as are necessary for a MEMS laser [9,13,14,18,26–28]. Single-chip integration represents the trend of development, because it has fast tuning and high compactness, and needs less work in packaging and batch production, though the optical quality of micromachined components is still not comparable with that of conventional ones. In brief, MEMS tunable lasers may require the real pivot or virtual pivot, depending on the target and working condition; they do not require a very large rotation angle, but impose a tight limit on the pivot shift. In addition, single-chip integration is preferable.

7.3.1 MEMS Virtual Pivot Designs

As discussed previously, the virtual pivot is of importance to the continuously tunable laser, especially for the Littman configuration. Several designs of the pivot structures

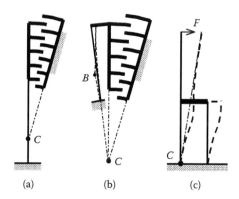

FIGURE 7.3
Design of virtual pivots that make use of the combined deformation of MEMS flexures: (a) virtual pivot by a cantilever beam, (b) virtual pivot by a folded cantilever beam, and (c) virtual pivot by a compound flexure.

have been demonstrated, as shown in Figure 7.3. A simple design is realized by attaching the rotary comb drive to the end of a cantilever beam, as shown in Figure 7.3a (called a *simple cantilever design*). It is well known that the end of a cantilever beam is rotated about a virtual pivot located at one-third the beam length (called the *1/3 rule*) when a force is applied at the free end [43]. This design was implemented to obtain a grating rotation of 1.6° in a MEMS tunable laser [28]. The 1/3 rule is also discussed in Reference 22. However, the actual pivot position is probably not at one-third beam length, as will be discussed in Section 7.4. A modified design is to fold the beams, as shown in Figure 7.3b (*folded cantilever design*). The movable comb is not at the extension of the cantilever beam. Instead, it is folded inside. The cantilever beam and the comb fingers are arranged with respect to a common center *C*, which is not at the one-third beam length. This type of design was used as an element in constructing several complicated pivot structures that were patented and implemented in commercialized MEMS lasers [10,21]. For a single element of such a design, the 1/3 rule still holds (as marked by the point *B*) because the driving force is applied at the free end of the cantilever beam. Using several elements that are jointed together will make the virtual pivot more complicated, and it may vary with the rotation angle rather than stay at *C*. For this reason, the comb fingers have to be tilted and offset so as to compensate for the pivot shift, as proposed in Reference 21, making the design and analysis more complicated. Another design realizes a virtual pivot by tailoring the deformation of a compound flexure (*compound flexure design*), which consists of a cantilever beam and a portal frame, as shown in Figure 7.3c [25]. When a concentrated force *F* is applied at the free end, the rotation angle comes from only the cantilever beam, although the displacement is contributed from both the cantilever beam and the portal frame. By proper choice of the relative lengths of the cantilever beam and the portal frame, the virtual pivot can be located at the root of the portal frame or at other remote positions. When the developed designs are examined under the requirements discussed previously, the simple cantilever design and the compound flexure design allow for single-chip integration, but both are unadjustable virtual pivots. In the presence of fabrication errors and environmental changes (such as temperature, humidity, etc.), the actual position may deviate from the desired position and may eventually limit the tuning range. The folded cantilever design is also an unadjustable virtual pivot. In addition, it does not allow for single-chip integration when many elements are used to build complicated driving structures [10,21].

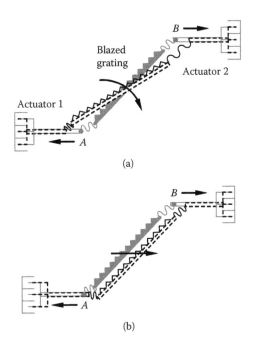

FIGURE 7.4
Design of the adjustable virtual pivot that employs two actuators to generate independent rotation and translation of the grating: (a) pure rotation of the grating and (b) pure translation of the grating.

Recently, a new design of an adjustable virtual pivot (*adjustable virtual pivot*) was demonstrated for solving the wavelength-tracking problem in the Littrow configuration [32]. The concept of the tuning method is shown schematically in Figure 7.4. The blazed grating is still used as the wavelength-tuning element in the tunable laser. It is not rotated about a fixed pivot. Instead, its two ends are mounted onto two independent MEMS actuators (commonly, electrostatic comb drives, or thermal actuators). In this way, the translation and rotation of the blazed grating can be controlled separately by properly shifting the two joints (*A* and *B*), driven by the two actuators. In principle, when two soft joints are moved in opposite directions, or they are shifted in the same direction but by different amounts, a grating rotation would occur, as shown in Figure 7.4a. When the two joints move in the same direction and at the same rate, a translation of the grating could occur, as shown in Figure 7.4b. Therefore, the independent control of the movement provides a separate adjustment of the cavity length and the grating incidence angle. Thus, it is easy to obtain any relationship of grating angle and cavity length, and this offers a fundamental way to eliminate the mode-hopping problem. Moreover, it is convenient to adjust the initial condition of tunable lasers so that discrepancies of gain chip size are tolerated. From the pivot's point of view, the instantaneous pivot position and rotation arm length can be adjusted to any desired values.

7.3.2 MEMS Real-Pivot Designs

A real pivot is by no means a key feature of MEMS tunable lasers. In a recent work, Yeh et al. proposed a real-pivot design (*symmetric spring design*) for a large rotation angle, as shown in Figure 7.5 [41]. It consists of four sets of bidirectional rotary comb drives and

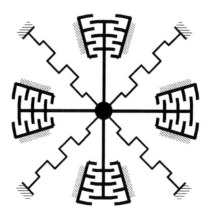

FIGURE 7.5
Design of a real pivot by a symmetric serpentine flexure.

four mechanical springs, all jointed to a floating but rigid central plate. The symmetrical arrangement of the actuators and springs ensures a fixed pivot regardless of the rotation angle, and the serpentine shape of the beams allows for considerable axial extension and thus large rotation angle. External optical components can be mounted onto the central plate. However, this design does not have room for other components such as mirrors, lenses, and gratings to be fabricated on the same chip, because the actuators and springs take up all the space. When this design is examined for the requirements discussed earlier, it has a small shift but a real pivot. However, single-chip integration is not possible.

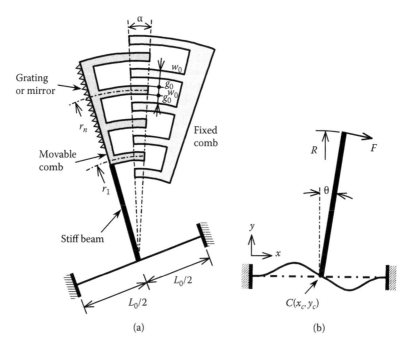

FIGURE 7.6
Real-pivot design using the double-clamped beam when actuated by a rotary comb drive: (a) the diagram, and (b) the analytical model.

A new real-pivot design has been proposed recently [40,44] that employs deformation of a double-clamped beam, as shown in Figure 7.6. This design is called a *double-clamped design*. A rotary comb drive is used as the actuator to rotate the external reflector (mirror or grating) attached to it. The movable comb is connected to the center of a soft double-clamped beam through a stiff beam. As the stiff beam is designed to be much stiffer than the double-clamped beam, its deformation is negligible. When a pure torque is applied at the midpoint of the double-clamped beam, it yields pure rotation and no position shift, thanks to the symmetry of the load and structure. As a result, the midpoint of the double-clamped beam provides an ideal real pivot. As can be observed from Figure 7.6a, a direct advantage of this design in comparison with the symmetric spring design is that one side of the grating/mirror is not occupied and thus can be used for external components (such as fibers, lenses, laser chips, etc.), facilitating single-chip integration. Many other advantages of the double-clamped design over the commonly used simple cantilever design will be given in the mechanical analyses that follow.

7.4 Analysis and Comparison of Real and Virtual Pivots

This section will analyze and compare the mechanical properties of the real and virtual pivots, and the aspects of beam deformation, pivot position, and pivot shifts. As the rotation angle is not large, the mechanical characteristics can be studied without involving large deflection behavior. The following analyses are based on the double-integration method and virtual-work method [43].

7.4.1 Real Pivot Using a Double-Clamped Beam

With the increase of the driving voltage, the electrostatic force would deform the double-clamped beam through the stiff beam, as shown in Figure 7.6b. As a result, the stiff beam would be rotated by a certain degree. In an ideal case, in which a pure torque M is applied to the midpoint $C(x_c, y_c)$ of the double-clamped beam, the deformation y of the beam can be expressed as

$$y = \frac{M}{8L_0 EI}(L_0 - 2x)(x^2 + L_0\langle L_0 - 2x\rangle) \tag{7.11}$$

where E is the Young's modulus, I is the momentum of inertia, and L_0 is the length of the double-clamped beam. Following the notation in Reference 43, here the point brackets $<\,>$ mean $<t> = t$ if $t > 0$, and $<t> = 0$ for $t \leq 0$. Under this torque, the rotation angle θ at the midpoint can be expressed as

$$\theta = \frac{ML_0}{16EI} \tag{7.12}$$

From Equation 7.11, $y|_{x=L_0/2}$ is always 0 for whatever value of θ. That is, the pivot does not move with change of rotation angle. This is easy to understand from the symmetry of the structure and load, as the midpoint of the beam is not allowed to have position shift.

However, the design in Figure 7.6 is actually driven by an off-centered load F generated by the rotary comb-drive actuator. It is possible to obtain a pure torque at the beam midpoint by putting another rotary comb drive in symmetry, but it would occupy more space and thus make it difficult to arrange other parts of the tunable lasers such as laser chips and coupling optical fibers. With only one side of the rotary comb drive, the off-centered force can be equivalent to a concentrated force F plus a torque $M = FR$ applied at the midpoint, where R is the equivalent arm length. The relationships can be derived as

$$F = \frac{N_0 \varepsilon_0 h_0}{g_0} V^2 \tag{7.13}$$

$$R = r_1 + (N_0 - 1)\Delta r / 2 \tag{7.14}$$

where N_0 is the finger number of the movable comb, g_0 the gap between adjacent fingers, h_0 the finger depth, ε_0 the permittivity of air, V the actuation voltage, r_1 the distance of the first movable finger to the rotation center, $\Delta r = 2(g_0 + w_0)$ the period of the fixed comb, and w_0 the width of the comb finger.

The position shifts (Δx_c, Δy_c) of the midpoint C due to the off-centered force F can be calculated by the virtual-work method [43]. After certain derivations, they can be expressed in term of the rotation angle θ as

$$\Delta x_c = \frac{t_0^2}{3R} \theta + \frac{2L_0^2}{105R} \theta^3 \tag{7.15a}$$

$$\Delta y_c = \frac{L_0^2}{12R} \theta^2 \tag{7.15b}$$

where t_0 is the width of the double-clamped beam. In Equation 7.15a, the first term is due to the axial extension or compression of the beam, whereas the second one comes from the shift of the deformed beam in the axial direction. Equations 7.15a and 7.15b imply that a small pivot shift requires a long arm length R, short beam length L_0, and narrow beam width t_0. This is because a longer arm length results in smaller force with a given torque, shorter beam length leads to larger stiffness for bending in the y direction and rotation about the beam center, and narrower beam width requires smaller torque to obtain the same rotation angle. Altogether, these factors contribute to the reduction of the pivot shift.

The preceding pivot shift is derived for an ideal case in which the width and depth of the double-clamped beam is uniform over the whole length; that is, the value of EI is kept constant. However, in real-world cases, the beam width and depth could be changed nonuniformly, especially for the deep-etched MEMS devices. This is because of the presence of undercut (in the width direction) and notching effects (in the depth direction) in the deep reactive ion etching (DRIE) process; both are dependent on the trench width and shape surrounding the beam. In the real pivot formed by the double-clamped beam, the trench and other structures are commonly symmetric about the beam midpoint, and thus the change of beam width and depth is also symmetric. Therefore, such fabrication error will not introduce additional pivot shift. In contrast, it would be a serious problem in virtual pivot designs.

7.4.2 Virtual Pivot Using a Cantilever Beam

Before comparing the real pivot with the virtual pivot, the relationships for the cantilever beam are derived. The design of the rotary comb drive, supported by the cantilever beam shown in Figure 7.7a, can be regarded as a soft cantilever beam subject to an offset force F applied through a stiff beam, as modeled in Figure 7.7b. The frame that supports the movable comb is represented by the stiff beam, whose deformation is neglected. It is assumed that the length of the soft beam is L_1 and that of the stiff beam is bL_1. For ease

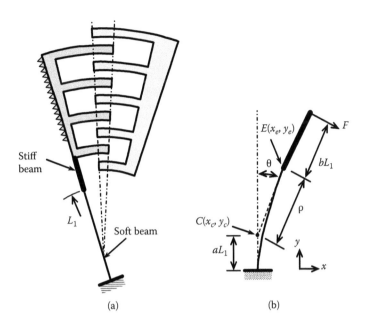

FIGURE 7.7
Virtual pivot by the cantilever beam: (a) schematic diagram and (b) mechanical model.

of comparison with the real pivot, the cantilever beam is placed in the y direction, and it bends toward the x direction.

The deformation x of the cantilever beam and the rotation angle θ of the stiff beam can be expressed as

$$x = \frac{Fy^2}{6EI}[3(1+b)L_1 - y] \quad \text{for} \quad y \leq L_1 \tag{7.16}$$

$$\theta = \frac{FL_1^2}{2EI}(1+2b) \tag{7.17}$$

With these equations, the position of the virtual pivot can be determined. As the end point $E(x_e, y_e)$ of the soft beam is the start point of the stiff beam, its rotation center is actually the pivot. The instantaneous rotation radius ρ can be expressed as

$$\rho = \left.\frac{x}{\theta}\right|_{y=L_1} = \frac{2+3b}{3(1+2b)}L_1 \tag{7.18}$$

Assuming that the instantaneous pivot is at a distance of aL_1 from the beam-clamped end as shown in Figure 7.7a,

$$a = \frac{1+3b}{3+6b} \tag{7.19}$$

The changes of a and ρ as a function of b are shown in Figure 7.8. With an increase of b, the value of a increases gradually, whereas the normalized rotation radius ρ/L_1 drops slowly. At $b = 0$, $a = 1/3$, which is commonly known; that is, the virtual pivot of a cantilever beam is at one-third of its length, as mentioned in Reference 22. When $b \to \infty$, the values of both a and ρ/L_1 are approximated to 1/2. Therefore, it provides the design flexibility of

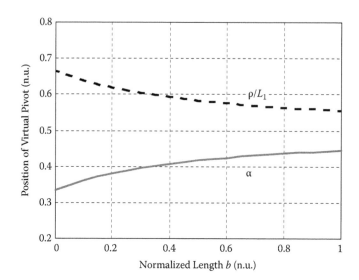

FIGURE 7.8
Position change of the virtual pivot formed by the cantilever beam (n.u. stands for normalized unit).

choosing the position and rotation radius of the virtual pivot by varying the length of the stiff beam, which is determined by the position and number of the movable comb fingers. It is observed from Equations 7.18 and 7.19 that a and ρ are kept constant regardless of the rotation angle. Therefore, the virtual pivot is stationary.

However, a deformation of the soft beam would shift the end point E as there is no force in the axial direction to elongate the soft beam. As a result, the virtual pivot C would be shifted. Assuming $\theta \ll 1$, after certain derivations, it has

$$\Delta x_c \approx \frac{(2+3b)L_1}{18(1+2b)}\theta^3 \tag{7.20a}$$

$$\Delta y_c \approx \frac{(1+5b+5b^2)L_1}{15(1+2b)^2}\theta^2 \tag{7.20b}$$

It can be seen from Equations 7.18–7.20b that the beam width does not affect the pivot position and shift. With respect to the analysis of the real pivot, these equations are valid only for the ideal case of uniform beam width and depth. In the presence of fabrication error, the virtual pivot would have significant shift. It is assumed that the width and depth of the cantilever beam are changed linearly along its length (see Figure 7.7), that is, $t(y) = t_1(1-k_0y/L_1)$ and $h(y) = h_1(1-k_0y/L_1)$, where k_0 is the rate of change, and t_1 and h_1 are the designed beam width and depth, respectively. By using the double-integration method to calculate the deformation and then comparing with the pivot position, the additional pivot shift can be expressed to the first order as

$$\Delta x_{ca} \approx \frac{2}{3}k_0L_1\theta \tag{7.21a}$$

$$\Delta y_{ca} \approx \frac{2(1+6b+6b^2)L_1}{9(1+2b)^2}k_0 \tag{7.21b}$$

Here, the x shift is linear with the rotation angle, but the y shift is independent.

7.4.3 Comparison of Real Pivot and Virtual Pivot

The double-clamped beam and the cantilever beam have different deformation and stiffness levels, as shown in Figures 7.6 and 7.7. To compare the real pivot with the virtual pivot, some standards should be used to make the conditions equivalent. First, the two types of beams are chosen to have the same material and same length, that is, $L_0 = L_1$. Second, the rotary comb drive should have the same parameters (finger number, width, gap, overlap, etc.). Last, the two types of pivots should have the same rotation stiffness; that is, they should have the same rotation angle under the same drive voltage. With these standards, it requires that $R = 8(1 + 2b)L_1 t_0^3 / t_1^3$.

The shifts of the real pivot and the virtual pivot are compared in Figure 7.9. The parameters for calculation are listed in Table 7.1; these are from fabricated devices [40]. When the rotation angle is increased from 0 to 4°, the shift in the x direction is increased from 0 to 0.19 nm for the real pivot, as shown in Figure 7.9a. In the ideal case of a uniform beam the virtual pivot is shifted by less than 0.03 nm. Assuming that fabrication introduces a nonuniformity to the beam width and depth, let $k_0 = 4\%$; the real-pivot shift does not change obviously, and is thus not drawn, but the virtual pivot shift increases by about nine times to 0.28 nm. From Figure 7.9a, the real pivot has a different shift from that of the virtual pivot in the x direction. However, the values are less than 1/10,000 of the wavelength 1.55 μm; such a difference is negligible. With regard to the pivot shift in the y direction shown in Figure 7.9b, the real pivot has about 7.4 nm at 4° in comparison to the value of 2 nm for the virtual pivot. Such values meet the requirements of 15 nm in the l direction and 40 nm in the y direction, as discussed in the previous section. Therefore, in an ideal case, the y shift is negligible for both the real and virtual pivots. However, in the presence of fabrication nonuniformity, the virtual pivot experiences a drastic increase of shift to about 1946 nm. Such a shift would significantly influence the wavelength tuning range of tunable lasers. In an example in [26], the tuning range is reduced from 30.3 to only

TABLE 7.1

Parameters of the Real and Virtual Pivot Designs

Comb Fingers

Finger number N_0	40
Finger separation g_0	2.5 μm
Finger width w_0	2 μm

Real Pivot by Double-Clamped Beam

Beam length L_0	150 μm
Beam width t_0	2 μm
Radius of the first movable finger r_1	1059 μm
Equivalent arm length R	1234.5 μm

Virtual Pivot by Cantilever Beam

Beam length L_1	150 μm
Beam width t_1	3 μm
Radius of the first movable finger r_1	160 μm
Normalized pivot position a	0.452 (normalized unit)
Normalized length of stiff beam b	1.236 (normalized unit)
Change of beam width/depth k_0	4%

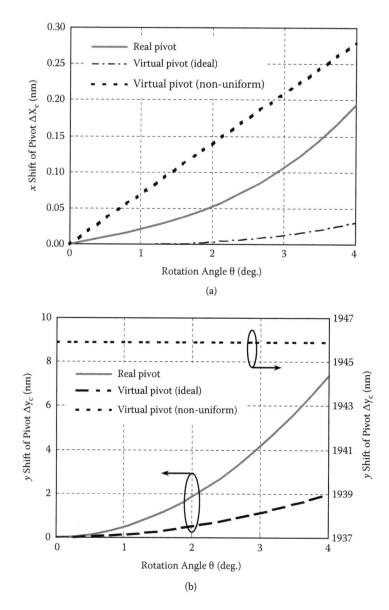

FIGURE 7.9
Comparison of the position shifts of the real pivot and virtual pivot at different rotation angles: (a) pivot shift in the *x* direction and (b) pivot shift in the *y* direction.

5.9 nm when the pivot position is changed by 2.1 μm (this value is calculated by substituting the rotation angle 1.13° into Equation 13 of Reference 26).

From the preceding discussions, it can be concluded that the double-clamped beam real-pivot design has an advantage over the simple cantilever virtual-pivot design in terms of the symmetric rotation beam and better tolerance for fabrication error. The reason is that the virtual pivot requires a proper combination of several parameters to maintain a stationary pivot and is thus sensitive to changes, as noted in [25]. In contrast, the real pivot is more robust. Although the real pivot has a larger shift in the *x* direction than the virtual pivot, because of the use of only one rotary comb drive and the consequent loss of load symmetry, the shift is very small and has negligible influence.

An experimental study on the real pivot and the virtual pivot was conducted by Zhang et al. (2007) [40]. The pivot designs are fabricated by deep etching on a silicon-on-insulator wafer and are released by a dry-etching method. The double-clamped beam has a depth variation of 16.1% over the whole beam but maintains the symmetry to the midpoint. As a result, it produces a continuous rotation angle of 4.7° until blocked by a stopper. In comparison, the cantilever beam has a depth reduction of 4% from the clamped end to the free end, resulting in a rotation angle of only 2.4°, limited by the side stability problem that originates from the pivot shift. Therefore, the real pivot is more suitable for obtaining a fixed pivot for MEMS tunable lasers owing to its advantages of large tolerance to fabrication error and environment change. More details will be given in Section 7.6.

7.5 Theory of Continuous Wavelength Tuning

As already discussed, different concerns apply to MEMS tunable lasers compared to conventional bulky ones. This section will reexamine the previous theoretical studies on the pivot position and wavelength tunability, and will then adapt them for MEMS tunable lasers.

7.5.1 Ideal Pivot Position

An analytical diagram of the Littrow laser is shown in Figure 7.10. It resembles the main parts of a Littrow laser, including a gain chip for optical amplification, a microlens for collimation and coupling, a blazed grating for wavelength selection, a pivot for grating rotation, and a dispersive element accounting for wavelength dependence. In this configuration, the laser-cavity resonance and the grating diffraction both impose their requirements on the oscillating wavelengths, as stated in Reference 7 for Littman lasers. By applying the same

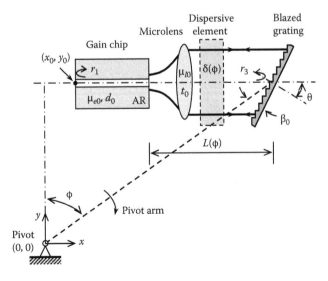

FIGURE 7.10
Analytical model of the Littrow external cavity laser.

requirements to the Littrow lasers,

$$N_0\lambda_m = 2[\mu_{e0}d_0 + L(\phi) + (\mu_{l0}-1)t_0 + \delta(\phi)] \qquad (7.22)$$

$$m_0\lambda_g = 2p_0\sin\theta \qquad (7.23)$$

where λ_m and λ_g represent the wavelengths determined by the laser cavity and the grating, respectively (subscript m for mode and g for grating); μ_{e0} is the refractive index of the gain chip; d_0 is the length of the gain chip; ϕ is the rotation angle of the pivot arm about the pivot point (clockwise direction is positive); $L(\phi)$ is the length of the external cavity in the free space (varies with ϕ); μ_{l0} is the refractive index of the lens (subscript l for lens); and t_0 is the effective thickness of the lens (different from the lens center thickness t_{l0} as it is a thick lens); θ stands for the diffraction angle that changes with ϕ by the relationship of $\theta = \phi - \beta_0$, where β_0 is the angle between the pivot arm and the grating surface. In most of the analyses and implementations of the ECTLs, the grating surface plane is chosen to pass the pivot point (i.e., $\theta \equiv \phi$) [24,34,37]. Here, we intentionally make the difference between θ and ϕ in order to determine their different influences on the wavelength-tuning range. The variables with the subscript 0 indicate that their values do not change with ϕ. All the other variables are functions of ϕ, directly or indirectly.

The quantity $\delta(\phi)$ in Equation 7.22 represents all the additional changes of the cavity length caused by the wavelength dependence of the gain chip, lens, and other dispersive components (not including the grating). As the ECTL wavelength is dependent on ϕ, the additional cavity length change is eventually a function of ϕ; this is why it can be expressed in a form of $\delta(\phi)$. In the gain chip, a change of ϕ also influences the cavity length by altering the carrier density of the gain medium (to be discussed later in this chapter). By introducing the term $\delta(\phi)$, theoretical analysis become more general for the ECTLs, regardless of the types of optics used for laser-beam coupling (e.g., lenses [10] or prisms [37]). From the definition it is easy to see that $\delta(\phi_0) = 0$, where ϕ_0 is the initial value of ϕ. To the first order, it can be approximated by $\delta(\phi) = \eta_0\Delta\phi$, where $\Delta\phi = \phi - \phi_0$, and η_0 is the dispersion coefficient, defined as $\eta_0 = \frac{d\delta}{d\phi}|_{\phi=\phi_0}$.

In Figure 7.10, the pivot point is chosen as the origin of the coordinate system (x, y), and the end facet of the gain chip is located at (x_0, y_0). From the geometry, it has

$$L(\phi) = y\tan(\phi) - x_0 - d_0 \qquad (7.24)$$

To obtain an ideal mode-hop-free wavelength tuning, $\lambda_m = \lambda_g$ should be always satisfied for all ϕ. Therefore

$$y = \cot(\phi)\left[\frac{N_0}{m_0}p_0\sin(\theta) + x'\right] \qquad (7.25)$$

where $x' = x_0 - (\mu_{e0}-1)d_0 - (\mu_{l0}-1)t_0 - \delta(\phi)$. To better understand the pivot position affected by the grating angle as given by Equation 7.25, we first study two simple and extreme cases before applying the equation to more practical cases. In an extreme case of normal incidence where $\theta = \phi \to 0$, y would be infinite if $x' \neq 0$, but kept constant at N_0p_0/m_0 if $x' = 0$. Normal incidence is not practical because it is difficult to maintain $x' = 0$ for all ϕ (the variation of $\delta(\phi)$ will be elaborated later). In addition, the zeroth diffraction order is used to form the laser oscillation, but this order is wavelength independent. Therefore, the grating is nothing but a common mirror in normal incidence. In another extreme case of the grazing incidence where $\theta = \phi \to 90$, $y \to 0$ and $x' \to -N_0p_0/m_0$; that is, the ideal pivot

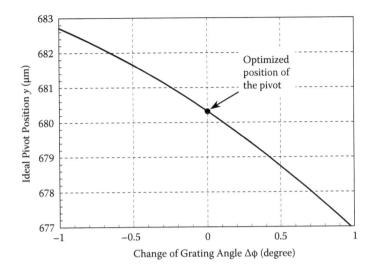

FIGURE 7.11
Variation of the ideal pivot position with grating rotation angle.

can approach to a fixed position. However, the grazing incidence to the blazed grating has very low diffraction efficiency owing to the large incident angle (nearly 90°).

To balance the resolving power and diffraction efficiency, in most applications the value of θ is chosen to be between 30 and 60° (e.g., 45° in References 24 and 25, and 51.81° in Reference 37). In such a range, it can be seen that y varies with ϕ (θ is also a function of ϕ), which implies that a moving pivot point is required for continuous tuning. An example of the relationship between y and ϕ is shown in Figure 7.11. When the grating is rotated around its initial position from –1 to +1° (corresponding to about 53 nm wavelength tuning), y should be changed from 682.7 to 676.9 μm. A displacement of 5.8 μm should be applied to the pivot to maintain an ideal tracking of λ_m and λ_g. The parameters are shown in Table 7.2, which are from a MEMS laser prototype [26].

However, for mechanical convenience, a fixed pivot is preferable in most of the tunable lasers [26–28,37]. A fixed value y_0 should be chosen according to the criterion $y_0 = y(\phi_0)$ to make $\lambda_m(\phi_0) = \lambda_g(\phi_0)$ (ϕ_0 is the original pivot angle). When $\phi \neq \phi_0$, the wavelength difference $\Delta\lambda_{mg1}$, owing to the fixed-pivot position, is proportional to the first order of the rotation angle incremental $\Delta\phi$, that is,

$$\Delta\lambda_{mg1} = (\lambda_m - \lambda_g)|_{y=y_0} \propto \Delta\phi \quad \text{for} \quad \phi \neq \phi_0 \tag{7.26}$$

It should be noted that y is the ideal pivot position and is a function of ϕ, whereas y_0 is a fixed value that represents the optimized position for a fixed pivot. If the pivot takes its position as y, according to Equation 7.25, $\lambda_m - \lambda_g$ will always be 0. However, when the pivot is fixed to y_0, $\lambda_m - \lambda_g$ is not equal to 0 except for $\phi = \phi_0$. To avoid confusion, the term *wavelength difference* will hereinafter always bear a precondition that the pivot position is fixed. The physical meaning of $\Delta\lambda_{mg1}$ is the difference between the nominal wavelengths determined by the laser cavity and the grating. Because the actual output wavelength is affected by both the grating and the laser cavity, the laser can be continuously tuned only when λ_m is able to follow λ_g for all ϕ. Otherwise, λ_m may become the output wavelength when ϕ varies over a certain range, and λ_g rises to dominance over another range, or even worse, the laser

TABLE 7.2

Parameters of the MEMS Littrow Tunable Laser

Pivot position y_0	680.3 μm
Pivot position x_0	50 μm
Grating period p_0	3.2 μm
Diffraction order m_0	3
Central wavelength λ_0	1.51 μm
Initial open angle ϕ_0	45°
Initial diffraction angle θ_0	45°
Length of gain chip d_0	250 μm
Initial refractive index of gain chip μe_0	3.375
Initial length of external cavity $L(\phi_0)$	380.3 μm
Mode number N_0	1661
Center thickness of microlens t_{l0}	30 μm
Initial refractive index of microlens μ_{l0}	3.482
Focal length of microlens f_0	179 μm
Effective thickness of microlens t_0	12.1 μm
Dispersion coefficient η_0	−106.5 mm/rad
Grating rotation angle $\Delta\phi$	0 to 1.13°
Actuation voltage V	0 to 30.2 V

output beats between λ_m and λ_g. To avoid such a mode hopping and competition associated with the choice of a fixed pivot, the value of $\Delta\phi$ must be limited, which in turn affects the tuning range. Therefore, a fixed-pivot point has a limited tuning range.

7.5.2 Optimization of Pivot Position for Maximum Tuning Range

From Equations 7.22 and 7.23,

$$
\begin{cases}
\dfrac{d\lambda_m}{d\phi} = \dfrac{2}{N_0}\left[\dfrac{y}{\cos^2(\phi)} + \dfrac{d\delta}{d\phi}\right] \\[3mm]
\dfrac{d\lambda_g}{d\phi} = \dfrac{2p_0}{m_0}\cos(\theta)
\end{cases}
\tag{7.27}
$$

$$
\begin{cases}
\dfrac{d^2\lambda_m}{d\phi^2} = \dfrac{2}{N_0}\left[\dfrac{2y\sin(\phi)}{\cos^3(\phi)} + \dfrac{d^2\delta}{d\phi^2}\right] \\[3mm]
\dfrac{d^2\lambda_g}{d\phi^2} = -\dfrac{2p_0}{m_0}\sin(\theta)
\end{cases}
\tag{7.28}
$$

To maximize the continuous tuning range, we further let $\frac{d\lambda_m}{d\phi} = \frac{d\lambda_g}{d\phi}$ for all ϕ (or equivalently, let $\frac{dy}{d\phi} = 0$). Under this condition, it yields

$$
y = \frac{\cos^2(\phi)[x' + \eta_0\tan(\theta)]}{\cos^2(\phi)\tan(\phi) - \tan(\theta)}
\tag{7.29}
$$

Commonly, in Equation 7.28, $\frac{d^2\delta}{d\phi^2}$ is negligible because $\frac{d^2\delta}{d\phi^2} \ll \frac{2y\sin(\phi)}{\cos^3(\phi)}$. From Equation 7.28, $\frac{d^2\lambda_m}{d\phi^2} > 0$, but $\frac{d^2\lambda_g}{d\phi^2} < 0$. Therefore, there is no way for $\frac{d^2\lambda_m}{d\phi^2} = \frac{d^2\lambda_g}{d\phi^2}$. It follows that if the pivot of the Littrow laser is fixed, the wavelengths given by the cavity mode and the grating can be matched only to the first order during wavelength tuning. This is why the Littrow laser has a limited continuous tuning range.

When the pivot position is chosen so that $y_0 = y(\phi_0)$, the wavelength difference $\Delta\lambda_{mg}$ can be expressed as

$$\Delta\lambda_{mg} = \lambda_m - \lambda_g \approx \frac{\Delta\phi^2}{2}\left(\frac{d^2\lambda_m}{d\phi^2} - \frac{d^2\lambda_g}{d\phi^2}\right)\bigg|_{\phi=\phi_0} = \Delta\phi^2\left[\frac{2y_0\sin(\phi_0)}{N_0\cos^3(\phi_0)} + \frac{2p_0}{m_0}\sin(\theta_0)\right]$$

$$= \lambda_0\Delta\phi^2[0.5 + \chi_0 \mathrm{tg}\phi_0\cot(\theta_0)]$$

(7.30)

where χ_0 is a coefficient standing for $\chi_0 = 1 + \eta_0/y_0$. From Equation 7.30, $\Delta\lambda_{mg}$ is proportional to $\Delta\phi^2$ and therefore allows a much larger rotation angle in comparison to that given by Equation 7.26. It is noted that both $\Delta\lambda_{mg}$ and $\Delta\lambda_{mg1}$ stand for the wavelength differences when the pivot position is fixed at y_0, and $\Delta\lambda_{mg}(\phi_0) = \Delta\lambda_{mg1}(\phi_0) = 0$. However, they differ in such a way that $\Delta\lambda_{mg}$ is the residual difference between λ_m and λ_g under the conditions that $\lambda_m(\phi_0) = \lambda_m(\phi_0)$ and $\frac{d\lambda_m}{d\phi}\big|_{\phi=\phi_0} = \frac{d\lambda_g}{d\phi}\big|_{\phi=\phi_0}$, whereas $\Delta\lambda_{mg1}$ is under the only condition $\lambda_m(\phi_0) = \lambda_m(\phi_0)$.

Let us choose half of the laser mode spacing $\delta\lambda_{FSR}$ as the tolerance for the wavelength mismatch; that is, $\Delta\lambda_{mg} \leq \frac{1}{2}\delta\lambda_{FSR} = \frac{\lambda_0^2}{4\mu_{e0}d_0}$. From Equation 7.23 we have $\delta\lambda = \lambda_0\cot(\theta_0)\Delta\phi$, where $\delta\lambda$ is the wavelength change. Substituting them into Equation 7.30, the maximum tuning range $\delta\lambda_{max}$ can be expressed as

$$\delta\lambda_{max} = \lambda_0\cot(\theta_0)\sqrt{\frac{2\lambda_0}{\mu_{e0}d_0[1 + 2\chi_0\tan(\phi_0)\cot(\theta_0)]}}$$

(7.31)

Here, $\delta\lambda_{max}$ includes the tuning in both the positive and negative direction rotations. It is worth noting that larger tuning ranges can be obtained with longer wavelength and shorter cavity. The reason is that the laser mode spacing would increase for longer wavelength and shorter cavity. As a result, a larger tolerance is allowed for $\Delta\lambda_{mg}$. It can also be seen that the grating angle and the arm open angle affect the tuning range differently. The former is more significant.

In the laser design, once the parameters m_0, p_0, μ_{e0}, d_0, ϕ_0, $\theta(\phi_0)$, and $x'(\phi_0)$ are chosen, there are only three variables to be determined: x_0, y_0, and N_0. As x_0 and y_0 are related by geometric relations, and Equation 7.24 links y_0 to N_0, only one variable, y_0 (or N_0), is left for choice. Physically, the design would require giving the positions of the grating and laser chip, and direction of the rotation arm; now, choose a point on the arm as the pivot. Larger y_0 (and N_0) means using a longer arm length for the grating rotation, and vice versa. However, there is one optimized value of y_0 that makes $\frac{dy}{d\phi}\big|_{\phi=\phi_0} = 0$, which can be proved to be identical to the condition $\frac{d\lambda_m}{d\phi}\big|_{\phi=\phi_0} = \frac{d\lambda_g}{d\phi}\big|_{\phi=\phi_0}$. Consequently, Equation 7.31 represents the optimized pivot position.

When the grating is rotated perpendicularly to the laser axis, it has $\phi_0 = 90°$. From Equation 7.31, it yields $\delta\lambda_{max} = 0$; that is, no wavelength change occurs. This agrees with the analysis of Trutna and Strokes [37].

7.5.3 Variation of Gain Medium Refractive Index with Grating Rotation

This section calculates the value of $\delta(\phi)$. Many studies on the proper pivot position do not consider the influence of dispersion properties of the gain medium and optical lens [34–37]. In a widely tunable external cavity laser, the effective refractive index of the gain medium may vary significantly with the tuning of wavelength, which eventually shifts the effective length of the light path from its nominal value. The refractive index change mainly arises from two factors: (1) the variation of the effective reflectivity of the external reflector, and (2) the wavelength dependence of the refractive index of the gain medium $In_{1-x}Ga_xAs_yP_{1-y}$, as given by

$$\Delta\mu_e = \frac{\partial\mu_e}{\partial\rho}(\rho - \rho_0) + \frac{\partial\mu_e}{\partial\lambda}(\lambda - \lambda_0) = \Delta\mu_{e1} + \Delta\mu_{e2} \tag{7.32}$$

where n is the refractive index of the gain medium, and ρ and ρ_0 are the carrier density and the threshold carrier density, respectively. Here, $\Delta\mu_{e1}$ and $\Delta\mu_{e2}$ stand for the index changes caused by the carrier density and the dispersion property, respectively.

7.5.4 Influence of Effective Reflectivity of the External Reflector

As explained previously, the microfabricated lens in a MEMS Littrow laser can generally collimate the light only in the plane parallel to the wafer surface. There is no collimation in the vertical plane. The rotation of grating would change the distance between the grating and the gain chip, and thus change the effective reflectivity of the grating, which in turn would alter the photon density in the gain chip and eventually adjust the refractive index of the gain medium.

The analysis here is based on the assumption that the scalar Gaussian beam optics is valid for dealing with the beam collimation by the lens, the diffraction by the blazed grating, and the coupling of the diffracted beam into the internal cavity of the gain chip. Assuming that the lens provides perfect coupling in the y direction, as shown in Figure 7.10, the effective reflectivity of the grating can be estimated on the basis of the coupling of the Gaussian beams:

$$r_3 = T_0\sqrt{\frac{2w_{z0}w_{z1}}{w_{z0}^2 + w_{z1}^2}} \tag{7.33}$$

where T_0 is the transmittance of the lens, w_{z0} is the light beam size emitted from the gain chip in the z direction, and w_{z1} is the light beam size returned from the grating. Both w_{z0} and w_{z1} represent the waist radii at the plane of the AR-coated facet and are related as follows:

$$w_{z1} = w_{z0}\sqrt{1 + \frac{\lambda_0^2 z(\phi)^2}{\pi^2 w_{z0}^4}} \tag{7.34}$$

where $z(\phi)$ is the effective light path between the AR-coated facet and the grating, as given by

$$z(\phi) = 2[y_0\tan(\phi) - x_0 - d_0 + (n_{10} - 1)t_0] \tag{7.35}$$

Using Equations 7.33–7.35, one can analytically obtain

$$\Delta r_3 = \zeta\Delta\phi \tag{7.36}$$

where $\zeta = \frac{dr_3}{dw_{z1}} \cdot \frac{dw_{z1}}{dz} \cdot \frac{dz}{d\phi}$ represents the coefficient. For notational simplicity, ζ is not expressed in detail.

In a stationary laser oscillation, the amplitude condition should be met [3], yielding

$$r_1 r_3 \exp[(g_m - \gamma)d_0] = 1 \tag{7.37}$$

where r_1 and r_3 are the reflectivities of the uncoated gain chip facet and the grating, respectively (as shown in Figure 7.10). The variables g_m and γ are the modal gain and loss of the gain medium. Therefore,

$$\Delta g_m = -\frac{1}{d_0 r_3} \Delta r_3 \tag{7.38}$$

The carrier density variation induces the change of gain Δg, as given by Reference 3:

$$\frac{\partial \mu_e}{\partial \rho}(\rho - \rho_0) = \alpha_0 \frac{\partial \mu_e''}{\partial \rho}(\rho - \rho_0) = -\frac{\partial g}{\partial \rho}\frac{\alpha_0 \lambda}{4\pi}(\rho - \rho_0) = -\frac{\alpha_0 \lambda}{4\pi}\Delta g = -\frac{\alpha_0 \lambda}{4\pi \Gamma_0}\Delta g_m \tag{7.39}$$

where α_0 is the line width enhancement factor, g is the material gain, and Γ_0 is the optical confinement factor. It should be noted that in Equation 7.37 the gain Δg_m is a modal gain, whereas in Equation 7.39 the gain Δg is a material gain. This difference significantly alters the value as Γ_0 is generally quite small (e.g., 0.1) [3,26]; however, this has long been neglected in earlier studies that dealt with similar cases [34–37,45].

Substituting Equations 7.36 and 7.38 into 7.39, and taking the values at $\phi \rightarrow \phi_0$,

$$\Delta \mu_{e1} = \frac{\alpha_0 \zeta_0 \lambda_0}{4\pi d_0 r_{30} \Gamma_0} \Delta \phi \tag{7.40}$$

where $\zeta_0 = \zeta|_{\phi=\phi_0}$ represents the initial grating reflectivity coefficient, and $r_{30} = r_3|_{\phi=\phi_0}$ is the initial reflectivity of the grating. In the experiment, the parameters are as listed in Table 7.2. The calculated value of $\zeta_0 = -0.32$ and $r_{30} = 0.10$. When the grating is rotated by 1.13°, the effective grating reflectivity is reduced to 0.09. Taking the typical parameters of the gain medium so that $\alpha_0 = 3$ and $\Gamma_0 = 0.05$, we have

$$\Delta \mu_{e1} = -0.092 \Delta \phi \tag{7.41}$$

7.5.5 Wavelength Dependence on Gain-Medium Refractive Index

The gain medium is also a dispersive material whose refractive index varies slightly with wavelength. When $x = 0.28$ and $y = 0.6$ (the parameters of the gain chip used in the experiment), the refractive index of $In_{1-x}Ga_xAs_yP_{1-y}$ is given as [46]

$$\mu_e = \left[10.452 + \frac{2.009}{\lambda^2} + \frac{0.245}{\lambda^4} \cdot \ln\left(\frac{\lambda^2 - 0.114}{\lambda^2 - 1.664} \right) \right]^{\frac{1}{2}} \tag{7.42}$$

where λ is the wavelength. At $\lambda = \lambda_0 = 1.51$ μm, $\mu_{e0} = 3.375$. After Taylor expansion at $\lambda_0 = 1.51$ μm, it yields

$$\Delta \mu_{e2} \approx -0.334 \Delta \lambda / \lambda_0 \tag{7.43}$$

From Equation 7.23, $\Delta \lambda / \lambda = \Delta \phi \cot(\theta_0)$; thus

$$\Delta \mu_{e2} = -0.334 \Delta \phi \tag{7.44}$$

In total,

$$\Delta \mu_e = -0.426 \Delta \phi \tag{7.45}$$

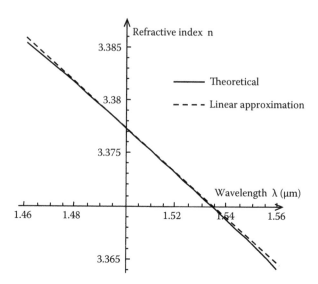

FIGURE 7.12
Change of the refractive index of $In_{1-x}Ga_xAs_yP_{1-y}$ with the wavelength.

Eventually, the additional cavity variation can be expressed as

$$\delta(\phi) = d_0 \Delta \mu_e = -106.5 \Delta \phi \text{ in } \mu m \tag{7.46}$$

Therefore, $\eta_0 = -106.5$.

The relationship between refractive index and wavelength is shown in Figure 7.12. The linear approximation of Equation 7.43 matches the theoretical curve, with marginal discrepancy over the wavelength range 1.48–1.54 μm. With the increase of the wavelength, the refractive index decreases gradually. A wavelength change of 30.3 nm would result in a refractive index change of −0.008. Using the same laser parameters in Table 7.2, the effective light path length of the gain chip will be changed by −2.0 μm. According to Equation 7.22, the cavity mode would be shifted from its expected position by a value of −2.46 nm, which is about twice the cavity mode spacing $\delta \lambda_{FSR} = 1.36$ nm. Here, the expected position refers to the wavelength value when the refractive index is considered to be independent of the wavelength (i.e., taking $\eta_0 = 0$ in Equation 7.29). In a Littrow laser, such a wavelength shift would cause unwanted mode hopping. Therefore, the refractive index change during wavelength tuning should be taken into account in the design of Littrow lasers.

7.5.6 Comparison of Optimized Design and Conventional Design

To further illustrate the influence of the refractive index change to the Littrow laser design, the variations of $\delta(\phi) + \Delta \lambda_{mg}$ and $\Delta \lambda_{mg}$ with wavelength tuning are compared and shown in Figure 7.13. If the pivot position is designed according to Equation 7.29 (in which the gain chip refractive index is considered dependent on the wavelength and the grating angle), the mismatch of λ_m and λ_g is only $\Delta \lambda_{mg}$ because $\delta(\phi)$ is counterbalanced (named as Case 1). In contrast, if the refractive index is regarded as a fixed value, the mismatch is $\delta(\phi) + \Delta \lambda_{mg}$ (named as Case 2) because $\delta(\phi)$ cannot be eliminated. It can be seen from Figure 7.13 that, within the limit of half-mode spacing, continuous wavelength tuning is between a and b for Case 1, and between c and d for Case 2 (a, b, c, and d are points on the horizontal axis; $a = -27.5$ nm, $b = 27.5$ nm, $c = -8.5$ nm, and $d = 11.1$ nm). It shows that, by taking into account

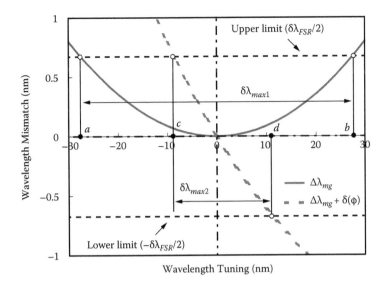

FIGURE 7.13
Comparison of wavelength tuning ranges predicted by the conventional design and the improved design.

the refractive index change, the continuous tuning range can be improved from $\delta\lambda_{max2}$ = 19.6 nm to $\delta\lambda_{max1}$ = 55.0 nm (see Figure 7.13). The difference in the tuning ranges showcases the advantage of the new theoretical analyses.

7.6 Experimental Studies and Discussions

This section will show an example of an implemented prototype of MEMS continuously tunable laser. The characteristics and experimental issues will be the focus.

7.6.1 Device Description

Experimental studies have also been conducted on the MEMS continuously tunable lasers. A prototype of the laser as shown in Figure 7.14 was demonstrated recently [26,27,47]. The laser design was based on optimization of the pivot position as discussed earlier. The parameters of the MEMS laser are listed in Table 7.2. The MEMS components, including a blazed grating, a rotary comb drive for actuating the grating, a microlens, and some trenches for containing the gain chip and coupling fiber, are fabricated simultaneously on an SOI wafer using the deep reactive ion etching (DRIE) process. The SOI wafer has a structural silicon layer 75 μm thick, an insulator silicon oxide layer 2 μm thick, and a substrate. After being patterned and etched, the MEMS structure is dry-released by the notching effect. In the final process step, it is selectively coated with a gold layer 0.5 μm thick. The gold coating improves the reflectivity of the grating and provides electrical connections. However, the microlens should not be coated; this is to allow the laser light to pass through. Such selective coating is implemented by a shadow mask that protects the microlens from being coated while exposing the other areas. The microlens has a focal length f_0 = 179 μm, which expands the laser beam in the horizontal direction from a beam diameter of about 3 to 211 μm. After this expansion, the laser beam covers about 96 grating teeth.

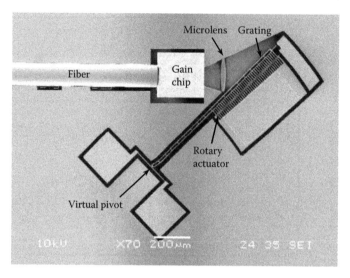

FIGURE 7.14
Overview of the MEMS Littrow laser.

The grating is etched on the sidewall of a rotary comb-drive structure, as shown in Figure 7.15a,b. The tooth profile is uniform, and the grating surface is smooth and clean, which guarantees good diffraction efficiency of the grating.

Once fabricated, the MEMS structure is assembled with the gain chip and the optical fiber. The MEMS Littrow laser has dimensions of 2.0 × 1.5 × 0.6 mm (not including the full length of the optical fiber). The gain chip with a contact on the top surface is placed face down into a trench in the silicon wafer (see Figure 7.14). A layer of indium foil, about 45 μm thick, is sandwiched between the gain chip and the trench bottom. The bonding is not self-aligned, and the chip and fiber are manually aligned to the markers that are fabricated. This foil serves two purposes. It bonds the gain chip to the trench, thereby providing better ohmic contact and heat transfer. It also raises the optical axis of the laser emission to align with the MEMS grating and optical fiber. Because in the MEMS fabrication the upper part of the etched grating (between 20 and 75 μm high) has good profile fidelity and surface quality, the optical axis is chosen at 45 μm height. The fiber is etched using 49% hydrofluoric acid to reduce the radius to 45 μm (etching rate 3.5 μm/s). The coupling loss is typically 2–4 dB for the cleaved fibers, and 0.5–0.8 dB for the lensed and tapered fibers. In the experiment, two probes are employed to apply the injection current. The negative contacts the exposed surface of the gain chip, whereas the positive touches a square sink on the wafer. The bottom of the square sink is connected to the gain-chip trench through gold wires.

The experimental setup include mainly a laser driver and some stages. For bare MEMS laser samples without the fixation of laser chips and optical fibers, in the experiment, one sample is placed on a 5-axis stage for positioning, and an optical fiber is aligned by another 5-axis stage for coupling of the output light. Two probes positioned by 3-axis stages are employed to apply injection current to the laser chip; and one more probe is employed for fine alignment of the laser chip to the MEMS lens and grating. A microscope is used to visualize the operation of the MEMS laser sample. Other equipment includes an oscilloscope (for monitoring the junction voltage of the laser chip), an optical power meter, and an optical spectrum analyzer. All the equipment is fixed to an isolation table. If the MEMS laser samples have laser chips and fibers packaged and wires bonded, the adjustment stages are not necessary.

(a)

(b)

FIGURE 7.15
SEM micrographs of the close-ups of the MEMS rotary grating: (a) side view of the grating surface and (b) top view of the grating tooth profile.

When a potential is applied to the comb drive, the comb fingers produce an electrostatic force between the moving fingers and the fixed ones, and thus generate an electrostatic torque to rotate the grating. The rotary comb drive follows the standard structure of the in-plane rotary actuators [40–42] except for an improved design of the pivot. Through a stiff frame structure that supports the movable comb, the comb drive is connected to the middle point of a clamped beam whose two ends are anchored to the substrate. Under pure torque from the frame, the center of the clamped beam generates a pure rotation, providing an ideal virtual pivot as shown in Figure 7.16. The beam has dimensions of $218 \times 75 \times 2 \ \mu$m and a stiffness of 5.6×10^{-7} N·m (the Young's modulus of the silicon is taken as 190 GPa). The rotary comb drive has 40 pairs of fingers spaced by gaps of $2 \ \mu$m, producing a torque of $1.2 \times 10^{-11} V^2$ N·m (V is the actuation voltage in volts). The comb drive provides a unidirectional

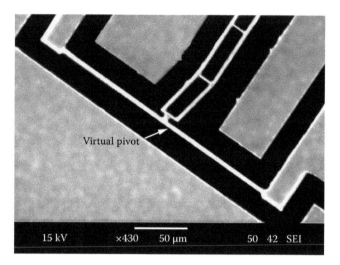

FIGURE 7.16
Close-up of the double-clamped beam for virtual pivot. The stiff beam is connected to the midpoint of the double-clamped beam through a soft joint, which avoids overstretching the beam during rotation.

pure rotation, measured as $\Delta\phi = 1.24 \times 10^{-3} V^2$ (designed $\Delta\phi = 1.19 \times 10^{-3} V^2$), where $\Delta\phi$ is in units of degrees. From Figure 7.17, the grating rotations in the "forward" stage (i.e., the angle increases from small to large) and "backward" stage (i.e., from large to small) have small hysteresis, and both of them match the designed value with marginal discrepancy.

7.6.2 Characteristics of Blazed Grating and Microlens

As the grating is the key component, its performance is first studied. Then, the microlens–grating set formed by the combination of the grating and the microlens is characterized.

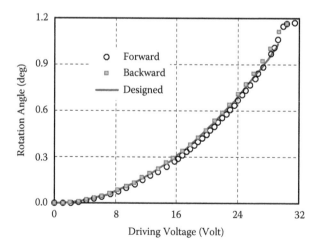

FIGURE 7.17
Rotation angle of the grating versus the driving voltage compared to the designed relationship. Little hysteresis is observed between forward and backward actuations.

The grating has a period of 3.2 μm and is designed to work at a blazed angle of 45° (corresponding to the third diffraction order), as shown in Figure 7.15. The grating sample is prepared by isolating and taking it out from a spare MEMS die. The far-field diffraction pattern is measured using a collimated laser beam illuminated at the designed angle. The diffracted power and pattern vary with change in the input wavelength, as shown in Figure 7.18a. The diffraction efficiency reaches its maximum at about 1511 nm. The fullwidth half-maximum (FWHM) bandwidth is about 5 nm.

The performance of the microlens–grating set is measured using another spare die, as shown in Figure 7.18b. An antireflection (AR)-coated lensed single-mode fiber (SMF,

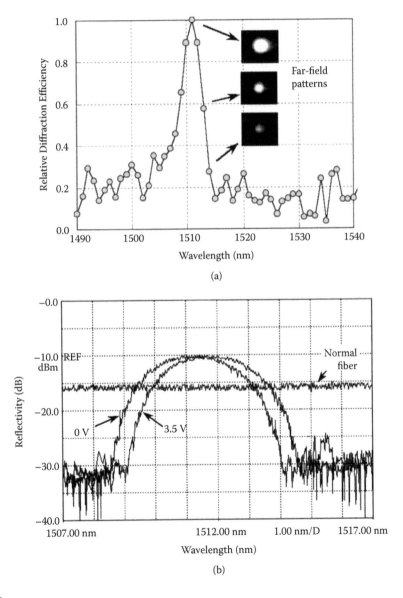

FIGURE 7.18

Characteristics of the grating and microlens: (a) the relative diffraction intensity and the far-field patterns of the grating and (b) the filtering spectra of the microlens–grating set in comparison to that of a normal fiber.

from Nanonics, Israel) is placed at the designed position of the gain chip to simulate optical input and output of the laser chip. The lensed fiber has a nominal focus diameter of 3.5 μm, which is close to the emitting spot size of the laser chip. In comparison to the normal single-mode optical fiber whose diameter is 10.4 μm, the lensed fiber can simulate the laser chip better in providing a Gaussian light beam to the microlens–grating, and in coupling the diffracted light back into the cavity or fiber core. In the experiment, a tunable laser provides the light source, and a synchronized optical spectrum analyzer measures the reflection. When no voltage is applied to the comb drive, the spectrum is centered at 1511.2 nm, with an FWHM bandwidth of 3.8 nm. The peak is at least 15 dB higher than the side lobes. As the voltage increases to 3.5 V, the center shifts to a longer wavelength by about 0.5 nm. As a reference, the reflection of a normally cleaved SMF under the same condition is also shown in Figure 7.18b. The peak reflectivity of the microlens–grating set is 5.3 dB higher than that of the SMF, corresponding to a value of 13.6% if the SMF reflectivity is taken as 4%. It shows that the microlens–grating set is functioning well. The 13.6% reflectivity is comparable to the estimated value of $r_{30} = 0.1$ in Section 7.5.4. The discrepancy of the bandwidth of 3.8 nm from the value of 5 nm may be attributed to the fact that the spot size of the lensed fiber is different from the designed size for the gain chip. The light is not well collimated, and therefore, the diffracted intensity decays rapidly when the light wavelength leaves the peak position, yielding a narrower bandwidth.

7.6.3 Characteristics of Laser Gain Chip

As the gain chip determines the laser tuning performance, it is also characterized in detail. The gain chip is a stripe-guide multiple-quantum-well InGaAsP/InP device having dimensions of $250 \times 300 \times 100$ μm (length × width × height). The gain curve of the laser gain chip is shown in Figure 7.19, which was obtained by measuring the amplified spontaneous

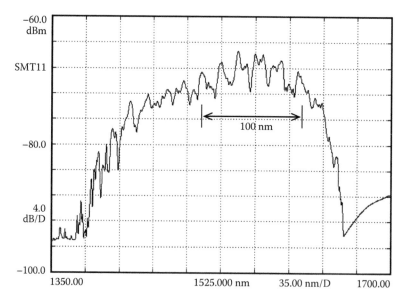

FIGURE 7.19
Gain curve of the laser chip.

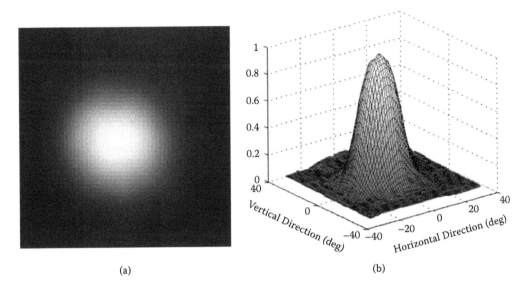

FIGURE 7.20
Far-field beam profile of the laser output: (a) intensity distribution of the laser spot and (b) 3-D view of the intensity distribution.

emission (ASE) spectrum. It has a broad wavelength range: the 3 dB bandwidth is ~100 nm (1490–1590 nm). It is observed that there is a certain oscillation in the curve. This is likely due to the interference of parasitic reflections. The beam quality is measured by a beam profiler after driving the MEMS laser over the threshold. The far-field beam profile is shown in Figure 7.20. It has a nearly circular cross section, with far-field divergence angles of 30.4° and 31.3° in the planes parallel and vertical to the contact surface, respectively (thus, the spot size is 1.63 × 1.58 μm). The beam has single transverse mode, and the intensity distribution is nearly Gaussian.

The gain-chip facet that faces the microlens is AR coated. The reflectivity is measured by monitoring the modulation index of the ASE spectra, as shown in Figure 7.21. The modulation index m is defined as $m = (P_{max} - P_{min})/(P_{max} + P_{min})$, where P_{max} and P_{min} are taken at the peak of the spectrum [48]. The modulation depth varies at different injection current levels. At a low level of $I = 10$ mA, m is only 0.17; at $I = I_t = 15$ mA, $m = 0.27$. As I rises to 25 mA, m reaches 0.63. The wavelength spacing of the maxima and minima of the modulation index corresponds to a round-trip phase shift. The round-trip amplification factor a of the gain chip has a relationship with m as given by $m = 2|a|/(1+|a|^2)$. The amplification factor a is also dependent on the injection current I as expressed by

$$\ln(|a|) = (\gamma L)\frac{I}{I_t} - \left(\gamma L - \frac{\sqrt{R_1 R_2}}{R_l}\right) \qquad (7.47)$$

where γ is the gain coefficient, L is the internal cavity length, and I_t and R_l are the threshold current and the facet intensity reflectivity of the original laser chip (before AR coating), respectively; R_1 and R_2 are the intensity reflectivity of the AR-coated facet and the other facet of the gain chip, respectively. Equation 7.47 suggest that $\ln|a|$ varies linearly with the normalized injection current I/I_t. In the curve of $\ln(|a|)$ versus I/I_t, the slope gives the gain coefficient γ, and the interception at $I/I_t = 0$ gives R_1.

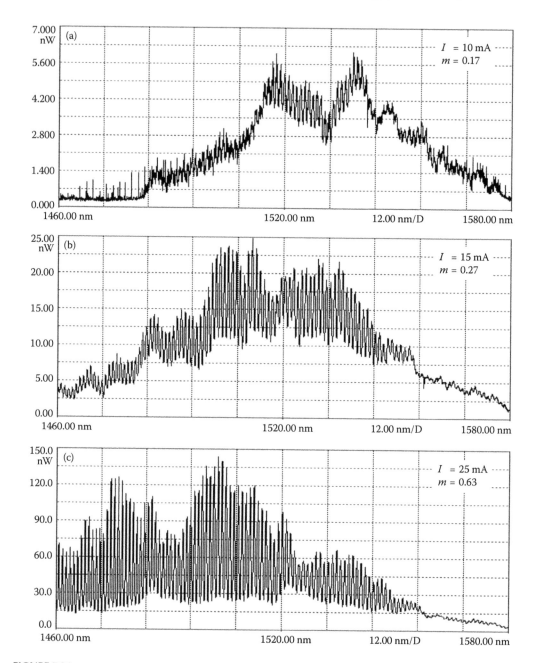

FIGURE 7.21

The ASE spectra of the gain chip at (a) $I = 10$ mA, $m = 0.17$; (b) $I = 15$ mA, $m = 0.27$; and (c) $I = 25$ mA, $m = 0.63$. For the original laser chip before antireflection (AR) coating, the threshold current is 15 mA.

The ASE spectra measured at various currents are used to obtain the curve of the round-trip amplification versus the normalized current, as indicated in Figure 7.22. At a high current level, the curve tends to saturate. However, at a low current level, the amplification $\ln(|a|)$ increases nearly linearly with the injection current I/I_t, as suggested by Equation 7.47. The trend line is $\ln(|a|) = 1.45I/I_t - 3.44$. With a Fresnel reflectivity $R_l = 0.31$, $R_2 \approx 0.9$, and $L = 250$ μm, it gives $\gamma = 58$ cm^{-1} and facet reflectivity $R_1 = 0.19\%$.

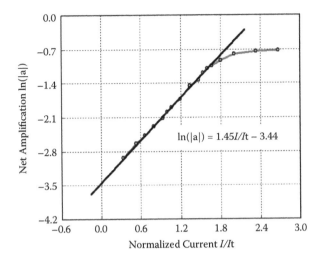

FIGURE 7.22
Round-trip amplification factor versus the normalized injection current.

7.6.4 Wavelength Tunability

The various output spectra at different actuation voltages are shown in Figure 7.23 at an injection current of 80 mA. The output maintains a single longitudinal mode over a wide range. The side-mode suppression ratio (SMSR) decreases with an increase of wavelength. Further wavelength tuning will result in a multimode output arising from other diffraction

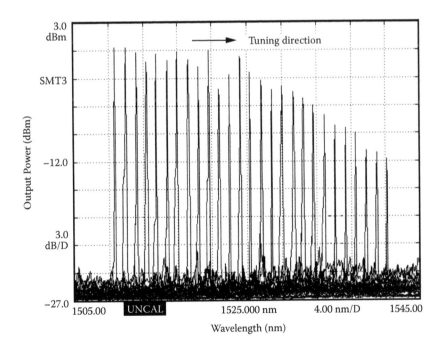

FIGURE 7.23
Superimposed spectra of the MEMS laser at different wavelengths.

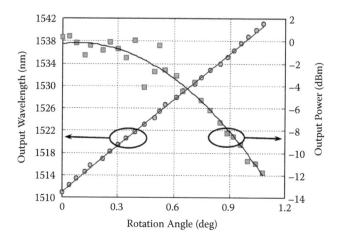

FIGURE 7.24
Variation of the output wavelength and power with the grating rotation angle.

orders. The variations of output wavelength and power are shown in Figure 7.24. The wavelength shifts linearly from 1510.9 to 1541.2 nm when the actuation voltage rises from 0 to 30.2 V, to rotate the grating by 1.13°. A tuning range of 30.3 nm is obtained. The tuning accuracy is about 0.03 nm/V². High tuning resolution can be obtained by fine adjustment of the actuation voltage. Wavelength repeatability and stability are measured to be about ±0.06 nm. The line width is below 0.01 nm, and does not change noticeably. The power initially has a maximum of 0.4 dBm, which drops to −11.8 dBm at the end. Its values fluctuate noticeably at the first half of the wavelength tuning range, which suggests that there is strong energy redistribution among the desired third order and other diffraction orders during grating rotation. In the second half, the power drops monotonically owing to the decrease of the diffraction efficiency with higher diffraction angle.

It is noted that the obtained tuning range of 30.3 nm is somewhat larger than $b = 27.5$ nm (as the grating of the MEMS laser is rotated to only one direction, the measured tuning range, 30.3 nm, should be compared with the predicted range $b = 27.5$ nm rather than the entire range, 55 nm). It can be attributed to the neglected dispersion of the lens. At 1541.2 nm, according to Equation 7.30, $\Delta\lambda_{mg} = +0.80$ nm, which is larger than half of the mode spacing, 1.36 nm, of the gain chip. However, the effective refractive index of the silicon lens decreases slightly with an increase in the wavelength. It can be approximated by $n(\lambda) = -0.0760\lambda + 3.5939$ within the wavelength range of 1.4–1.8 μm. According to this relation, a wavelength change from 1510.9 to 1541.2 nm would reduce the refractive index by 2.28×10^{-3}, which in turn would shorten the light path by 0.069 μm. Equivalently, the wavelength difference is compensated by −0.08 nm, so $\Delta\lambda_{mg} = + 0.72$ nm, which is quite close to half of the mode spacing. The remaining discrepancy may be due to the omitted higher-order term of $\delta(\phi)$.

The light-current (LI) curves of the laser under different actuation voltages are shown in Figure 7.25. For comparison, the original LI curve of the gain chip in the free space is also illustrated, which is very close to the axis of the injection current. This means the laser output is very small when there is no grating to form the resonant cavity. It also indicates that the AR coating of the gain-chip facet is quite successful. Otherwise, the residual reflectivity would let it emit laser light when the injection current becomes quite high. With the presence of the grating, which acts as the reflector, obvious laser oscillation can be observed when the injection current goes beyond a certain threshold. However, the LI

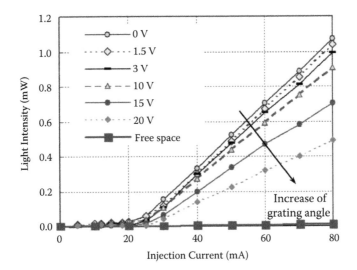

FIGURE 7.25
Light–current curve of the laser under different actuation voltages.

curves shift to the right-bottom corner with the increase in voltage. The threshold current rises from 21.9 to 24.7 mA, and the differential quantum efficiency (slope of the LI curve) drops from 0.019 to 0.009 mW/mA, as shown in Figure 7.26. This is due to the reduction of the effective diffraction efficiency when the grating rotation angle increases. A clear inflection can be observed in the external efficiency curve at about 3 V, which coincides with the light power change in Figure 7.23. Theoretically, the efficiency would decline monotonically. Because the laser structure is the same during the LI test, the inflection is unlikely due to the gain chip. It must be from the grating side; or it could be due to the energy redistribution among other diffraction orders, or due to the slight misalignment of the gain chip to the grating.

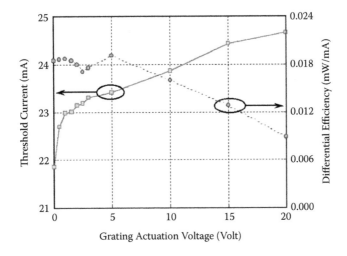

FIGURE 7.26
Variation of the threshold current and differential quantum efficiency with grating actuation voltage.

7.6.5   Comparison and Discussions

The 30.3 nm tuning range of this MEMS Littrow laser is smaller than that of the 40 nm range obtained in Reference 10, which was in a Littman configuration. As the Littman laser has a longer optical path, it tends to be less sensitive to the dispersion issue. However, because the laser in Reference 10 is not miniaturized, it is not appropriate to directly compare these tuning ranges. However, ignoring the technical issues that originate from deep etching, under the same conditions, the MEMS Littman lasers should have a larger tuning range than the Littrow ones but with smaller output power. To improve the tuning range and output power, the 3-D optical coupling system should be introduced to the external cavity so as to obtain high coupling efficiency while still maintaining the advantage of single-chip integration [30]. More details of the 3-D coupling system are given in Section 6.2.

A Littrow external cavity laser was designed and fabricated according to the conventional pivot design as in Reference 34, using the DRIE process. The design and fabrication are the same as the one discussed in Figure 7.14 except for the pivot position. In this laser, the pivot position is not optimized (i.e., the refractive index change during the wavelength tuning is not considered). Its pivot position is given by $y_0 = 680.0$ μm, whereas that of the optimized laser e is given by $y_0 = 680.3$ μm. One more difference is that the laser is tuned to a short wavelength owing to the arrangement of the rotation direction difference. The measured output spectra are superimposed, as shown in Figure 7.27, and the wavelength can be tuned continuously only from 1503.2 to 1497.3 nm (tuning range 5.9 nm). Further tuning would result in mode hopping. Theoretically, it can be seen from Figure 7.14 that $c = -8$ nm; that is, the wavelength tuning is at the most 8 nm when the laser is tuned toward the short wavelength. The obtained wavelength of 5.9 nm is close to the predicted value. From this it can be seen that a small change of pivot position (by 0.3 μm) affects the tuning range significantly. By pivot optimization, the tuning range can be increased about five times; this well demonstrates the effectiveness of pivot optimization.

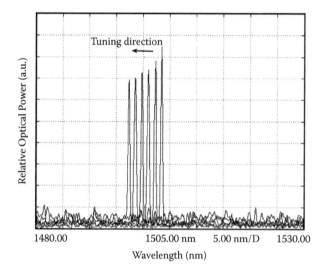

FIGURE 7.27
Superimposed spectra of a MEMS laser designed by conventional theory. It has a very limited tuning range of 5.9 nm (a.u. stands for arbitrary unit).

7.7 Summary

In conclusion, this chapter has described laser designs, theoretical analyses, and experimental studies of MEMS continuously tunable lasers, with a special focus on the new designs, features, and issues that MEMS technology brings to tunable lasers. The MEMS structures facilitate various real and virtual pivot designs, making it possible to construct new laser configurations in addition to the standard Littrow and Littman ones, and also enabling new features such as the elimination of the mode-hopping problem associated with conventional Littrow lasers. Reexamination of tuning theory shows that pivot position should be optimized to account for the refractive index change of the gain medium during the tuning, whose influence is negligible in conventional long-cavity-length lasers but is significant in MEMS lasers. An experimental study exemplifies an integrated laser system that demonstrates an improvement of wavelength-tuning range from 5.9 to 30.3 nm by optimizing the position of the fixed pivot and verifies the effectiveness of the design analysis.

As in the case of discretely tunable lasers, the application of MEMS technology to continuously tunable lasers not only produces many engineering improvements such as very short cavity and superior performance but also enables design opportunities at the physical level, such as introducing new designs, new mechanisms, and new principles.

References

1. Liu, A. Q., Zhang, X. M., Lu, C., and Tang, D. Y., Review of MEMS external-cavity tunable lasers, *J. Micromech. Microeng.*, 17, R1, 2007.
2. Liu, A. Q. and Zhang, X. M., Photonic MEMS: from laser physics to cell biology, In *Proc. 14th Int. Conf. Solid-State Sens. Actuators Microsyst. (Transducers '07)*, Lyon, France, 2007, 2485.
3. Petermann, K., *Laser Diode Modulation and Noise*, Kluwer Academic, London, 1988.
4. Bradley, D. J., Durrant, A. J. F., Gale, G. M., Moore, M., and Smith, P. D., A-3–characteristics of organic dye lasers as tunable frequency sources for nanosecond absorption spectroscopy, *IEEE J. Quantum Electron.*, QE-4, 707, 1968.
5. Crowe, J. W. and Craig, R. M., Jr., GaAs laser linewidth measurement by heterodyne detection, *Appl. Phys. Lett.*, 5, 7274, 1964.
6. Capelle, G. and Philips, D., Tuned nitrogen laser pumped dye laser, *Appl. Opt.*, 9, 2742, 1970.
7. Liu, K. and Littman, M. G., Novel geometry for single-mode scanning of tunable lasers, *Opt. Lett.*, 6, 117, 1981.
8. Harvey, K. C. and Myatt, C. J., External-cavity diode laser using a grazing-incidence diffraction grating, *Opt. Lett.*, 16, 910, 1991.
9. Ukita, H., Uenishi, Y., and Tanaka, H., A photomicrodynamic system with a mechanical resonator monolithically integrated with laser diodes on gallium arsenide, *Science*, 260, 786, 1993.
10. Berger, J. D. and Anthon, D., Tunable MEMS devices for optical networks, *Opt. Photon. News*, 14, 42, 2003.
11. Uenishi, Y., Tsugai, M., and Mehregany, M., Hybrid-integrated laser-diode micro-external mirror fabricated by (110) silicon micromachining, *Electron. Lett.*, 31, 965, 1995.
12. Uenishi, Y., Honma, K., and Nagaoka, S., Tunable laser diode using a nickel micromachined external mirror, *Electron. Lett.*, 32, 1207, 1996.
13. Liu, A. Q., Zhang, X. M., Murukeshan, V. M., and Lam, Y. L., A novel device level micromachined tunable laser using polysilicon 3D mirror, *IEEE Photon. Technol. Lett.*, 13, 427, 2001.
14. Zhang, X. M., Liu, A. Q., Tang, D. Y., and Lu, C., Discrete wavelength tunable laser using microelectromechanical systems technology, *Appl. Phys. Lett.*, 84, 329, 2004.

15. Liu, A. Q. and Zhang, X. M., Widely and discretely tunable laser using MEMS technology, In *Proc. IEEE/LEOS Int. Conf. Optic. MEMS Their Applic. (Optical MEMS 2003)*, Hawaii, 2003, 73.

16. Zhang, X. M., Liu, A. Q., Tang, D. Y., Lu, C., Hao, J. Z., and Asundi, A., Tunable external-cavity laser using MEMS technology, In *Proc. 12th Int. Conf. Solid-State Sens. Actuators Microsyst. (Transducers '03)*, Boston, MA, 2003, 1502.

17. Zhang, X. M., Liu, A. Q., Murukeshan, V. M., and Chollet, F. A., Integrated micromachined tunable lasers for all-optical network applications, In *Proc. 11th Int. Conf. Solid-State Sens. Actuators (Transducers '01)*, Munich, Germany, 2001, 1314–1317.

18. Sidorin, Y., Blomberg, M., and Karioja, P., Demonstration of a tunable hybrid laser diode using an electrostatically tunable silicon micromachined Fabry–Pérot interferometer device, *IEEE Photon. Technol. Lett.*, 11, 18, 1999.

19. Aikio, J. K., *Extremely Short External Cavity (ESEC) Laser Devices*, VTT publication 529, ESPOO 2004, http://www.vtt.fi.

20. Aikio, J. K. and Karioja, P., Wavelength tuning of a laser diode by using a micromechanical Fabry–Pérot interferometer, *IEEE Photon. Technol. Lett.*, 11, 1220, 1999.

21. Jerman, J. H. and Grade, J. D., U.S. Patent 6,469,415, 2002.

22. Huang, W., Syms, R. R. A., Stagg, J., and Lohmann, A., Precision MEMS flexure mount for a Littman tunable external cavity laser, *IEE Proc. Sci. Meas. Technol.*, 151, 67, 2004.

23. Cai, H., Zhang, X. M., Wu, J., Tang, D. Y., Zhang, Q. X., and Liu, A. Q., MEMS tunable dual-wavelength laser with large tuning range, in *Proc. 14th Int. Conf. Solid-State Sens. Actuators Microsyst. (Transducers '07)*, Lyon, France, 2007, 1433.

24. Lohmann, A. and Syms, R. R. A., External cavity laser with a vertically etched silicon blazed grating, *IEEE Photon. Technol. Lett.*, 15, 120, 2003.

25. Syms, R. R. A. and Lohmann, A., MOEMS tuning element for a Littrow external cavity laser, *IEEE J. Microelectromech. Syst.*, 12, 921, 2003.

26. Zhang, X. M., Liu, A. Q., Lu, C., and Tang, D. Y., Continuous wavelength tuning in micromachined Littrow external-cavity lasers, *IEEE J. Quantum Electron.*, 41, 187, 2005.

27. Liu, A. Q., Zhang, X. M., Tang, D. Y., and Lu, C., Tunable laser using micromachined grating with continuous wavelength tuning, *Appl. Phys. Lett.*, 85, 3684, 2004.

28. Liu, A. Q., Zhang, X. M., Li, J., and Lu, C., Single-/multi-mode tunable lasers using MEMS mirror and grating, *Sens. Actuators A*, 108, 49, 2003.

29. Zhang, X. M., Cai, H., Liu, A. Q., Hao, J. Z., Tang, D. Y., and Lu, C., MEMS Littman tunable laser using curve-shaped blazed grating, In *Proc. Int. Conf. Solid State Sens. Actuators Microsyst. (Transducers '05)*, Seoul, South Korea, 2005, 1, 804.

30. Zhang, X. M., Cai, H., Lu, C., Chen, C. K., and Liu, A. Q., Design and experiment of 3-dimensional micro-optical system for MEMS tunable lasers, In *Proc. 19th IEEE Int. Conf. Micro Electro Mechan. Syst. (MEMS 2006)*, Istanbul, Turkey, 2006, 830.

31. Cai, H., Zhang, X. M., Zhang, Q. X., and Liu, A. Q., MEMS tunable coupled-cavity laser, In *Proc. 14th Int. Conf. Solid-State Sens. Actuators Microsyst. (Transducers '07)*, Lyon, France, 2007, 1441.

32. Cai, H., Zhang, X. M., Yu, A. B., Zhang, Q. X., and Liu, A. Q., MEMS tuning mechanism for eliminating mode hopping problem in external-cavity lasers, In *Proc. 20th IEEE Int. Conf. MEMS (MEMS 2007)*, Kobe, Japan, 2007, 159.

33. Geerlings, E., Rattunde, M., Schmitz, J., Kaufel, G., Wagner, J., Kallweit, D., and Zappe, H., Micro-mechanical external-cavity laser with wide tuning range, In *Proc. 20th IEEE Int. Conf. Micro Electro Mechan. Syst. (MEMS 2007)*, Kobe, Japan, 2007, 731.

34. McNicholl, P. and Metcalf, H. J., Synchronous cavity mode and feedback wavelength scanning in dye laser oscillator with gratings, *Appl. Opt.*, 24, 2757, 1985.

35. de Labachelerie, M., Sasada, H., and Passedat, G., Mode-hop suppression of Littrow grating-tuned lasers: erratum, *Appl. Opt.*, 33, 3817, 1994.

36. Lotem, H., Mode-hop suppression of Littrow grating-tuned lasers: comment, *Appl. Opt.*, 32, 3816, 1994.

37. Trutna, W. R., Jr. and Strokes, L. F., Continuous tuned external cavity semiconductor laser, *J. Lightwave Technol.*, 11, 1279, 1993.

38. Lerner, J. M. and Laude, J. P., New vistas for diffraction gratings *Electro-optics*, 15, 77, 1983.

39. Kwon, O. K., Kim, K. H., Sim, E. D., Kim, J. H., and Oh, K. R., Monolithically integrated multi-wavelength grating cavity laser, *IEEE Photon. Technol. Lett.*, 17, 1788, 2005.
40. Zhang, X. M., Liu, A. Q., Lu, C., and Tang, D. Y., A real pivot structure for MEMS tunable lasers, *IEEE J. Microelectromech. Syst.*, 16, 269, 2007.
41. Yeh, J. A., Chen, C.-N., and Lui, Y.-S., Large rotation actuated by in-plane rotary comb-drives with serpentine spring suspension, *J. Micromech. Microeng.*, 15, 201, 2005.
42. Berger, J. D., Zhang, Y. W., Grade, J. D., Lee, H., Hrinya, S., and Jerman, H., Widely tunable external cavity diode laser based on a MEMS electrostatic rotary actuator, In *Proc. Opt. Fiber Commun. Conf. Exhibit (OFC 2001)*, Washington, 2001, 2, 198.
43. Pytel, A. and Singer, F. L., Strength of Materials, 4th ed., Harper & Row, New York, 1987.
44. Zhang, X. M., Liu, A. Q., Tamil, J., Yu, A. B., Cai, H., Tang, D. Y., and Lu, C., Real pivot mechanism of rotary comb-drive actuators for MEMS continuously tunable lasers, In *Proc. 14th Int. Conf. Solid-State Sens. Actuators Microsyst. (Transducers '07)*, 2007, Lyon, France, 1437.
45. Liu, A. Q., Zhang, X. M., Murukeshan, V. M., Lu, C., and Cheng, T. H., Micromachined wavelength tunable laser with an extended feedback model, *IEEE J. Sel. Topics Quantum Electron.*, 8, 73, 2002.
46. Broberg, B. and Lindgren, S., Refractive index of $In_{1-x}Ga_xAs_yP_{1-y}$ layers and InP in the transparent wavelength region, *J. Appl. Phys.*, 55, 3376, 1984.
47. Liu, A. Q., Zhang, X. M., Cai, H., Tang, D. Y., and Lu, C., Miniaturized injection-locked laser using microelectromechanical systems technology, *Appl. Phys. Lett.*, 87, paper 101101, 2005.
48. Kaminow, I. P., Eisenstein, G., and Stulz, L. W., Measurement of the modal reflectivity of an antireflection coating on a superluminescent diode, *IEEE J. Quantum Electron.*, QE-19, 493, 1983.

8

MEMS Injection-Locked Lasers

Xuming Zhang and Ai-Qun Liu

CONTENTS

8.1 Introduction 318
    8.1.1 Technological Origin 318
    8.1.2 Rate Equations 319
    8.1.3 Development of MEMS Injection-Locked Lasers 321
8.2 Design of Laser Configurations 323
8.3 Theoretical Study of Injection Locking 324
    8.3.1 Generalized Model Based on the Concept of Injection Seeding 325
    8.3.2 Locking Property under Single External Injection 328
        8.3.2.1 Governing Equations 328
        8.3.2.2 Threshold Injection Power and Critical Detuning 330
        8.3.2.3 Comparison of the GM with the SSM 331
    8.3.3 Locking Property under Multiple External Injections 334
8.4 Experimental Studies and Discussions 335
    8.4.1 Device Description and Experimental Setup 335
    8.4.2 Locking Property under Single External Injection 337
    8.4.3 Locking Property under Multiple External Injections 340
    8.4.4 Locking Property under Weak Injection 342
        8.4.4.1 Lockable Spectrum Range 343
        8.4.4.2 Influence of Injection Power 343
        8.4.4.3 Tolerance for Alignment 346
        8.4.4.4 Influence of Polarization 347
8.5 Summary 349
References 350

This chapter focuses on the theoretical development and experimental study of micro-electromechanical systems (MEMS) injection-locked laser (ILL), which employs two laser chips, one as the master laser to lock the state of the other, the slave laser. The role of MEMS is to provide an integration platform and tuning elements for easy adjustment and control of the locking states. Chapters 6 and 7 provide a technological base for the ILL device from the standpoints of MEMS laser designs and wavelength-tuning methods. From the engineering point of view, the two types of tunable lasers developed in previous chapters are stand-alone devices; each uses only one laser chip. In contrast, the ILL device consists of two lasers. From the physical point of view, the ILL device is more complicated as it has to take into account many laser modes and their interactions, which are not necessary with the previously mentioned lasers.

The theoretical and experimental studies are inspired by the prospect of applying the MEMS ILLs to emerging applications such as injection-locking-based all-optical switching, wavelength conversion, coherent communications, high-power lasers, portable atomic clocks, etc. They typically make use of multiple injections to lock multilongitudinal mode lasers, which have been demonstrated successfully using conventional free-space and fiber components, but the application is limited by their bulky size and their need for complicated control. Moreover, theoretical analysis is not available as the previous small-signal models (SSMs) in literature are mostly for simple conditions such as weak injection power, and single-injection and single-mode slave lasers. Therefore, it is necessary to develop a new model (termed the *generalized model*, i.e., GM) to cover broader conditions including single and multiple injections, weak and strong injection power, and single-mode and multimode slaves, which are the typical working conditions of MEMS ILLs.

On the basis of these considerations, this chapter is organized as follows. First, an introduction in Section 8.1 will cover the fundamentals of injection-locking technology and the development of MEMS ILLs. Then, the typical configurations of MEMS ILLs will be presented in Section 8.2. Next, the GM will be derived and applied to the cases of single injection and multiple injections in Section 8.3. The GM will also be compared with the SSM under both weak injection and strong injection conditions. Finally, the issues in experimentation will be elaborated in Section 8.4, including the experimental setup, the implementation of MEMS ILLs, the locking characteristics under single and multiple injections, and the verification of the GM by the experimental data.

8.1 Introduction

This section will follow briefly the road of development from conventional ILLs to MEMS ILLs, covering also the working principles and their applications.

8.1.1 Technological Origin

Injection locking is a technology for locking the oscillation state of a resonant object (i.e., oscillator, termed *slave*) by injecting an external coherent radiation (termed *master*). It is believed to have been discovered in the 17th century by the Dutch scientist Christian Huygens, who is known for inventing the pendulum clock. Huygens noticed that the pendulums of two clocks that were hung close to each other on the wall moved in unison [1]. The application of injection locking for electrical oscillators was pioneered by Van der Pol in 1927 [2], who injected an external signal into a self-sustained RLC oscillator and found that, when the injection frequency gets close to the RLC frequency, the oscillation suddenly occurs only at the injection frequency, and the beating between the injection and the intrinsic RLC oscillation disappears correspondingly. Injection locking soon found applications in microwave oscillators [3], and its use was later extended to laser oscillators [4,5] soon after the invention of the laser. The first ILL was demonstrated by Stover and Steier in 1966 using one HeNe laser to lock another one [4]. Since then, tremendous efforts have been made in the injection locking of various types of lasers, especially semiconductor lasers [6]. More reviews on the applications of injection locking will be discussed later in the chapter. The theoretical study of injection-locking phenomena also has attracted much attention. In 1964, Lamb built up a semiclassical theory to describe the operation of a multimode laser parametrically and to investigate various laser behaviors such as threshold condition, frequency pulling and pushing, and mode competition [7]. Although

Lamb borrowed the idea of external injection in explaining the frequency locking of laser modes, he did not really study laser behavior under external injection. However, Lamb's laser theory provided a base for most of the later theoretical investigations of injection locking. Early theoretical studies of injection locking focused mostly on the general theory [5,8–10], whereas recent studies tend to target analyses for specific applications [11–16]. Tang and Statz [5] applied the Van der Pol equation in lasers, but the inversion density was eliminated. Pantell took into account the inversion density and presented a more general analysis using a field equation and a rate equation, but neglected the refractive index change with the carrier density [8]. The same problem occurred in some other studies that were developed specifically for the injection locking of semiconductor lasers [9]. A milestone work was presented in 1982 by Lang, who synthesized the previous studies and developed a small-signal theory for the multimode semiconductor lasers [10]. Lang's theory has been extensively used and proved in many theoretical and experimental studies [12–14,17,18]. However, the small-signal approximation and the simplification of considering all side modes as one would limit its validity in many cases, such as in large-signal and strong side modes. Further studies were also reported that focused on aspects such as strong injection, locking conditions/stabilities [17], locking of the single-mode slave lasers [18], lockable range [19], etc. Later theoretical studies targeting specific applications have attracted considerable attention. For example, Cassard and Lourtioz developed the analysis for injection-locked high-power pulsed lasers by treating the locking as a competition between the injection- and noise-driven fields [16]. Injection locking for Q-switched lasers was also investigated [11]. Recently, injection of multiple lasers has been studied broadly for optical networks. Hörer and Patzak presented a large-signal analysis for wavelength conversion using two injection lasers simultaneously [12]. Cai et al. proposed the generation of dense wavelength division multiplexing (DWDM) channels using multiline optical injection, and studied the associated cross talk, noise, and stability [13]. All-optical regenerators and optical transmitters were also investigated using the two-side-mode locking method [14]. The theory for chaos synchronization was established by Murakumi, who by treated the slave as a driven damped oscillator [15].

8.1.2 Rate Equations

The conceptual model of injection locking is shown in Figure 8.1. The external injection couples an electrical field E_{ext} at a frequency v into the internal cavity of the Fabry–Perot (FP) laser diode. E_0 stands for the electrical field of the laser cavity, and Ω is the oscillation frequency. $\omega(n, \Omega)$ is the resonant frequency of the cavity mode, which is dependent on the excited carrier density n and the oscillation frequency Ω. It should be noted that although v, ω, and Ω

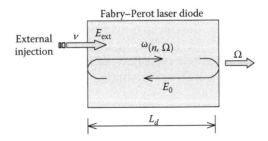

FIGURE 8.1
Conceptual model of injection locking to a semiconductor laser.

stand for the frequencies related to the laser cavity, they are different in such a way that v is the frequency of external injection, Ω is the real frequency of the locked output, and ω is the nominal frequency defined by the refractive index $\eta(n, \Omega)$ and the cavity length L_d as

$$\omega = \frac{\pi M c_0}{\eta L_d} \tag{8.1}$$

where M is an integer representing the mode number, and c_0 is the velocity of light in vacuum.

Based on Lamb's semiclassical laser theory [7], the state inside the FP cavity can be expressed by the field and rate equations as [10]

$$\frac{dE_0(t)}{dt} = \frac{1}{2}[G(n) - \Gamma]E_0(t) - j[\omega(n) - v]E_0(t) + \kappa E_{ext} \tag{8.2}$$

$$\frac{dn}{dt} = -\gamma_s n - G(n) \cdot [\,|E_0(t)|^2 + N_u(t)] + P \tag{8.3}$$

$$\frac{dN_u(t)}{dt} = [G(n) - \Gamma]N_u(t) + \gamma C_{sp} n \tag{8.4}$$

where $E_0(t) = E_0 \exp(-jvt)$ is the time-varying electrical field in the laser cavity, $G(n)$ stands for the modal gain that is dependent on the carrier density, and Γ is the cavity loss that is regarded as a constant in the locking. κ is a proportionality constant for the external injection given by $\kappa = \frac{c_0}{2\eta L_d}$, which represents the average contribution of the external field E_{ext} to the propagating wave inside the laser cavity. It is noted that E_0 and E_{ext} are slow-varying fields in contrast to the fast-varying term $\exp(-jvt) \cdot \gamma_s$ is the inverse carrier lifetime, and C_{sp} is the ratio for the spontaneous emission into the oscillating mode. P is the carrier injection rate per unit volume determined by the injection current density J, active layer thickness d, and electric charge e as $P = \frac{J}{ed}$. $N_u(t)$ denotes the photon density of the possible unlocked neighboring mode.

Equation 8.2 is the rate equation accounting for amplitude and phase (i.e., the frequency and wavelength) of the electrical field. The first term on the right side stands for the net gain of the electrical field after a round trip, the second for the frequency pulling between the injection and the laser mode, and the last for the contribution of the external injection. Equation 8.3 is the rate equation for the excited carrier density. On the right side, the first term is the decay of the carrier due to limited lifetime, the second is the carrier loss that is converted into photons, and the last one is the current injection. Equation 8.4 is the rate equation for the photon density of the unlocked neighboring mode. The first term accounts for the round-trip gain, and the second for the contribution from the spontaneous emission.

These rate equations describe the interactions inside the laser cavity and provide profound physical insight into the injection-locking mechanism. Many theoretical studies are based on the standard form or certain modification of such equations [10,12–14,17,18]. However, there are some drawbacks in these equations:

- In Equation 8.2, the external injection at v (named the *seeded mode*) is considered to have degenerated into the laser mode at Ω (named the *targeted mode*); that is, there is no seeded mode in the laser output. However, it is observed in the experiment that the seeded mode always presents in the output and becomes dominant when the slave is well locked.

- The degeneration of the seeded mode makes it impossible to study the progress of locking from the unlocked state to the fully locked state (e.g., by changing the wavelength of the injection).
- The rate equation (Equation 8.2) is valid only for small-signal conditions (i.e., the injection current is slightly over the laser threshold, and the laser injection is weak). In real-world applications of high-power lasers and all-optical switching, the lasers are commonly far beyond the threshold, and the injection is strong.
- The rate equation assumes only one external injection. It cannot deal with the case of multiple external inputs injected into a multimode slave.

In the MEMS ILLs, the slave is commonly a multimode FP laser chip, the injection power can be quite high in comparison to the output power of the slave in the free-running state (i.e., no external injection), and they may need multiple external injections. Therefore, Equations 8.2–8.4 are no longer suitable. Detailed derivation of the new GM for the MEMS ILLs will be given in Section 8.3.

8.1.3 Development of MEMS Injection-Locked Lasers

The fusion of MEMS technology and external cavity tunable lasers has led to the innovation of many laser devices with promising specifications as described in Chapters 5 and 6. The success of these stand-alone laser devices has eventually led to the construction of many complicated laser systems. Liu et al. demonstrated in 2005 the first miniaturized ILL using MEMS fabrication and integration [20]. The device consists of an MEMS grated-tuned laser and an FP multimode laser with dimensions of $3 \times 2 \times 0.6$ mm. Single injection and multiple injections to the slave laser are both tested; it has achieved a side-mode suppression ratio (SMSR) of 55 dB, a range of fully locked state (FLS) of 0.16 nm, and a rate of all-optical switching at 100 MHz. New theoretical analysis is also developed in this work. More details will be given in Sections 8.3 and 8.4. Recently, Cai et al. demonstrated a more advanced version of MEMS-based laser system [21], in which the two laser chips lock each other at the same time and form a tunable coupled-cavity laser (CCL; a type of mutually ILL). A deep-etched parabolic mirror actuated by an electrostatic comb drive is used to adjust the gap of the CCL for optical coupling and optimal phase match. Stable single-mode laser output over 6.5 nm is demonstrated. More details are given in Section 8.2.

The MEMS ILL possesses some clear advantages when compared with traditional ILLs that are mostly based on bulky, free-space optical components (lenses, prisms, etc.) [5,21] and complicated fiber optic components [22]. One advantage is its integration. Many high-performance MEMS components can be integrated into a single system, making it more functional and powerful. Especially, the adjustment components can be fabricated and integrated on the same chip, making it easy to tune the parameters and to build up some special working conditions. For example, most of the conventional CCLs have fixed air gap between the two laser chips and require careful design of working conditions. In contrast, the MEMS CCLs offer an additional degree of freedom to adjust the air gap to the optimal tuning state [21]. Another clear advantage is its high compactness. The whole system can be packaged into a volume of several cubic millimeters or even smaller, making it portable for field applications and also easy for forming large arrays. The compactness also leads to better mechanical stability, easier encapsulation, and longer lifetime. These features make the MEMS ILLs very promising for use in many emerging applications.

Injection locking has a number of applications in lasers. Semiconductor lasers are naturally advantageous for injection locking as they have a low Q factor cavity with a broad passband (tens of gigahertz) and an even broader gain region (~10 THz), which makes injection locking easy and stable. Its function in semiconductor lasers can be roughly categorized into two types: (1) improving the characteristics of the slave, and (2) synchronizing the master and the slave. For the former, locking is able to improve the properties of the slave laser in spatial, spectral, temporal, and polarization domains. In the spatial domain, the locked laser can have a beam shape close to the ideal Gaussian distribution and thus ensure better propagation and optical coupling. In the spectral domain, the locked laser can obtain purely single wavelength, narrow line width, high SMSR, low wavelength shift or chirping when subjected to external disturbance and direct modulation, etc. In the temporal domain, injection locking significantly improves response speed and modulation bandwidth. In the polarization domain, the locked laser has a polarization state identical to the master laser and is thus able to avoid polarization switching instability. These improvements are especially useful for many applications. For example, high-power lasers are commonly operated in multiple longitudinal mode and have wide diverging angles, making coupling and targeting difficult. By the use of a low-power single-mode laser as the master, high-power lasers are able to obtain a much higher power level, but they have single wavelength, high SMSR, narrow line width, and diffraction-limited beam quality [11,16,21,24]. Tuning of the master can lead to the wavelength tuning of the slave (though tuning may not cover the full wavelength range). Locking an array of high-power laser chips has attracted particular interest [22,24]. Before locking, the total peak power is increased by the rule of N (i.e., the number of the laser chips). In contrast, after locking, the total peak power shoots up by the rule of N^2. Injection locking provides a simple way to obtain large peak power. For high-speed modulation (i.e., time domain), injection locking can be used to ensure single-mode operation [25] and to reduce the mode partition noise. Additionally, it can significantly broaden the modulation bandwidth and flatten the modulation response [26]. Such performance improvement has caused injection locking to be widely applied in high-power lasers and optical communications [11,16,22,24,27]. The locking of polarization has been widely used in vertical-cavity side-emitting lasers (VCSELs), which do not discriminate between the transverse electric (TE) and transverse magnetic (TM) modes and thus suffer from polarization instability and power fluctuation. On the other hand, the synchronization of master and slave in wavelength, phase, and chaos state has resulted in broad applications of injection locking in coherent communications [28].

One of the latest applications of injection locking is all-optical switching [22], which is achieved by simultaneous injection of three signals: a data signal, a control signal, and a continuous wave (CW) stabilizer signal, into an FP slave laser. As the CW signal is for stabilizing the slave when there is no control signal or data signal, it is not involved in the optical switching process. The wavelength and power of signals are chosen so that the slave is always injection locked by one of the signals in the following priority: the control signal, the data signal, and the CW stabilizer signal. That is, in the presence of the control signal, the slave will be fully locked by the control signal despite the presence or absence of the data signal. Therefore, the data signal is suppressed; that is, it is turned off. In the absence of a control signal, the data signal will lock the slave, that is, it is turned on. In this way, the optical data signal can be switched by the control signal in the optical form itself without the need for the conventional optical-electronic-optical conversion. The function of the CW stabilizer signal is to suppress the FP modes of the slave when the intensities of both the data and control signals are low. A similar principle can be used for wavelength conversion [12,27].

Another novel application is the micromachined atomic wristwatch [29–31]. Commonly, a ^{133}Cs atomic clock needs a pair of laser sources to excite the atoms between two hyperfine energy levels through a nonlinear mechanism called *coherent population trapping* (CPT) [32].

The laser source is modulated by a local oscillator and then shone through a gas cell of ^{133}Cs. When the laser wavelength and the modulation frequency are properly adjusted, the absorption of the gas cell is minimized due to CPT. A photodetector measures the laser power and gives feedback to the oscillator so as to maintain the minimum absorption. As the oscillation frequency is directly related to the transition between hyperfine energy states, it is used as the timing signal. The lasers should have stable wavelength, very narrow line width (~10 MHz), and very high modulation rate (~10 GHz). Injection locking makes it easy to meet these requirements. As the vapor cell of the atomic clock is miniaturized [31] and the overall size of the atomic clock is downscaled, the laser sources should also be shrunk accordingly (on the order of millimeters). However, conventional ILLs are bulky and require complicated driving and adjusting mechanisms that make them unrealistic for use in atomic wristwatches. A good solution is to miniaturize the ILL and to integrate it on a single chip using the MEMS technology.

8.2 Design of Laser Configurations

The MEMS ILLs developed so far have two typical designs. The first design (termed *direct coupling design*) was demonstrated in 2005 [20]. It employs direct coupling between the two laser chips, as shown in Figure 8.2. A MEMS continuously tunable laser is used as the master to provide the external locking light for an FP laser, which acts as the slave. In between, a micromachined optical component is used as the coupling control element to align and adjust the optical coupling between master and slave. An optical fiber can be used to couple the output and also to inject extra external lasers in case of multiple injections (e.g., in all-optical switching, an extra signal is needed from the external data [22]). The direction-coupling design of MEMS ILLs is merely a miniaturized replication of the conventional CCL. The improvements are single-chip integration of MEMS tunable lasers (for locking state control) and the coupling control element (for coupling efficiency control).

The other design (termed *parabolic mirror design*), proposed recently, was inspired by the successful demonstration of a deep-etched parabolic mirror for micro-optical coupling systems used in MEMS variable optical attenuators [33]. The parabolic mirror consists of a pair of parabolic sections arranged symmetrically; each section is a part of the parabola (see Figure 1 in Reference 33). According to geometric optics, the rays from the focus of a parabola can be collimated ideally after reflection, and vice versa. The end facets of the laser chips are located at the two foci of the parabolas, C_1 and C_2, respectively. Therefore, the incoming rays from C_1 are collimated by the first mirror and are then converged to

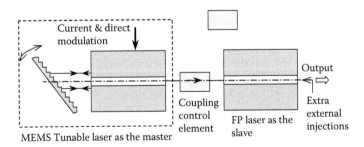

FIGURE 8.2
Direct-coupling design of the MEMS injection-locked lasers, in which the air gap between the two chips are fixed but the coupling efficiency can be controlled.

C_2, ensuring a good coupling. The gap between the two chips can be adjusted by simply translating the parabolic mirror, making it convenient to select the initial working conditions (e.g., cavity length, wavelength, mode stability, SMSR, etc.). One beauty of this design is that, with the translation of the parabolic mirror, the coupling efficiency does not drop obviously. It was found in the experiment that the coupling efficiency is reduced by only 5% when the parabolic mirror is translated by 20 μm. Because a gap change of about a half-wavelength is large enough for optimizing the working conditions [21], the drop of coupling efficiency during the adjustment is negligible. In this way, the locking state can be adjusted at a constant injection power, making it convenient for the experiment and application. Strictly speaking, the parabolic mirror design was not originally intended for use in an ILL [21]. In the experimental demonstration, one chip (i.e., the lasing chip) is driven above the threshold to provide the laser light, whereas the other (i.e., the tuning chip) is driven below the threshold as the phase control. Wavelength tuning is realized by injecting an electrical current into the tuning chip so as to change its refractive index, cavity length and, eventually, the oscillation mode. As a result, the wavelength tuning speed can reach nanosecond level, and thus this design breaks the limits of milliseconds imposed by the micromechanical movement in other MEMS tunable lasers [34]. Nevertheless, the parabolic mirror design can also be used for ILL if the tuning chip is driven above the threshold. In this way, the two chips will be mutually locked and can exhibit stable single-mode and other tuning behaviors under proper conditions [21,35].

8.3 Theoretical Study of Injection Locking

The theoretical analyses of injection locking have been extensively recorded in literature [8–15,17–19,36], as reviewed in Section 8.1.1. An important small-signal analysis was presented by Lang [10] and was later simplified by Li to include mathematical expressions [36]. Their model has been extensively used and proved in many theoretical and experimental studies [12–14,17,18]. In this chapter, this SSM will be used as a reference to verify the newly developed GM under the condition of weak injection power. Similar to the drawbacks described in Section 8.1.3, most of the previous analyses are simplified based on the following assumptions: (1) *quasi-locking condition*, that is, the slave laser is tuned within a small wavelength range near to the locking state [9,10,13,15,19,36]; (2) *single-mode slave condition*, that is, the slave laser is in single longitudinal mode [9,13,15,17–19]; (3) *single-injection condition*, that is, there is only one external injection and that is injected at one wavelength position [9,10,13,15,17–19,36]; and (4) *small-signal condition*, that is, the injection current of the slave is just over the threshold and the optical injection is weak [8–10,16,17–19,36]. Although the previous analyses are capable of explaining some observed phenomena such as threshold injection power [9,10,15,17–19,36], asymmetric locking range [10,17,36], and locking instability [10,12,36], their applicability is limited. For example, quasi-locking limits the validation of the analyses to a very narrow wavelength range near to the FLS (i.e., only one locked mode exists, whereas all others are suppressed). Therefore, it cannot be used to study the locking property far from the FLS, especially the progression from the unlocked state (ULS) to the FLS. The conditions of single-mode slave and single injection make the previous analyses valid only for the case of single injection to a single-mode laser (SISL). However, it has become more common to use multiple injections to lock a multimode laser in some emerging applications, for example, in all-optical switches and wavelength converters [22]. In the case of multiple injections to multimode lasers (MIMLs), theoretical study is continuing. Further, in practice, it is more common to drive the slave at large injection current and strong optical injection for better performance (in terms of high optical powers, better stability, and narrow line width). The validity of the previous SSMs remains uncertain. Although there are some attempts at large-signal analysis, they rely more on numerical simulation [12] rather than analytical form. To tackle these problems, the GM will be described later in this chapter to target broad conditions such as either weak or strong injection power, single injection or multiple injections, single-mode or multiple-mode slaves, and near to the FLS or in the ULS.

8.3.1 Generalized Model Based on the Concept of Injection Seeding

To simplify the notation, several terms related to laser modes in injection locking are defined in Figure 8.3. The external injections at different wavelengths are termed *injected modes,* and are shown in Figure 8.3a. Of all the multiple longitudinal modes of the slave in the free-running state (i.e., no external injections), those close to the injected modes are termed *targeted modes* (i.e., targets for locking), whereas the others are termed *side modes,* as shown in Figure 8.3b. In the presence of the external injections, the injected modes would appear in the optical spectrum of the slave as shown in Figure 8.3c. As these modes are seeded by injections, they are termed *seeded modes.* The wavelengths of the seeded modes are exactly the same as those of the injected modes. Under the injection, the slave may

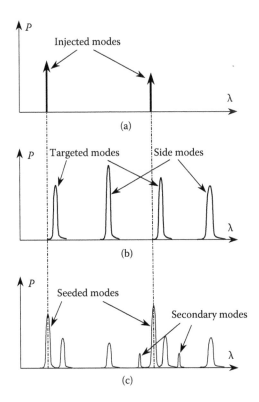

FIGURE 8.3
Definition of terms related to laser modes for injection-locked multimode lasers. (a) External optical injections, (b) laser modes of the slave in the free-running state, and (c) laser modes of the slave under external injections.

also present some secondary modes owing to the interaction between the modes (e.g., the wave-mixing effect [27]). These modes are termed *secondary modes*. Because the targeted modes and side modes are determined by the phase condition of the FP resonant cavity in the slave laser, they are collectively called *cavity modes*. All modes that are in oscillation under external injections are termed *oscillation modes*, including cavity modes, seeded modes, and secondary modes. It is worth noting that injected modes are generated from the master laser, whereas oscillation modes are generated within the slave laser.

The key point regarding SSMs is the concept of linear superimposition of electromagnetic (EM) fields; that is, the external injection is treated as an EM field to be linearly superimposed onto the original EM field inside the slave cavity [10,13,17,36]. More specifically, the amplitude of the injected mode is considered to contribute to the amplitude of the target mode, whereas the detuning (i.e., the wavelength–frequency difference between the injected and targeted modes) contributes to the phase of the targeted mode, similar to a vector summation. This linear superimposition method has become a standard treatment for both the rate equation method and the propagation method since Adler developed the so-called "Adler equation" for electrical injection in 1946 [3] and later Lang adopted it for semiconductor lasers in 1982 [10]. This method is based on an implicit assumption that the injected mode transfers all its power to the target mode and that it would disappear from the final optical output spectrum of the slave (i.e., non-seeded modes exist). It might be true in the FLS, but it is well observed that, out of the FLS, the seeded mode would exist, and it could be the dominant mode of the slave output [17,19]. The linear method also implies that the locking properties (power, wavelength drift, line width, etc.) are influenced by the initial phase of the injected mode; for example, the injected mode would have no contribution if it has a phase difference of $\pi/2$ from the initial electrical field in the FP cavity of the slave. This prediction is contrary to the experimental observation. Moreover, in the MIML case, the inclusion of phase would complicate the analysis because of multiple injections, whose relative phases are difficult to control and monitor.

The key point regarding the GM is the concept of seeding. The idea originates from the understanding that, in a free-running laser, the spontaneous emission actually provides the seed photons, which are amplified and then selected by the FP cavity condition to form intrinsic laser modes. In other words, the laser modes are seeded by spontaneous-emission photons. Similarly, it can be considered that, in ILLs, the external injections provide the seed photons, which are then amplified to form the seeded modes. It is noted that the term *injection seeding* was once suggested by Siegman to describe the case when a weak external injection is used to establish the initial condition for a pulsed oscillator during the turn-on period [1]. In this chapter, injection seeding means that the external injections seed the new oscillation modes in the slave. As the concept of injection seeding leaves out the phase term, the propagation method cannot be used any longer because the phase is essential for electrical field interference. Consequently, the GM is based on the rate equation method.

The analytical model of the ILL is shown in Figure 8.4. The external injections provide seed photons U_j to be injected into the resonant cavity of the slave, in which all the oscillation modes S_i are amplified (or suppressed) and then form the final output (S_i and U_j are the photon numbers, and i and j are integers referring to the i-th and j-th modes, respectively). Therefore, the laser output has all the cavity and seeded modes. In addition to the seeding effect, there are some other physical processes taking place inside the laser cavity: (1) The oscillation modes interact with one another to produce secondary modes through nonlinear effects (e.g., wave mixing [27]) and (2) all the oscillation modes compete with each other and deplete the carriers. Based on the multimode laser rate equation (Equation 3.1 in Reference 37), a generalized mode for injection locking is conceived by including the

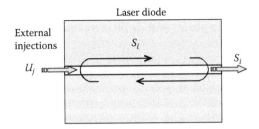

FIGURE 8.4
Analytical model of the injection-locked laser (ILL) for derivation of the generalized model (GM).

injections and interactions, as given by

$$\frac{dS_i}{dt} = S_i \left(R_{st,i} - \frac{1}{\tau_{ph,i}} - \sum_j \theta_{i,j} S_j \right) + R_{sp,i} \kappa_{tot0} + \sum_j \xi_{i,j} S_j + \sum_j \Xi_{ij} U_j \qquad (8.5)$$

where t is the time, $R_{st,i}$ is the stimulated emission coefficient of the i-th mode, $\tau_{ph,i}$ is the photon lifetime, $\theta_{i,j}$ is the cross-mode gain saturation coefficient of the j-th to the i-th mode, $R_{sp,i}$ is the spontaneous emission coefficient, κ_{tot0} is the total enhancement factor of spontaneous emission, $\xi_{i,j}$ is the cross-mode optical pumping coefficient of the j-th to the i-th mode, and Ξ_{ij} is the seeding coefficient. The rate equation (Equation 8.5) takes into account factors such as photon amplification and saturation (the first term on the right side of Equation 8.5), seeding of spontaneous emission (the second term), cross-mode pumping (the third term), and the seeding of the external injections (the fourth term). Equation 8.5 can also be used to describe the case of single injection by ignoring some coefficients. In dealing with the secondary generated modes, the coefficients $\xi_{i,j}$ and Ξ_{ij} may not be constant but depend on the photon numbers of the other modes. As Equation 8.5 treats the laser modes individually, it is easy to deal with the case of multiple injections and multimode slave. In this sense, the model represented by Equation 8.5 is generalized. This is where the name *generalized model* comes from. With respect to polarization, it is assumed that the seeded mode is identical to the targeted mode. Otherwise, the seeded mode will be divided into TE and TM components; each is considered a separate injection to lock a targeted mode with a corresponding polarization state.

As the external injections cause power redistribution among all the modes, the carrier density and frequency positions of the cavity modes are changed, as given by

$$\frac{dn}{dt} = \frac{I_0 - I_{th0}}{e_0 V_0} - \frac{n - n_{th0}}{\tau_{s0}} - \frac{1}{V_0} \sum_i S_i R_{st,i} \qquad (8.6)$$

$$\frac{d\phi}{dt} = 2\pi(f - f_0) = \frac{1}{2} \alpha_0 v_{g0} \frac{\partial g}{\partial n} (n - n_{th0}) \qquad (8.7)$$

where n and n_{th0} are the carrier density and the threshold carrier density, respectively. Similarly, I_0 and I_{th0} are the injection and the threshold currents, respectively. e_0 is the electron charge, V_0 is the volume of active medium (optical confinement factor Γ_0 should be taken into consideration), and τ_{s0} is the carrier lifetime. In Equation 8.7, ϕ is the phase of the slowly varying complex field (following the definition in Equation 2.77 of Reference 37), and f_0 and f are the initial frequency and the frequency after injection, respectively. The variables f_0 and f refer to any of the FP cavity modes. It is assumed that all the cavity modes are shifted by nearly the same frequency. α_0 is the line-width enhancement factor, v_{g0} is the group velocity in the active medium, and $\delta g/\delta n$ is the gain coefficient.

8.3.2 Locking Property under Single External Injection

To prove the validation of the GM, it is first used to deal with a simple case of single external injection. When an external signal is injected into a multimode slave laser, the study can be focused on three separate laser modes: the seeded mode T, the targeted mode S_1, and the side mode S_2 that is nearest to the seeded mode. The governing equations will be presented first, followed by the derivation of the threshold injection power and critical detuning, and finally, the numerical comparison between the GM and the SSM in cases of strong and weak injections.

8.3.2.1 Governing Equations

In the case of single injection, the rate equations for photon numbers, carrier density, and frequency shift can be simplified from Equations 8.5–8.7 to have the following forms:

$$\frac{dS_1}{dt} = S_1\left(R_{st1} - \frac{1}{\tau_{ph0}}\right) + R_{sp1}\kappa_{tot0} - \xi_1 S_1 + \xi_2 U_0 \tag{8.8}$$

$$\frac{dS_2}{dt} = S_2\left(R_{st2} - \frac{1}{\tau_{ph0}}\right) + R_{sp2}\kappa_{tot0} \tag{8.9}$$

$$\frac{dT}{dt} = T\left(R_{stT} - \frac{1}{\tau_{phT}}\right) + \xi_1 S_1 + (1-\xi_2)U_0 \tag{8.10}$$

$$\frac{dn_e}{dt} = \frac{J_0}{e_0 V_0} - \frac{n_e}{\tau_{s0}} - \frac{S_1 R_{st1} + S_2 R_{st2} + TR_{stT}}{V_0} \tag{8.11}$$

$$\Delta f_{drift} = \frac{1}{4\pi}\alpha_0 v_{g0} a_{10}\chi\Delta n_e \tag{8.12}$$

where n_e is the net carrier density defined by $n_e = n - n_{th0}$, Δn_e is the change in the net carrier density, J_0 is the net injection current given by $J_0 = I_0 - I_{th0}$, and Δf_{shift} is the frequency drift of the FP cavity comb due to injection. Because the external injection consumes more carrier, $\Delta n_e \leq 0$. As a result, $\Delta f_{shift} \leq 0$; that is, external injection always leads to a red shift. U_0 is the photons number of the injected mode given by $U_0 = P_{in0}/(h_0 v_T)$, where P_{in0} is the injected power, h_0 is the Planck's constant, and v_T is the frequency of the seeded mode, which is always identical to that of injection. R_{st1}, R_{st2}, and R_{stT} are the stimulated emission coefficients given by $R_{st1} = g_1 v_{g0}$, $R_{st2} = g_2 v_{g0}$, and $R_{stT} = g_T v_{g0}$, where g_1, g_2, and g_T are mode gains that are related to the net carrier density as $g_1 = a_{10}n_e\chi$, $g_2 = a_{20}n_e\chi$, and $g_T = a_{T0}n_e\chi$. a_{10}, a_{20}, and a_{T0}, are the linear gain coefficients, which are determined by the gain curve of the slave; they remain nearly constant during injection locking [37]. Because in most cases the seeded mode is close to the targeted mode, it is assumed that $a_{T0} \approx a_{10}$. χ is the gain suppression factor given by

$$\chi = \frac{1}{1 + \frac{S_\Sigma}{S_{sat0}}} \tag{8.13}$$

where S_Σ represents the total photon number as expressed by

$$S_\Sigma = S_1 + M_0 S_2 + \frac{\tau_{ph0}}{\tau_{phT}} T \qquad (8.14)$$

$$S_{sat0} = \frac{w_0 t_0}{\tau_{s0} v_{g0} \alpha_{m0} \frac{\partial g}{\partial n}} \qquad (8.15)$$

where M_0 is the number of side modes that should be taken into account. S_{sat0} is the saturation photon number α_{m0} is the modal facet loss given by $\alpha_{m0} = -\ln(r_1 r_2)/L_0$; r_1 and r_2 are the amplitude reflectivities of the two laser facets; and w_0, t_0, and L_0 are the width, depth, and length of the active medium of the slave, respectively. As the lifetime of the seeded mode could be much smaller than that of the cavity mode at a certain detuning (e.g., when $\Delta v = 0.3\ \delta_{FSR}$, $\tau_{phT} \approx 0.5\ \tau_{ph0}$), the ratio τ_{ph0}/τ_{phT} in Equation 8.14 normalizes the photon number of the seeded mode to the same time scale as that of cavity modes. Otherwise, energy conservation will be breached because the output power P_i of the i-th individual mode is related to the photon number S_i and the photon lifetime $\tau_{ph,i}$ by $P_i \propto S_i/\tau_{ph,i}$. This treatment makes it feasible not to analyze the slave in the FLS, especially under strong injections.

In Equations 8.8–8.11, R_{sp1} and R_{sp2} are the effective spontaneous-emission coefficients of the targeted mode and the side mode, respectively. They are related to the stimulated emission coefficients by $R_{sp1} = n_{sp0} R_{st1}$ and $R_{sp2} = n_{sp0} R_{st2}$, with n_{sp0} being the inversion factor [37]. The variable ξ_1 is the cross-mode pumping coefficient of the targeted mode to the seeded mode, whereas ξ_2 is that of the seeded mode to the targeted mode. As optical pumping requires high-energy photons to pump out low-energy photons, $\xi_1 = 0$ and $\xi_2 > 0$ when v_T is greater than that of the targeted mode v_1, and $\xi_1 > 0$ and $\xi_2 = 0$ when $vT < v_1$. It is noted that only the cross-mode pumping between the targeted mode and the seeded mode is considered as they are very close in the frequency (or wavelength) domain. The side modes are quite far from the seeded mode in the frequency domain, and thus the interaction is weak.

The variables τ_{ph0} and τ_{phT} are the photon lifetimes for the cavity modes and the seeded mode, respectively. τ_{ph0} is defined by $1/\tau_{ph0} = \alpha_{s0} + \alpha_{m0}$, with α_{s0} being the modal scattering loss of the FP cavity. The seeded mode suffers one more loss, in addition to the scattering loss and the facet loss, because its frequency may not match the FP cavity resonant condition. Therefore,

$$\frac{1}{\tau_{phT}} = \alpha_{s0} + \alpha_{m0} + \alpha_\psi \approx \frac{1}{\tau_{ph0}}(1 + \beta_0 \Delta v^2) \quad \text{for } \Delta v \ll \delta v_{FSR} \qquad (8.16)$$

$$\beta_0 = \frac{2\pi^2}{L_0(\alpha_{s0} + \alpha_{m0})\delta v_{FSR}^2} \qquad (8.17)$$

where the phase mismatch loss α_ψ is given by $\alpha_\psi = -\ln[\cos(\psi)]/L_0$, ψ is the phase mismatch such that $\psi = 2\pi\Delta v/\delta v_{FSR}$, $\Delta v = v_T - v_1$ is the actual frequency detuning (i.e., the real frequency difference between the seeded mode and the targeted mode), $\delta v_{FSR} = c_0/2\bar{\mu}_e L_0$ is the spacing of the FP laser modes, c_0 is the velocity of light in vacuum, and $\bar{\mu}_e$ is the effective group refractive index. β_0 is a phase mismatch factor that directly links the photon lifetime to the frequency detuning.

In a steady state, all the terms $d/dt = 0$ in Equations 8.8–8.11. After some derivation, the governing equations for the static locking properties can be expressed as

$$S_1 = \frac{n_{sp0}\kappa_{tot0} + \xi_2\tau_{ph0}U_0}{1 - G_1 + \xi_1\tau_{ph0}} \tag{8.18}$$

$$S_2 = \frac{n_{sp0}\kappa_{tot0}}{1 - G_2} \tag{8.19}$$

$$T \approx \frac{U_0\tau_{ph0}}{1 + \beta_0\Delta v^2 - G_1} \tag{8.20}$$

$$n_e = \frac{\tau_{s0}J_0}{e_0V_0} - \frac{\tau_{s0}}{V_0\tau_{ph0}}(S_1G_1 + M_0S_2G_2 + TG_T) = \frac{\tau_{s0}}{V_0}\left(\frac{J_0}{e_0} - \frac{S_\Sigma}{\tau_{ph0}}\right) \tag{8.21}$$

$$\Delta f_{shift} = -\rho_0(S_\Sigma - S_{\Sigma0}) \tag{8.22}$$

where G_1, G_2, and G_T are the normalized round-trip gains of the targeted mode, side mode, and seeded mode as given by $G_1 = v_{g0}\tau_{ph0}g_1$, $G_2 = v_{g0}\tau_{ph0}g_2$, and $G_T = v_{g0}\tau_{ph0}g_T$, respectively. ρ_0 is the frequency shift coefficient given by $\rho_0 = \tau_{s0}\alpha_0v_{g0}a_{10}\chi/(4\pi\tau_{ph0}V_0)$. Here it is assumed that the main side modes are identical in terms of power change, which comes from the experimental observation that all the main side modes decrease with the same trend. This treatment is not accurate, but it is not far from reality as compared to the simplification treatment in the SSM [10,36] that treats all the side modes as one mode. When the slave is driven far over the threshold current, the sum of the side modes could give a value much higher than the targeted mode (though any single side mode may be small), which in turn requires a very high gain and thus invalidates the predicted data. Therefore, the SSM is limited to slaves driven slightly over the threshold. In contrast, our treatment allows for strong injection current to drive the slave far over the threshold.

8.3.2.2 Threshold Injection Power and Critical Detuning

The threshold injection power for the FLS can be determined from Equations 8.18–8.22. To lock the slave with a minimum injection power, the injection should be well aligned to one of the FP cavity modes; that is, $\Delta v = 0$. Usually, the threshold power of the injection is weak compared to the original output power of the slave. It is observed in the experiment that under weak injection, the total output power (S_Σ) does not change too much in comparison with the drastic increase or drop of T, S_1, and S_2 (e.g., T is 0 in the free-running state but can be $>10^4$ after locking, whereas S_1 and S_2 change from large values to nearly 0). Therefore, the changes in photon numbers ΔS_1, ΔS_2, and ΔT; change in carrier density Δn_e; and frequency shift can be related solely to the change in total photon number ΔS_Σ, as expressed by

$$\Delta S_1 = \frac{n_{sp0}\kappa_{tot0} + \xi_2\tau_{ph0}U_0}{(1 - G_{10} + \xi_1\tau_{ph0})^2}\Delta G_1 = -\zeta_1\Delta S_\Sigma \tag{8.23}$$

$$\Delta S_2 = \frac{n_{sp0}\kappa_{tot0}}{(1 - G_{20})^2}\Delta G_2 = -\zeta_2\Delta S_\Sigma \tag{8.24}$$

$$\Delta T = -\zeta_T\Delta S_\Sigma \tag{8.25}$$

where changes in normalized gains ΔG_1 and ΔG_2 are given by $\Delta G_1 = -a_{10}\gamma\Delta S_\Sigma$ and $\Delta G_2 = -a_{20}\gamma\Delta S_\Sigma$, and γ is the gain slope such that $\gamma = v_{g0}\tau_{ph0}\chi[\frac{\tau_{s0}}{\tau_{ph0}V_0} + \frac{n_{e0}\chi}{S_{sat0}}]$, where n_{e0} is the initial net carrier density in the active gain medium. ζ_1, ζ_2, and ζ_T are the gain suppression factors as given by $\zeta_1 = \frac{a_{10}\gamma(n_{sp0}\kappa_{tot0}+\xi_2\tau_{ph0}U_0)}{(1-G_{10}+\xi_1\tau_{ph0})^2}$, $\zeta_2 = \frac{a_{20}\gamma n_{sp0}\kappa_{tot0}}{(1-G_{20})^2}$, and $\zeta_T = \frac{a_{10}\gamma U_0\tau_{ph0}}{(1-G_{10})^2}$. G_{10} and G_{20} are the initial values of G_1 and G_2, respectively, in the free-running state. It is noted that ΔT is the change of the seeded photon from its nominal value $T_0 = U_0\tau_{ph0}/(1-G_{10})$.

In fact, the change of total photon number before and after locking is due to the decrease of side and target modes and the increase of the seeded mode; that is,

$$\Delta S_\Sigma = \Delta S_1 + M_0\Delta S_2 + T_0 + \Delta T \tag{8.26}$$

As $\Delta v = 0$, the ratio of $\tau_{ph0}/\tau_{phT} = 1$ and thus does not appear in Equation 8.26. By substituting Equations 8.23–8.25 into Equation 8.26, it can be determined that the increment of the total photon number $\Delta S_{\Sigma th}$ at the threshold of locking state is given by

$$\Delta S_{\Sigma th} = \frac{S_{\Sigma 0}}{1 + \zeta_1 + M_0\zeta_2} \tag{8.27}$$

Let the condition $T \geq k_0 S_{\Sigma 0}$ be set as the criterion for the FLS, where k_0 is the decision level (e.g., 0.8). From Equation 8.25, the threshold injection power U_{0th} can be expressed as

$$U_{0th} = \frac{b_1 S_{\Sigma 0}}{1 - b_2 S_{\Sigma 0}} \tag{8.28}$$

where b_1 and b_2 stand for $b_1 = \frac{k_0(1-G_{10})}{\tau_{ph0}}$ and $b_2 = \frac{k_0 a_{10}\gamma}{(1-G_{10})(1+\zeta_1+M_0\zeta_2)}$, respectively. The threshold injected power P_{0th} can be expressed as $P_{0th} = h_0 v_1 U_{0th}$.

The range for the frequency detuning can be derived from Equation 8.20 as

$$0 \leq \Delta v \leq \Delta v_{up} = \sqrt{\frac{U_0 S_{10}\tau_{ph0} - n_{sp0}\kappa_{tot0}S_{\Sigma 0}}{\beta_0 S_{10}S_{\Sigma 0}}} \tag{8.29}$$

where Δv_{up} is the upper limit of frequency detuning (termed *critical detuning*), and S_{10} is the initial value of S_1 in the free-running state. The lower limit of the detuning is set to 0 rather than $-\Delta v_{up}$ according to the experimental observation. It is also noted that Δv refers to the actual detuning because the frequency of the targeted mode is shifted by Δf_{shift} under external injection as given by Equation 8.22. In the experiment, it is more convenient to use the nominal detuning Δv_{nom} (i.e., the difference between the current injection frequency and the initial frequency of the targeted mode). Therefore, $-\Delta f_{shift} \leq \Delta v_{nom} \leq \Delta v_{up} - \Delta f_{shift}$. However, as Δf_{shift} is also dependent on Δv, the closed-form expression is complicated.

8.3.2.3   *Comparison of the GM with the SSM*

The relationships between the photon numbers and frequency detuning predicted by the GM are shown in Figure 8.5 at two different injection power levels of –10 dBm and –4 dBm, which represent the cases of weak and strong injections, respectively. For comparison, the data predicted by the SSM developed by Li (Equations 8.20–8.22 of Reference 36) are also shown as a reference because Li's work has been widely accepted and proved [12–14,17,18]. The parameters of ILL are listed in Table 8.1. Other data are $S_{10} = 8955$, $S_{20} = 2997$, $S_{\Sigma 0} = 2.69 \times 10^4$, $M_0 = 6$, $n_{e0} = 2.62 \times 10^{24}$, $1-G_{10} = 2.9 \times 10^{-4}$, and $1-G_{20} = 8.7 \times 10^{-4}$, where S_{20} is the initial value of S_2 in the free-running state. These data are measured or calculated from the

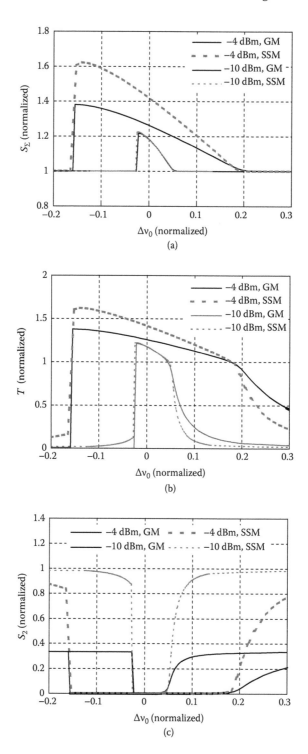

FIGURE 8.5
Comparison of the photon numbers predicted by the generalized model (GM) and the conventional small-signal model (SSM) at strong and weak injection power levels. (a) Total output photon numbers S_Σ (normalized by $S_{\Sigma 0}$), (b) the seeded mode T (normalized by $S_{\Sigma 0}$), and (c) the side mode S_2 (normalized by S_{10}). The nominal detuning frequency Δv_0 is normalized by δv_{FSR}.

TABLE 8.1

Parameters of the MEMS Injection-Locked Laser for Theoretical Analysis and Experimentation

Peak wavelength of the slave λ_0	1533.890 nm
Length of the active gain medium L_0	267.5 μm
Width of the active gain medium w_0	1.30 μm
Thickness of the active gain medium t_0	0.5 μm
Optical confinement factor Γ_0	0.1
Group refractive index of InGaAsP ($\bar{\mu}_e$)	3.375
Amplitude reflectivities of the slave r_1, r_2	0.54
Mode spacing of the slave $\delta\nu_{FSR}$ ($\delta\lambda_{FSR}$)	166 GHz (1.30 nm)
Line width enhancement factor α_0	3
Scattering loss inside the cavity α_{s0}	36 cm^{-1}
Photon lifetime τ_{ph0}	1.38 ps
Total enhancement factor of spontaneous emission κ_{tot0}	1.3
Inversion factor n_{sp0}	2
Carrier lifetime τ_{s0}	2 nm
Saturation photon number of slave S_{sat0}	1.0×10^6
Initial gain suppression factor χ_0	0.80
Phase mismatch factor β_0	3.27×10^{-22} Hz^{-2}
Frequency drift coefficient ρ_0	1.20×10^6 Hz
Threshold current of the slave I_{th0}	15 mA
Injection current of the slave I_0	30 mA

experiment in Reference 20. They will also be used in the experimental study in Section 8.4. For simplicity, the axes S_Σ, T, S_2, and $\Delta\nu_0$ in Figure 8.5 are normalized with respect to $S_{\Sigma0}$, $S_{\Sigma0}$, S_{10}, and $\delta\nu_{FSR}$, respectively. In Figure 8.5a, the total photon number does not change significantly when the frequency detuning is larger than $0.18\delta\nu_{FSR}$ for −4 dBm (or $0.06\delta\nu_{FSR}$ for −10 dBm), and then increases gradually with the decrease of frequency detuning. Further reduction of detuning to a certain value (about $-0.16\delta\nu_{FSR}$ and $-0.02\delta\nu_{FSR}$ for −4 dBm and −10 dBm, respectively) would cause the photon number to drop abruptly to its initial value of $S_{\Sigma0}$. This detuning corresponds to the critical detuning defined in Equation 8.29. The curves given by the two models are quite close at a weak injection of −10 dBm. At a strong injection of −4 dBm, the critical detunings predicted by the two models are nearly identical; however, the SSM predicts a larger total photon number than the GM. From Figure 8.5a, the output power would be increased by 38 and 60% as predicted by the GM and the SSM, respectively. Similarly, in Figure 8.5b, the photon numbers of the seeded modes calculated by the two models match well at weak injection, but a larger value is predicted by the SSM at strong injection. When the detuning is larger than critical detuning, the photon number of the seeded mode predicted by the GM decreases more slowly than that given by the SSM, especially under strong injection. This might be due to the fact that the SSM takes into account the phase (directly related to the detuning), and the increase of detuning rapidly makes the injection field and the laser cavity field out of phase. In contrast, the phase is omitted in the GM. In Figure 8.5c, a clear difference can be seen in the photon numbers of the side mode predicted by the two models. Out of the locking range, the side model given by the SSM is close to S_{10}. This is not realistic as the side mode should be comparatively weak in comparison to the targeted mode ($S_{20} \approx 0.33\ S_{10}$). In contrast, the GM handles the side model properly. It can be seen from Figure 8.5 that the SSM is valid only under conditions of weak injection power, close to the FLS, and nearly single-mode slave, whereas the GM works well for both weak and strong injection power, single and multimode slave, FLS, and ULS. The reason for this is that the SSM involves linearization process, whereas the GM does not.

8.3.3 Locking Property under Multiple External Injections

The GM in Equation 8.5 can be used to describe the laser property under multiple injections. Here, a simple case of two external injections a and b will be analyzed because, in many applications, only two injections are employed [22]. Similar to Equation 8.20, the two seeded modes can be expressed as

$$T_a = \frac{U_a \tau_{ph0}}{1 + \beta_0 \Delta v_a^2 - G_a} \tag{8.30}$$

$$T_b = \frac{U_b \tau_{ph0}}{1 + \beta_0 \Delta v_b^2 - G_b} \tag{8.31}$$

where T_a and T_b are the photon numbers of the two seeded modes, respectively; U_a and U_b are the photon numbers of the injections; Δv_a and Δv_b stand for the detunings; and G_a and G_b represent the normalized gain of the corresponding seeded modes.

Assume that the laser is initially locked by only the injection b (with a power of P_{bin}) at the position of one side mode. When another injection, a, is present at the position of the peak side mode, the power will be redistributed among the oscillation modes. Figure 8.6 shows the relationship between the photon numbers and the seeded photon U_a when the other injection (b) is kept constant. In Figure 8.6, all the photon numbers are normalized by $S_{\Sigma 0}$; in the horizontal axis, the seeded photon U_a is also normalized by U_b. The parameters are the same as those for Figure 8.5 by letting the injection a target the peak mode and b the side mode. When $P_{bin} = -10$ dBm (i.e., $U_b = 7.72 \times 10^{14}$), T_a rises gradually with the increase of U_a/U_b, whereas T_b keeps on decreasing, and the total photon number S_Σ is close to $S_{\Sigma 0}$. At $P_{bin} = -4$ dBm (i.e., $U_b = 3.07 \times 10^{15}$), T_a increases rapidly over $S_{\Sigma 0}$, and S_Σ rises nearly linearly. At the same time, T_b is dropped. When $U_a/U_b = 1.68$ (i.e., 2.3 dB), T_a reaches $S_{\Sigma 0}$ and becomes dominant (this power difference is termed the *gating threshold*). This relationship underlies one of the important applications of multiple injections—all-optical switching. In Figure 8.6, T_b is dominant in the absence of U_a, but is well suppressed in the presence of a strong U_a. Let U_a be the control signal and U_b be the data signal; the optical pulse of U_a

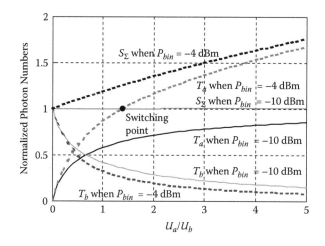

FIGURE 8.6
Changes of photon numbers (normalized by $S_{\Sigma 0}$) with normalized injection power difference under two external injections.

would be able to turn on or off the optical pulse train of U_b. This phenomenon has been used in recent applications [22] but has not been analyzed theoretically.

8.4 Experimental Studies and Discussions

This section will use a real prototype of MEMS ILL as the example to discuss the experimental setup, laser characteristics under different conditions, and verification of the developed GM.

8.4.1 Device Description and Experimental Setup

In a real MEMS ILL demonstrated recently [20], the MEMS structure is fabricated on a 6 in. silicon-on-insulator (SOI) wafer using the deep reactive ion etching (DRIE) process. The structure layer is silicon material of 35 μm thickness the oxide layer is 1 μm thick, and the handling wafer is 400 μm thick. The structure layer is deeply etched to form the gratings, actuators, microlenses, and trenches for laser chips and optical fibers. The floating structure is released by patterning and etching from the bottom side of the handling wafer. A gold-coating process is used by a shadow mask to provide bond pads, electrical routing, and grating reflection surfaces while keeping the microlenses uncoated. After fabrication, the wafer is diced into small pieces of dies to expose the fiber grooves to the edge. During the laser assembly and packaging, the laser chips are put into the MEMS trenches with their top surfaces facing down. A layer of indium foil about 10 μm thick is sandwiched between the laser chips and the trenches on the dies. The foil lifts up the optical axis, and also provides good electrical and heat transfer. The alignment of the laser chips to the MEMS gratings is done coarsely by the marks on the dies, and finely by driving the laser to oscillate and then monitoring the output. Fixation of the laser chips is accomplished by applying thermal conductive epoxy. After that, the single-mode optical fibers are aligned to the fiber grooves on the dies and then fixed by the UV-curable epoxy.

The scanning electron micrograph (SEM) of an integrated MEMS ILL is shown in Figure 8.7, which follows the direction coupling design as discussed in Section 8.2. After

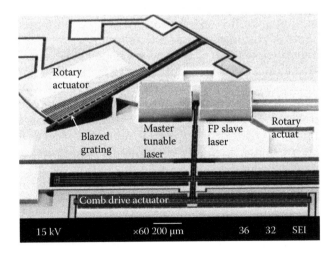

FIGURE 8.7
Scanning electron micrograph (SEM) of a single-chip-integrated injection-locked laser.

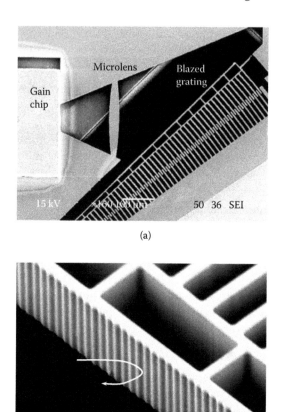

(a)

(b)

FIGURE 8.8
Close-ups of the key components of the MEMS injection-locked laser: (a) rotary comb drive and microlens for the tunable laser and (b) blazed grating.

fabrication and integration, the ILL has dimensions of $3 \times 2 \times 0.6$ mm. In the real prototype as shown in Figure 8.7, the master laser is a continuously tunable laser in the Littrow external cavity configuration (similar to the one developed in Chapter 6). The wavelength tuning is obtained by rotating the blazed grating using the MEMS rotary comb actuator as shown in Figure 8.8a. A close-up of the grating is shown in Figure 8.8b. It is etched on the sidewall of the structure layer and is then coated with a gold layer of 0.2 μm. The grating has 60 teeth at intervals of 3.3 μm. The third-order diffraction angle is 45°. The actuator can rotate the grating by 1.0° with a voltage of 30 V. A prism is used as the coupling control element. It maintains the mutual coupling of oscillating mode between master and slave while preventing the other modes of the slave from entering the master. In this sense, it acts as an isolator. The position of the prism also affects the coupling efficiency.

In the master laser part, the gain chip of the master is a stripe-guided multiquantum-well InGaAsP/InP laser chip having a dimension of $250 \times 300 \times 100$ μm (length $\times$ width $\times$ height). One facet of the gain chip is antireflection coated (intensity reflectivity < 0.1%), whereas the other is high-reflection coated (intensity reflectivity ≈ 85%). The output wavelength can be tuned continuously from about 1510 to 1540 nm at a resolution of 0.001 nm. The slave is an FP laser chip coming from the same wafer as the gain chip, but its facets

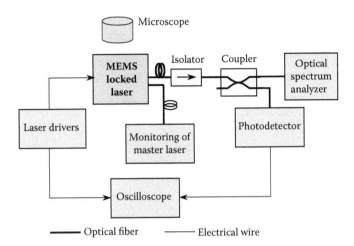

FIGURE 8.9
Experimental setup for characterizing the MEMS injection-locked laser.

are just normally cleaved. The internal cavity length is 267.5 μm. The threshold current is 15 mA. In the following test, the slave laser is driven at 30 mA (DC) and gives an output optical power of about 1.5 mW.

The experimental setup is illustrated in Figure 8.9. The MEMS ILL is placed under a microscope, on which a charge-coupled device (CCD) camera is mounted. This is to visualize the integration, assembly, and operation of the MEMS components, the laser chips, and the optical fiber. The master and slave lasers are driven by laser drivers (SDL-820 Laser Diode Driver, SDL, and 236 Source Measure Unit, Kaithley, respectively). The output of the slave laser is connected to a 50/50 2 × 2 optical coupler through an optical isolator. The isolator is to avoid back reflection into the slave, which would deteriorate the laser characteristics and complicate the analysis. If multiple injections are needed, external injections may be made from the fiber. In this case, the isolator has to be removed. The coupler has one output branch connected to an optical spectrum analyzer (Q8384 Optical Spectrum Analyzer, Advantest) for spectrum measurement, and the other one to a photodetector (2832-C Dual-Channel Power Meter + 818-IR detector head, all from Newport) for optical power measurement. The electrical signals of the injection current from the laser drivers and the optical power from the photodetector are both fed to a digit oscilloscope (TDS 360, Tektronix) to measure the waveforms. At the same time, the output of the master laser is also monitored.

8.4.2 Locking Property under Single External Injection

From the engineering point of view, the quality of injection locking is dependent on many parameters of the master laser such as injection power, wavelength detuning, polarization direction, and even the misalignment relative to the slave. The locking property under weak injection has been extensively studied, both theoretically and experimentally. Therefore, this study focuses on strong injection (i.e., the injection power is comparable to the output power of the slave in the free-running state). The parameters of the slave are listed in Table 8.1 and in the Section 8.3. In the free-running state, the slave output has more than 30 modes with a total output power of −2 dBm as shown in Figure 8.10a. In the window of observation, there are four modes. The peak mode at 1533.890 nm is 4.8 dB stronger

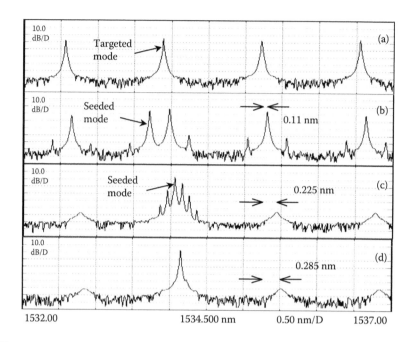

FIGURE 8.10
Output spectra of the slave at different locking states. (a) Free-running state; (b) $\Delta\lambda_0 = -0.19$ nm, far from locking; (c) $\Delta\lambda_0 = 0.065$ nm, close to locking, but have wave mixing; and (d) $\Delta\lambda_0 = 0.255$ nm, fully locked state.

than the second highest mode, and is chosen as the targeted mode. According to Equation 8.28, the injection power should be $\geq$−23.1 dBm (the SSM predicts $\geq$−74 dBm). The actual threshold injection power for locking is −21.9 dBm. The GM gives a close estimation in comparison to the SSM. Table 8.2 summarizes some key parameters predicted by the GM as compared to the experimental data.

The external injection is from the master laser with an estimated power of −4 dBm. Therefore, it is a strong injection. The output spectra of the slave are illustrated in Figure 8.10b–d when the detuning is varied. In Figure 8.10b, the external injection is far from the targeted mode (the nominal detuning $\Delta\lambda_0 = -0.085$ nm and the actual detuning $\Delta\lambda = -0.19$ nm), and it is simply added up to the spectrum of the free-running state. However, the cavity modes are red shifted by 0.105 nm. In addition, secondary modes are excited on both sides of every cavity mode. In Figure 8.10c, as the seeded mode moves closer to the targeted mode ($\Delta\lambda_0 = 0.065$ nm and $\Delta\lambda = -0.16$ nm), the red shift is increased to 0.225 nm, and most of the output power is concentrated on the seeded mode and the targeted mode, whereas the side modes

TABLE 8.2

Comparison of the Parameters Predicted by the Generalized Model (GM) with the Experimental Data

Laser Characteristics	Generalized Model	Experiment
Critical detuning	0.310 nm	0.295 nm
Increase of total output power	38%	30%
Maximum red shift	0.27 nm	0.29 nm
Threshold injection power	−23.1 dBm	−21.9 dBm
Gating threshold	2.3 dB	2.0 dB

become suppressed. At the same time, the secondary modes become more obvious. These secondary modes may be due to the wave-mixing effect between the seeded mode and the cavity modes. In previous studies, the mechanism for new wavelength components generation is not taken into account, and thus cannot be simulated. However, this kind of new mode could be important in some applications such as coherent communications and chaos synchronizations [38]. In contrast, the GM of Equation 8.5 allows production of new modes by treating them as separate and then including them in the nonlinear interactions inside the laser cavity following Equations 8.6 and 8.7. However, in this experiment the influence of the secondary modes is neglected as they are relatively weak (>20 dB power difference) and is not among our focus areas. In Figure 8.10d, as the seeded mode comes very close to the targeted mode ($\Delta\lambda_0 = 0.255$ nm and $\Delta\lambda = -0.03$ nm), the laser is fully locked. The seeded mode dominates the other modes by an SMSR of 53.4 dB. At this time, the cavity mode is shifted by 0.285 nm. It is observed that the total output power in the locking state is increased by 30%, which is comparable to the 38% estimate given by the GM (see Table 8.2) but significantly smaller than the 60% estimate given by the SSM (see Figure 8.5a). Further increase of $\Delta\lambda_0$ up to 0.295 nm would push the cavity mode to a longer wavelength, but the actual detuning is always close to but smaller than 0. Once the actual wavelength detuning becomes larger than 0, the slave suddenly loses the locking state. This is why the boundary condition $\Delta v \geq 0$ (i.e., $\Delta\lambda \leq 0$) should be applied to the GM.

The relationship between red shift $\Delta\lambda_{shift}$ and nominal wavelength detuning $\Delta\lambda_0$ is measured and compared with the simulation of the GM as shown in Figure 8.11. It can be seen from the test data that the cavity mode has a red shift of 0.065 nm even when the amount of detuning is large (< −0.3 nm). This phenomenon is termed as the *red shift baseline*. It may be due to the fact that the injected photons consume the carrier and thus reduce the carrier density even if they do not satisfy the FP cavity condition. In the case of weak injection, this baseline of red shift is negligible. As the GM does not take this type of red shift into account, a value of 0.065 nm is added to the simulation result. After compensation, the GM curve has a reasonable agreement with the experiment data. The critical detuning predicted by the GM is 0.31 nm, whereas the measured value is 0.295 nm. According to Equation 8.29, the critical detuning is 0.255 nm, and becomes 0.32 nm after compensating for the baseline. The maximum red shift is 0.27 nm by the GM and 0.29 nm by the experiment (see Table 8.2).

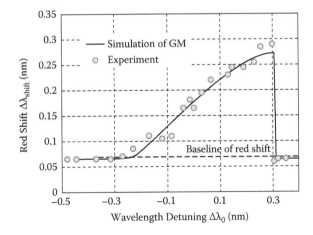

FIGURE 8.11
The relationship between red shift and nominal detuning.

8.4.3   Locking Property under Multiple External Injections

The locking property under multiple external injections is also investigated. For simplicity and also for the sake of most applications, two injections are used: one from the master (termed the *control signal*) and the other from another laser source injected through the optical fiber (termed the *data signal*). The injection wavelength positions of the control and the data are shown in Figure 8.12a when the DC injection current of the slave is not turned on. The control has a wavelength of 1535.39 nm, and the data has a wavelength of 1534.06 nm. The data signal is fixed at −4 dBm. It is able to lock the slave to an SMSR of 51.4 dB when there is no control injection as shown in Figure 8.12b. When the control signal is increased, the seeded control mode rises gradually whereas the seeded data mode decreases, following the same trend as shown by Figure 8.6. When the control is −2.0 dBm, the slave is locked by the control to an SMSR >50 dB even in the presence of the data signal as shown in Figure 8.12c. The gating threshold is 2.0 dB (i.e., the injection power of the control is 1.58 times that of the data). From Figure 8.6, the GM predicts a value of 2.3 dB

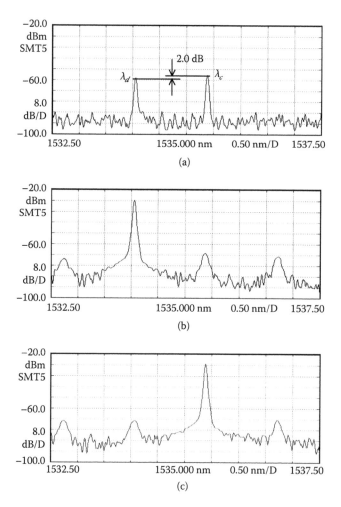

FIGURE 8.12
Output spectra of all-optical switching. (a) Injected control and data, (b) locked only by the data, and (c) locked by the control in the presence of data.

(i.e., 1.68 times). Reasonable agreement has been obtained (see Table 8.2). Such a power difference is relatively small compared to the values used in the literature, but the obtained SMSR is much higher than the typical value of 20 dB [22].

To demonstrate the switching between the control and the data, the response of the slave laser to a modulated injection is measured as shown in Figure 8.13. In the experiment, the injection wavelength is determined by obtaining the FLS in the static state (i.e., the injection is DC). Keeping the wavelength unchanged, the master laser is modulated and injected to

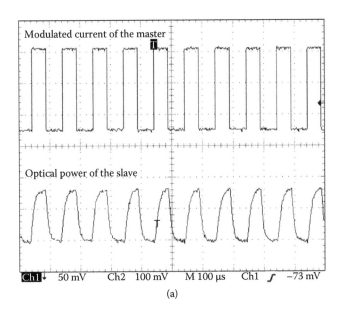

(a)

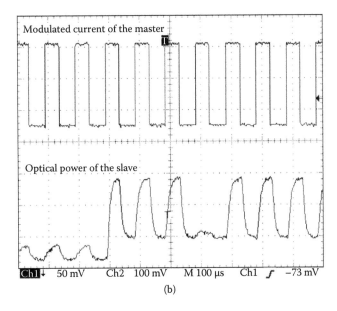

(b)

FIGURE 8.13
Waveforms of the slave output under the pulsated injection signal. The injection wavelength is preadjusted for locking under static injection. (a) The slave can follow the injection modulated at 9.99 KHz and (b) locking becomes unstable when injection is modulated at 10.2 KHz.

the slave. At a low modulation rate of 9.9 kHz, the slave can follow the power change of the injection signal as shown in Figure 8.13a. The waveform of the slave output comes from the fact that the power of the slave in the locked state is about 30% higher than that in the free-running state. With the increase of the modulation rate to 10.2 kHz, locking becomes unstable as shown in Figure 8.13b. The slave cannot follow the injection change and therefore misses some pulses. This instability has been mentioned by Lang [10]. In his study, the slave power should vary periodically; however, in the experiment, the loss of pulse is nonperiodic.

As shown in Figure 8.13, the modulation limit is about 10 KHz, which is too low compared with values such as 5 and 10 Gbs reported in the literature [22,39]. The problem is solved by further adjustment of the injection wavelength. After that, the slave can easily respond to the injection modulation rate up to 100 MHz (limited by the available equipment) while maintaining a high-quality waveform. The influence of detuning on the maximum response speed might be attributed to the fact that the pulsated injection makes slave parameters slightly different from those under static injections. As the theoretical study in this chapter focuses mainly on the static state, the dynamic property is not analyzed any further. However, the GM in Equation 8.5 should also be valid as it is actually a dynamic model.

When the control is pulsated at 1 MHz and the data at 50 MHz, the waveforms are shown in Figure 8.14. It can be seen that the data is turned off once the control is on, and vice versa. As the control and the data are all in their original optical form, Figure 8.14 demonstrates the function of all-optical switching as implied by Figure 8.6. Such a switching speed of 50 MHz is much higher than that of conventional micromechanical switches (~1 KHz), and has potential applications in packet optical switching.

8.4.4 Locking Property under Weak Injection

The locking properties discussed previously are measured under the strong injection power of −4 dBm. The MEMS ILL is also measured under a weak injection of −10 dBm (larger than the threshold injection power −21.9 dBm, but much smaller than the initial total

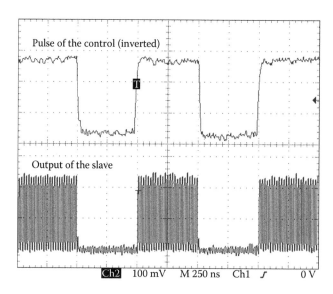

FIGURE 8.14
Waveforms of all-optical switching.

output power of the slave –2 dBm). Various locking properties are investigated, including the lockable spectrum range, influence of injection power, tolerance for misalignment, and polarization dependence.

8.4.4.1 Lockable Spectrum Range

The experimental result of the lockable spectrum range is shown in Figure 8.15. In the free-running state, that is, with no external injection, the output spectrum is multimode as shown in Figure 8.15a. It is noted that the power reading in the spectrum does not reflect the actual value as it is found that, in this OSA, the reading changes with the sweep mode, wavelength resolution, and sampling point number. However, it reflects the relative relationship. Figure 8.15b shows the lockable mode at the minimum wavelength of 1520.08 nm (termed the *minimum mode*) when the power of single injection is fixed at –15 dBm. Modes with shorter wavelengths cannot be locked. Figure 8.15c illustrates the locking of the mode at 1521.41 nm, which is just next to the minimum mode. All the longitudinal modes on the right side of the minimum wavelength are lockable up to a maximum wavelength of 1570.13 nm, as shown in Figure 8.15d (termed the *maximum mode*). Modes with longer wavelengths cannot be locked. As shown in Figure 8.15e, the mode at 1571.44 nm, which is adjacent to the maximum mode, loses the locking if the external injection power keeps constant. A lockable spectrum range of 50.05 nm is obtained, covering 39 longitudinal modes. In the locking states in Figure 8.15b–d, the SMSR is > 30 dB; that is, most of the optical power is concentrated on the locked modes. The peak power after locking is about 10 dB higher than that of the free-running state; but the total power does not increase significantly, unlike in the case of strong injection in which the total power is increased by 30%. It can be seen from Figure 8.15e, that, even after amplification in the laser cavity, the injected light in the unlocked state is still about 10 dB lower than the peak power of the free-running state. Therefore, the red shift baseline does not exist under weak injection, which is opposite to the case of strong injection. It is also noted that the lockable spectrum range is shorter on the left side than on the right side. This may be due to the asymmetry of the gain curve.

If the injection power decreases, the lockable range becomes smaller as shown in Figure 8.16. At the critical value of –21.9 dBm, only the peak wavelength of the multilongitudinal FP spectrum can be locked; in other words, the lockable spectrum range becomes zero. Further reduction of injection power makes none of the modes lockable. The theoretical prediction is also shown in Figure 8.10. The theoretical curve follows the trend of the experimental one.

8.4.4.2 Influence of Injection Power

To study the influence of injection power on mode locking, a mode next to the peak mode of the free-running state is locked with an injection power of –10 dBm. The mode is well locked as shown in Figure 8.17a. The SMSR reaches a value as high as 42 dB, in contrast to the original 4.8 dB of the free-running state. To obtain the FLS, the wavelength is adjusted by a step of 0.001 nm to maximize the SMSR. With the decrease of injection power, locking remains stable until injection drops to –21.9 dBm. The slave laser loses locking, and the spectrum becomes multimode as shown in Figure 8.17b. At a still lower level of injection power, locking cannot be seen. An interesting observation in Figure 8.17b is that when the locking state disappears, the powers of the side modes return to near-original values, whereas the power of the targeted mode is reduced. One possible reason is that the targeted mode transfers part of its power to the seeded mode through a process similar to

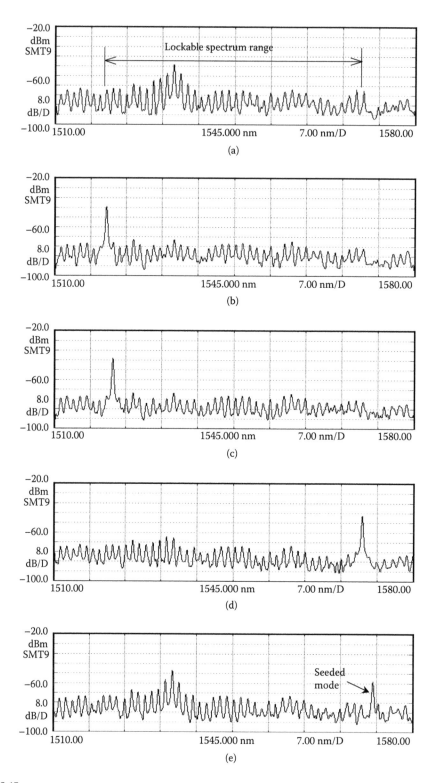

FIGURE 8.15
Lockable spectrum range. (a) Free-running spectrum, (b) the minimum lockable wavelength 1520.08 nm, (c) locking at 1521.41 nm, (d) the maximum lockable wavelength 1570.13 nm, and (e) loss of locking at 1571.44 nm.

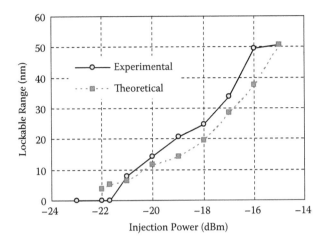

FIGURE 8.16
Theoretical and experimental data of the lockable spectrum range with change in injection power.

optical pumping because the wavelength difference is small (0.06 nm). As the wavelength of the targeted mode is longer than that of the seeded mode, the latter does not reversely pump the former. This is an experimental evidence of the optical pumping coefficients $\xi_{i,j}$ in Equation 8.5, and ξ_1 and ξ_2 in Equation 8.8. The pumping coefficients are quite small and dependent on the wavelength difference. It can be neglected in most cases.

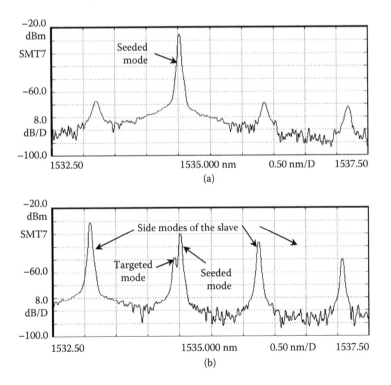

FIGURE 8.17
Change of the spectrum with effective injection power. (a) Locking state at –10 dBm and (b) loss of locking at –21.9 dBm.

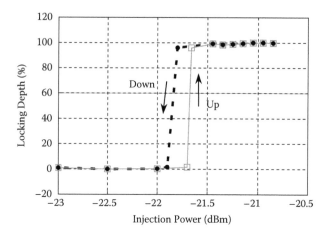

FIGURE 8.18
Hysteresis of the locking state with the change of injection power.

When injection power increases from very low to high values, locking does not emerge until –21.7 dBm. There is a hysteresis of 0.2 dB. The relationship between locking quality and injection power is shown in Figure 8.18. Here, the term *locking depth* is introduced to quantitate the locking quality as defined by

$$\zeta = \frac{\text{SMSR} - \text{SMSR}_0}{\text{SMSR}_{\text{max}} - \text{SMSR}_0} \times 100\% \tag{8.32}$$

where ζ is the locking depth (in percentage), SMSR is the current side-mode suppression ratio determined by the power of the mode of interest minus the maximum power of all the other modes, SMSR_0 is the initial ratio in the free-running state, and SMSR_{max} is the maximum ratio for the perfect locking state. All the power values are in units of dBm. Note that the variables SMSR and SMSR_0 can be negative if the mode of interest is not the peak mode of the free-running state. Locking depth ζ is valid only for longitudinal modes of the slave laser, not for the master. According to definition, ζ always falls in the range 0–100%; 0 for the free-running state and 100% for the FLS. The hysteresis can be seen clearly in Figure 8.18. Either in the up or down state, the locking depth keeps flat initially and changes abruptly on reaching the critical values. After the abrupt change, the increase of injection power brings the locking depth closer to 100%.

The hysteresis of threshold injection can be regarded as a history-dependent property of injection locking; that is, the current locking property is influenced by the previous state, similar to Markov chain. It can be explained qualitatively using the GM. From Equations 8.18–8.22, it can be seen that the final locking property is influenced by the initial parameters S_{10}, S_{20}, T_0, and ne_0. When injection changes from strong to weak to pass the threshold point, the slave is initially locked and tends to maintain this locking state. Therefore, it requires the injection power to go to a lower value. Conversely, when injection increases from the unlocked state, it requires higher injection.

8.4.4.3 Tolerance for Alignment

The alignment between the master and slave lasers is very important when the lasers are assembled within the MEMS structure. As the lateral misalignment would reduce the

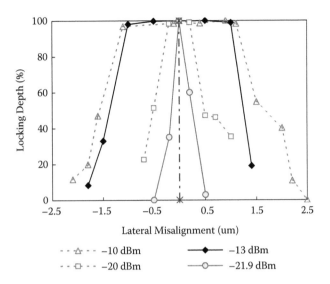

FIGURE 8.19
Influence of the lateral misalignment on locking depth.

coupling from master laser to the slave, it should be within a tolerance value. The misalignment tolerance at different injection power levels is shown in Figure 8.19. When the effective injection power is −10 dBm, the slave has a locking depth close to 100% within the misalignment range of −2.1 to 2.5 μm. Out of this range, the locking depth drops sharply to a very low level. When the injection power decreases to −13 dBm, the tolerance shrinks to a range from −1.8 to 1.4 μm. At −20 dBm, it is from −0.7 to 1 μm. At the critical value of −21.9 dBm, the depth rate is close to 100% only when the lateral misalignment is negligibly small. In other words, the alignment tolerance is very small. From Figure 8.19, the curves of locking depth are not very symmetrical to the line of misalignment, which is equal to zero. It can also be observed in Figure 8.19 that the locking quality experiences sharp changes in all the curves. It suggests that the transition is abrupt once the injection power drops to the threshold level.

The misalignment tolerance can be explained in terms of injection power. At the critical injection, it allows no misalignment. This explains the sharp drop of the curve for the injection of −21.9 dBm in Figure 8.19. At high injection, the actual injection can be maintained over the threshold even in the presence of misalignment. The higher the injection, the larger the tolerance. The coupling efficiency drops rapidly with the increase of misalignment; this may explain why the tolerance does not increase linearly with the injection power as illustrated in Figure 8.19.

8.4.4.4 Influence of Polarization

Because the stripe waveguide of the active layer in the slave laser has high modal gain for the TE mode than that for the TM mode, the polarization of external injection should affect the locking quality. In the analyses and experimental results stated previously, the polarization of injection locking is always identical to the slave. In all the experiments citer earlier, optical outputs of the master and slave are always linearly polarized in the TE mode. This section will study the influence of injection polarization. The experimental observation is shown in Figure 8.20. In the measurement, the polarization state of injection from an external laser is adjusted by a polarization controller.

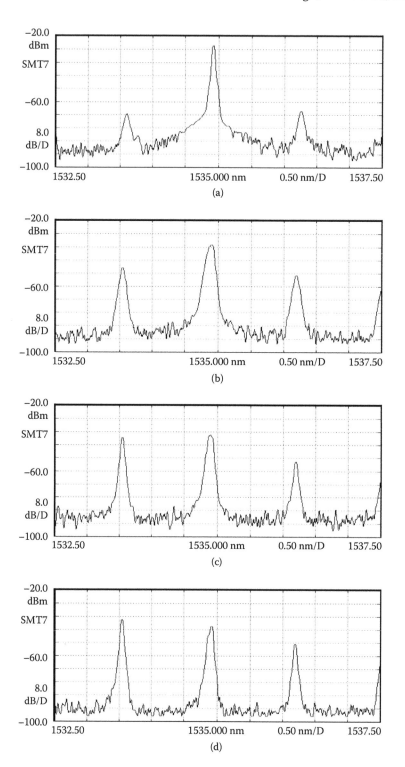

FIGURE 8.20
Degrading of locking quality with the change of polarization direction: (a) The fully locked state, (b) polarization angle at 30°, (c) polarization angle at 45°, and (d) polarization angle at 60°.

When the polarization direction of the injection light is well aligned, the slave laser is fully locked at 1534.92 nm with a maximized SMSR of 42 dB, as shown in Figure 8.20a. The peak is high and sharp, the line width is only about 0.02 nm (limited by the resolution of the (optical spectrum analyzer) OSA). When the polarization angle is changed to about 30°, the locking state becomes deteriorated as shown in Figure 8.20b. The output becomes multimode with an SMSR of only 14 dB. The power level of the peak is dropped by 8.7 dB. The spectral shape of the modes is rounded and broadened. The line width is increased to 0.1 nm. At the same time, the total output power is reduced by 20%. As the polarization direction is rotated to 45°, the slave output is further degraded (see Figure 8.20c). The SMSR becomes as low as 1.7 dB, and the peak power is dropped by another 4 dB. When the polarization angle is changed by 60° or higher, the slave loses the locking entirely as shown in Figure 8.20d. The slave output spectrum is very similar to that of its free-running state. It is also measured that the FP mode comb is shifted by −0.005 nm for every step of increase of the polarization angle. This *blue shift* during the degrading of the locking state agrees with the red shift observed in Figure 8.16, in which the locking quality (or locking depth) is gradually improved.

Based on the GM, the degradation of the locking state due to polarization change can be attributed to the combination of reduced effective injection power and increased noise. The injection can be decomposed into TE and TM modes. In the FLS, the polarization of injection is aligned with the TE direction. With the deviation of the polarization angle, the effective injection power in the TE mode decreases, and therefore, the locking state is gradually degraded (i.e., the side modes increase) as shown in Figure 8.20a–d. At the same time, the effective injection power in the TM mode is increased. Just outside the FLS, the TM mode is too weak to oscillate, and therefore it acts as a noise source to the TE mode. As a result, the line width is broadened significantly as shown in Figure 8.20b. Further increase of the TM mode will make it strong enough to oscillate, and thus the line width becomes narrow as shown in Figure 8.20c,d.

8.5 Summary

This chapter presents the theoretical and experimental studies of MEMS ILLs with the primary focus on the development of a new GM based on the concept of injection seeding, which treats external injections as the seed photons and leaves out the phase. Compared with the previous SSMs of injection locking, which treat injections as linear superimposition of EM fields and is valid only for weak injection and single-mode slave laser, the GM has many advantages; for example, it can handle the cases of multiple injections to a multimode slave laser and strong injection power, which are the common conditions when the MEMS ILLs are applied to some emerging applications such as all-optical switching and wavelength conversion. For experimental study, a prototype of MEMS ILL has been used as the example to discuss the issues. The prototype mainly consists of an MEMS Littrow tunable laser as the master, an FP laser chip as the slave, and an optical fiber for optical coupling; all are integrated onto a SOI wafer within a volume of $3 \times 2 \times 0.6$ mm. The locking properties are measured under strong injection power with single and multiple external injections. Locking states with a side mode suppression ratio >50 dB have been obtained, and all-optical switching up to 50 MHz has also been demonstrated. Other static properties such as the influence of injection power, polarization, and misalignment tolerance have also been investigated. The GM predicts the laser characteristics such as critical detuning, increase of total output power, maximum red shift, threshold injection power, and gating threshold to be 0.31 nm, 38%, 0.27 nm, −23.1 dBm, and 2.3 dB, respectively, compared to the corresponding measured data of 0.295 nm, 30%, 0.29 nm, −21.9 dBm, and 2.0 dB. The reasonable agreement validates the effectiveness of the GM.

References

1. Siegman, A. E., *Lasers*, University Science Books, Mill Valley, CA, 1986.
2. Van der Pol, J., Forced oscillations in a circuit with nonlinear resistance, *Phil. Mag.*, 3, 65, 1927.
3. Adler, R., A study of locking phenomena in oscillators, *Proc. Inst. Radio Eng.*, 34, 351, 1946.
4. Stover, H. L. and Steier, W. H., Locking of laser oscillators by light injection, *Appl. Phys. Lett.*, 8, 91, 1966.
5. Tang, C. L. and Statz, H., Phase locking of laser oscillators by injection signal, *J. Appl. Phys.*, 38, 323, 1967.
6. Sung, H.-K., *Strong Optical Injection Locking of Edge-Emitting Lasers and its Applications*, Ph.D. thesis, Department of Electrical Engineering and Computer Science, University of California at Berkeley, USA, 2006.
7. Lamb, W. E., Jr., Theory of an optical maser, *Phys. Rev.*, 134, A1429, 1964.
8. Pantell, R. H., The laser oscillator with an external signal, *Proc. IEEE*, 53, 474, 1965.
9. Kobayashi, S. and Kimura, T., Optical phase modulation in an injection locked AlGaAs semiconductor laser, *IEEE Trans. Microwave Theory Tech.*, MTT-30, 1650, 1982.
10. Lang, R., Injection locking properties of a semiconductor laser, *IEEE J. Quantum Electron.*, QE-18, 976, 1982.
11. Ibrahim, M. M. and Alhaider, M. A., Injection-locked passively Q-switched lasers, *IEEE J. Quantum Electron.*, QE-18, 109, 1982.
12. Hörer, J. and Patzak, E., Large-signal analysis of all-optical wavelength conversion using two-mode injection-locking in semiconductor lasers, *IEEE J. Quantum Electron.*, 33, 596, 1997.
13. Cai, B., Johansson, L. A., Silva, C. F. C., Bennett, S., and Seeds, A. J., Crosstalk, noise, and stability analysis of DWDM channels generated by injection locking techniques, *J. Lightwave Technol.*, 21, 3029, 2003.
14. Kashima, N., Yamaguchi, S., and Ishii, S., Optical transmitter using side-mode injection locking for high-speed photonic LANs, *J. Lightwave Technol.*, 22, 550, 2004.
15. Murakami, A., Phase locking and chaos synchronization in injection-locked semiconductor lasers, *IEEE J. Quantum Electron.*, 39, 438, 2003.
16. Cassard, P. and Lourtioz, J.-M., Injection locking of high-power pulsed lasers—part I: monochromatic injection, *IEEE J. Quantum Electron.*, 24, 2321, 1988.
17. Mogensen, F., Olesen, H., and Jacobsen, G., Locking conditions and stability properties for a semiconductor laser with external light injection, *IEEE J. Quantum Electron.*, QE-21, 784, 1985.
18. Hadley, C. R., Injection locking of diode lasers, *IEEE J. Quantum Electron.*, QE-22, 419, 1986.
19. Lidoyne, O., Gallion, P., Chabran, C., and Debarge, G., Locking range, phase noise and power spectrum of an injection-locked semiconductor laser, *IEE Proc.*, 137, 147, 1990.
20. Liu, A. Q., Zhang, X. M., Cai, H., Tang, D. Y., and Lu, C., Miniaturized injection-locked laser using microelectromechanical systems technology, *Appl. Phys. Lett.*, 87, paper 101101, 2005.
21. Liu, Y., Liu, H. K., and Braiman, Y., Injection locking of individual broad area lasers in an integrated high power diode array, *Appl. Phys. Lett.*, 81, 978, 2002.
22. Chan, L. Y., Wai, P. K. A., Lui, L. F. K., Moses, B., Chung, W. H., Tam, H. Y., and Demokan, M. S., Demonstration of an all-optical switch by use of a multiwavelength mutual injection-locked laser diode, *Opt. Lett.*, 28, 837, 2003.
23. Cai, H., Zhang, X. M., Zhang, Q. X., and Liu, A. Q., MEMS tunable coupled-cavity laser, In *Proc. 14th Int. Conf. Solid-State Sens. Actuators Microsyst. (Transducers '07)*, Lyon, France, 2007, 1441.

24. Abbas, G. L., Yang, S., and Chan, V. W. S., Injection behavior of high-power broad-area diode lasers, *Opt. Lett.*, 12, 605, 1987.
25. Kobayashi, S., Yamada, J., Machida, S., and Kimura, T., Single-mode operation of 500 Mbit/s modulated AlGaAs semiconductor laser by injection locking, *Electron. Lett.*, 16, 746, 1980.
26. Simpson, T. B., Liu, J. M., and Gavrielides, A., Bandwidth enhancement and broadband noise reduction in injection-locked semiconductor lasers, *IEEE Photon. Technol. Lett.*, 7, 709, 1995.
27. Paiella, R., Hunziker, G., Ziari, M., Mathur, A., and Vahala, K. J., Wavelength conversion by cavity-enhanced injection-locked four-wave mixing in a fiber-bragg-grating coupled diode laser, *IEEE Photon. Technol. Lett.*, 10, 802, 1998.
28. Chen, H. F. and Liu, J. M., Open-loop chaotic synchronization of injection-locked semiconductor lasers with gigahertz range modulation, *IEEE J. Quantum Electron.*, 36, 27, 2000.
29. Wynands, R., The atomic wrist-watch, *Nature*, 429, 509, 2004.
30. Knappe, S., Shah, V., Schwindt, P. D. D., Hollberg, L., Kitching, J., Liew, L.-A., and Moreland, J., A microfabricated atomic clock, *Appl. Phys. Lett.*, 84, 2694, 2004.
31. Knappe, S., Schwindt, P. D. D., Gerginov, V., Shah, V., Liew, L., Moreland, J., Robinson, H. G., Hollberg, L., and Kitching, J., Microfabricated atomic clocks and magnetometers, *J. Opt. A: Pure Appl. Opt.*, 8, S318, 2006.
32. Jau, Y.-Y., Post, A. B., Kuzma, N. N., Braun, A. M., Romalis, M. V., and Happer, W., Intense, narrow atomic-clock resonances, *Physical Review Letters*, 92, paper 110801, 2004.
33. Zhang, X. M., Liu, A. Q., Cai, H., Yu, A. B., and Lu, C., Retro-axial VOA using parabolic mirror pair, *IEEE Photon. Technol. Lett.*, 19, 692, 2007.
34. Liu, A. Q., Zhang, X. M., Lu, C., and Tang, D. Y., Review of MEMS external-cavity tunable lasers, *J. Micromech. Microeng.*, 17, R1, 2007.
35. Coldren, L. A. and Koch, T. L., Analysis and design of coupled-cavity lasers-Part I: Threshold gain analysis and design guidelines, *IEEE J. Quantum Electron.*, QE-20, 659, 1984.
36. Li, L., Static and dynamic properties of injection-locked semiconductor lasers, *IEEE J. Quantum Electron.*, 30, 1701, 1994.
37. Petermann, K., *Laser Diode Modulation and Noise*, Kluwer Academic Publishers, London, 1988.
38. Liu, Y., Chen, H. F., Liu, J. M., Davis, P., and Aida, T., Synchronization of optical-feedback-induced chaos in semiconductor lasers by optical injection, *Phys. Rev. A*, 63, 03, paper 031802, 2001.
39. Weich, K., Patzak, E., and Hörer, J., Fast all-optical switching using two-section injection-locked semiconductor lasers, *Electron. Lett.*, 30, 493, 1994.

9

Deep Etching Fabrication Process

Jing Li

CONTENTS

9.1 Crystal Structure of Silicon...354
9.2 Silicon Wet Etching...357
    9.2.1 Silicon Isotropic Wet Etching..357
    9.2.2 Silicon Anisotropic Wet Etching..358
9.3 Introduction to Dry Silicon Etch...358
9.4 Chemistry and Physics of DRIE...359
    9.4.1 Parameters of the Etching Process 362
        9.4.1.1 Effect of Platen Power 363
        9.4.1.2 Effect of Gas Flow with Source Power...............................364
        9.4.1.3 Effect of Duration on the Cyclic Etch/Passivation Cycles.................368
        9.4.1.4 Effect of Process Pressure...368
9.5 Loading Effect...370
    9.5.1 Loading Effect Dependence on Design of Various Features.............370
    9.5.2 Etch Rate versus Trench Width...370
    9.5.3 Etch Rate versus Trench Position on the Wafer................................373
    9.5.4 Etch Rate versus Exposed Area of the Wafer 373
    9.5.5 Etch Rate Along the Process..375
    9.5.6 Beam Verticality..376
9.6 Notching Effect..377
    9.6.1 Mechanisms of the Notching Effect..377
    9.6.2 Elimination of Notching Effect..380
    9.6.3 Dry Release by Notching Effect...384
        9.6.3.1 Uniform Beam Release...384
        9.6.3.2 Nonuniform Beam Release 385
9.7 Process of DRIE Fabrication on SOI Substrate 390
9.8 Summary...391
References 392

In this chapter, a universal and reliable deep reactive ion etching (DRIE) process on a silicon-on-insulator (SOI) wafer to achieve high-performance microelectromechanical system (MEMS) devices is described. As the performance of MEMS devices is strongly process dependent and MEMS fabrication is not standardized, it is very important and relevant to develop a flexible process that is applicable for different systems.

    Among the numerous process steps involved, etching is the most essential, especially for bulk micromachined microstructures with high aspect ratio, in which deep, highly directional etching and release from beneath are necessary but challenging to achieve. Dry gaseous etching is more advanced compared to wet etching because of its merits of independence with respect to substrate crystalline orientation, accurate transfer of lithographically defined patterns into underlying layers, high resolution and cleanliness, ease of automation, and better control. Release is a special etching process employed to remove a specific layer to obtain movable MEMS structures, which is never used in integrated circuit (IC) technology or where new problems such as stiction may occur. An SOI-wafer-based process simplifies the process, reducing the number of masks necessary at the expense of the notching effect, because of the buried insulation layer and microloading effect. These issues have been addressed in this chapter and proved by process results.

The chapter begins with a section on the crystal structure of silicon followed by a brief discussion on wet etching. Then, dry etching is discussed, with the focus on chemical and physical aspects of the alternative etch and passivation DRIE process. The effects of process parameters, including operating power, gas flow rates, durations, and chamber pressure, on the structures are analyzed individually in terms of etch rate, etch profile, and etch selectivity and are compared with the experimental results to optimize process conditions.

Paying special attention to high-aspect-ratio structures, loading effects are experimentally studied. Etch rate is analyzed, first as a function of the feature size, location, exposed area, and variations within the process itself. Then, the verticality of microbeams as a function of trench width is investigated with the purpose of obtaining a better design.

Last, a new spacer-oxide thin-film technique is developed to eliminate the severe notching produced during the processing of an SOI wafer, based on analysis of the origin of the notching phenomenon. Moreover, this problem is used to selectively release microstructures, thus realizing a reliable dry release.

9.1 Crystal Structure of Silicon

A crystal can always be divided into fundamental units having a characteristic shape, volume, and contents. In many crystals, a cube may be chosen as the unit cell, with atoms placed at various points within it. Cubic structures can be categorized by the location of atoms into simple cubic (sc), body-centered cubic (bcc), face-centered cubic (fcc), sodium chloride (NaCl), cesium chloride (CsCl), zinc blende (ZnS), and diamond (c), as shown in Figure 9.1.

Silicon has a diamond structure in which the atoms are bonded tetrahedrally with covalent bonds, the lattice planes and directions being described by a Miller index. This index allows specification, investigation, and discussion of distinct planes and directions of a crystal. The most widely used planes are {100}, {110}, and {111}, as shown in Figure 9.2. The angles between different planes are listed in Table 9.1.

Wafer flats are employed to identify the wafer orientation for automatic equipment and to indicate the type and orientation of crystal. For large crystals, no flats are ground. Instead, a notch is machined for positioning and orientation purposes.

Generally, a primary flat and secondary flat are used. The primary flat is the flat of longest length located in the circumference of the wafer that has a specific crystal orientation relative to the wafer surface. The secondary flat indicates the crystal orientation and doping of the wafer, whose location varies.

For <100> wafers, the flats are shown in Figure 9.3 for n- and p-type wafers. The secondary flat is at 180° for n-type and at 90° for p-type.

For <111> wafers, the flats are illustrated in Figure 9.4 for n- and p-type wafers. There is no secondary flat for p-type wafer, whereas the secondary flat is at 45° for the n-type. However, very few 6 in. or 8 in. <111> wafers are manufactured.

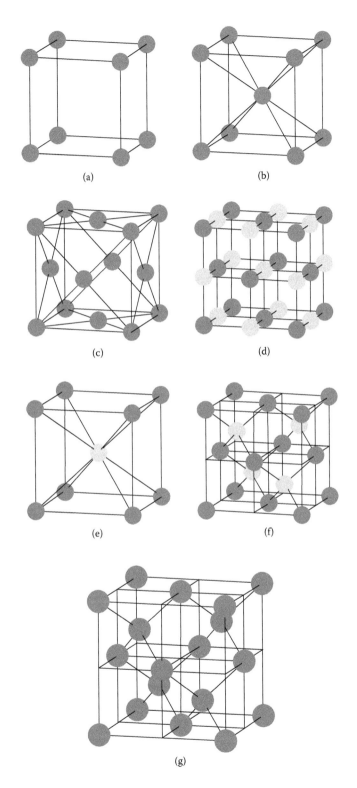

FIGURE 9.1
The various cubic structures are (a) sc, (b) bcc, (c) fcc, (d) NaCl, (e) CsCl, (f) ZnS, and (g) diamond.

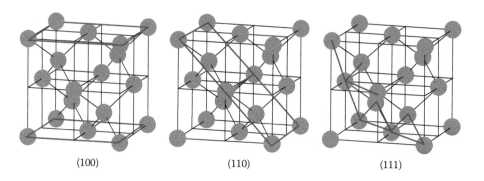

(100)   (110)   (111)

FIGURE 9.2
Most widely used planes of silicon.

TABLE 9.1

Angles Between Different Crystal Orientations

Angle (°)	100	110	010	001	101
100	0.00	45.00	90.00	90.00	45.00
011	90.00	60.00	45.00	45.00	60.00
111	54.74	35.26	54.74	54.74	35.26
211	35.26	30.00	65.91	65.91	30.00
311	25.24	31.48	72.45	72.45	31.48
511	15.79	35.26	78.90	78.90	35.26
711	11.42	37.62	81.95	81.95	37.62

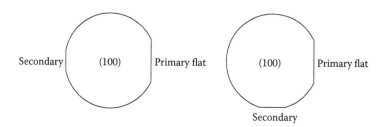

FIGURE 9.3
Wafer flats for <100> wafers.

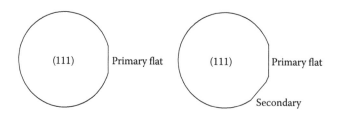

FIGURE 9.4
Wafer flats for <111> wafers.

FIGURE 9.5
Cross-sectional view of isotropic silicon wet etching (a) with agitation and (b) without agitation.

9.2 Silicon Wet Etching

A wide variety of silicon wet etchings are available, which are classified according to differences in properties, costs, and levels of compatibility with the other process steps. Employing the wet etch technique, various subtractive structures can be achieved through appropriate selection of masking materials, etchants, and process conditions. Wet etch can be generally categorized into two classes, isotropic and anisotropic, the latter being obtained as a result of different etching rates along different crystal orientations.

9.2.1 Silicon Isotropic Wet Etching

Isotropic silicon etching is illustrated in Figure 9.5a,b for etching with and without agitation, respectively. Pits and cavities with rounded surfaces can be achieved. Agitation is performed to speed up the transport of reactants and products and to keep the transport more uniform, so that the same etching rate can be obtained along different directions.

The most common etchant is $HF/HNO_3/CH_3COOH$ (hydrofluoric/nitric/acetic [HNA] acids). The reaction is,

$$18HF + 4HNO_3 + 3Si \rightarrow 2H_2SiF_6 + 4NO(g) + 8H_2O \qquad (9.1)$$

The etching rate is a function of the chemical mixture ratio and silicon doping, which is listed in Table 9.2 along with the etching rate of the mask materials.

TABLE 9.2

Silicon Isotropic Wet Etching: Etchants and Etch Rates

	Etchant	Reagent Ratio	Temperature (°C)	Etch Rate (μm/min)	Etch Ratio (100)/(111)	Dopant Dependence	Mask Etch Rate (μm/min)
1	HF HNO_3 CH_3COOH	1 3 8	22	0.7~3.0	1:1	Etch rate reduces 150 times for $<10^{17}$cm^{-3}	SiO_2, 30 μm/min
2	HF HNO3 CH_3COOH	1 2 1	22	4	1:1	No dependence	Si_3N
3	HF HNO_3 CH_3COOH	3 25 10	22	7	1:1	Not known	SiO_2, 70 μm/min

FIGURE 9.6
Cross-sectional view of wet silicon anisotropic etch for (a) (100) substrate and (b) (110) substrate.

9.2.2 Silicon Anisotropic Wet Etching

Silicon anisotropic etching is orientation dependent, which means that the etch goes much faster in one direction than in another. For silicon, the slowest plane is (111) regardless of the wafer type, as shown in Figure 9.6 for different types of substrates. The typical etching-rate ratio along the different orientations is 1 (111), 300 to 400 (100), and 600 (110), which relies on the chemical composition, concentration, and process temperature. The most commonly used etchants are potassium hydroxide (KOH) and tetramethyl ammonium hydroxide (TMAH). Further information on the mask materials and selectivity can be obtained in Reference 1.

9.3 Introduction to Dry Silicon Etch

Dry silicon etching is a process in which the material removal reactions occur in the gas phase, including both nonplasma- and plasma-based methods. The nonplasma-based etch is isotropic, using a spontaneous reaction of the appropriate reactive gas mixture that typically contains fluorine. This etch provides high selectivity to many masking layers including Al, SiO_2, Si_3N_4, photoresist (PR), and phosphosilicate glass (PSG), and can be precisely controlled by temperature and partial pressure of the reactants. One of the most widely employed reactant gases is XeF_2, which chemically reacts with silicon as follows:

$$2XeF_2 + Si \rightarrow 2Xe + SiF_4 \tag{9.2}$$

The other method of dry silicon etching is plasma-based anisotropic etch, which is used more widely in MEMS fabrication because the directional etch is independent of silicon crystal orientation and unintentional prolongation of etching can be easily prevented. The plasma-based dry etch can be categorized into four types: chemical etch, physical etch, reactive ion etch (RIE), and DRIE, the latter two being more frequently employed. The RIE process is a chemical etch accompanied by ionic bombardment (i.e., ion-assisted etch). The plasma is the source of both ions and chemical etchant agents. The combined effect of etchant atoms and energetic ions in producing etch products can be much larger than that produced by either pure chemical etch or by sputtering alone. The etch is a chemical reaction that occurs in nature without directionality, whereas the reaction rate is determined by energetic ion bombardment. The anisotropic property results from an ion-enhanced reaction. Although RIE can produce a good directional etch, the reaction rate and anisotropic profile properties are obtained at a trade off between the two. This is because increase in etch rate needs a higher reactive species concentration, which in turn leads to higher gas pressure and more collisions which consequently compromises anisotropy. Another conflict between high etch rate and anisotropic profile exists because of the higher kinetic energy required for a good profile, which will decrease the etch rate. In other words, the etch depth is constrained by this process. DRIE is introduced for high-aspect-ratio micro-structure fabrication, which employs lower-energy ions that affect the etch rate less and have higher selectivity, the plasma maintaining a pressure of 0.5–3 mtorr.

9.4 Chemistry and Physics of DRIE

The DRIE process uses a high-density low-pressure (HDLP) system, which reduces the probability of ion collision because sheath thickness decreases at higher ion density and ion mean free path increases at lower pressure. Ion directionality is improved, and this in turn enhances the control of anisotropy. Two main mechanisms are used in reactive dry silicon etching to realize high-aspect-ratio structures:

1. Reduction in reaction probability at the sidewall to below a critical value, which implies a reduction of the spontaneous lateral etching rate to zero.
2. Sidewall passivation-based protection from radical, atomic, and neutral etching species, with or without addition of a polymerizing chemistry. This anisotropy can be enhanced by reducing the reaction probability, depending on the precursor gases.

Simultaneous etch and sidewall passivation processes have been successfully employed in COMS and in some very limited MEMS applications. Suitable passivation is formed during etch. This technique can be described by a simple model of an SF_6/O_2 process. The ion and radical species are generated by electron impact dissociation:

$$SF_6 + e^- \rightarrow S_X F_Y^+ + S_X F_Y^* + F^* + e^- \tag{9.3a}$$

$$O_2 + e^- \rightarrow O^+ + O^- + e^- \tag{9.3b}$$

The energetic O is used to passivate the silicon surface by forming an oxide film:

$$O^* + Si(s) \rightarrow SiO_n(sf) \tag{9.4}$$

where (s) and (sf) denote surface and surface film, respectively. Because the bottom silicon is covered by the passivation layer, it must be removed directionally to etch silicon with F further:

$$SiO_n(sf) + F^* \rightarrow SiO_n(sf) + F \tag{9.5a}$$

$$SiO_n(sf) + F \overset{IonEnergy}{\rightarrow} SiF_x(ads) + SiO_x F_y(ads) \tag{9.5b}$$

where (ads) means that F adsorbs onto the surface and the ion enhances the adsorption, reaction, and desorption. The silicon is now exposed and the etch with F can proceed with product formation and desorption as a gas (g):

$$Si + F^* \rightarrow Si + nF \tag{9.6a}$$

$$Si + nF \overset{IonEnergy}{\rightarrow} SiF_x(ads) \tag{9.6b}$$

$$SiF_x(ads) \rightarrow SiF_x(g) \tag{9.6c}$$

Obviously, the ion flux provides directionality, and hence, the anisotropy. The removal of passivation and continuation of silicon etch occur simultaneously, and they must be balanced to maintain vertical profile. If not, the predominance of passivation results in slow etching or even stopping of the etch owing to surface residue buildup. On the other hand, insufficient passivation leads to a loss in anisotropy owing to increase of lateral etch.

Cryogenic enhancement is one of the possibilities used to address the passivation-related problem. Fluorine-based gas is selected for high etch rate because it has the highest electronegativity, which is required for silicon halide volatility. However, the lateral etch also increases as the sidewall passivation is chemically attacked by high F concentration. Thus, the cryogenic system provides substrate temperatures as low as $-110°C$ to utilize high fluorine concentrations because the etch product volatility is reduced. These drawbacks are critical when using PR because of the superlow temperature, which induces thermally cracking. Also, surface contamination (leading to formation of black silicon grass) resulting from impurity condensation onto the wafer has to be considered. Sidewall roughness is also an obstacle to application in optical devices.

Inductively coupled plasma (ICP) sources offer the most stable and widest operating window and are used in surface technology systems–ICP (STS-ICP) systems operating at 13.56 MHz (both the excitation and bias frequency). The schematic of an STS Multiplex ICP deep etch machine used to obtain high-aspect-ratio microstructures is shown in Figure 9.7.

This equipment employs a process based on the technique invented by Larmer and Schilp [2]. This passivation technique is different from simultaneous etch and passivation, in which passivation is deliberately separated from etch by sequentially alternating etch and deposition steps rather than integrating sidewall protection with the etching process. The main reaction gases are SF_6 for etching and octafluorocylcobutane (C_4F_8) for passivation. The process involves introduction of a short polymer deposition cycle immediately after each etch cycle. The formal polymer-forming step deposits a layer of passivation (C_xF_y) on

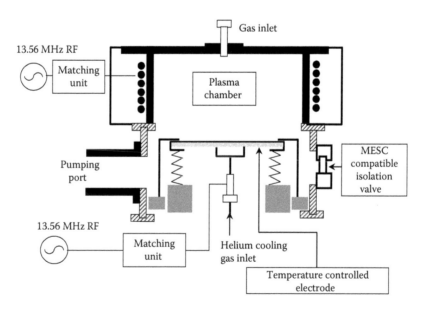

FIGURE 9.7
STS Multiplex ICP Process Module.

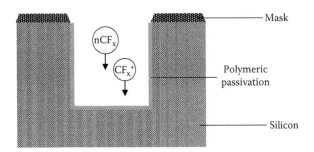

FIGURE 9.8
Schematic of the passivation polymer film deposition.

the sidewall and base of the feature by ionization and dissociation of a precursor gas, i.e., C_4F_8. The deposition gas is dissociated by the plasma to form ion and radical species first:

$$C_4F_8 + e^- \rightarrow CF_x^+ + CF_x^{\cdot} + F^* + e^- \tag{9.7a}$$

The species then undergo polymerization reactions to generate a polymeric layer:

$$CF_X^{\cdot} \rightarrow nCF_2(ads) \rightarrow nCF_2(f) \tag{9.7b}$$

where (*ads*) means CF_2 adsorbs onto the surface and (*f*) denotes that the passivation layer is a deposited film. This is an isotropic process, which means that the passivation layer is deposited on the trench surfaces including sidewall and bottom, as shown in Figure 9.8.

$$nCF_2(f) + F^* \overset{IonEnergy}{\rightarrow} CF_x(ads) \rightarrow CF_x(g) \tag{9.8}$$

Then, the etch cycle starts with dissociation of reactive gas SF_6, as expressed in Equation 9.5a, followed by the bottom passivation layer removal [3]. The silicon to be etched is exposed and the etch can proceed as described by Equations 9.6a–c. A schematic of this alternative process is shown in Figure 9.9.

The polymer deposition and silicon-etching steps are repeated successively in passivation and etch cycles. Thus, the anisotropy is enhanced by protecting the sidewall from radical, atomic, and neutral etch species, and the vertical etch is controlled mainly by the ion bombardment aiding removal of the surface polymer.

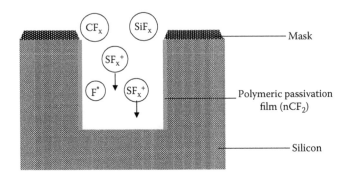

FIGURE 9.9
Etching cycle of the alternative passivation process.

In such a process, reaction chemistry selection is critical to finely control the balance between etch and passivation. First, the deposited polymer film must have good coformality irrespective of whether it is for a positive- or negative-tapered trench, providing adequate sidewall protection. Second, this film must be easily removed by etching plasma. Polymeric layers meet both these requirements; in particular, they need only low levels of surface ion bombardment in the presence of a reactive chemical to be completely removed from the surface. Therefore, high selectivity to PRs without compromising anisotropy is obtained as low self-bias potentials allow the directionality of the etching to be easily controlled. Compared to the cryogenic system, this process can be carried out at room temperature, avoiding the need for supercooling. The reaction gases SF_6 and C_4F_8 as well as other options such as NF_3, CHF_3, and C_2F_6 are all nontoxic.

Rapid etching of deep trenches with high aspect ratio and the ability to stop on oxide (because of the high selectivity) are very important aspects of a trench process, for various applications. Precise trench profile and taper control, which minimize critical dimension (CD) loss, along with low-stress highly conformal oxide refill are therefore essential to enable elimination of voids in the trench refill. Minimizing the undercut at the silicon dioxide interface is essential as substantial footing will cause nonuniform dielectric films to be deposited, which will reduce the breakdown field between adjacent silicon islands.

9.4.1 Parameters of the Etching Process

Based on the principle of DRIE as described in the proceeding section, etching can be carried out to obtain high-aspect-ratio microstructures. However, the characteristic etching performance parameters such as etch rate and etch selectivity, as well as surface morphology and post-etch mechanical behavior (including pattern profile and anisotropy), exhibit a strong dependence on processing variables [4,5]. For instance, the profile and etch surface are not a concern in realization of the dicing line, whereas a fast etch rate is important to obtain the deep trenches. Achieving a perfectly vertical microstructure profile is much more important for the micromirror compared to having a fast etch rate. Verticality and surface smoothness are two key requirements that need to be achieved to ensure a high-performance micromirror. Therefore, understanding the effects of the individual processing parameters allows for optimizing both etch and deposition rates, so that when both are combined, the overall DRIE efficiency is maximized. There are six major steps in etching: reactive species (ions and radicals such as F^*) generation, diffusion to surface, adsorption at surface, reaction at surface, desorption of reactive cluster, and diffusion of by-products into bulk gas (refer to Figure 9.10).

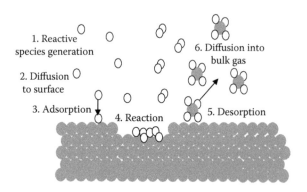

FIGURE 9.10
Schematic of the etch steps.

Accordingly, factors that influence the process include concentration of etchant species generated in the plasma, flux of etchant species to the silicon surface, reaction rates at the surface, temperature of the silicon, removal rate of reaction products, and consumption of etchant by other species in the plasma or by-products formed during etching. Consequently, macroparameters to be adjusted and monitored during the process are gas flow rate, coil and platen RF power, chamber pressure, wafer temperature, and individual cycle times. The effects of these parameters will be discussed on the basis of experiments conducted on 6 in. diameter silicon wafers, in terms of etch rate, etch profile, and etch selectivity. After the loading of experimental wafers into the reaction chamber, an initial gas stabilization is carried out, during which the electronic-grade, highly purified SF_6 gas is also loaded into the chamber at a preset flow rate, which is adjustable using a mass flow controller. The gas is pumped away by means of a turbopump. Then, the helium cooling gas flow rate is checked to confirm that there is not much warp or some other deformation in order to guarantee good wafer cooling. The desired pressure is maintained by using a throttle valve. The plasma is generated by a 600 W 13.56 MHz RF coil generator and the platen 13.56 MHz RF power is applied to accomplish the wafer chuck, the energy of the ions being controlled by this power supply.

9.4.1.1 Effect of Platen Power

Platen power plays an important role in this deep etching process—it removes the base passivation polymeric layer. This RF power is used to accelerate ions to a sufficiently high potential (typically greater than 20 eV) to bombard the trench bottom while maintaining the sidewall passivation intact, thereby promoting the anisotropy. Therefore, the etch rate, etch selectivity to the mask material, and etch profile are functions of the platen power. Platen power increases etch rate because the energy of the reactive ions is higher and also speeds the mask erosion rate, thus degrading the selectivity. A comparison of these effects is shown in Table 9.3, under the following etch conditions: etching cycle of 8.0 s and passivation cycle of 5.0 s; gas flow during etching of 30 sccm of C_4F_8, 100 sccm of SF_6, and 10 sccm of O_2 and gas flow for passivation of 160 sccm of C_4F_8; RF coil power of 600 W; process pressure of 20 mtorr; and platen power ramped from 14 to 23 W. The table shows that etch rate is increased and etch selectivity to hard mask (silicon dioxide) is reduced with higher platen power. The profile of the etched trench also alters from positive to vertical to negative with the ramping platen power. The feature profile with high aspect ratio is even closer to the base when the platen power is low. Undercut is another phenomenon relying on platen power, as shown in Table 9.3 (the relevant definition of undercut is schematically shown in Figure 9.11). The beam profile is presented by the angle θ between the contour of

TABLE 9.3

Comparison of Etch Rate, Etch Selectivity, Feature Profile, and Undercut for Various Platen Power

Specification	14 W	18 W	20 W	23 W	Remark
Undercut (μm)	0.37	0.32	0.28	0.24	Decrease
Beam top width (μm)	1.02	1.02	1.08	1.10	Increase
Beam bottom width (μm)	1.77	1.54	1.15	1.08	Decrease
Beam depth (μm)	53.6	56.0	55.3	55.8	Deeper
Beam profile (θ°)	0.80	0.53	0.07	−0.01	More vertical
Etch rate (μm/min)	1.34	1.40	1.38	1.398	Increase
Selectivity (Si:SiO$_2$)	160:1	165:1	143:1	94:1	Lower

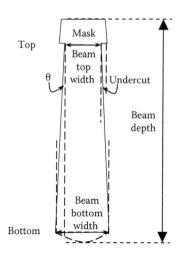

FIGURE 9.11
Schematic of the undercut in a DRIE-etched beam.

the etched side and the ideal vertical edge. The etching results for platen power of 14 and 18 W are shown in Figures 9.12 and 9.13, respectively, where (a) overview of etching profile, (b) top (for the undercut), and (c) bottom details are demonstrated by scanning electron microscope (SEM) micrographs.

Etch rate, undercut, and profile angle are plotted in Figure 9.14, which verifies that the etch rate increases and undercut decreases with higher platen power.

One of the most important parameters for optical devices is the verticality, which changes from a positive slope to nearly 90° when the power is set at 23 W, while simultaneously, the etch selectivity between silicon and the silicon dioxide hard mask is sacrificed with higher platen power, as shown in Figure 9.15.

9.4.1.2   Effect of Gas Flow with Source Power

The precursor (SF_6) of the reactive species during the etch cycle is dissociated in the plasma, providing neutral particles, reactive ions, and other highly reactive compounds, whereas the alternative passivation cycle uses C_4F_8. The gas flow rates and source power significantly influence the deep etching process.

When the flow rate of etching gas, SF_6, is fixed at either 100 or 130 sccm and RF coil power ramps from 600 to 1000 W, the silicon etch rate is limited by source power, as evidenced by Figure 9.16. The figure shows that etch rate increases with coil power as a consequence of more reactive etchant species (F^*) being generated. The relationship between the etch rate and gas flow rate can be deduced from this figure, which shows that etch rate is limited by precursor gas flow when insufficient precursor gas is available. In both cases, the main etch-rate-limiting mechanism is insufficient generation of etchant species (F^*) due to either an insufficient gas flow rate or a too low source power. However, it does not mean that etch can be speeded up infinitely by increasing gas flow or coil power.

This phenomenon can be explained by a kinetic scheme for the generation and loss of atomic fluorine. The generation of F^* species from SF_6 is the result of the dissociation reaction characterized by a reaction speed of k_G:

$$SF_6 \xrightarrow{k_G} F^*$$

(9.9)

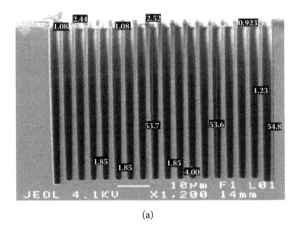

(a)

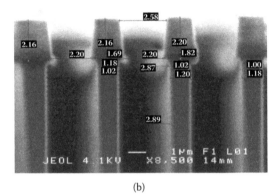

(b)

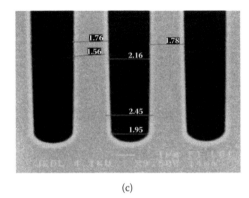

(c)

FIGURE 9.12
Deep-etched trenches with platen power of 14 W: (a) overview, (b) top, and (c) bottom.

Several mechanisms for the consumption of F^* species are listed as follows:

1. Recombination with SF_n (n is 3, 4, or 5) species: As expressed by Equation 9.3, SF_n species are generated with reactive etchant F^*. In the process chamber, it is possible for these species to recombine with F^* at the combination rate of k_R.

$$SF_n^* + F^* \xrightarrow{k_R} T \qquad (9.10a)$$

where T denotes the termination species.

2. Recombination with other F^* species: When the reactive F^* meet each other, they will combine to from F_2 at the rate of k_F.

$$F^* + F^* \xrightarrow{k_F} F_2 \qquad (9.10b)$$

3. Reaction with silicon (effective consumption): The reactive species react with surface silicon at the etch reaction rate of k_E.

$$F^* + Si \xrightarrow{k_E} SiF \qquad (9.10c)$$

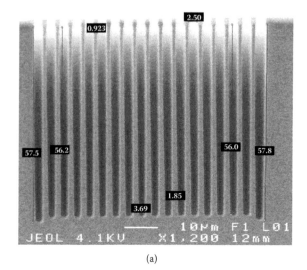

(a)

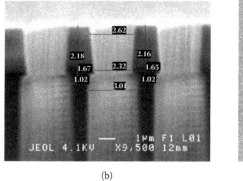

(b)

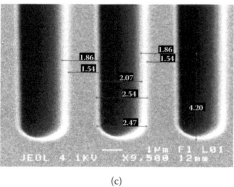

(c)

FIGURE 9.13
Deep-etched trenches with the platen power of 18 W: (a) overview, (b) top, and (c) bottom.

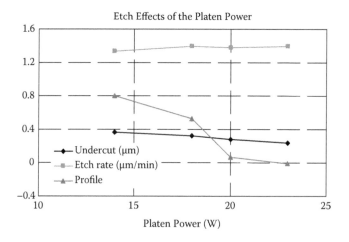

FIGURE 9.14
Effects of platen power on etch rate, undercut, and profile of the deep-etched beam.

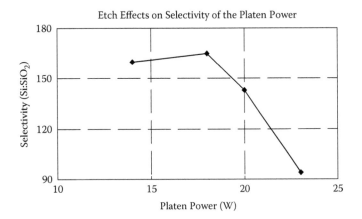

FIGURE 9.15
Effects of platen power on etch selectivity.

Because the generation and consumption of the etchant species must be balanced, the overall process can be expressed as

$$k_G[SF_6] = k_R[SF_n][F^*] + k_F[F^*][F^*] + k_E[F^*][Si] \tag{9.11}$$

where [] represents the concentrations of various species. It is noted that the contributions of the three mechanisms to the consumption of the species are different. The limiting silicon etch rate is due to the equilibrium between etchant generation from SF_6 and its recombination with dissociated SF_n, as illustrated in Figure 9.17. The recombination between reactive species F^* plays a marginal role, if any. The third contribution is related to the exposed area of silicon $[Si]$, which does not affect the absolute value of the etch rate, although it does make the etch rate saturate at a lower $[SiF_6]$ level. The fluorine species concentration $[F^*]$ is an increasing function of $[SiF_6]$ when the generation constant k_G is several orders of magnitude greater than the recombination constant k_R. In other words, etch rate is limited by etchant species $[F^*]$ loss due to its recombination with SF_n. Therefore, reducing the possibility of recombination by using a precursor gas that has a lower recombination rate with atomic fluorine species is one approach to reducing this loss. The other

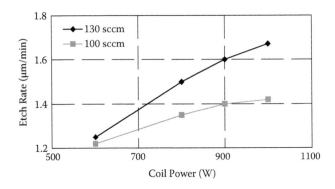

FIGURE 9.16
Variation in silicon etch rate as a function of coil power and SF_6 flow rate.

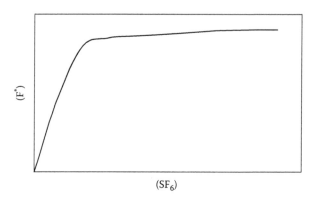

FIGURE 9.17
The equilibrium between the generation and recombination of $[F^*]$.

method to increase the etch rate is to quickly remove potential recombination particles from the process chamber.

The passivation cycle is another key step in realizing a high-aspect-ratio microstructure. This cycle influences the anisotropy, aspect ratio, undercut, and overall etching rate. Therefore, achieving a fast polymeric film deposition rate is essential in order to improve the profile control. This speed relies on the passivation gas flow rate and source power. Extra energy is necessary to dissociate the passivation precursor gas C_4F_8 to accomplish the deposition, as expressed by Equation 9.7a; for this reason, the source power is critical because it provides the kinetic power starting and maintaining all the chemical processes. However, increasing source power to obtain a higher deposition rate of the polymer film also increases the wafer surface temperature, which will decrease the polymer deposition rate. Therefore, better wafer cooling is necessary to optimize the overall etching rate.

In conclusion, coil RF power raises the etching rate because of more etching species. Also, the insufficient precursor gas flow rate decreases etch rate.

9.4.1.3 Effect of Duration on the Cyclic Etch/Passivation Cycles

DRIE is a cyclic process of etch and passivation. The various combinations of etch and passivation cycles affect the overall etch rate, trench profile, and etched surface roughness. A longer-duration etch cycle and a shorter passivation cycle definitely increase the etch rate as expected, while the anisotropy is sacrificed, resulting in bowing sidewalls, which is not desirable. Another more serious problem is the sidewall scallops. Because lateral etch is restricted by the polymeric film deposited during the separate passivation cycle, scallops invariably arise, as schematically shown in Figure 9.18a and in the SEM micrograph picture of Figure 9.18b. The etched beam sidewall surface is roughened by this effect.

Therefore, the durations of etch and passivation cycles influence the quality of the sidewall. A trade-off between the etch rate, beam profile, and sidewall quality must be carefully selected.

9.4.1.4 Effect of Process Pressure

The process pressure affects the overall etch results, including etch rate, etch selectivity, and etch profile. Generally, the number of fluorine radicals increases initially when the pressure is higher because more reaction gas leads to higher etch rate. However, when

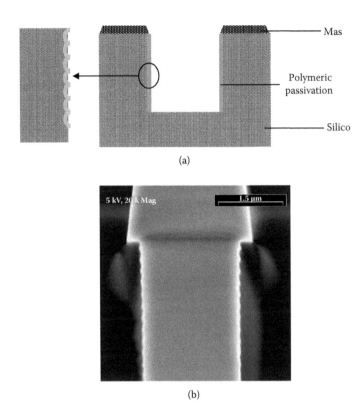

(a)

(b)

FIGURE 9.18
Scallops on the trench sidewall: (a) schematic representation and (b) SEM micrograph of an actual structure.

the pressure increases further, the DRIE etch rate does not improve anymore because the dissociation efficiency for the plasma is lower and the number of reactive species reduces. Eventually, when the pressure continues to increase, the reactive ions suffer more collisions because their mean free distance is drastically reduced, leading to a higher recombination in the plasma, which results in a lower etch rate.

The process pressure also affects the profile of the etched feature. Bowing and closing up of the features toward the bottom in high-aspect-ratio trenches are observed as a result of high pressure. This is because more collisions between the reactive species at higher pressure induce more significant isotropy. It can also be explained by the decrease in ion bombardment directionality due to higher pressure, resulting in lower-energy ions arriving at the trench bottom normally because of increased scattering. Therefore, reducing the chamber pressure can provide a more vertical trench at the expense of a reduced etch selectivity between silicon and the mask material because the mask etch rate increases as a consequence of increased mean free distance at lower pressures. Higher pressure can reduce the PR etch rate, but sputtering and redeposition of the mask material promotes the formation of the so-called silicon grass, that is, microscopic needles etched in silicon owing to localized micromasking by the deposited mask material. Therefore, a compromise between the trench profile and the etch rate and selectivity to mask has to be made.

This ICP DRIE is a very complicated process, in which etching of silicon is the result of chemical and physical reactions, and the anisotropy is ensured by the separated passivation deposition on the sidewalls combined with lateral polymer removal. The process conditions, including RF coil power, RF platen power, gas flow rate, process pressure, durations of the etching and passivation, and substrate temperature, affect the macroscopic etch characteristics such as etch rate, etch profile, etch selectivity, and etch surface. The dependencies of the characteristics on the process conditions are not independent and some may even conflict with one another; therefore, compromises have to be found to optimize the specific process.

9.5 Loading Effect

Apart from process conditions that affect deep etching results, the microstructure design and the distributions of overall structures may lead to different results under the same etching conditions. In this section, the etching issues related to feature design are studied.

The loading effect is a well-known phenomenon originating from the nonuniform plasma distribution, nonvertical pattern profile of the soft and hard masks, and various pattern densities. It can be categorized into microloading and macroloading effect. Microloading is the result of local variations in the exposed area on the wafer, which causes faster etching at the larger exposed features. This phenomenon is widely known as the RIE lagging effect. It is generally caused by the local depletion of etchants at high aspect ratio etch features on the wafer [6]. The etching is also a function of global variation in the exposed surface area of the wafer. This means that the etch rate is dependent on the overall exposed area and pattern distribution on the wafer. This is known as the macroloading effect when it is caused by the global etching difference.

9.5.1 Loading Effect Dependence on Design of Various Features

To illustrate the loading effects, a specific design was employed with 24 trenches of different widths ranging from 0.4 to 200 μm, and the process conditions were fixed as in Table 9.4.

In this section, the hard mask is a 2.0-μm-thick plasma-enhanced chemical vapor deposition (PECVD) oxide with precursor of tetraethyl silicate (TEOS). After patterning, the designed trenches with different mappings are etched in the 6 in. silicon wafers. The results are obtained by taking SEM pictures in several positions on the wafer, which are labeled as in Figure 9.19.

9.5.2 Etch Rate versus Trench Width

The deep trenches obtained after 30 min DRIE under the process conditions listed in Table 9.4 are measured at the center of the wafer. Very significant microloading effect was observed among the narrow trenches, as can be seen in SEM micrographs shown in Figure 9.20a. The etch depth varies dramatically among the trenches with widths from 0.4 to 20.0 μm, whereas the variation is negligible for wider trenches, as plotted in Figure 9.21.

TABLE 9.4

Process Conditions Used to Illustrate the Loading Effect

Gas Flow (sccm)			Power (W)		
C_4F_8	SF_6	O_2	Coil	Platen	Process Time (s)
30	100	10	600	14	8
160	0	0	600	0	5

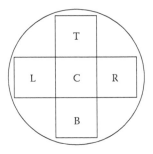

FIGURE 9.19
Position labels on the 6 in. wafer for the loading effect experiments.

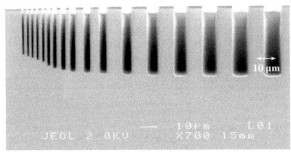

(a)

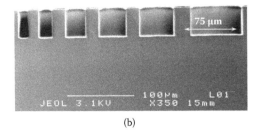

(b)

FIGURE 9.20
SEM micrographs of various trenches at the wafer center: (a) trenches with widths from 0.4 to 10.0 μm and (b) trenches with widths from 10.0 to 75.0 μm.

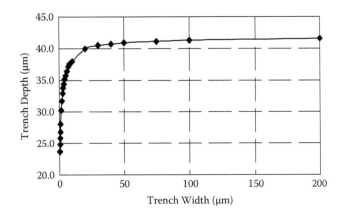

FIGURE 9.21
Etch depth as a function of trench width.

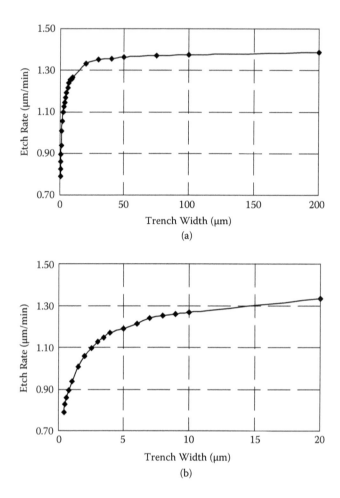

FIGURE 9.22
Etch rate as a function of trench width: (a) trench widths between 0.4 to 200.0 μm and (b) trench widths from 0.4 to 20.0 μm.

The etch depth increases from 23.70 μm for the 0.4 μm wide trench to 31.7 μm for the 2.0 μm wide trench, and to 40.0 μm for 20.0 μm wide trench. Thereafter, the depth increases by only 1.6 μm for larger widths, up to 200.0 μm.

The overall average etch rate in this process is plotted in Figure 9.22a. The rate increases from 0.80 μm/min for the 0.4 μm wide trench to 1.39 μm/min for the 200.0 μm wide trench. However, the acceleration of etch rate is not constant or linearly proportional to the trench width. The etch rate increases dramatically for the narrow trenches, that is, when the widths change from 0.4 to 20.0 μm (Figure 9.22b), whereas the difference can be ignored as the trenches become wider. This lagging effect and the tampered bottom of narrower trenches are due to the narrow openings that hinder the arrivals of reactive ions and diffusion of reaction by-products. It is also related to higher aspect ratios because it is more difficult for the plasma etch species to physically reach the bottom of the trench.

This microloading effect makes MEMS fabrication a tricky task, as it produces different depths for different features. Additionally, the microloading effect changes for different etch conditions, which makes it harder to control the aspect ratio needed in MEMS. Changing the process conditions, for example, increasing gas flow rate, increasing chamber pressure to optimum, or decreasing electrode power during the etch cycle, may be helpful in reducing the microloading effect. It would be a breakthrough if the same depth could be achieved for features of different size, especially those in the range 0.4–10 μm, which are widely employed in MEMS or nanoelectromechanical system (NEMS) devices and for which the loading effect is the most significant.

9.5.3 Etch Rate versus Trench Position on the Wafer

Etch rate is not only a function of feature design, but also of pattern distribution on the wafer that result from nonuniform plasma distribution in the process chamber. The etching results of the trenches at five different positions, labeled C (center), R (right), T (top), L (left), and B (bottom) in Figure 9.19, are compared in Figure 9.23. It can be seen that the lagging effect exists universally across the wafer, implying that it is the result of localized distribution of the patterns. The narrower the trench to be etched, the slower the etch rate, because of the insufficient reactive species and difficulty for the by-product to escape. However, the lagging effect varies from one position to another. The sequence of loading effect (difference between maximum and minimum etch depth) in ascending order of magnitude is 14.40 μm (T), 16.00 μm (L), 16.30 μm (R), 16.60 μm (B), and 17.90 μm (C). The trench at the wafer center lags the most because it is the most difficult area for the by-products inside the narrow trenches to diffuse outward through and be evacuated even if the reaction species are uniformly distributed.

For trenches with the same width, etch rate changes according to trench position. For example, the etching rate of a 2.5 μm wide trench at the five different positions is 1.10 μm/min (C), 1.05 μm/min (R), 1.12 μm/min (T), 1.13 μm/min (L), and 1.12 μm/min (B).

Therefore, overall microstructures on the wafer and localized patterns with different widths have various etch rates under the same process conditions. These effects lead to more difficulties for MEMS devices with varying trench widths, especially for high-aspect-ratio structures.

9.5.4 Etch Rate versus Exposed Area of the Wafer

Etch rate is affected by local variations in the exposed surface area on the wafer, which are due to trenches with different widths. The pattern at a specific position has its own etching

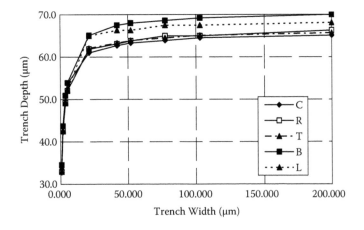

FIGURE 9.23
Etch depth of trenches located at different positions.

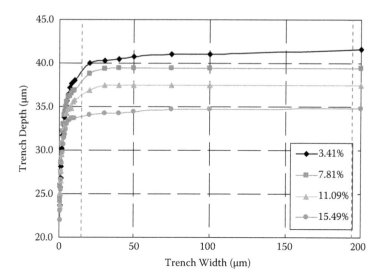

FIGURE 9.24
Etch depth in relation to the exposed area (%) for various trenches.

characteristics, as discussed in the previous section. The overall exposed area is another important factor that influences the etch profile, etch depth, and etch rate because the consumption of reactive species differs according to the entire exposed area. To show this effect, the same mask is mapped differently onto the 6 in. wafer to obtain various exposed areas. For instance, the exposed areas are 3.41, 7.81, 11.09, and 15.49%. They are etched under the same conditions and the center sample is taken as the reference gauge. The etch depths of various trenches with different exposed areas are shown in Figure 9.24. The figure illustrates that smaller the overall exposed area, faster the etch rate and, therefore deeper the trench. In general, the depth of trenches decreases when the percentage of exposed area on the wafer increases. This phenomenon is very significant for trenches with widths greater than 20.0 μm. For example, the depth differences between 3.41 and 15.49% exposed areas for trenches 20.0 and 200.0 μm wide are 5.0 and 7.8 μm, respectively, after 30 min. This is due to the depletion of reactive etching ions, because the larger exposed area uses up more etchant ions and, thus, etch depth reduces and etch rate decreases.

The deviation of narrower trenches is not as significant as wide trenches. Figure 9.25 compares the etch depth of trenches narrower than 20.0 μm under the same conditions. The difference increases as trench width increases, because wider trenches deplete more reactive species than narrower ones.

The microloading effect is affected by the overall etching areas as well. Loading decreases as the percentage of the exposed area increases. The difference of etch depths between the narrowest trench (0.4 μm wide) and the widest one (200 μm wide) is listed in Table 9.5. It can be seen that the difference reduces from 17.90 μm for the smallest exposed area to 11.83 μm for the largest exposed area. This is expected as the widest trenches depend most on the percentage of exposed area. When the overall etch area increases, the etch rate of wider trenches decreases more than that of narrower trenches, thereby decreasing the microloading effect. This effect is more significant when deeper trenches are required, owing to the etch rate decrease. The average etch rate of the narrow trenches (2.5 μm wide) decreases slowly as the exposed area increases, from 1.10 μm/min for a 3.41% exposed area to 1.02 μm/min for a 15.49% exposed area. However, for the 200 μm wide trench, the

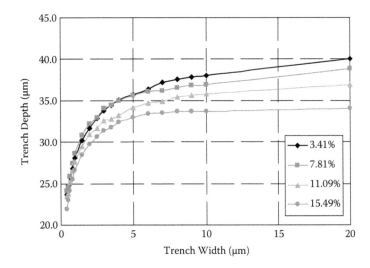

FIGURE 9.25
Etch depth versus exposed area for trenches with widths less than 20 μm.

difference is considerable, from 1.39 μm/min for 3.41% to 1.13 μm/min for 15.49%. Less than 30% exposed area is recommended for the 6 in. wafer process.

9.5.5 Etch Rate Along the Process

In the previous sections, the stated etch rate value is considered to be the average during the entire deep etching process. However, in practice, the rate decreases during the course of the process. The deep etch process is divided into several steps and the average rate for each step is calculated. Thus, the variation in time of the etch rate during the entire process is obtained. The targeted final etch depth is 75 μm and the testing process of deep etching is divided into three steps of 40, 20, and 40 min, respectively. The etching depths and individual average etch rates for different trench widths are shown in Table 9.6 and plotted in Figure 9.26, which verifies that the etch rate for a specific trench decreases with trench depth. At the first step (40 min), the average etch rate for a 2.5 μm wide trench is 1.00 μm/min and drops to only 0.67 μm/min in the third step (also 40 min). This deceleration is more significant for narrower trenches. The microloading effect also varies with the process. The deeper the trench, the greater the difference in etch rate. This is because etching depth increase is accompanied by change in the rate of successive etch and passivation cycles; due to variations in the passivation step coverage and transport of reactive and product species, into and out of the trench. There is relatively less change

TABLE 9.5

Etch Results for the Different Exposed Areas Under the Same Conditions

Exposed area (%) specifications	3.41	7.81	11.09	15.49
Depth of 0.4 μm trench (μm)	23.70	24.12	23.48	21.97
Depth of 200 μm trench (μm)	41.60	38.70	36.90	33.80
Maximum depth difference (μm)	17.9	14.58	13.42	11.83
Average etch rate of 2.5 μm trench (μm/min)	1.10	1.10	1.06	1.02
Average etch rate of 200 μm trench (μm/min)	1.39	1.29	1.23	1.13

TABLE 9.6

Etch Rate During the Etching Process of Trenches with Different Widths

Trench Width Time	2.5 μm Trench		7.0 μm Trench		30.0 μm Trench		120.0 μm Trench	
	Depth (μm)	Rate (μm/min)	Depth (μm)	Rate (μm/min)	Depth (μm)	Rate (μm/min)	Depth (μm)	Rate (μm/min)
40 min (first step)	40	1.00	43.2	1.08	45.2	1.13	50.2	1.26
Estimate after 20 min	20	—	21.6	—	22.6	—	25.1	—
20 min (second step)	16.1	0.81	17.7	0.89	21.5	1.08	24.0	1.20
First + second step	56.1	0.94	60.9	1.015	66.7	1.11	74.2	1.24
40 min (third step)	26.8	0.67	33.5	0.84	44.6	1.12	47.7	1.19
100 min (overall)	82.9	0.83	94.4	0.94	111.3	1.11	121.9	1.22

in the reactive species motion in and out of the wide trenches. Therefore, the fabricated pattern must deviate from the designed trench depth under various etch conditions and duration, especially for narrow trenches. Because etch rate decreases with the process, the estimated depth based on a previous step is deeper compared to the fabricated result and vice versa.

9.5.6 Beam Verticality

The etching profile is highly dependent on process conditions and pattern characteristics. The influence of trench width on the sidewall profile was investigated experimentally. The SEM micrographs of trenches with widths between 0.4 and 200.0 μm obtained after etching continuously for 30 min are shown in Figure 9.20. The verticality of trenches with different widths was analyzed and the results are shown in Figure 9.27. The figure indicates that the beam sides have a positive angle when the trench widths are less than 10.0 μm and have a negative angle when the trenches are more than 10.0 μm wide, under the same etching conditions. For trenches with a width of 10.0 μm, the deviation from vertical

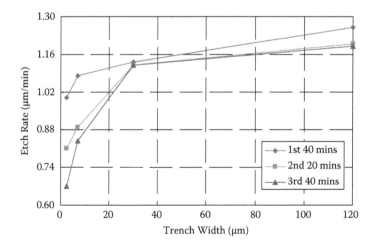

FIGURE 9.26
Etch rate versus trench width for the three steps during the etch process.

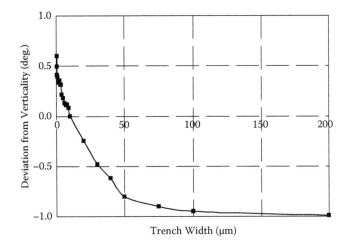

FIGURE 9.27
The verticality of the beam versus the trench width.

is the least. Therefore, some dummy microstructures are employed to fix the trench width at this particular value to guarantee the verticality of the microstructure.

9.6 Notching Effect

SOI substrate is becoming more and more commonly adopted in MEMS fabrication because it simplifies the process and maintains precise control on the depth of the device as a result of the device silicon–buried oxide–handle silicon structure of the substrate. However, the buried oxide layer, which is a dielectric material, introduces new problems apart from those in the DRIE etching of monolithic silicon, as discussed earlier. A severe undercutting (notching/footing effect) of the silicon is unavoidable at the interface of the device silicon and buried oxide. Another common problem for SOI microstructure fabrication is the stiction caused by the wet release process. In this section, the mechanisms of the notching effect are discussed, a new process (multiple-step deep etching associated with thin oxide-spacer layer deposition and anisotropic etching) is introduced in order to eliminate this problem, and experimental results are shown to verify this process. Additionally, this notching effect is utilized to dry-release the desired movable structures, thus eliminating stiction.

9.6.1 Mechanisms of the Notching Effect

Performing DRIE etch through the device silicon layer is an essential step in the fabrication of microstructures on an SOI substrate. However, plasma etching of the silicon over the insulator layer has long been recognized as the cause of the notching problem at the silicon–insulator interface [7–9], as schematically shown in Figure 9.28a,b is the SEM micrograph of the notching phenomenon on an SOI wafer. The poor profile caused by the notching may result in resonant frequency variations in the microstructure, leading to degraded performance. The notching effect is dependent on the aspect ratio, so the profiles

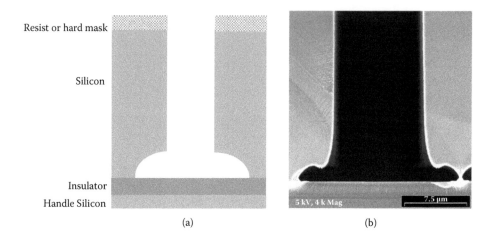

FIGURE 9.28
Notching effect: (a) schematic representation and (b) SEM micrograph.

and characteristics of the final devices may further vary across the wafer, affecting the repeatability and reliability, especially for the thick device wafer.

Charge accumulation is the main reason for the notching. Its mechanism can be explained by three phenomena (see Figure 9.29):

1. An electric field effect due to the charging transient, when the impinging ions change direction
2. The etching reactions of energetic ions impinging on the bottom surface of the exposed silicon
3. The forward-scattering effects

At the onset of lateral etching, more and more insulator layer is exposed and the notching deepens as a result of the forward scattering of ions. Moreover, the oxide surface charges up, leading to forward deflection of ions because more oxide is exposed as etching proceeds [10]. Though the trench sidewall is covered by the polymeric thin film, these deflected ions quickly sputter away the thin passivation layer, exposing the silicon to be etched at the device silicon–buried oxide interface. Additionally, the significant issue that

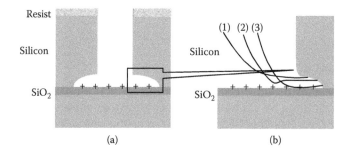

FIGURE 9.29
Schematic of the notching effect: (a) charge accumulation and (b) mechanisms of the notch formation.

is unavoidable in a DRIE process is the microloading effect of aspect-ratio-dependent etching (ARDE), which worsens the microstructures because different wide trenches have different etch rates under the same conditions. For SOI-substrate-based fabrication, the designed features may be widened or even destroyed because of the insulator layer under the device silicon. This phenomenon can be explained by the deflection, capturing, and subsequent depletion of ions.

The notching effect depends on the aspect ratio, layout design, whether the silicon lines are electrically connected [11] or not, thickness of the mask layer [12], and the electron temperature [8]. The aspect-ratio-dependent charging is due to the directionality difference between ions and electrons when approaching the wafer surface. The positively charged ions arrive nearly normal to the wafer plane because of the effect of the sheath field, whereas the electrons face a decelerating field. The higher-energy electrons tend to be deflected and the lower-energy electrons are turned back, though both are initially uniformly distributed in all directions at the sheath edge. Monte Carlo simulations verified that electrons arrive at an uncharged wafer surface with a near-isotropic distribution, but in a high-aspect-ratio structure, most of the directional ions could reach the bottom of the trenches, whereas the electrons mostly end up near the top of the device layer [11]. This mechanism was verified by sputtering a metal layer on the dielectric glass wafer, which was electrically connected to the silicon substrate [7]. The notching effect was eliminated, though the incident positive ions still bombarded the surface. These charges could not be accumulated on the glass substrate because they were evacuated from this metal-covered bottom to the silicon substrate because of the electric connection. Hence, no ion trajectory bending occurred in the direction of the silicon surface. However, it is impossible to continue this premetal covering for the SOI process as the buried oxide layer is sandwiched by two silicon layers. Figures 9.30 and 9.31 show the cross-sectional views of trenches with widths ranging from 0.4 to 200 μm for the DRIE results of a 35 μm thick SOI wafer etched continuously for 30.3 min under the process conditions listed in Table 9.4. It is clear that the 4.0 μm trench is just etched down to the buried oxide layer because the time is set by the average etching rate on this specific trench. However, trenches narrower than 4.0 μm are not etched through to the buried oxide layer, whereas the trenches with widths ranging from 5.0 to 10.0 μm encountered serious notching at the interface between device silicon and insulator layer. For trenches with widths larger than 20.0 μm (see Figure 9.31), no prominent notching or other side effects are noted. This is attributed to the dependence of the notching on aspect ratio.

This lateral attack etch is a serious problem for SOI-based MEMS fabrication because it makes it hard, if not impossible, to have only one trench width in the device- or system-level design. Even if this criterion is met, the uniformity around the entire wafer may

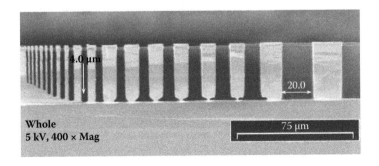

FIGURE 9.30
Trenches with widths from 0.4 to 20.0 μm etched in an SOI wafer.

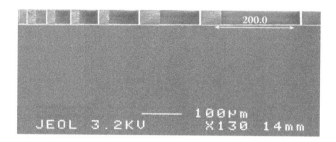

FIGURE 9.31
Trenches with widths from 20 to 200 μm etched in an SOI wafer.

not be satisfactory. As a result, the notching problem cannot be addressed if continuous DRIE etching is employed. This problem is worsened if the thickness of the device silicon increases, as there will be an even more serious loading effect due to accumulation of more positive ions.

Figure 9.32 is an example of the severe notching effect when trenches are etched on an SOI wafer with device silicon 75 μm thick, in which the widths of the trenches and silicon beams are 7.0 and 2.0 μm, respectively. The actual final height of the silicon beam is only 46.5 μm, including the seriously damaged bottom. The actual intact beam height is only 33.2 μm, which is less than half of the corresponding designed value because of the serious notching effect.

9.6.2 Elimination of Notching Effect

There are several ways to address the notching effect [8–12], but each method has its own limitations: it could be time consuming or difficult to control [11], it could be constrained by hardware or have difficulty in detecting the end point, it could be suitable only for bonded wafers [10], and it could exhibit degraded profiles [12]. A new spacer-oxide thin-film technique has been developed in this study, and the process flow is outlined in Figure 9.33 for trenches of three different widths.

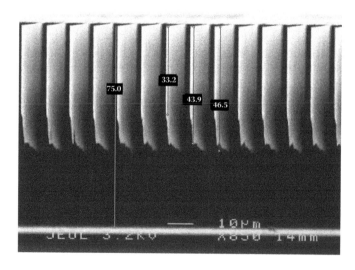

FIGURE 9.32
Notching of trenches etched in an SOI wafer with 75 μm device silicon layer.

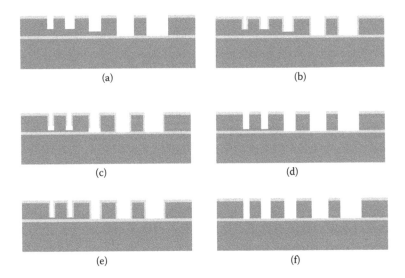

FIGURE 9.33
DRIE process flow that eliminates the notching effect: (a) first DRIE etch of the widest trenches till the buried oxide layer, (b) 5000 Å PECVD oxide deposition and anisotropic etch to remove the lateral deposited oxide, (c) DRIE etch of the second-widest trenches up to buried oxide layer, (d) etch back the deposited oxide on the sidewalls, (e) redeposit and lateral-etch another 5000 Å PECVD oxide layer, and (f) repeat the deep etching and remove the deposited oxide after process completion.

In this technique, a PECVD oxide layer of 3.5 μm thickness is first deposited as the hard mask. In general, its thickness has to be sufficient to protect the device silicon from being etched. After patterning the trenches, the major etch step of the device layer is divided into several subetch steps according to the etch rate of trenches of various aspect ratios. The first subetch step focuses on the fastest-etched trenches. As illustrated in Figure 9.33a, the widest trenches are etched through to the buried oxide layer, whereas the narrow trenches are still not completely etched. If this etching regime were to continue, notching would occur at the bottom of the wide trenches owing to the dielectric buried oxide layer. To prevent this from happening, PECVD oxide is then employed to cover the surface of the trenches, including their sidewalls, top surfaces, and trench bottoms. To continue with the etching for the other trenches, an anisotropic oxide etching step has to be employed (e.g., using an Applied Materials P5000 Mark II), which removes the lateral deposited oxide, in a CHF_3-based etchant. The result is shown in Figure 9.33b.

After this, the STS-ICP DRIE is used again until the device layer is etched through for the second-widest trenches (see Figure 9.33c). For the narrowest trenches with the slowest etch rate, the oxide conformal coverage, lateral SiO_2 removal, and DRIE etch-through steps are all repeated again. However, to ensure vertical profiles and smooth sidewalls for the beams, the previous spacer oxide is removed before the second deposition, as illustrated by Figure 9.33d. In this way, the deposited oxide will not accumulate to a thick layer, which may induce a step at the interface between the various etchings. Then, a new oxide layer is deposited and laterally etched to protect the sidewall, as shown in Figure 9.33e. Finally, the last etching is carried out followed by spacer-oxide removal using an isotropic oxide etch (refer to Figure 9.33f).

Using this technique, trenches with different widths can be obtained on an SOI wafer without the notching problem. This is accomplished by dividing the deep etching into more steps and associating it with auxiliary oxide deposition and lateral oxide removal steps, which are intercalated in between the DRIE steps in order to protect the sidewalls. This process was experimentally verified on an SOI wafer with a device layer 35 μm thick, using a predesigned mask with trench widths varying from 0.4 to 200.0 μm. The focus is on the trenches with widths from 2.0 to 10.0 μm, for which notching is the most serious problem, but which are typically most widely used in MEMS devices.

The relationship between etch rate and trench width is used as the basic data in deciding the proper time distribution of multiple-step etching on the SOI wafer. Based on the etch rate obtained on etching the monolithic silicon wafer, the multiple-step etching process was carried out to etch trenches with different widths in order to fully eliminate the notching effect. The first etching targets the 10.0 μm wide trench, whose etch time is expected to be 27 min and 30 s according to the rate obtained from the initial silicon wafer etching. After this step, all of the trenches are etched to different depths, whereas only the 10.0-μm-wide trench is etched through. Then, a PECVD TEOS oxide deposition is employed to cover all the exposed surfaces, followed by the anisotropic etch of silicon oxide on the lateral surfaces, leaving only the protection oxide layer on the sidewalls. In this way, the sidewalls of the 10.0 μm trench are covered by oxide and the trench bottom is the buried thermal oxide. For the narrower trenches, the sidewalls are also covered by oxide, whereas the bottom bare silicon is exposed.

The second DRIE step is carried out to etch through the trenches with widths ranging from 5.0 to 9.0 μm. Oxide deposition and anisotropic oxide etching expose the bottom silicon, and the third deep etching step is repeated, this time targeting trenches with widths of 2.0, 2.5, 3.0, 3.5, and 4.0 μm, before which the previous sidewall oxide layer is removed by isotropic oxide etching. The final cross-sectional views of the specially designed trenches are presented in Figure 9.34a. It clearly shows that trenches narrower than 2.0 μm are not etched through, whereas trenches with widths of 2.0, 2.5, and 5.0–10.0 μm are etched through without the notching problem. However, undercutting still occurs at the foot of trenches with widths of 3.0, 3.5, and 4.0 μm, respectively. This is the result of pronounced etch rate variations among patterns with critical dimensions of 2.0, 2.5, 3.0, and 3.5 μm. However, this phenomenon can be eliminated if more etching steps are employed.

Another observed phenomenon is the rounded shape of the trench bottom (see Figure 9.34b), for which there could be two causes. The first one is the overetch that occurs during the anisotropic etching of deposited oxide (for securing a very well-exposed single-crystal silicon (SCS) surface at the bottom to subsequently DRIE etch), which results in etching some of the buried oxide. The second cause could be the multiple-step

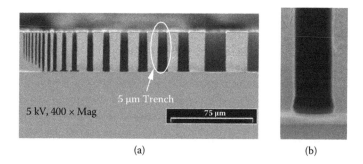

(a)   (b)

FIGURE 9.34
Improved etching results on a 35 μm SOI wafer: (a) all trenches (from 0.4 to 20 μm wide) and (b) bottom of the 5.0 μm wide trench.

silicon etching. Although the etching selectivity between silicon and oxide is larger than 65, under such conditions, the exposed buried oxide would be deeply etched because of the long exposure to the directional high-energy plasma. It is noted that the bottom of the DRIE-etched trench is cone shaped instead of being as flat as the top surface. However, the loss of a little buried oxide does not affect the process or properties of the devices, because the next process step in most cases is oxide etching to release the microstructures.

One of the most important steps in this process is the PECVD oxide coverage. This particular deposition must be conformal over all of the surfaces, especially the sidewalls and bottom of every trench. Cross-sectional SEM micrographs are used to verify that oxide depositions in this process provide a good coverage. However, the dielectric oxide film induces charging of the electrons during scanning, making the inspection of the thin oxide layer extremely difficult. This problem can be eliminated in the following manner. First, an additional polysilicon layer is deposited on top of the oxide layer. The sample is then dipped in buffered oxide etchant (BOE) to etch the PECVD SiO_2 thin film after cutting the cross-sectional sample in order to obtain a hollow layer between the SCS and deposited polycrystalline silicon. Figure 9.35a,b shows the hollowed oxide layers at the bottom and sidewalls that were previously covered by the polysilicon, verifying that the bottom and sidewalls of all the trenches are well covered by the PECVD oxide. This confirms that the process is reliable.

Another concern in this process is the thickness of the hard-mask oxide layer. According to the etch rate selectivity between silicon and oxide under the conditions listed in Table 9.4, the 2.0 μm oxide is thick enough to be used as a hard mask for etching through the 75.0 μm device silicon layer. Also, the hard mask must be thick enough to safely resist the multiple etching process. The lateral oxide removal (refer to Figure 9.33b) that exposes the bottom silicon requires a certain amount of overetching. Thus, the previously deposited top hard mask is invariably consumed. The variations of hard-mask thickness for different process steps measured at the large nonpatterned areas are listed in Table 9.7. It clearly shows that oxide depletion at the edge is faster than that at the center. The residual oxide thickness at the edge after triple DRIE is only 8822 Å. The measured oxide layer in this table is thicker than that in the localized patterned area because of the nonuniform distribution of the reaction gases. Therefore, a thicker hard-mask layer is required to accomplish multiple DRIE etching.

This fabrication approach serves as a baseline for MEMS device fabrication on SOI wafers as it provides more flexibility for microstructure design and realization. Several varying feature sizes of microstructures can be designed and fabricated at the same

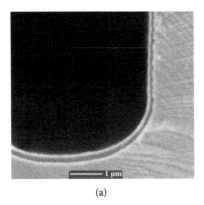
(a)

(b)

FIGURE 9.35
SEM micrographs of the deposited oxide: (a) bottom and (b) sidewall.

TABLE 9.7

Hard-Mask Oxide Thicknesses Obtained During the Multiple Etch Process

Process Step	1 Original	2 1st DRIE	3 1st oxide Deposition	4 1st Oxide Etch	5 2nd DRIE	6 2nd Oxide Deposition	7 2nd Oxide Etch	8 3rd DRIE
Center μm	3.5627	2.7511	3.3497	2.5705	2.3819	2.8051	2.1798	1.4587
Edge μm	3.4818	2.5029	3.0723	2.4430	2.0050	2.4070	1.7635	0.8822

time, eliminating the restriction of designing only a single uniform size for each process. Another benefit derived from this process is the significantly enhanced profile of the deep microstructure, which is the key requirement for MEMS optical devices.

9.6.3 Dry Release by Notching Effect

The permanent stiction due to the capillary forces that occurs during the rinsing and drying of wet release is another common problem observed in a conventional SOI process [13]. This effect has been studied theoretically and experimentally [14,15], and several modified release approaches have been reported, with their respective drawbacks. To solve these problems, special attention is paid to dry release.

Here, a more flexible fabrication method is described and demonstrated to release movable structures on an SOI wafer without stiction. A thin spacer-oxide film is utilized and the notching effect is exploited beneficially in order to provide a dry chemical release. This novel solution entirely eliminates any problems associated with wet chemical release and other reported dry gas releases. The notching effect or undercutting is employed because this specific etching takes place laterally, and hence is a very good candidate for a release process. The depth of notching depends on many factors, such as the overetch time, type of material, thickness of sidewall passivation, and size of the feature. Other factors include electron temperature, ion energy, and the ion/electron current at the surface.

DRIE makes use of separate etch and passivation process to attain the deep trenches. Though the sidewall is covered by the polymerized $(CF2)_n$, this thin polymer-like layer cannot protect the silicon trench sidewall from being etched by the high-energy ions repelled from the exposed silicon dioxide layer, resulting in the development of the notching effect. However, a thin sidewall spacer-oxide film could be used to eliminate the notching problem, as discussed previously in Section 9.6.2. More significantly, the release depth can be controlled by adjusting the depth covered by the spacer-oxide layer, even if overetching for release occurs.

9.6.3.1 Uniform Beam Release

Beams with uniform trenches can be released by employing the notching effect. The process flow is briefly outlined in Figure 9.36. After patterning the trenches on the hard-mask oxide layer on the SOI wafer, DRIE is employed to etch the device silicon layer to within 1.0–2.0 μm from the buried oxide layer, as shown in Figure 9.36a. The exposed surfaces are then covered by the PECVD oxide (see Figure 9.36b). To etch the remaining silicon and to release the beams, the lateral deposited oxide is removed by anisotropic oxide etching whereas the sidewall oxide remains intact, as in Figure 9.36c. Then, DRIE is repeated. This is a critical step for uniform dry release as it plays two important roles: first, it etches through the device layer until the buried oxide layer is reached, and second, it releases the

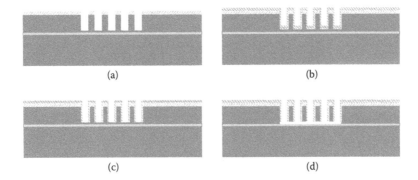

FIGURE 9.36
Process flow for dry release employing the notching effect: (a) first DRIE etch, (b) PECVD oxide deposition, (c) anisotropic etch of the lateral deposited oxide to expose the bottom silicon, and (d) second DRIE etch up to the buried oxide layer and lateral release of the beams by notching effect.

features by the notching effect (see Figure 9.36d). Compared to other dry-release methods, the time tolerance of this process is much more flexible. In conventional dry etching, overetching can result in sacrificing of beam depths. However, because of the sidewall oxide coverage, in this process the structures can endure longer overetch as the selectivity between silicon and silicon oxide is high.

This process is verified by using SOI wafers with a device silicon layer 35 μm thick [16]. After patterning the structures, DRIE is carried out for 25 min to a depth of about 33.0 μm followed by the thin PECVD oxide layer deposition. The lateral oxide is anisotropically etched using the Precision 5000 Mark II (Etch MXP®) etcher from Applied Materials. Overetching is required to ensure exposure of the bottom silicon for the subsequent vertical etching step. Finally, a DRIE etch is performed for 12 min to etch through the trenches and to release the microbeams. The released structures are shown in Figure 9.37a,b, which illustrates that the beams are released well without stiction or other side effects.

In the event of insufficient etching, undercutting occurs only at the foot of both sides of the beam and the beams cannot be totally released, as is evident in Figure 9.38a. When the process time is exact, the foot of the released beam will be shaped like an upright cone that is pure SCS without oxide coverage. As for the top part of the beam, the PECVD oxide coverage remains intact. If the DRIE continues for a longer period, the cone part will be etched off and the top part will remain to give the final released beams, as shown in Figure 9.38b.

The depth of the released microstructures can be adjusted by combining the DRIE etch, and oxide thin-film deposition, and anisotropic etching steps. By allowing sufficient etch time for the second DRIE etch, the achieved beam height is equal to the depth of the initially etched trench, whose sidewall is protected by the oxide film. Therefore, by adjusting the first etching time, suspended microstructures of various depths can be obtained.

9.6.3.2 Nonuniform Beam Release

The notching effect is strongly related to the aspect ratios of the patterns (see Figure 9.30). Under the process conditions stated in Table 9.4, the notching effect is significant for trenches with aspect ratios greater than 1.75. No apparent undercutting is noted for features with aspect ratios of about 1.5 or less, as electrons are able to bombard the lower part of the etched structures and the bottom of the surface. The charge buildup and the

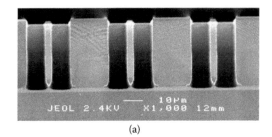

(a)

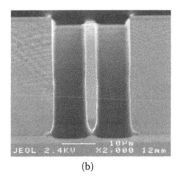

(b)

FIGURE 9.37
SEM micrographs of the dry-released beams: (a) released group of beams and (b) released single beam with an upright cone bottom.

resulting notching effect are precluded. The extreme high-aspect-ratio patterns (greater than 20) are too narrow for the etchant gas to enter deeply and for the product gas to escape. The etch rate then drops significantly and the interface between the conductive silicon and isolative thermal oxide is not exposed. The notching effect for features with aspect ratios between 1.75 and 20 (with widths ranging from 2.5 to 20 μm on an SOI wafer with a device layer 35 μm thick) strongly depends on the pattern. This can be observed from the SEM micrographs of Figure 9.39, in which various trench dimensions are illustrated. The notching is so small in Figure 9.39a that it can be neglected for the 20.0-μm-wide trench. However, the notching depth, defined by the extent of the lateral etch, increases as the aspect ratio increases. A notching depth of 4.335 μm for the trench with an aspect ratio of 3.50 can be observed in Figure 9.39a, and a depth of 4.230 μm can be observed for the

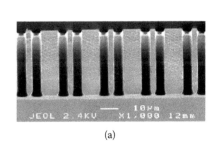

(a)

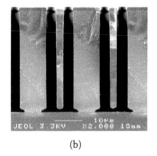

(b)

FIGURE 9.38
SEM micrographs of insufficiently released and overreleased beams: (a) insufficiently released beams and (b) overreleased beams.

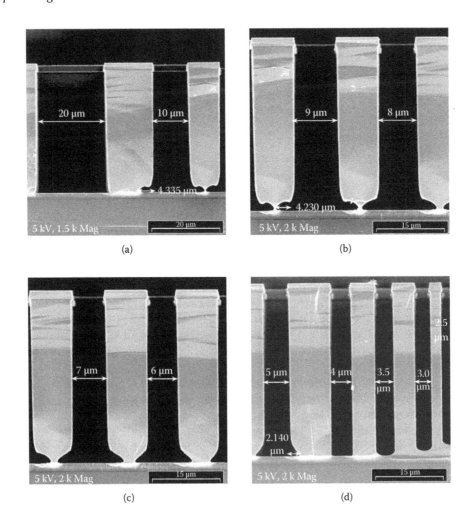

FIGURE 9.39
SEM micrographs of notching on trenches with various widths: (a) 20 and 10 μm wide trenches, (b) 9.0 and 8.0 μm wide trenches, (c) 7.0 and 6.0 μm wide trenches, and (d) 5.0, 4.0, 3.5, and 3.0 μm wide trenches.

trench with an aspect ratio of 3.89 in Figure 9.39b. The depth reduces to 2.140 μm when the aspect ratio increases to 7.00 (see Figure 9.39d).

The dependence of notching depth on the trench width is plotted in Figure 9.40. Accordingly, this result can be used to easily control the notching depth by tailoring the aspect ratios of the etched features.

Different etch durations are required for different trenches in order to etch through the 35 μm SOI device layer. Trenches with different aspect ratios under the same etch conditions present different lateral undercutting, as shown in Figure 9.30. The lateral etch rate (ratio of notching depth over notching time) for trenches with widths ranging from 4.0 to 10.0 μm are plotted in Figure 9.41, which is uniform within this range. This uniform notching rate is attributed to the consistent charge distribution in this aspect-ratio range. Table 9.8 gives the notching depth, overetching time, and notching rate for a series of trenches.

The duration of overetch time is another important contributing factor for notching depth. The etch results with longer overetch times are depicted in Figure 9.41 for a series

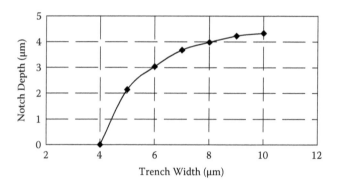

FIGURE 9.40
Notching depth versus trench width.

of trenches with widths of 2.5, 3.0, 3.5, and 4.0 μm under the same etch conditions, except for an additional 5.0 min of etching so that the obtained results can be compared with the previous ones shown in Figure 9.39. This figure shows that the 2.5 μm wide trench is etched through with the additional 5 min of DRIE etch, whereas the final etch depth is less than 35 μm (see Figure 9.39d) for the trench etched without this additional 5.0 min. However, the other three wider trenches suffer from notching during this additional 5 min etch, as shown in Figure 9.42.

Therefore, structures with different trench widths will exhibit different notching depths if a single continuous etching approach is employed. By using the proposed multistep plasma etching process, such nonuniform features can be etched and dry-released to an equal depth. Therefore, this new improved SOI wafer fabrication technique allows a large flexibility in the design of MEMS devices with possible variation in microstructure dimensions while simultaneously presenting the geometrical and operational properties of the devices.

For a nonuniform design that includes both wide and narrow trenches, an oxide-spacer coverage is employed to avoid excessive notching. The process can be described as follows. First, all trenches are patterned in the top mask layer, as shown in Figure 9.43a. Then, the trenches are etched, but not entirely down to the silicon and oxide interface for even the widest trenches (see Figure 9.43b). All the patterns are then conformally covered by PECVD TEOS oxide, followed by plasma etching of the oxide deposited on the lateral bottom in order to continue vertical etching and to realize the lateral release, as illustrated in Figure 9.43c. The last step is crucial as it plays several key roles. It etches through the widest trenches and deepens the narrow trenches. As etching progresses, release of the

TABLE 9.8

Notching Depths and Rates for Trenches of Various Widths

Trench Width (μm)	Etch Rate (μm/min)	Time for 35 μm (min) Layer Etch	Overetch Time (min)	Notch Depth (μm)	Notch Rate (μm/min)
4.0	1.155	30.300	0	0	—
5.0	1.222	28.651	1.466	2.140	1.460
6.0	1.242	28.178	1.939	3.035	1.565
7.0	1.256	27.856	2.261	3.675	1.625
8.0	1.277	27.408	2.709	3.980	1.469
9.0	1.283	27.277	2.840	4.230	1.489
10.0	1.296	26.996	3.121	4.335	1.389

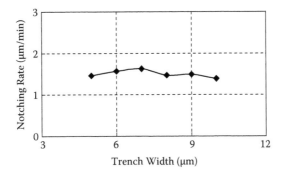

FIGURE 9.41
Notching rate versus trench width.

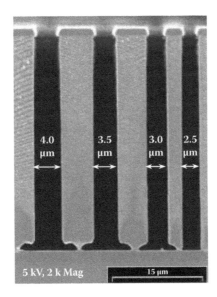

FIGURE 9.42
SEM micrograph of the notching effect on trenches with widths of 2.5, 3.0, 3.5, and 4.0 μm.

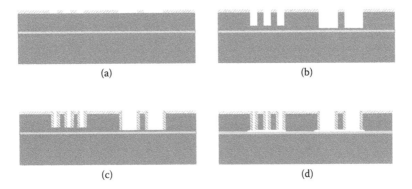

FIGURE 9.43
Process flow of dry release of nonuniform beams using notching effect: (a) trench patterning, (b) first DRIE, (c) sidewall oxide coverage and lateral bottom etch, and (d) notching release.

beams with wide trenches begins to take place because of the notching effect, while the narrow trenches become deeper. When the narrow trenches are etched through to the interface with the buried oxide, the beams with wide trenches will have suffered some degree of notching or lateral etching. Hence, beams with wide trenches are released completely while notching just begins for beams with narrow trenches. Continuing the undercutting of the narrow beams, repelled positive ions cannot react with the spacer oxide of the beams, so that the depth of the beams with wide trenches will not be etched further. DRIE etching is stopped when the narrow beams are completely released, as shown in Figure 9.43d. Thus, both DRIE anisotropic etching and dry gas release that make use of the notching effect are realized.

9.7　Process of DRIE Fabrication on SOI Substrate

The universal process flow for MEMS optical components, including microfunctional structures, electrodes for the actuation, fiber grooves for passive alignment, and dicing line for separation, is outlined in Figure 9.44.

The electrodes are patterned and sintered to obtain good ohmic contact with silicon, as shown in Figure 9.44a. Then, the hard mask (e.g., silicon dioxide) is deposited and the microstructures, including the fiber grooves, are patterned by employing lithography and a hard-mask etching process (see Figure 9.44b). Multilayer patterning is employed to pattern the dicing line onto the hard-mask layer, followed by the first DRIE etch through the device while protecting the microstructures patterned on the hard-mask oxide layer with PR (see Figure 9.44c). In this way, lithography onto deep trenches is prevented and accuracy of the functional structures is guaranteed. The dicing line is realized by a buried oxide etch followed by a second DRIE etch of the handle silicon, as shown in Figure 9.44d.

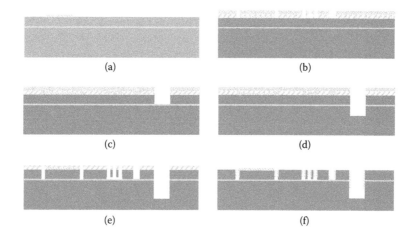

(a)　　　　　　　　　　　　　　　(b)

(c)　　　　　　　　　　　　　　　(d)

(e)　　　　　　　　　　　　　　　(f)

FIGURE 9.44
Universal DRIE process flow for MEMS optical devices: (a) electrode patterning and sintering, (b) microstructure patterning on a hard-mask layer, (c) dicing-line patterning and first DRIE etch through the device silicon layer, (d) buried oxide etch and second DRIE etch of the handle silicon, (e) microstructure DRIE and release of the movable structures, and (f) sidewall metal coating.

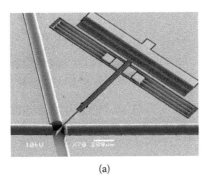
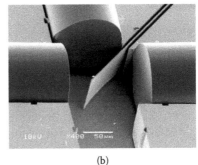

(a)   (b)

FIGURE 9.45
SEM micrograph of the optical switch fabricated using our novel SOI DRIE fabrication process: (a) overview and (b) close-up view of the mirror with optical fibers.

After this the PR is stripped, exposing the patterned hard mask to proceed with microstructure fabrication, including another deep etch to the expected depth and release of the movable microstructures, as shown in Figure 9.44e. It should be noted that the release is achieved by using a gaseous dry approach as explained in Section 9.6.3 instead of the classical wet method. Finally, the metal is deposited onto the functional optical components, such as mirror or gratings, to realize a highly reflective surface, as shown in Figure 9.44f.

By using this process, the MEMS optical switch can be easily fabricated, and the SEM micrographs of functional devices are shown in Figure 9.45. The overview of the MEMS optical switch is shown in Figure 9.45a. The trench width covers 2.5 μm of the overlapped fingers, 7.0 μm of the nonoverlapped fingers, 50.0 μm of the folded beams, and 127.0-μm fiber grooves. All of these trenches with different widths are fabricated at the same time on an SOI wafer with a device silicon layer 75 μm thick. Figure 9.45b shows the central mirror together with the assembled optical fibers.

9.8 Summary

This chapter described the crystal structure of silicon material followed by a brief discussion on wet silicon etch, focusing on the DRIE process on SOI substrates. The principle of DRIE was first described from the chemistry and physics viewpoints. The effects of process conditions, such as gas flow rate, coil power, platen power, process pressure, and duration of the separated etch and passivation cycles, on the macroscopic characterizations of etch results (etch rate, etch profile, and etch selectivity to mask layer) have been analyzed. Then, the loading effects were discussed, after which the specific problems (notching effect) occurring on SOI wafer due to the buried oxide layer were explored in detail; the solution involved thin oxide film deposition and multiple DRIE process. Further, the notching effect has been used beneficially to achieve a nondefect dry release for both uniform and nonuniform beams.

This process, based on a spacer-oxide protection, has been used for the fabrication of the 2 × 2 optical switch. From a broader perspective, the process can be used to fabricate any high-aspect-ratio suspended structures for many other MEMS device applications, such as a variable optical attenuator (VOA), a tunable laser (TS), an RF switch, and so on.

References

1. Gregory, T. and Kovacs, A., Micromachined Transducers Sourcebook, McGraw-Hill.
2. Larmer, F. and Schilp, A., Method of anisotropically etching silicon, German Patent DE4241045, 1994.
3. Gormley, C., Yallup, K., Nevin, W. A., Bhardwaj, J., Ashraf, H., Huggett, P., and Blackstone, S., State of Art Deep Silicon Anisotropic Etching on SOI Bonded Substrates For Dielectric Isolation and MEMS Applications, 5th International Symposium on Semiconductor Wafer Bonding Science, Technology and Applications, The Fall Meeting of the Electrochemical Society, Hawaii, October 17–22, 1999.
4. Chen, K. S., Zhang, A. A. X., and Spearing, S. M., Effect of process parameters on the surface morphology and mechanical performance of silicon structures after deep reactive ion etching (DRIE), *J. Microelectromech. S.*, 11(3), 264, June 2002.
5. Ashraf, H., Bhardwaj, J. K., Hall, S., Guibarra, E., Hopkins, J., Hynes, A. M., Johnston, I., Lea, L., Mccauley, S., Nicholls, G., and O'Brien, P., Defining Conditions for the Etching of Silicon in an Inductive Coupled Plasma Reactor, *Proceedings of the Materials Research Society Fall Meeting*, Boston, Massachusetts, 299, November 29–December 3, 1999.
6. Chabloz, M., Jiao, J., Yoshida, Y., Matsuura, T., and Tsutsumi, K., A method to evade microloading effect in deep reactive ion etching for anodically bonded glass silicon structures, *J. of MEMS*, 283, 2000.
7. Tsujimoto, K., Kumihashi, T., Kofuji, N., and Tachi, S., High-Gas-Flow-Rate Microwave Plasma Etching, *IEE of Jpn., Proc. 14th Dry Etch Sym.*, II-1 49, Tokyo, 1992.
8. Toshihisa, N., Takashi, K., Tetsuya, N., Akira, N., Takayoshi, I., and Akimitsu, N., The Electron Charging Effects of Plasma on Notch Profile Defects, *Jpn. J. Appl. Phys.*, 34, 2107, 1995.
9. Fujiwara, N., Maruyama, T., and Yoneda, M., Pulsed Plasma Processing for Reduction of Profile Distortion Induced by Charge Buildup in Electron Cyclotron Resonance Plasma, *Jpn. J. Appl. Phys.*, 35, 2405, 1996.
10. Sun, Y., Piyabongkarn, D., Sezen, A., Nelson, B. J., and Rajamani, R., A high-aspect-ratio two-axis electrostatic microactuator with extended travel range, *Sensor. Actuat. A-Phys.*, 102, 49, 2002.
11. McAuley, S. A., Ashraf, H., Atabo, L., Chambers, A., Hall, S., Hopkins, J., and Nicholls, G., Silicon micromachining using a high-density plasma source, *Appl. Phys.*, Vol. 34, No. 10, 2001, pp. 2769–2774.
12. Letzkus, F., Butschke, J., Hofflinger, B., Irmscher, M., Reuter, C., Springer, R., Ehrmann, A., and Mathuni, J., Dry Etch Improvement in the SOI Wafer Flow Process for IPL Stencil Mask Fabrication, *J. Microelectron. Eng.*, 53, 609, 2000.
13. Fujitsuka, N. and Sakata, J., A new processing technique to prevent stiction using silicon selective etching for SOI-MEMS, *Sensor. Actuat. A-Phys.*, 97–98, 716, 2002.
14. Legtenberg, R., Tilmans, H. A. C., Elders, J., and Elwenspoek, M., Stiction of surface micromachined structures after rinsing and drying: model and investigation of adhesion mechanisms, *Sensor. Actuat. A-Phys.*, 43, 230, 1994.
15. Buhler, J., Steiner, F.-P., and Baltes, H., Silicon dioxide sacrificial layer etching in surface micromachining, *J. Micromech. Microeng.*, 7, R1, 1997.
16. Li, J., Zhang, Q. X., Liu, A. Q., Goh, W. L., and Ahn, J., Technique for preventing stiction and notching effect on silicon-on-insulator microstructure, *Journal of Vacuum Science and Technology B*, 21(6), 2530, November, December, 2003.

10

Deep Submicron Photonic Bandgap Crystal
Fabrication Processes

Selin Hwee Gee Teo and Ai-Qun Liu

CONTENTS

10.1 Introduction 394
10.2 Design of Fabrication Process Flow 394
10.3 Subwavelength Lithography 397
10.4 Lithography Setup and Methods 400
    10.4.1 Lithographic Parameters Manipulation 401
    10.4.2 Optical Proximity Error Model 402
    10.4.3 Partial Coherence and Pattern Density Control 406
    10.4.4 Chemically Amplified Resist Deactivation 408
    10.4.5 Photoresist Trimming for Critical Dimensions Reduction 408
10.5 Post Lithographic Processings 415
    10.5.1 Unique Considerations for Photonic Crystal Rods Etch 416
    10.5.2 Development of Deep Reactive Ion Etching (DRIE) 416
    10.5.3 Dense Deep-Submicrometer Critical Dimensions (CDs) Etching 420
    10.5.4 RIE Lag and Notching at Buried Oxide Interface 422
    10.5.5 Etch Postprocessing 423
10.6 Summary 425
References 427

This chapter begins with a presentation of the process flows designed for planar optical photonic crystal (PhC) structures. Following this, the optical lithography (OL) process is introduced to initiate and explain the strategies taken to push the lithographic process beyond its conventional limits while keeping photomask fabrication cost and system simulation time cost low. With this foundation, the lithographic experiment setup is presented and discussed to enable further presentations of the details involved to achieve high-resolution PhC patterning. Here, rigorous design of experiments yielded necessary PhCs devices with precise lattice dimensions even near regions of "defect structures" designed for device operations.

Completion of the OL process manipulations then allows development of the etching process and other postprocessing modules. Especially challenging in this phase was the development of the etching module, where the stringent requirement of etch angle needed for successful realization of such superdense array of submicron-sized PhC lattice had to be satisfied before any working PhC device could be achieved. Deep reactive ion etching (DRIE) was the method of choice developed. Based on the developed methodology, etch sidewalls of high verticality and record-high aspect ratios (HARs) with ultralow scallop-depths were obtained (i.e., minimum etch sidewalls undulations).

10.1 Introduction

The objective of this chapter is to develop the fabrication technologies required to realize deep submicrometer-sized optical PhC devices, which are sensitive to the canonical communication wavelengths centered at 1550 nm. Especially because the performances of PhC devices are heavily process dependent—and PhC fabrication technologies are not mature—it is therefore crucial to develop a flexible fabrication template suitable for realization of different PhC types (such as the two-dimensional (2D) hole-/rod-type PhC). Among the many processing steps required to fabricate working PhC devices, patterning and pattern transfer into device material are especially critical and challenging because PhC devices are typically submicrometer in size and arranged in contrasting superdense and nearly isolated densities of bulk and defect regions, respectively. To successfully demonstrate working PhC devices, the difficulties attributed by these unique characteristics of PhC must be overcome. To do so, the issue of drastically reduced process window of deep ultraviolet (DUV) lithography, DRIE, and post-processing modules for a superdense array of submicron-sized rods or holes features with typical diameters of 90–230 nm and spacing of 340 nm apart is specifically treated in this chapter.

10.2 Design of Fabrication Process Flow

In the experiments to achieve sublithographic wavelength OL, only conventional binary chrome-on-glass photomasks without phase shift features were used. Positive DUV chemically amplified resist (CAR) was coated atop a bottom antireflection coating (BARC) using a Tokyo Electron ACT8 wafer-track system. A DUV Nikon S203B laser step-and-scan system was used for controlled 248-nm-wavelength exposure [1]. The silicon-on-insulator (SOI) wafers were first prepared using standard steps of cleaning, particle counting, and surface profiling before a blanket deposition of 5000 Å undoped silicate glass (USG) using a Novellus Concept Two Sequel plasma enhanced chemical vapor deposition (PECVD) system. Uniformity of hard mask and other characterizations were then checked for before commencement of the lithography process.

To allow independent manipulation of the lithographic and time-multiplex etch processes, different requirements for masking, together with corresponding recipes, are needed. These are essential to realize drastically different mask openings and etch depths for the PhC array of rods (~ 13-μm deep with 340-nm-wide openings) and for deep trenches with more than 50 μm depth and 125 μm wide openings required to couple standard single-mode-fibers (SMFs).

A rigorous design of the fabrication process for single-crystal silicon wafers without any buried oxide layers is as follows. An initial mask for the deep-submicron-sized PhC structures were first defined through a thin layer of hardmask, using a similarly thin layer of photoresist spun on top of an antireflection coating. As the pattern was transferred onto the predeposited USG hardmask using magnetic resonance reactive ion etching (MERIE), a photoresist strip was carried out for a refill of the even thicker resist needed for the longer-duration deep-fiber grooves etching, lithographically exposed and aligned to the first layer of the PhC lattice already defined on the wafer. When the required depth of the fiber grooves was achieved, any remaining photoresist was then stripped to reveal the PhC structures in the hardmask for the final device etch. Upon achieving sufficient HAR, the PhC rod-type 2-D device was formed. In another process, a thinner 2-D PhC device comprising an SOI wafer perforated by a slab of holes may be obtained using the process flow shown in Figure 10.1.

Figure 10.1 also demonstrates that the design of process flow is a critical factor in the fabrication of PhC with other structures. This is not only for the effective development of each process module such that different requirements with corresponding recipes may be

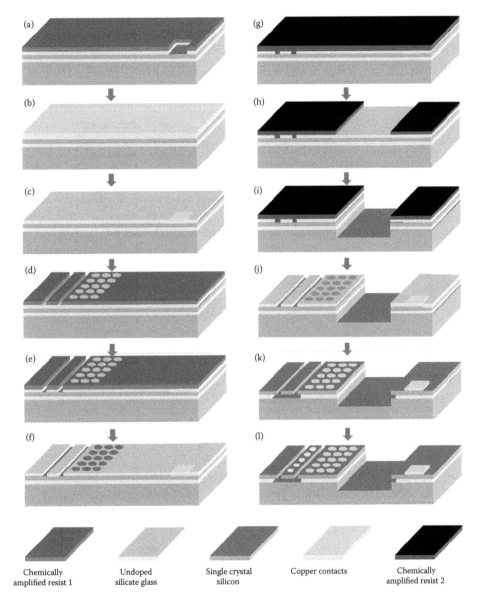

Chemically amplified resist 1   Undoped silicate glass   Single crystal silicon   Copper contacts   Chemically amplified resist 2

FIGURE 10.1
Fabrication process flow for two-dimensional hole-type photonic crystal in silicon-on-insulator wafers.

discovered through a rigorous design of experiments, but the proper sequence of processing steps also prevents elemental contamination of reaction chambers, material degradations, and process overruns. Here, the fabrication process flow for control of the lithographic, deep-etching processes with disparate requirements are shown as a schematic of the integrated fabrication process flow for PhC with optical testing structures (OTS), which may be described in four main parts: Steps (a)–(c) describe the deposition of the metal electrodes and oxide hard mask layers; steps (d)–(f) describe the PhC and comb-drive CD pattern definition and transfer process; steps (g)–(i) describe the opening of large alignment structures; steps (j)–(l) depict the oxide release and also postprocessing modules.

The film material deposition processes was initiated with USG on cleaned SOI wafers. Following electrode pattern opening in hard mask (Figure 10.1a), copper seed deposition and metallization plating (Figure 10.1b) were done. Planarization was then carried out using chemical mechanical polishing (CMP). To avoid contamination in subsequent steps, another layer of USG film was deposited to cover the defined electrode (Figure 10.1c). The deposition of this protective USG layer therefore completes the first part of the thin-film

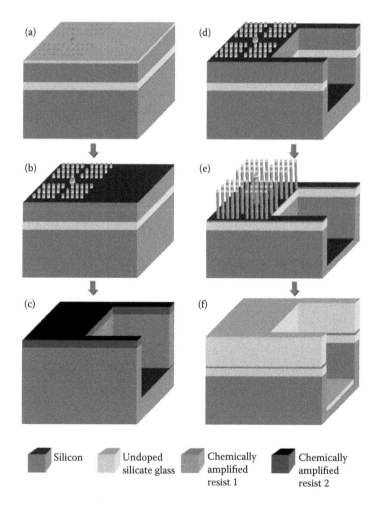

FIGURE 10.2
Fabrication process flow for rod-type photonic crystal in silicon-on-insulator wafers.

processes and lays the foundation for the second set of steps necessary for critical dimensions (CDs) definition. Here, PhC, together with the optical testing structures, were defined using photoresist on a layer of antireflection coating (Figure 10.1d) prior to MERIE etching of PhC for good profile transfer of CDs onto the hardmask layer (Figure 10.1e). Complete resist strip after this step (Figure 10.1f) in the second phase then allows the third phase of OTS fabrication to begin.

Here, a different photoresist of higher viscosity was spun on the wafer to form a thicker resist film without first coating a BARC layer (Figure 10.1g). Due to the shallow depths of the hardmask patterns defined in the previous phase, the resist coating recipe may be developed to successfully planarize the surface, enabling exposure of the second photomask after alignment to at least five underlying marks patterned during the first exposure (Figure 10.1h). The testing structures exposed may therefore be sequentially etched for the: hardmask oxide, SOI device silicon, SOI buried oxide and then SOI handle silicon again till an etch depth of nearly 60 μm (Figure 10.1i). This step completes the construction of OTS, and the wafers undergo the final phase of PhC and OTS, etching and other postprocessing modules.

In Figure 10.1j, photoresist strip and polymer cleaning were performed prior to device etching so as to enable simultaneous etching of both PhC structures and the OTS, using etch recipe developed to optimize the etch quality of the PhC devices. After device sidewalls optimization, chemical buffered oxide etching (BOE) was then performed to release both the movable structures and the air-bridge type PhC devices (Figure 10.1k). In this step, not only was peeling of the metallic electrodes prevented, precaution was also taken to insure no further rework in case of contamination arising from the exposed electrodes.

Figure 10.2 shows the fabrication process flow for rod-type PhC fabrication on SOI wafers for buried oxide-layer cladding. In this process, PhC rods realized in the device layer may be filled in between with silicon dioxide for waveguiding-mode symmetry. Here, due to the introduction of an insulating oxide layer and reduced-radius PhC rods waveguide, out-of-plane control of radiation was introduced, and the PhC rods need not be etched for increased HAR. For SOI wafers of device thickness 3 μm and reduced-radius PhC rods of radii less than 50 nm, special techniques are needed for subwavelength lithography, smooth-etch sidewall profiles, etc.

10.3 Subwavelength Lithography

In the past, the most straightforward way to improve resolution was to decrease the wavelength of radiation, as illustrated in Figure 10.3. However, reduction of wavelength beyond Argon Fluoride (ArF) lasers emitting at 193 nm would require a drastic redesign of the lithographic system because shorter wavelengths are simply absorbed by the quartz lenses that direct source light in current systems with laser wavelengths between 193 and 432 nm. Alternatively, resolution enhancement techniques (RETs) [2–4] such as PSM (phase shift masking), optical proximity correction (OPC), and off-axis illumination (OAI) may be used for exposure systems of specific wavelengths to improve their respective resolution limits (as indicated by the line marked with round symbols in Figure 10.4).

Although each of these techniques improves baseline resolutions, they are limited mainly by their respective drawbacks, which for PSM (Figure 10.3a) and OPC are the cost and computational efforts required for implementation, whereas for the various OAI schemes illustrated in Figure 10.3b, it is the discrimination needed for varying pattern densities. Therefore, alternative approaches to achieving subwavelength lithography resolutions have been widely investigated in recent years [5,6].

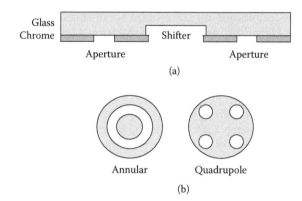

FIGURE 10.3
Illustration of (a) phase-shift masking technique with use of shifters within glass mask and (b) off-axis illumination of the annular and quadrupole schemes.

To develop OL, a key consideration was that, out of the major components used in a set of optical exposure systems, the radiation source could not be easily changed as compared to the parameters of the condenser, mask, wafer, pattern design, resist chemistry, coating mechanisms, etc. The various parameters of the imaging system such as the lens-set numerical aperture (*NA*) and partial coherence of the condenser system design can also be varied for different lithographic performance. Although an ideal lithographic projection tool should be coherent with unity *NA* (a measure of how well a lens is able to collect diffracted light from a photomask and in turn project the image onto the wafer, as illustrated in Figure 10.5), real systems without immersion technology can neither provide unity *NA*,

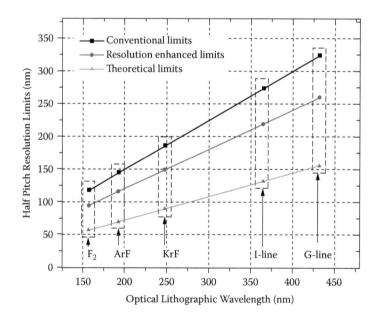

FIGURE 10.4
Limits in resolution improvement through reduction of optical lithographic wavelengths.

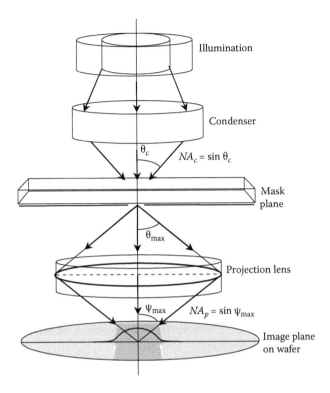

FIGURE 10.5
Schematic of an on-axis illumination lithographic system for depiction of maximum cone angles in the condenser and projection lens.

nor purely coherent illumination because the total image is the sum of all the intensity images from all source points. The degree of coherence factor is usually expressed as

$$\sigma = \frac{NA_c}{NA_p} \tag{10.1}$$

where $NA_c = n\sin\theta_c$ and $NA_p = n \sin \psi_{max}$ represents the NA of the condenser and the projection lens, which is in turn determined by n, the refractive index at the image plane; and ψ_{max} and θ_c, which are the maximum cone angle of rays subtended by the maximum pupil diameter at the image and condenser plane, respectively (see Figure 10.5).

Based on the theoretical resolution limits of an otherwise ideal imaging system, the Rayleigh equation for the resolution of the imaging system is given by

$$R = k_1(\lambda/NA) \tag{10.2}$$

where λ is the wavelength, and k_1 is the process-dependent constant. This equation not only provides the limits of OL as shown in Figure 10.3 but also indicates the latitude of processing windows for different CDs. Hence, for 100, 200, and 300 nm resolution, the k_1 process constant needed can be calculated as shown in Table 10.1.

Because controllable processes typically operate in the range of $k_1 \cong 0.6$ to 0.7, it can be deduced from the table that, although resolution near 200 nm is difficult to achieve without high NA, RETs are essential for CDs below 200 nm. At the same time, although lens with higher NA would be able to collect more diffracted orders and consequently more

TABLE 10.1

Theoretical Resolution Limits of Line Widths by
Deep Ultraviolet Optical Lithography

Numerical Aperture	300 nm	200 nm	100 nm
0.30	**0.36**	**0. 29**	*0. 12*
0.40	**0. 48**	**0. 39**	*0. 16*
0.50	*0. 60*	**0. 49**	*0. 20*
0.60	*0. 73*	*0. 59*	*0. 24*
0.70	*0. 85*	*0. 68*	**0. 28**
0.80	*0. 97*	*0. 78*	**0. 31**
0.90	*1.09*	*0. 88*	**0. 35**
1.00	*1.21*	*0. 98*	**0. 39**

Resolution enhancement techniques needed

Typical optimized process parameters

Process parameter without optimization

Process constant beyond theoretical limits

information to correctly reproduce the required image, there would be an inherent trade-off with depth of focus (*DOF*), because

$$DOF = k_2 \frac{\lambda}{NA^2} \qquad (10.3)$$

where k_2 is the second process constant. It is clear, therefore, that for high resolution, increases in lens *NA* quickly decreases the *DOF*, which should have preferably a large value (because tolerances for wafer flatness, resist coating, etc., should be more than a few hundred nanometers). Therefore, alternative control of photomask pattern densities and lithographic partial coherence factors should be used to achieve such similar resolution improvements. For partially coherent systems, the resolution of the imaging system can be described with some correction to Equation 10.2 such that

$$R = k_1 \frac{\lambda}{(1+\sigma)NA} \qquad (10.4)$$

Therefore, the increase of the coherence factor can be used to derive improved resolutions.

10.4 Lithography Setup and Methods

The lithographic experiments were carried out using a resist track, a DUV scanner based on the Krypton Fluoride (KrF) laser of 248 nm, and a binary mask of four times magnification without phase shift features with positive DUV resist (Shipley UV210 and JSR 221). Prior to resist coating, it is important that wafers are cleaned and primed to minimize the effect of surface oxides, which form long-range hydrogen bonds with water adsorbed from air (resist adheres to water vapor rather than to wafer surface), resulting in poor resist adhesion to wafer surface. Therefore, in the experiments, cleaning and priming with hexamethyldisilazane (HMDS) was carried out using the liquid priming process of spinning solvent-diluted HMDS. Resist can then be spun onto primed wafer surface with the target

of the greatest possible uniform thickness. Although the physics of spinning is complicated, it can be determined to be strongly dependent on the evaporation rate of the solvent
used together with the assistance of spin speed charts. A preexposure bake then drives the
solvent from the resist, enabling high-quality profile when appropriately carried out, but
decreasing the sensitivity of the process if overdone. Postexposure, another bake is needed
to reduce standing waves on resist profiles. Development of the resist was then carried out
following a final postdevelopment bake to improve the resist's etch resistance.

10.4.1 Lithographic Parameters Manipulation

Although the resolution of the lithographic process is not a fixed number but depends on
the factors of pattern type, resist thickness, and wafer topography, it is generally easier
to print an isolated feature than a dense pattern such as that of a PhC structure. At the
same time, it is important that the focus of a system is always checked and compensated
for prior to exposure by adjustment of the lens to the wafer distance. For the projection
method of lithography used, the "resolution" that determines the narrowest line/space
of a repetitive pattern that can be correctly reproduced is given by Equation 10.2, and
the other critical factor, DOF, is given by Equation 10.3. These factors effectively form the
limiting determinants for the spread functions that control the size of isolated and dense
patterns. When these parameters are not carefully controlled, typical failure modes that
occur include uprooted resist (Figure 10.6a) due to diffracting wave defocusing, missing
patterns (Figure 10.6b) due to illumination overdose, and distorted shapes of resist profile
(Figure 10.6c) due to overlapping wave-functions interference.

According to Equation 10.2, it is clear that, for good resolution, the value of the proportionality constant k_1 and λ should be the smallest possible, whereas the value of NA is the
largest possible. Technologically, the reduction of λ hits a barrier beyond 193 nm because
shorter wavelengths are simply absorbed by the quartz lenses that, direct source light onto
the wafer and therefore cannot be used without a drastic redesign of lithographic systems.
At the same time, because the DOF of Equation 10.3 is inversely proportional to the square
of the NA, increasing the aperture to improve resolution quickly decreases the DOF, which
should have a value greater than 500 nm because wafers cannot be perfectly flat, and
photoresist usually has thickness variation of several hundred nanometers. To increase
the value of NA for better resolution, lens designs of larger diameters and greater lens
element count are used to increase the amount of light collected through the photomask.

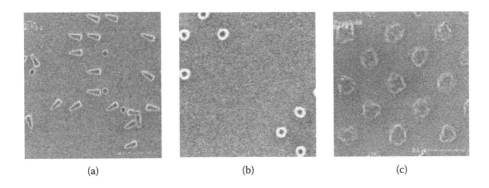

(a)   (b)   (c)

FIGURE 10.6
Lithographic problems that arise from unoptimized process conditions include (a) uprooted resists, (b) missing
patterns, and (c) distorted shapes.

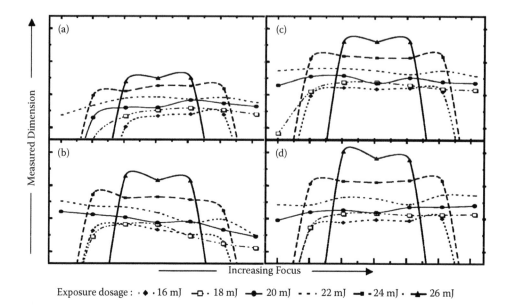

Exposure dosage : ◆ · 16 mJ ▫ · 18 mJ ● 20 mJ · · · 22 mJ ■ · 24 mJ ▲ 26 mJ

FIGURE 10.7
Plots of focus exposure matrix for (a) low-, low-; (b) high-, low-; (c) low-, high-; and (d) high-level of resist thickness and antireflection coating, respectively.

On the other hand, to reduce the trade-off between R and DOF, k_1 can be reduced through improvement of the illumination coherence, antireflection coatings, and also resist parameters. Hence, two-level, multiple two-factorial experiments were carried out by running the experiments with different levels of BARC and CAR thickness to yield results of process windows summarized by the focus exposure matrixes (FEMs) of Figure 10.7.

From the plots, it can be seen that the poorest exposure latitude was obtained for the case of high level of CAR and low level of BARC (Figure 10.7b), while an increase in level of BARC still resulted in slanted profiles of the process window (Figure 10.7d). On the other hand, decrease in the level of CAR showed improvement in latitude of process window for both high and low levels of BARCs (Figure 10.7a,c). The plot of the FEMs with low level of CAR and high level of BARC can be seen to yield the greatest exposure latitude and also focus range (Figure 10.7c). Hence, the effect of CAR thickness can be significant and a smallest k_1 can be obtained for a lower level of CAR thickness and higher level of BARC. These results supported the fact that reduced aberrations result from minimum resist thickness although presence of BARC aids reduction of the effect of lithographic dimensional errors due to minimization of scattering from substrate material. Furthermore, it has also been identified that these two parameters have a minimum interaction effect as indicated by the non-twisting of their response surface when the results of Figure 10.7 are plotted in three-dimensions (3-D).

10.4.2 Optical Proximity Error Model

Beyond the improvement of Equations 10.2 and 10.3 as discussed, other methods for resolution enhancement include techniques such as OPCs, use of PSM, and use of OAI systems to manipulate the amplitude, phase, and propagation direction of the illuminating electromagnetic waves, respectively. Although there are high-computing-powered programs available for OPC, these are often costly and in state-of-the-art integrated circuits (IC) layout tools. On the other hand, PSM is also relatively much more costly (as compared to a conventional binary photomask). Finally, although the OAI of annulus, dipole, and quadrupole types had been demonstrated to work well for dense areas of repetitive patterns, the technique faces limitation for isolated features, which will often be encountered in our application to PhC devices in the form of tunneling defects, etc.

In this study, a binary mask and nonshifted axis illumination exposure system were used for achieving the required PhC defect structures within a dense matrix of otherwise similar-sized PhC rods. Here, the achievement of designed tunneling defect size is critical for successful fabrication of PhC devices with useful properties. The current challenge in this area lies in the fact that PhC devices are superdense arrays of periodic structures with feature size close to the lithographic wavelength. Hence, by the effect of diffraction, dense neighboring features illuminate to interfere with each other either constructively or destructively to print smaller and larger rods, respectively, as opposed to being isolated to give rise to the OPE effects. This unavoidable phenomenon affects the fundamental dispersion characteristic of a designed PhC drastically as can be seen in Figure 10.8, in a single-line-defect

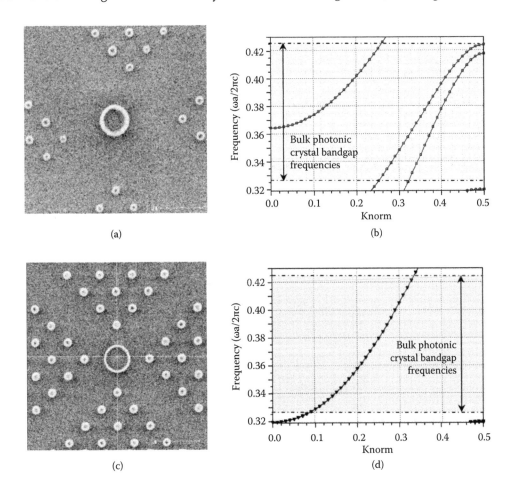

(a)   (b)

(c)   (d)

FIGURE 10.8
(a) Top view scanning electron micrograph (SEM) of photonic crystal array with proximity errors along the periphery rods of the intersecting line-defect waveguides and (b) its corresponding multimode dispersion simulation result. (c) Photonic crystal after resolution enhancement by proximity correction, and (d) the single-mode dispersion property within the bandgap frequency range.

waveguide where the rods near the waveguide defects patterned (Figure 10.8a) much smaller than designed (Figure 10.8b), such that the originally single mode (Figure 10.8d) waveguide becomes multimodal within the bandgap frequencies (Figure 10.8c).

To overcome this problem of OPE between bulk, semiisolated structure and isolated features, test patterns with mask biases of varying degrees and shapes were designed to yield CD measurements for varying exposure energy (E), focus (F), "correction factor" (CF) and "shape factor" (SF). Here, CFs were varied with different sizes of the periphery PhC rods and defined as their diameter ratio to that of the intended bulk rods' diameter as designed on the optical mask. The SF used corresponds to the circular and square shapes of the rods drawn with width of squares equated to the diameters of the circular rods needed (i.e., for the square-shaped rods drawn, the same definition of correction factor applies with simply replacement of square width with circular diameter).

Based on these two designed experiment control parameters, the measured results are recorded as "experiment factor" (EF) giving the measured rods' diameter normalized by the actual bulk PhC rod diameter. For the two sizes of rods of diameters 300 and 360 nm investigated, two types of SFs (square and circular shape) and eight sets of CFs were tested using varying F and E between 0 and 0.2 and 18 and 22 mJ/cm^2, respectively. Each treatment combination of SF and PhC rod size was varied against the other parameters as described to yield 43 sets of measurements each, resulting in a total of 172 data sets.

By first studying the result of these experiments through scatter matrix plots, correlations of the experiment factors with respect to CD errors measured can be deduced. Therefore, based on initial hypothetical multilinear models, with regressors of CF, E, and F deduced accordingly, the OPE measured for the different structures can be used to obtain fitting parameters for each of the regressors such that prediction errors by the models are minimized. In this way, much-simplified OPC models were derived for both the semiisolated and isolated structures, where the former was fitted with singular power factors of CF, E, and F, whereas the latter was fitted with a double power factor for CF and singular power factor for E and F.

In Figure 10.9, the exaggerated effects of variations in CFs (as ratios of periphery PhC rod diameters to bulk ones) for values of 1.1 to 1.8 are shown as designed on the photomask (with actual CF values used ranging from 1 to 1.07 in steps of less than 0.01).

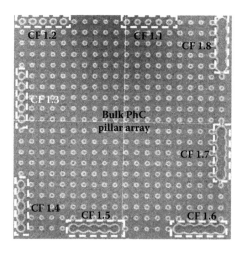

FIGURE 10.9
SEM of varying CFs (in dashed-line boxes) along the periphery of bulk photonic crystal rods in exaggerated scale.

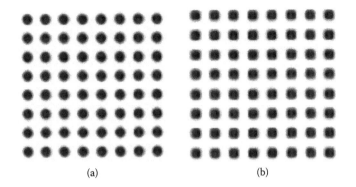

(a)   (b)

FIGURE 10.10
Micrographs of fabricated photomask with chrome-on-glass patterns of photonic crystal rods (a) without shape factor correction and (b) with shape factor correction.

Also, variations in *CF*s were designed as different shapes of circles and squares drawn on the layout of the schematic. The micrograph photographs of the fabricated photomask with these *SF*s are given in Figure 10.10. By plotting scatter plots of lithographic errors (*LE*) for each of these experiment conditions, trends for dependencies can be identified and expressed as a function of *E*, *F*, and *CF* such that

$$LE_k = a_0 + a_1 E_{k,1} + a_2 F_{k,2} + a_3 CF_{k,3} + \varepsilon_k \tag{10.5}$$

where a_0, a_1, a_2, and a_3 constitute a set of the best estimate modeling parameter, and ε is the error of estimation, which can be reduced [7] to

$$\hat{\underline{a}} = (\underline{X}^T \underline{X})^{-1}(\underline{X}^T \underline{Y}) \tag{10.6}$$

where $\underline{X}$ is a 43-row matrix of ones with the second and subsequent columns replaced by coefficients of *E*, *F*, and *CF*, respectively, and $\underline{Y}$ is given by the measured values of *LE*. The results of the regression model were therefore obtained and summarized by Table 10.2. Using these model parameters, the prediction errors for each treatment can be shown to have median errors of less than 1% and standard F-numbers of 14 to 67 indicating significance of regression at above 97.5% confidence level. This OPC model therefore allows for proximity corrected PhC device structures with appropriate defect sizes, and average measured deviations below 3% target value.

TABLE 10.2

OPC Regression Model Parameters Obtained Through Minimization of Regression Modeling Errors for Different Shape Factors and Photonic Crystal Rods Diameters

Shape Factor/Model	Constant	Energy	Focus	Correction Factor
Square (width = 360 nm)	−0.4301	−0.0012	−0.0151	1.4711
Square (width = 300 nm)	−1.1162	−0.0045	−0.0089	2.1917
Circle (diameter = 360 nm)	−0.5868	−0.0001	0.0084	1.5873
Circle (diameter = 300 nm)	−0.5901	−0.0203	−0.0143	2.0161

An OPC model was therefore developed based on such a design of experiment for experimental compensations, which allowed for proximity-corrected PhC device structures with appropriate defect sizes and average measured deviations below 3% target value. Thus, mask bias corrections can be made in anticipation of OPE, improving resolution of near subwavelength DUV OL without use of expensive PSM or high-end computations to yield compensated results of Figure 10.8c, where PhC device structures near defect structures are obtained with desired dimensions.

10.4.3 Partial Coherence and Pattern Density Control

Based on this, critical line-width experiments may be carried out. The optimized resist coating recipe was applied, first with wafer surface priming by HMDS for adhesion preparation, followed by a layer of BARC (Shipley AR3) that was spun at a combination of rotation speeds between 1.5 and 4.5 krpm for 40 s before bake at 205° for 60 s. Following this, the spun layer of BARC with thickness 600 Å was then baked at 130° for 60 s. For application of CAR with thickness 4800 Å, the CAR was spun at rotation speeds between 1.5 and 2.5 krpm for 40 s. After a preexposure bake, the wafers undergo a wafer edge exposure for removal of edge bead formed in the resist coating spiral stage (after approximately a thousand revolutions). Line widths experiments were then carried out using patterns of lines with widths 200 nm and duty factors (line/space ratio) of 1:3 and 1:1, as shown in Figure 10.11. CDs of each set of duty factor experiments were taken using a Hitachi 9200 CD SEM imaging system. Based on a lithographic imaging system with a calibrated zero focus and *NA* of 0.68, the designed line widths of 200 nm were used in OL experiments for coherence factor values 0.38, 0.45, and 0.51.

The results of the partial coherence variation experiments were plotted as measured CDs with respect to the exposure dosage used, as shown in Figure 10.12. From the CD plots of the line-width experiments, it can be seen that the 1:3 density patterns are much more sensitive to coherence factor variations than that of the 1:1 density patterns. Especially, the 1:3 density patterns have measured line widths, which were drastically reduced for the exposure with coherence factor 0.51 as compared to the exposures with coherence factors 0.38 or 0.45. On the other hand, the higher-density 1:1 patterns had similarly reduced line widths for all three coherence factors used. From further such experiments, it was obtained that reduction of original line widths to less than 200 nm allowed for further reduction of CDs obtained. However, such progressive reduction of designed line widths

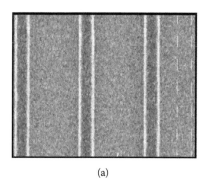 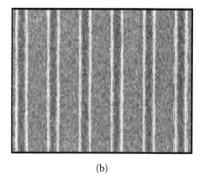

(a)　　　　　　　　　　　　　　　　　(b)

FIGURE 10.11
Lithography experiment patterns with constant line widths and different pattern density of line/space duty factors (a) 1:3 and (b) 1:1, respectively.

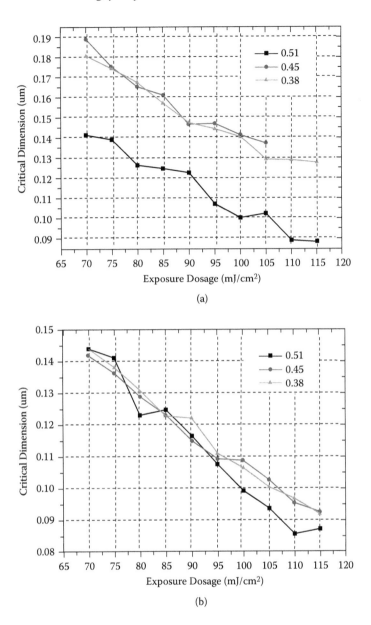

FIGURE 10.12
Critical dimension plot for lithography experiments with a constant numerical aperture of 0.68 and varying coherence factors of 0.38, 0.45, and 0.51 for line patterns with duty factors of (a) 1:3 and (b) 1:1, respectively.

for corresponding reduction of CDs approaches its limit at a drawn line width of ~130 nm. This is due to the reduced process window of the smaller CDs, where a small deviation in exposure energy or focus leads to lithographic failures.

Examples of typical exposure failures brought about by such constraints are as shown in Figures 10.13 and 10.14. In Figure 10.13a, unresolved features results from insufficient exposure dosage in spaces between dense line patterns, and in Figure 10.13b, resist peeling occurs at narrow ends of line patterns due to overexposure. In Figure 10.14, there is the necking effect, where biasing of patterns near line terminations are required to prevent neck breakages at

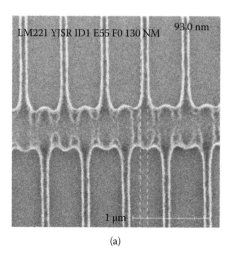 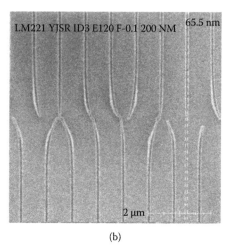

(a)   (b)

FIGURE 10.13
SEM images of sublithographic wavelength exposure failures for (a) unresolved dense structures and (b) lifted-off resist.

line ends. The layout and exposure lithographic variations are shown for a narrow line end (Figure 10.14a,b) where there is the pull-back mechanism, and in a connected line end (Figure 10.14d,e) where there is the flaring mechanism at the end of the connected lines. Combined together, these two effects constitute the mechanisms for necking such that pull-back coupled with flaring in the mild case leads to patterns of critical line widths with constricted "neck" interfaces as shown in Figure 10.15a. In the unbridled case of necking, extreme constrictions lead to critical line breakages at line-base interface as shown in Figure 10.15b.

10.4.4 Chemically Amplified Resist Deactivation

DUV radiation exposure on the CAR results in the formation of minute catalytic amounts of acids in the exposed areas of CAR as determined by the openings on the chrome photomask. Subsequent processing by acid chemical-amplification processing then allows the CAR to become soluble in the alkaline-developer solution, allowing the removal of the exposed resist. If, however, the minute acid formed during exposure is unintentionally quenched by the base impurities on the wafer or in the environment, chemical amplification can be deactivated, causing erroneous patterning where CAR are not effectively removed by the developer where an opening should be. This type of "resist poisoning" is illustrated in Figure 10.16a, where a PhC lattice of holes was patterned, and some holes are not properly developed to result in positions of missing holes.

Such defects install optical states within bandgap frequencies, which can become microcavities that act as unintended resonators (Figure 10.16b). To deter the formation of such deactivated CAR, the presence of weak-base pyrimidine ($C_4H_4N_2$) resulting from treatment with the solvent should be avoided with minimization of resident time after the solvent-cleaning process for successful lithographic patterning [8]. It is important to avoid such instances of resist poisoning because reworks are not only complicated but also often ineffective.

10.4.5 Photoresist Trimming for Critical Dimensions Reduction

Beyond the many RETs presented so far, there is still an overriding need to decrease the size of the smallest feature printable using OL. Although the development of

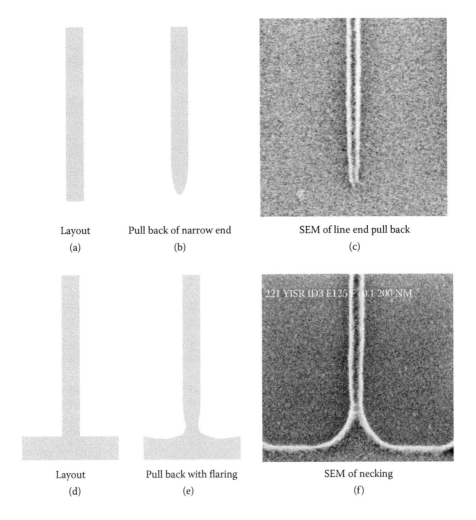

Layout (a)   Pull back of narrow end (b)   SEM of line end pull back (c)

Layout (d)   Pull back with flaring (e)   SEM of necking (f)

FIGURE 10.14
From left to right are (a) schematic illustration of layout, (b) lithographic variations, (c) SEM example due to pull-back mechanism at line end; and (d) schematic illustration of layout, (e) lithographic variations, (f) SEM example of necking for flaring mechanism at the end of connected lines.

shorter-wavelength lithography together with its accompanying lens systems, resist chemistries, and mask material, etc., are continuously being explored, developmental progressions take considerable time. Hence, it is very useful to make use of the technique of dry plasma etching to further reduce the size of the smallest feature printed by resolution-enhanced OL. Such a method, known as "photoresist trimming" (or "photoresist ashing," "oxide lateral etching," and "resist thinning") [9], makes use of oxygen-based plasmas to etch photoresist polymers isotropically. Halogens or fluorocarbon gases are also added to provide control over the ratio of the lateral and vertical etch rates.

In this work, the resist trimming recipes were developed for a thin layer of CAR, deposited atop a layer of organic BARC. An ICP etcher was used to generate tetrafluoromethane (CF_4) and oxygen (O_2) plasma in this process. Whereas the O_2 plasma was used mainly to isotropically etch a developed resist pattern, CF_4 fluorocarbon was used mainly as a polymerizing gas to ensure a vertical resist profile after trimming. In such photoresist

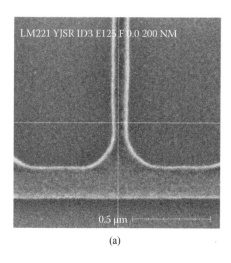
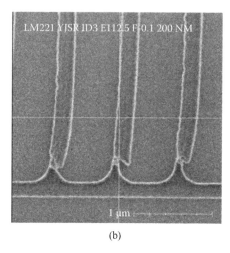

(a)   (b)

FIGURE 10.15
(a) Patterns of critical line widths with constricted "neck" interfaces and (b) extreme constrictions leading to critical line width breakages at structure interface.

trimming process, the key parameters to be controlled are gas composition, wafer temperature, bias voltage, chamber pressure, source power, and total gas flow.

First, in terms of gas composition, an initial addition of polymerizing CF_4 gas helps accelerate the hydrogen extraction process from the resist pattern, which therefore increases the trim rate. However, this trim rate increment effect saturates with further supply of CF_4 gas, because the increased concentration of polymerizing gas deposits around the resist patterns, hindering the etch process by the reduced concentration of O_2 plasma. Second, the trim rate increases as wafer temperature increases, as can be expected because higher temperature enhances chemical reaction rates and reduces sidewall polymer deposition. Third, the lateral trim rate, on the other hand, generally decreases with increases in bias voltage because with higher-bias voltages, the ions are more energetic and are less likely

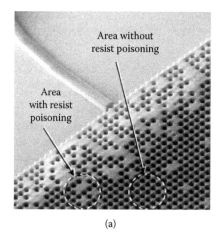
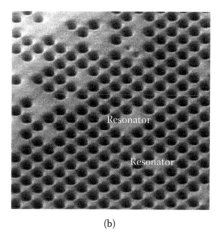

(a)   (b)

FIGURE 10.16
(a) Cubic photonic crystal holes array patterns with areas of resist poisoning (missing holes in a uniform holes array) and (b) unintended resonators form within the photonic crystal holes lattice.

to be deflected laterally to aid the trim chemical reaction. Fourth, increasing the chamber pressure serves to decrease ion energy because an increased number of collisions arises. However, it also increases the scattering of reactive species onto the resist sidewall and therefore enhances the trimming reaction. Further, increases in pressure also increase the neutral chemical flux, which also serves to increase the trim rate. Fifth, the trim rate also increases with source power because the enhanced dissociation of O_2 plasma reactive species aids the resist etching process. Finally, the effect of increasing total gas flow only increases trim rate as much as it is possible to increase the number of chemical reactions around the photoresist pattern surfaces.

An optimized resist-trimming recipe has high uniformity across the wafer and maintains good pattern integrity (e.g., no line bending, etc.). This can be obtained using only O_2 and Argon (Ar) gas, with flow rate of 5 and 45 standard cubic centimeters per minute (sccm), respectively. For a chamber pressure of 2 mT, platen power of 25 W, and coil power of 600 W, a trim rate of 400 Å/min was obtained in the vertical direction, together with a trim rate of 500 Å/min in the horizontal direction for isolated resist patterns. For highly dense PhC resist patterns, however, the trim rate tends to be correlated with the pattern density. For PhC resist patterns (created in a stack of 600 Å BARC [Shipley AR3] and 2800 Å CAR [JSR M221] and exposed using the same track and DUV OL techniques), the uniformity and quality of the trimming process were measured using test patterns with varying CDs and densities. Figure 10.17 shows the trim result of PhC resonators with radius ranging from 100 to 250 nm, subjected to trimming process time of 55 s.

For each CD resonator placed within a similar single-resonator structure surrounded by bulk PhC rods such as that shown in Figure 10.18b, five measurement points were taken at the middle, left, right, top, and bottom positions of the wafer. These readings were plotted on the vertical axis of Figure 10.17 for each corresponding point of measurement (plotted along the horizontal axis). Based on these readings, it was shown that an average radius trimmed was 256 Å—corresponding to a lateral trim rate of 559 Å/min for each PhC rod CD. Although very similar trim rates were obtained for PhC resonators of different CDs placed within a similar bulk PhC lattice, it was found that the trim rates were varied by the

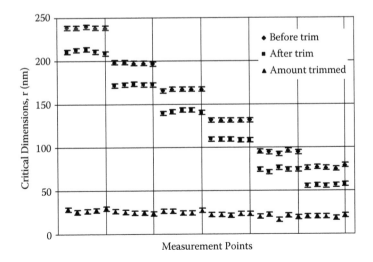

FIGURE 10.17
Critical dimension measurements of resonators with radius ranging from 100 to 250 nm, before and after the trimming process.

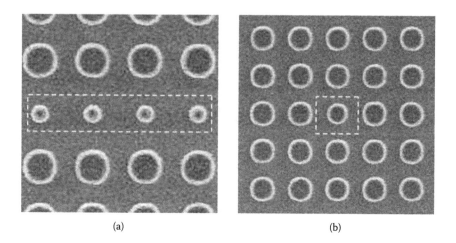

(a)   (b)

FIGURE 10.18
(a) Single-line reduced-radius rods within a line defect photonic crystal waveguide and (b) single resonator rod surrounded by bulk photonic crystal rods.

surrounding structures (such as that of single-line defects shown in Figure 10.18a), bulk PhC of an unperturbed lattice, or single resonators of Figure 10.18b.

Figure 10.19 shows the results of patterning and trimming found for different PhC pattern densities, specifically of the single-resonator and bulk-PhC lattice rods. Although bulk rods are patterned to become much larger than their reduced-sized counterparts drawn within a single-resonator pattern on the photomask, they trimmed down by nearly the same amount to give similar CAR etch rates. This is in contrast with the trim bias, found for PhC rods, patterned in a single resonator versus a single-line defect pattern, as plotted in Figure 10.20. Here, it can be seen that, although the differences in pattern density did

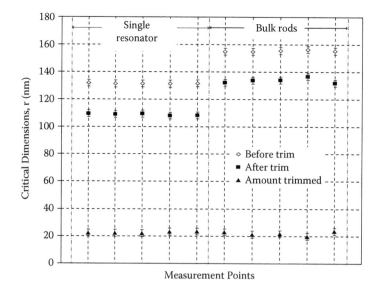

FIGURE 10.19
Comparison of trim result for photonic crystal rods of the same critical dimension, placed within the bulk lattice and within a single-resonator structure.

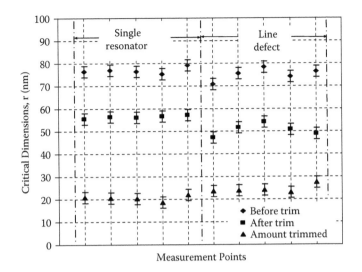

FIGURE 10.20
Difference in trim rate for varied pattern densities of single-line and resonator structures.

not result in significant differences in the initial lithographic process, the line-defect PhC rods were trimmed down to a smaller size than those found in a single-resonator configuration. Hence, higher corresponding etch rates were found for the single-line structures than that for the resonator ones. This may be attributed to the increased opening size along a "line" rather than a "point" defect, which would enable greater transport of trim reactants towards the trim sites.

Figure 10.21 shows SEMs of the CAR pattern in a single-line PhC waveguide before and after the trimming process. Here, it can be seen that the isotropic trim process not only

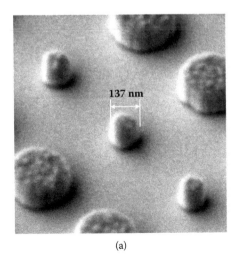
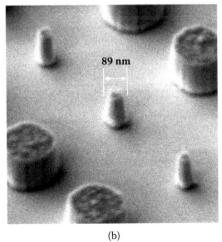

(a)   (b)

FIGURE 10.21
SEM of line defect rods flanked by bulk photonic crystal lattice. (a) Photoresist patterned by deep ultraviolet lithography and (b) trimmed photoresist with sub-100 nm critical dimensions.

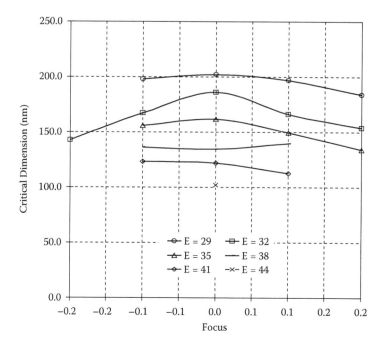

Figure 10.22
Focus exposure matrix for reduced-radius photonic crystal critical dimensions.

reduces the CDs of the PhC rods laterally but also serves to etch away vertically the BARC up to a thickness of 60 nm. From the profiles of the patterned photoresist as shown in Figure 10.20a (which are tapered from the base up) and the FEMs of Figure 10.22 (where small exposure latitudes were found), it can be seen that the limits of the process would impede further reduction of CDs using lithographic RETs. Hence, to obtain sub-100-nm CDs, resist trimming is an important technique. Figure 10.23 shows the top-down SEM

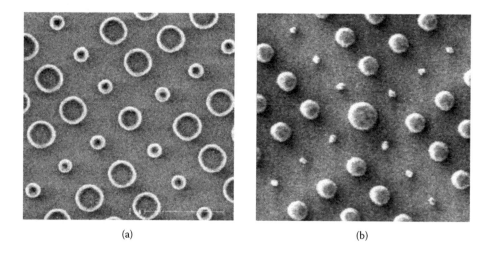

(a)   (b)

FIGURE 10.23
Top-down SEM view of photonic crystal optical intersections with critical dimensions (a) before and (b) after trim processing.

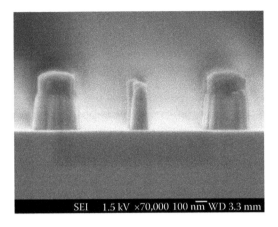

SEI   1.5 kV ×70,000 100 nm WD 3.3 mm

FIGURE 10.24
Cross-sectional SEM of trimmed photonic crystal rods resist pattern, with the reduced radius line-defect rod flanked by two bulk rod pattern.

image of some PhC optical intersections, with single-line reduced-radius PhC rods, resonator rod, and the surrounding PhC bulk crystal prior to and after the optimized trim process. Figure 10.24 shows the cross-sectional counterpart to Figure 10.23b.

10.5 Post Lithographic Processings

After achieving of high-resolution PhC lattices in lithography, the development of etching formulations are necessary to derive an acceptable etch process window for very deep etching of PhC lattices of HAR deep-submicron-sized rods (aspect ratio beyond 50) previously unattainable by the DRIE method for such a length scale. Together with the suppression of etch sidewall "scallops"—a critical shortcoming of time-multiplexed etching—in the form of sidewall undulations causing surface roughness, an atypically low value of scallop depth of only 12 nm was obtained.

Using an STS deep etcher with ICP sources for plasma generation (coil) and biasing (platen), DRIE was carried out using the Bosch process [10] in which the time-multiplexed scheme of etching and passivation enabled by well-controlled gas inputs facilitate the achievement of vertical sidewalls in deep etching that consumes relatively lesser protection material. This is due to the switching between etching and passivation steps that are repeated as necessary. First, a silicon opening unprotected by masking material is etched using fluorinated gas chemistry comprising mainly of sulphur hexafluoride (SF_6) to give a slightly isotropic profile (caused by lateral physical etch effect of impacting ions at etch surface). The action of reactive ions results mainly by the formation of ion and radical species by electron impact dissociation

$$SF_6 + e^- \rightarrow S_xF_y^+ + S_xF_y^* + F^* + e^- \tag{10.7}$$

so that the ion-assisted fluorine radicals then react with silicon openings to form a volatile silicon by-product gas, which is then removed. Here, by mixture of oxygen (O_2) in the etching gas chemistry, suitable sidewall passivation can be formed

$$O_2 + e^- \rightarrow O^+ + O^* + e^- \tag{10.8}$$

$$O^* + Si \rightarrow Si - nO \rightarrow SiO_n \tag{10.9}$$

even during the etching phase such that the passivation layer derives from the formation of SiO_n, which adsorbs onto the surface of exposed silicon as an oxide layer (with passivation properties because the subsequent etching reaction requires its removal prior to action by F). Therefore, after the etching phase, the chamber is pumped out, and passivation gas octafluorocyclobutane (C_4F_8) is applied for deposition onto the wafer (coating both base surfaces and also the sidewalls of trenches and other openings).

These two steps are then repeated as necessary to obtain the required amount of etch and passivation cycling. Generally, due to the assistance of platen power for the vertical etch direction (plane perpendicular to wafer surface), the directionality of reacting ions in the vertical direction is high, compared to the rate at which fluorine radicals are bombarded onto laterally exposed surfaces (plane parallel to wafer surface) because of impact scattering and also chemical reactions. Hence, the vertical etch component is much more significant such that of the etch rate in the lateral direction, is often considered negligible. At the same time, the formation and removal of passivation compounds together with the etching of silicon must be designed to act in balance so as to maintain etch profile anisotropy. Otherwise, dominance by passivation would lead to low etch rates or even etch termination by surface residue buildup. On the other hand, insufficient passivation leads to loss in anisotropy, and increased lateral etching occurs.

10.5.1 Unique Considerations for Photonic Crystal Rods Etch

General RIE etch concepts that are established for standard micrometer-sized devices have to be applied with much more stringent requirements because a reduction in feature size results in a corresponding decrease in the process window and tolerance. Hence, because the PhC used at the optical communication wavelength of 1.55 μm has CDs of only a few hundred nanometers, stepwise isotropy and lateral etch rates need to be carefully monitored to prevent fabrication failure, as shown in Figure 10.1.

At the same time, etch scallops of undulations in sidewalls are inevitably formed during the etch/passivation cycling process due to the existence of isotropy in the etch species bombardment phase. Past results of DRIE [11] has demonstrated scallop depths ranging between 50 and 300 nm. However, such depths of undulations on PhC rods can be unacceptable, as diffused scattering on the side walls caused by such drastic scalloping (i.e., macroscopic roughness), can obliterate precisely designed Bragg reflections (ordered reflections of the designed PhC lattice structure) that constitutes the photonic band gap (PBG) effect. Therefore, to ensure that the PhC properties are preserved, it is important that etch scalloping (resulting in etch profile undulations) be kept at a minimum to reduce the effect of diffused scatterings.

10.5.2 Development of Deep Reactive Ion Etching (DRIE)

The time-multiplexed action of etching and passivation gases applied with independent bias control of the substrate was studied using an initial recipe of manual pressure control at 50%, etching chemistry comprising of C_4F_8, SF_6, and O_2 gases, and passivation chemistry comprising only the C_4F_8 gas. The coil power was maintained at 600 W for both passivation and etch phase, and platen power was applied only for the etch phase. For dominating

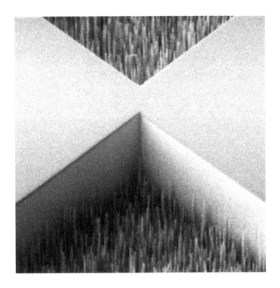

FIGURE 10.25
Formation of "black silicon" in areas of etch openings due to a passivation phase overdominance during time multiplexed, etch/passivation cycling.

the passivation processes, a reentrant etch sidewall profile was obtained with a lower etch rate, often resulting in self-induced etch stops. This is because plasma polymerization reactions require relatively high ion bombardment energies of several hundred electron volts (eVs) for their complete removal. At the same time, besides termination by the surface residue build up, the dominance of passivation also results in the incomplete removal of passivation polymers, which leads to the formation of grass-like residues commonly known as "black silicon" (Figure 10.25). Conversely, insufficient passivation causes loss in anisotropy such that lateral etching increases to result in the PhC rods being severed (Figure 10.26). Often, such lateral etch effect can easily dominate the small diameters of the PhC rods causing the rods to sever along their length, leading to "runaway etching." This

FIGURE 10.26
SEM image of fabrication failure by the effect of etching isotropy caused by the inappropriate reduction of passivation.

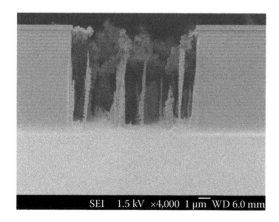

FIGURE 10.27
SEM image of runaway etch over the entire area that was previously a lattice of photonic crystal rods.

results in the etch openings becoming extended, and the etch species take over etching in the entire area that was previously filled with rods, leading to a "released hard mask" situation (Figure 10.27).

To overcome such undesirable defective etch situations, etching gas flow rates and durations can be reduced, and the corresponding values increased, for the passivation phase. At the same time, other methods of reducing the dominance of the etch include decreasing the chamber pressure and the platen power applied. Such processes and their respective parameters are compared in Table 10.3. Through the balance of etch and passivation dominance and a reduced platen power, an acceptable process window for the survival of deep-submicron-sized rods can be found, based on an etch chemistry of SF_6 (100 sccm), C_4F_8 (30 sccm), and O_2 (10 sccm) at a pressure of 50%, using platen power of 23 W, together with a passivation chemistry of C_4F_8 (160 sccm) at the same plasma power but without the application of platen bias to yield the results shown in Figure 10.28.

As can be seen, although the surviving PhC rods are obtained, the sidewall profiles of this etch process has rough surfaces due to large undulations arising from the etch scalloping effect. Here, it can be measured from cross-sectional SEM that the scallop depth is 106 nm, whereas the scallop height is about 150 nm. Intuitively, the removal of etch-passivation cycling in DRIE process should result in the minimization of sidewall scallops. However, it was found that, although a continuous etch with simultaneous plasma polymerization yields consistently smooth sidewalls, it was achieved at the expense of isotropic

TABLE 10.3

Process Parameters of the Etch and Passivation Phases for the DRIE Processes with Different Etching Profiles

Condition	Process Cycles	Gas Flow (sccm)			Power (W)		Process Time (s)
		C_4F_8	SF_6	O_2	Coil	Platen	
Reentrant etch profile	Etch	30	100	10	600	23	8
	Passivation	160	0	0	600	0	5
Lateral etch dominant	Etch	30	100	10	600	23	9
	Passivation	110	0	0	600	0	5
Continuous etch	Etch	90	40	0	800	20	—

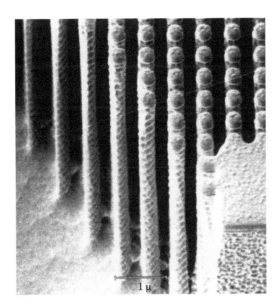

FIGURE 10.28
SEM image of surviving photonic crystal rod array without catastrophic fabrication failures but with rough sidewalls due to etch scallop undulations.

sidewall angles, leading to lateral undercutting, which, once again, led to the etch failures shown in Figures 10.26 and 10.27.

Therefore, to optimize the recipe for the reduction of scallop depths so as to achieve smoother sidewalls, a further reduction of etch isotropy was carried out to balance the reduction of etch dominance and increase in etch duration. Based on this principle, a large improvement in sidewall etch profile was obtained, as can be seen in Figure 10.29, where a

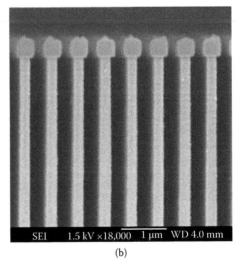

(a)   (b)

FIGURE 10.29
SEM image of comparison between (a) previous scallop depth of 102 nm and (b) smoothed scallop depth of 12 nm.

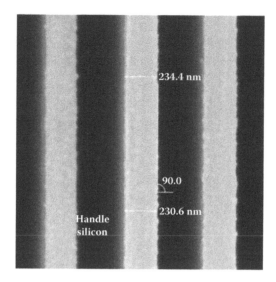

FIGURE 10.30
SEM cross-sectional image of fabricated photonic crystal lattice with controlled parameters, yielding nearly vertical sidewall profiles.

scallop depth of only 12 nm was obtained, which is a drastic improvement from an initial depth of 106 nm. Further, the development of such ICP etch processes may be developed to yield the conditions of sloped, retrograde, or vertical sidewall profiles. Here, with an optimized RIE condition, the etch profile can be moderated to obtain nearly vertical sidewalls as shown in the cross-sectional SEM of the etched PhC rod lattice in Figure 10.30.

At the same time, it was found from the experiments that, as opposed to the conventional assumption that lateral etching of DRIE structures are negligible beyond typical scalloping effects, progressive etch durations of deep-submicron-sized rods (unlike for larger structures) consistently resulted in reductions of PhC rod CDs, especially near the surface of the wafer. Based on this, it can be deduced that the reduced-sized geometries resulted from a significant etch action that is more dominant for the parts of the structures nearer the surface of the wafer than deeper into the wafer as the etch process progresses. Such a phenomenon can be attributed to the trapped nature of the etch species along the narrow etch paths of the PhC that resulted in the sidewalls being further etched laterally. Consequently, the incorporation of mask design bias (in addition to that for OPC derived in Section 10.3) is required for the successful realization of targeted devices, which would otherwise turn out to be smaller.

10.5.3   Dense Deep-Submicrometer Critical Dimensions (CDs) Etching

In the experiments to develop DRIE conditions that allow realization of superdense PhC arrays, the coil power of the RF source was maintained at 600 W, and varied platen power was applied for bias of ions energetic in the vertical direction, normal to the wafer surface. For a chemistry dominated by passivation, a sloped sidewall profile tends to result in gradually reduced etch openings that often leads to eventual self-induced etch stops.

Conversely, insufficient passivation causes loss in anisotropy such that lateral etching increases, resulting in retrograde sidewalls. To avoid such sidewall profiles, etch and passivation process parameters can be delicately balanced in such-high density, deep-submicrometer CD arrays to obtain sloped, retrograde, or vertical sidewall profiles. Therefore, the etch profiles of the CD structures can be moderated to obtain nearly vertical sidewalls as shown in the cross-sectional SEM of the etched PhC holes lattice shown in Figure 10.31.

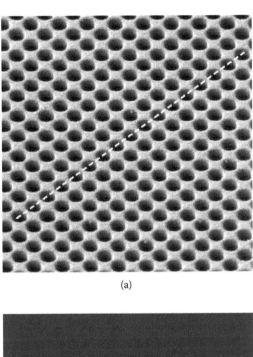

(a)

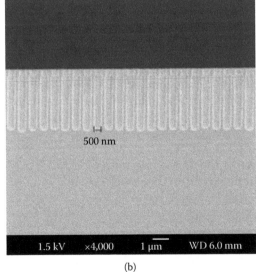

(b)

FIGURE 10.31
(a) Top view SEM image of photonic crystal holes lattice with dashed line showing where the cross section was taken to yield the (b) cross-sectional SEM of photonic crystal with controlled etch/passivation parameters giving vertical sidewall profiles.

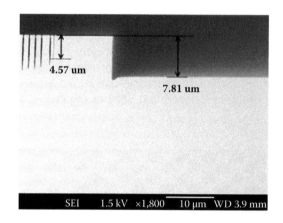

FIGURE 10.32
SEM of reactive ion-etch lag by aspect-ratio-dependent effects in plasma etching as shown for the high-density pattern on the left and the large optical testing structure opening on the right.

10.5.4 RIE Lag and Notching at Buried Oxide Interface

For etch patterns with different opening sizes, etch rates vary according to the effect of the aspect ratio, resulting in RIE lag. For etch openings of different line widths, the etch rates vary drastically. This can be seen from the cross-sectional SEM image of Figure 10.32, where the most drastic difference in etch depth was obtained for the OTS structures with ultrawide etch area, whereas, for the more narrow line patterns on the left, correspondingly increased etch depths were observed with progressively increasing line widths. Such variations in etch rates with respect to aspect ratios and etch opening sizes arise as a result of several etch mechanisms. Some of them are the ion and kinetic neutral shadowing effect; the transport of etch reactants toward, and away from, the etch surface; and the effect of electrical charges on the etch surface.

One of the most prominent among these mechanisms would be the transportation of etch reactants, which becomes progressively inhibited for narrower trenches. This is also true of the charge effect on the etch surface, with the charge distribution on the etched surfaces tending to be more highly concentrated in smaller etch surface opening areas than in larger ones. Hence, higher charge density at the bottom of smaller trenches results in higher probabilities of incoming charged reactive species being repelled, which acts to reduce or even prohibit further etch actions eventually.

Such charge accumulation effects also come into play in another circumstance—notching of device silicon at the device and buried cladding oxide interface. This critical RIE (reactive ion etching) effect is an important consideration in SOI device etching. An SEM of this type of etch problem is shown in Figure 10.33a. Here, there is severe undercutting of the silicon PhC near the bottom of the device at the interface of the buried oxide layer. This is due to the accumulation of positively charged RIE species, pulled down deep onto the etch surface by the vertical platen bias power.

Such residual charge accumulation on etch surfaces acts to deflect further incoming positive species when etching proceeds until the etch-stop buried-oxide charge insulating layer is reached. This results in the lateral etch of the PhC structures near the device-cladding interface, as the etch species are deflected onto the sidewalls, which can be seen in Figure 10.33b. In the case of mild notching, the performance of the fabricated PhC structures would be adversely affected. However, if notching due to overetching was allowed to become extensive, then total destruction of the designed device would result.

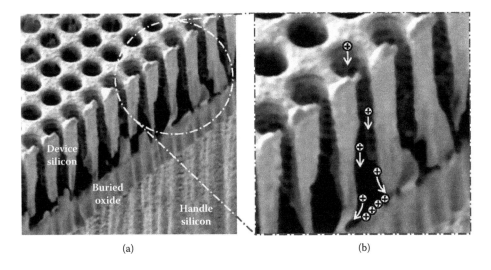

(a)   (b)

FIGURE 10.33
Notching of a silicon-on-insulator wafer, as shown in the (a) SEM of photonic crystal holes array etched into silicon and "etch-stopped" at the cladding buried oxide interface and (b) schematic illustration of charge buildup effect causing further incoming ions to be deflected onto etch sidewalls.

To prevent this problem, the process was designed as described in Section 10.2 such that only similar etch opening sizes and etch depths were required for each etch step. In this way, drastic etch-lag effects would not result in the undercutting of larger etch features at the etch-stop layer due to the overetch time required to complete etching for the narrowest openings. Otherwise, the use of the periodic discharging technique to remove accumulated charges leading to side-walls-deflected etch species need to be used, which is another tool-related method to ameliorate notching effects.

10.5.5 Etch Postprocessing

In the postprocessing of PhC and OTS devices, thermal oxidation was performed to improve the optical quality of device sidewalls through smoothening of the sharp edges. To enable the high-temperature furnace process, a total strip of CAR was necessary to avoid organic contamination of the oxidation chamber. Through a wet clean recipe of piranha bath consisting of sulfuric acid (H_2SO_4) and hydrogen peroxide (H_2O_2) at the high temperature of 130°C, the polymers and remaining photoresist leftovers in the lithographic and etch process were removed and wafers were inspected under a defect review SEM with energy dispersive x-ray analysis to ensure complete stripping. The cleaned wafers were then oxidized in the furnace at 1050°C for 90 min, consuming silicon at a rate of about 0.4 times the rate of oxidation. For oxidation of 1000 Å thickness, the sidewall edges of the time-multiplexed DRIE-etched sidewalls (Figure 10.34a) were then rounded with reduced scallop depths (Figure 10.34b). Thus, the optical quality of the fabricated device was improved, and optical scattering due to sidewall roughness was reduced.

Next, in the integration of electrode metallization, copper (Cu) seed deposition was carried out first using physical vapor deposition (PVD) for a conformal film of 250 Å Tantalum (Ta), before the actual deposition of Cu on top of the Ta layer, which proceeded until a conformal coating was formed. Following seeding, the actual Cu film thickness required was then applied by electrochemical plating (ECP). In the plating cell, the Cu plating

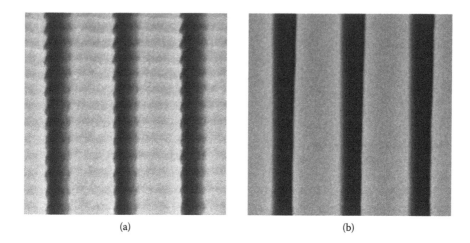

(a)   (b)

FIGURE 10.34
Postprocessing for (a) device sidewalls with undulating, time-multiplexed sidewall scallop roughness. After thermal oxidation, (b) the sharp edges of roughness on device sidewalls are visibly reduced.

process was carried out with a copper sulphate ($CuSO_4$), H_2SO_4, and distilled (DI) water solution bath. After plating, the wafer surface was cleaned by DI water spray in the post-plating module. Postannealing at 200°C was then carried out before the Cu film underwent CMP to achieve the planarized stack thickness required and also to remove unwanted Cu over areas that should not have been coated. This planarizing step enables subsequent processing of high-resolution lithography after another USG (undoped silicate glass) deposition was carried out to cover over the metal electrodes (thereby preventing elemental contamination in subsequent processing).

Upon completion of subsequent clean-room processes, the defined Cu electrodes were exposed in the final phase using BOE (buffered oxide etch) release, which removes silicon oxide at a rate of 2500 Å/min. The wafers released with the wet etch bath (comprising ammonium fluoride and hydrofluoric acid) are then cleaned with DI water.

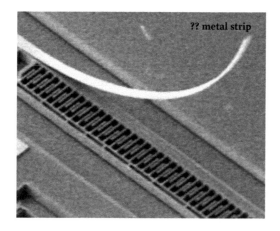

FIGURE 10.35
SEM of peeling of electrode copper metallization after buffered oxide etch and spin drying.

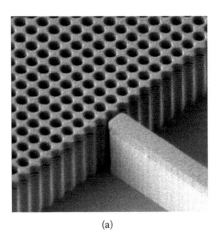 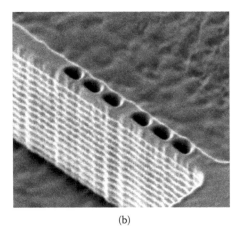

(a) (b)

FIGURE 10.36
(a) Photonic crystal of dense holes array together with fine waveguide structures prior to sidewall optimizations and (b) released monorail structure with focus ion-beam-etched photonic crystal holes.

Spin drying of clean wafers at a standard qualified recipe of 1800 rpm results in peeling of the thinner metal electrodes with less area of adhesion to the underlying device (Figure 10.35). To prevent such defects, two approaches may be taken: reduce the spin speed of the dryer to 300 rpm, or, alternatively, allow the wafers to be evaporated dry after an isopropyl alcohol (IPA) bath rinse without undergoing spin dry. Although the latter method was more time consuming and less precise, its advantage over spin drying is the reduced likelihood of the occurrence of stiction (the permanent sticking of released movable structures).

Thus, following sidewall optimization and buried-oxide release, the movable structures—the PhC air bridge structure and OTS—were realized. A final FIB milling was then carried out on the CD of a monorail PhC waveguide as shown in Figure 10.36a. Here, the FIB system uses liquid metal ion sources to form very small probes with high current densities. Where such high-density ion beam strikes the required points on the silicon air bridge, the silicon material was removed through a physical sputtering process. To carry out FIB milling on a CD structure, it was important that the surface oxide was removed prior to processing so that the effects of charging could be minimized for sharper SEM imaging and therefore more precise spatial milling. Optimized conditions were found for a magnification of 35k times and beam duration of 8 s for each PhC hole milling. For the higher order of magnifications used, greater imaging precision was adversely affected by the increased specimen charging effect that reduced the resolution of the patterned PhC holes The FIB-etched PhC holes on the monorail waveguide structure are as shown in Figure 10.36b, completing the synthesized processing of disparate PhC structures and the etch-depth process.

10.6 Summary

Based on the identified advantage of batch processing in OL and DRIE, insights into the optimization of these techniques have been presented. Here, simplified models for OPC were derived heuristically for both semiisolated and isolated defect structures, allowing for the enhancement of resolution without the use of expensive PSM or high-end compensation computation technique. At the same time, the DRIE experiments revealed many interesting effects, which were previously unemphasized because they were quite negligible for larger-dimensioned structures (such as that of the amplified etch angle, process time effect on lateral etch and sidewall scallops, etc). By careful consideration of such experimental effects, HAR (45–57) deep-submicron-sized rods with small scallop depths (12 nm) are the deepest and smoothest PhC by far achieved by DRIE technique used at the optical communication wavelength of 1.55 μm.

To conclude, in the fabrication of PhC structures with disparate patterns and device etch depths, many techniques were developed to overcome the unavoidable challenges. The foremost include the design of a feasible processing flow that allowed for both PhC CD definition and etch together with actuator electrode metallization and also suspended released structures. Through the investigation of each process module to satisfy the demands of the device-operating specifications (such as the PhC device requirement to operate at canonical optical communication wavelengths), the insights derived were applied to each processing module.

In the first part of the process, an alternative method for the achievement of the sublithographic wavelength line-width patterning, not commonly used in RETs such as PSM, OAI, and OPC, was derived. Using low lithographic system requirements, such as that of a DUV scanner working at a wavelength of 248 nm, CAR with BARC, and conventional chrome-on-glass photomasks, without the expensive phase-shift features or OAI systems, sublithographic wavelength CDs were obtained with the aid of pattern and partial coherence control. Here, based on the key factors of pattern density bias and imaging system partial coherence variations, the smallest resolution of 36% of mask value could be obtained for a resolution of 75 nm using a 200-nm binary photomask pattern. Although the limit in the reduction of design line widths and types of pattern together with constricted patterning at the line/base interface by line-end necking are the disadvantages of this technique, its advantages are its simplicity and cost effectiveness in enhancing resolution without the use of expensive or high-end peripherals, by means of simplified mask-making requirements and reduced demand on exposure hardware sophistication. Such a derived method may be used not only for PhC but also for applications such as metal–semiconductor–metal photodetectors, nanoelectromechanical structures, etc.

At the same time, besides a description of the issue of resist poisoning in the patterning experiments, etch-lag effects, the notching of devices at the interface of device silicon and buried-oxide layer was also presented for the simultaneous implementation of PhC with optical testing structures. The techniques of FIB milling in the finalization step of the processing following buried-oxide release for PhC structures on CDs enabled a wide processing window for both lithography and etch phases, avoiding the severe loading effects caused by disparate pattern dimensions (such as etch lag and notching). Besides realizing deep-etching TMRIE for the wide OTS grooves, PhC structures were optimized through the delicate etch and passivation multiplexing balance, and also through thermal oxidation. By the careful development of each experiment process module, high-resolution tunable PhC for use at an optical communication wavelength of 1550 nm was yielded, such that a monolithically integrated tunable PhC was fabricated simultaneously in mixed densities of superdense PhC holes, dense line patterns, isolated waveguide structures, and ultrawide OTS grooves that were defined aligned to the deep-submicron-sized PhC devices with sub-100-nm resolution CDs.

References

1. Selin, H. G. Teo, Liu, A. Q., Sia, G. L., Lu, C., Singh, J., Yu, M. B., and Sun, H. Q., Deep UV lithography for pillar type nanophotonic crystal, *International Journal of Nanoscience*, 4, 559, 2005.
2. Wong, A. K., Ferguson, R., Mansfield, S., Molless, A., Samuels, D., Schuster, R., and Thomas, A., Level-specific lithography optimization for 1-Gb DRAM, *IEEE Transactions on Semiconductor Manufacturing*, 13, 76, 2000.
3. Aoyagi, M., Nakagawa, H., Sato, H., and Akoh, H., Precise patterning technique for Nb junctions using optical proximity correction, *IEEE Transactions on Semiconductor Manufacturing*, 11, 381, 2001.
4. Smith, S., McCallum, M., Walton, A. J., Stevenson, J. T. M., and Lissimore, A., Comparison of electrical and SEM CD measurements on binary and alternating aperture phase-shifting masks, *IEEE Transactions on Semiconductor Manufacturing*, Vol. 16, no. 2, pp. 266–272, 2003.
5. Smith, B. W., Fan, Y. F., Zhou, J. M., Bourov, A., Zavyalova, L., Lafferty, N., Cropanese, F., and Estroff, A., Hyper NA water immersion lithography at 193 nm and 248 nm, *Journal of Vacuum Science and Technology B*, 22, 3439, 2004.
6. Cheng, M., Yuan, L., Croffie, E., and Neuretuther, A., Improving resist resolution and sensitivity via electric-field enhanced post exposure baking, *Journal of Vacuum Science and Technology B*, 20, 734, 2002.
7. Montgomery, D. C. and Runger, G. C., *Applied Statistics and Probability for Engineers*, 3rd edition, John Wiley & Sons, New York, 2003.
8. Bosch Gmbh, R. B., U.S. Pat. 4855017, 4784720, 1994.
9. Sin, C.-Y., Chen, B.-H., Loh, W. L., Yu, J., Yelehanka, P., See, A., and Chan, L., Resist trimming in high-density CF4/O2 plasmas for sub-0.1 um device fabrication, *Journal of Vacuum Science Technology B*, 20, 1974, 2002.
10. Meade, R. D., Rappe, A. M., Brommer, K. D., and Joannopoulos, J. D., Nature of the photonic band gap: some insights from a field analysis, *Journal of Optical Society of America B*, 10, 328, 1993.
11. Li, Z. Y., Gu, B. Y., and Yang, G. Z., Large Absolute Band Gap in 2D Anisotropic Photonic Crystals, *Physical Review Letters*, 81, 2574, 1998.

11

Control Strategies for Electrostatic MEMS Devices

Bruno Borovic and Frank L. Lewis

CONTENTS

11.1 Introduction 430
11.2 MEMS Optical Switch...431
11.3 Mathematical Modeling of Electrostatic Actuators................................432
      11.3.1 Mechanical Model 432
      11.3.2 Electrostatic Model...435
      11.3.3 Optical Intensity Model..436
11.4 Open-Loop and Closed-Loop Control 437
      11.4.1 Open-Loop Control 437
      11.4.2 Closed-Loop Control...438
      11.4.3 Discussion: Comparison and Issues 441
11.5 Summary...444
References 445

Increasing demands on the dynamical behavior of MEMS devices are reaching a point where mechanical design and simple-signal, open-loop driving cannot provide further improvements. Alternative approaches based on control theory, such as open-loop or closed-loop driving strategies, must be used instead to provide further enhancements in device performance. Simple input signals can be made more complex to fit MEMS dynamics better, therefore yielding better dynamic performance. Such control techniques are commonly referred to as preshaped open-loop driving and control. The ultimate step in improving precision and speed of response is use of feedback or closed-loop control. Unlike macromechanical systems, where the implementation of the feedback is relatively simple, in the MEMS case the feedback design may be quite problematic. Limited availability of sensor data, presence of sensor dynamics and noise, and the typically fast actuator dynamics all pose challenge in MEMS feedback design. The purpose of this chapter is to explore both open- and closed-loop strategies and to address the comparative issues of driving and control for MEMS devices. An optical MEMS switching device is used as an example in this study. On the basis of both experimental results and computer simulations, advantages and disadvantages of the different control strategies are discussed.

11.1 Introduction

MEMS devices are mechanically operated and, therefore, require an actuator to move them. Five basic on-chip actuators technologies have been developed [1]: magnetic, piezo-electric, thermal, optical, and electrostatic. Regardless of the applied actuation technique, MEMS devices are typically driven directly in an open-loop fashion by applying simple input control signals. Straightforward and simple actuation provides the MEMS designer with the improved device designs as a single choice to achieving better dynamical behavior. Hence, MEMS actuators have traditionally been gradually modified and improved in terms of mechanical design, suitable open-loop driving signals, and better area-efficiency [2–4].

On the other hand, the requirements for better dynamical behavior of the MEMS devices in terms of both speed of response and precision have resulted in the gradual introduction of improved actuation approaches. If the simple input signal is made more complex, taking care of the system dynamics, the approach results in the so-called "preshaped control" [5–7]. The dynamic model of the device is used to construct a preshaped input signal that enables the device to achieve better and faster dynamical performance.

However, preshaped actuation schemes are sometimes not enough. The lack of accurate models, fabrication inconsistencies, and lack of repeatability of the device parameters, compounded by special requirements on the dynamical behavior, all call for the use of closed-loop control design [7–12]. The first MEMS devices incorporating feedback were closed-loop sensors, with the objective of enhancing measurement accuracy [1]. An increase in complexity, device integration, and sophistication level of MEMS devices demands equally sophisticated integrated control systems. Unlike macromechanical systems where the implementation of the feedback is relatively simple, it is quite problematic in the MEMS case. The presence of sensor dynamics, fast high-frequency system dynamics, and requirements for the integration of the control system on the actual MEMS device have introduced additional challenges for feedback control design.

Both input shaping and closed-loop approaches significantly improve the dynamical behavior of MEMS and both strategies have their own advantages and disadvantages. The choice of driving strategy depends on several factors—the purpose of the device, complexity of the sensor implementation, available space, complexity of the electronic circuitry, dynamics of the device, and sensitivity of the dynamical response to the device parameters.

The purpose of this chapter is to compare and contrast open-loop and closed-loop design of MEMS control systems, detailing the design issues and choices. The experimental results, obtained from implementation of the preshaped open-loop and closed-loop control methods, were used to compare the two approaches and point out their advantages and disadvantages. It is found that with minimum additional implementation complexity, the closed-loop approach speeds up the system and improves its dynamical response. Criteria for choosing between the two control approaches is also established. As a case study, an optical MEMS device actuated by the electrostatic comb drive was used. The actuator shuttle has a light-modulating shutter attached to it. Optical feedback was used to reconstruct the position of the shuttle, which cannot be directly measured. The device can be used either as a variable optical attenuator (VOA) [13–15] or as an optical switch [9,14].

This chapter starts with the description of the device used as an example, followed by the development of an appropriate mathematical model. The second part discusses the open- and closed-loop approach and controller designs. At the end, some issues specific to the small-scale feedback are discussed.

11.2 MEMS Optical Switch

A MEMS optical switch, used as a platform on which the control strategies are to be discussed is shown in Figure 11.1. Detailed geometry of the device is given in Figure 11.2 [14]. The device was fabricated using deep reactive ion etching (DRIE) [16] on SOI wafers with a 75-μm-thick structural layer.

The device consists of two electrostatic comb-drive actuators: a suspension mechanism (the body of the device is also called a shuttle) and a shutter. A voltage applied to the comb-drive actuator generates a force that moves the shuttle. The shutter, which is attached to the shuttle, then cuts and modulates a light beam.

The shuttle consists of the shutter and backcone, 508 μm and 1430 μm long, respectively. Support frames are made lighter by creating cavities in the structures (see Figure 11.2). The widths of all features of the shuttle are 2 μm. The shutter itself is 3 μm wide and 190 μm long. There are 158 fingers on the comb drive, each having a width of 2 μm and a length of 27 μm. The gap between the fingers is 2.5 μm, with an initial overlapping of 5 μm. The width of the folded beams of the suspension is 3 μm with lengths 862.5 μm and 854.5 μm for the outer and inner beams, respectively.

The interface to the device is shown in Figure 11.3. The voltage applied to the actuator is separated and distributed to both forward (V_f) and backward acting (V_b) combs. The movement of the actuator modulates the light generated by a laser diode, and the light is sensed by photodetector. The voltage from photodetector V_{PD} is processed to determine the position x.

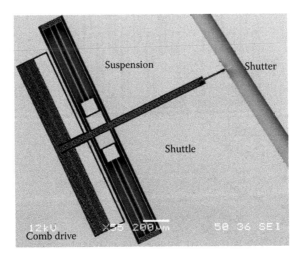

FIGURE 11.1
SEM image of the MEMS VOA device used as an example in this chapter.

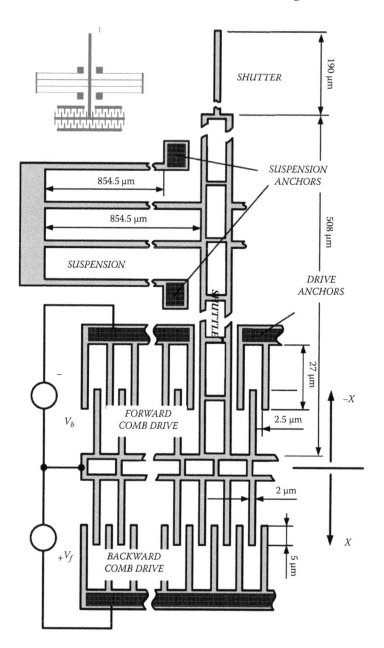

FIGURE. 11.2
Design and geometry of the VOA.

11.3 Mathematical Modeling of Electrostatic Actuators

11.3.1 Mechanical Model

The first step in modeling the mechanical part of the MEMS device is to determine its dynamics. The glimpse on the device from Figures 11.1 and 11.2 discloses that MEMS switch is used in one degree of freedom (DOF). Focusing only on the main DOF, the

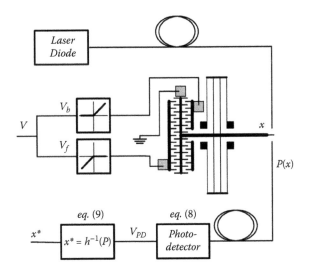

FIGURE 11.3
Interfacing the MEMS VOA.

device has effective mass, effective stiffness, and effective damping. All these parameters are lumped. This means it can be represented with the simplified model as shown in Figure 11.4.

The model contains effective moving mass of the switch in the x direction, effective stiffness, and effective damping. The switch is moved by applying force, F, to the mechanical structure. The governing differential equation for the system from Figure 11.4 is given as

$$m\frac{d^2x}{dt^2} + d\frac{dx}{dt} + kx = F \tag{11.1}$$

First, the effective moving mass has to be investigated, and to do that, one simple example is given. Large mass is suspended with two beams as shown in Figure 11.5. Thickness of the device is assumed to be uniform.

Motion of the device in x direction of the mass is observed. Deflected structure is also shown in Figure 11.5. The mass is assumed to be rigid, meaning all its parts deflect similarly. Each beam is chopped into three parts, and each part is assigned its deflection. Effective moving mass is given as

$$m_{eff} = \frac{1}{x^2}\left(m_{PM}x_{PM}^2 + m_{b1}x_{b1}^2 + \cdots + m_{b6}x_{b6}^2\right) \tag{11.2}$$

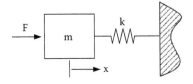

FIGURE 11.4
Simplified lumped model of the switch.

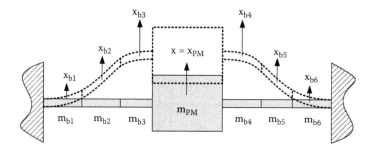

FIGURE 11.5
Main degree of freedom—effective mass and stiffness.

Choosing the main DOF deflection to be that of the mass, i.e., $x = xPM$, Equation 11.2 can be rewritten and effective moving mass calculated as

$$m_{eff} = m_{PM} + \left(\frac{x_{b1}}{x}\right)^2 m_{b1} + \cdots + \left(\frac{x_{b6}}{x}\right)^2 m_{b6} \qquad (11.3)$$

Effective stiffness can be calculated from the ratio of the force applied to the main DOF and deflection of the main DOF, x, as

$$k_{eff} = \frac{F}{x} \qquad (11.4)$$

If different x is chosen, the effective moving mass will be different. Therefore, the effective moving mass depends on the choice of the main DOF. However, the stiffness also changes to keep the characteristic frequency, i.e., modal frequency, constant

$$\omega = \sqrt{\frac{k_{eff1}}{m_{eff1}}} = \cdots = \sqrt{\frac{k_{effi}}{m_{effi}}} \qquad (11.5)$$

Or, more generally, Equation 11.2 can be rewritten as

$$m_{eff} = \frac{1}{x^2} \sum_{i=1}^{K} m_i x_i^2 \qquad (11.6)$$

Figure 11.5 represents one vibration mode of the device. Therefore, calculated mass in Equation 11.3 or the more generalized Equation 11.6 also represents effective modal mass. Effective modal stiffness is easily calculated from Equation 11.5. This is useful when lumped models are derived from, for instance, modal finite element analysis (FEA).

Using Equations 11.4 and 11.6, effective mass and stiffness can be determined from the shape of the mode, easily done by FEA. For examples given in Figures 11.1 and 11.2, the effective mass of the system turns out to be $m = 7.75 \times 10^{-9}$ kg. Stiffness is determined as $k = 1.05$ Nm^{-1}.

The last parameter in Equation 11.1 to be determined is damping. Considering the number of different complex mechanisms that cause it, including friction, viscous forces, and drag, etc., the discussion about damping is well beyond the scope of this chapter. For the device from Figure 11.1, damping was estimated experimentally. Assuming damping

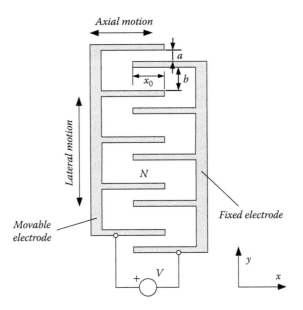

FIGURE 11.6
Capacitance of the electrostatic actuator.

coefficient to be constant, it turns out to be $d_0 = 8 \times 10^{-5}\ kgs^{-1}$. For interested readers, more detailed discussion on damping is provided in [17,18].

11.3.2 Electrostatic Model

Generated force of the comb structure depends on the structure of the capacitor. To derive force between two electrodes, the model of basic comb-drive unit is shown in Figure 11.6. The capacitance, as a function of position, x, is calculated as a sum of parallel capacitances among pairs of comb electrodes and is given as

$$C(x) = 2\varepsilon_0 Tn \frac{(x_0 + x)}{d_G} \tag{11.7}$$

where n is the number of finger electrodes of the each comb drive, ε_0 is the permittivity of the vacuum, d_G is the gap between fingers, T is the thickness of the structure, and x_0 is initial overlapping between the fingers. The capacitance of the comb drive calculated at the rest position is $C(0) = 0.42\ pF$. It increases as shuttle moves forward and decreases as it moves backward.

The partial derivatives of the capacitance calculated by Equation 11.7 with respect to x and y are given as

$$\frac{\partial C}{\partial x} = \frac{2\varepsilon_0 Tn}{d_G} \tag{11.8}$$

If voltage V is applied between two electrodes, the force generated in x and y direction is given as

$$F_x = \frac{1}{2} \frac{\partial C}{\partial x} V^2 = \frac{\varepsilon_0 Tn}{d_G} V^2 = k_e V^2 \tag{11.9}$$

where k_e is electrostatic constant. Calculated for the device from Figure 11.2, the value of the electrostatic constant is 41 nNV^{-2}. Note that the electrostatic force of the comb drive does not depend on its deflection. This is a typical property of comb drives.

The electrostatic constant in Equation 11.9 can be verified by conducting static experiments, i.e., $\frac{d^2x}{dt^2} = \frac{dx}{dt} = 0$). Static conditions reduce Equations 11.1 and 11.9 to $x = (k_e/k)V_f^2$ yielding the experimental result 81 nNV^{-2}! The analytically obtained value of k_e/k turns out to be half of the experimental value, i.e., -0.039 μmV^{-2}. As the stiffness represented by Equation 11.4 can be determined very accurately, the calculation for k_e in Equation 11.9 seems to be inaccurate. The reason for this can be attributed to the finite aspect ratio of the silicon structure fabricated by DRIE, which may increase the value of the capacitance calculated by Equation 11.5 several times [19]. Consequently, the experimental results for electrostatic is used in this chapter.

The MEMS device discussed here actually has two comb-drive actuators generating force in opposite direction. This modifies Equation 11.9 as

$$F_x = k_e\left(V_f^2 - V_b^2\right) \tag{11.10}$$

where V_f and V_b are the voltages applied to both forward- and backward-actuating electrodes, respectively. From the application point of view, these voltages are not supposed to act at the same time. Therefore, if these voltages are assumed to be $V_f \geq 0$ and $V_b \leq 0$, and they do not overlap in time, i.e., when $V_f > 0 \Rightarrow V_b = 0$ and $V_b < 0 \Rightarrow V_f = 0$, voltage $V = V_f + V_b$ can be defined. Equation 11.10 can be rewritten to accommodate proper orientation of generated forces as

$$F_x = k_e|V|V \tag{11.11}$$

This nomenclature is used further throughout this chapter. It is also worth mentioning that maximum static voltages that can be applied to the electrodes before they exhibit lateral pull-in are approximately $V = V_b = 10$ V.

11.3.3 Optical Intensity Model

In general, the position measurement can be implemented either through capacitive sensing or optical sensing. In this example, the optical sensing is used, and a model that relates deflection to the light intensity is developed (see Figure 11.7).

As it was mentioned before, the light beam is intercepted by the shutter, increasing and decreasing the throughput of the light. Analytical techniques were developed to theoretically determine the relationship between optical intensity and displacement in [7,9,13]. However, due to dissimilarities between the predicted model and an actual experiment, further discussion relies on the experimental results. The experimentally determined relationship, $V_{PD} = h(x)$, is given as

$$V_{PD}[V] = a_4x^4 + a_3x^3 + a_2x^2 + a_1x + a_0 \tag{11.12}$$

for $0 \leq x \leq 7$ μm, with the following parameters: $a_4 = 5.73 \times 10^{-5}$ $V\mu m^{-4}$, $a_3 = -0.0024$ $V\mu m^{-3}$, $a_2 = 0.018$ $V\mu m^{-2}$, $a_1 = 0.03$ $V\mu m^{-1}$, and $a_0 = -0.069$ V. The light intensity is linearly related to the V_{PD} with its minimum corresponding to $V_{PD} = -800$ mV and its maximum corresponding to 0 mV.

In order to reconstruct the position from the optical power, which is necessary to measure deflection, the inverse of Equation 11.12 has to be determined. The inverse of

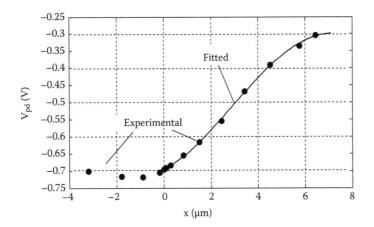

FIGURE 11.7
Position—optical intensity characteristics.

Equation 11.12 inside the interval $-0.7\,V \le V_{PD} \le -0.3\,V$, and for $x \ge 0$, can be approximated using a third order polynomial as

$$x^{*}[\mu m] = b_{3}V_{PD}^{3} + b_{2}V_{PD}^{2} + b_{1}V_{PD} + b_{0} \tag{11.13}$$

with parameters determined as $b_3 = 116\ \mu mV^{-3}$, $b_2 = 175.5\ \mu mV^{-2}$, $b_1 = 100\ \mu mV^{-1}$, and $b_0 = 23.6\ \mu m$.

Summarized, the model of the MEMS VOA is given as

$$m\frac{d^2x}{dt^2} + d\frac{dx}{dt} + kx = k_e|V|V \tag{11.14}$$

$$V_{PD} = h(x) \tag{11.15}$$

with values of parameters for the particular device discussed following Equations 11.1, 11.2, 11.4, 11.11, and 11.12.

11.4 Open-Loop and Closed-Loop Control

11.4.1 Open-Loop Control

Direct open-loop driving of the actuator is straightforward. An applied voltage step causes deflection of the actuator. The resulting step response is shown in Figure 11.8. The rise time is 190 μm and the overshoot is 17%. The settling time is approximately 550 μs.

Next, the simple step input signal is modified resulting in preshaped open-loop driving. The idea behind preshaping is to obtain a faster, aperiodic dynamic response. Here, different voltage pulses are combined to obtain a signal with a high voltage spike at the beginning and a trailing steady-state voltage as shown in Figure 11.9. A zero-voltage period exists between the initial spike and the steady-state voltage. The signal is defined by the amplitude of the initial spike, the steady-state voltage, and the values of the three triggering instants. Triggering instants define the beginning and the end of the initial spike as well as the beginning of the steady-state voltage value.

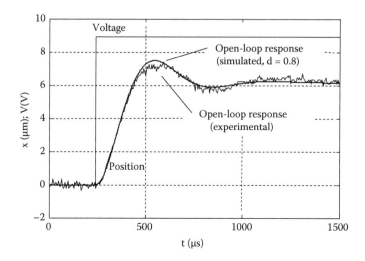

FIGURE 11.8
Open-loop response of the simulated model by Equations 11.14 and 11.15, compared with the experimental results.

For a specified steady-state voltage and a given maximum amplitude of the voltage spike, this input signal will give the fastest possible aperiodic response. The rise time of the response is determined by the difference between the third triggering instant. Detailed discussion of a similar signal shaping technique is given in Reference 20.

The simulated results for different amplitudes of initial spikes are shown in Figure 11.9. To speed up the response, the amplitude of the initial spike is increased. Consequently, the responses grow faster and all triggering instants are moved closer to the first one. The increase in rise time with respect to the amplitude of the initial spike is larger at low voltages than it is for higher voltages. On the other hand, when both the amplitude of the spike and the steady-state voltage are the same (e.g., 8.9 V), the rise time is the same as the direct open-loop's rise time; however, the response is aperiodic. The advantage of uniform voltage levels is that they enable simple implementation.

Experimental results are compared with those obtained from simulations and are given in Figure 11.10. The simulated open-loop step response is shown for reference. Both simulated and experimental results are well matched, having a similar rise time (i.e., ~100 μs). As can be seen, unexpected residual oscillations are present in the experimental response.

To suppress residual oscillations while keeping the faster rise time, the preshaped input signal was reshaped once again by adding two more multivibrators. The results are shown in Figure 11.11. Unfortunately, the situation was not significantly improved and the oscillations were not eliminated.

11.4.2 Closed-Loop Control

The next step is the implementation of the feedback controller. The controller contains feed-forward (FF) and feedback proportional derivative (PD) loop, and is shown in Figure 11.12. The detailed design of the controller is provided in Reference 9. Realization was accomplished using a fast control prototyping dSpace system. The sampling time was

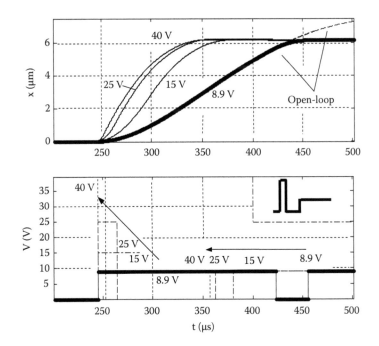

FIGURE 11.9
Preshaped open-loop response. All signals have the same final voltage level (simulation results).

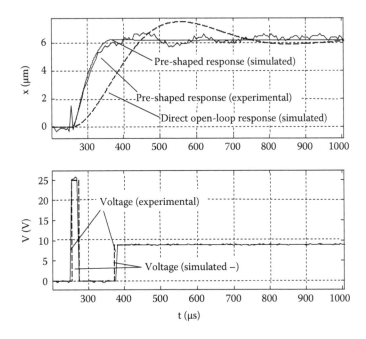

FIGURE 11.10
Preshaped open-loop response—different voltage levels and dependence of the speed on maximal available voltage (experimental and simulated results).

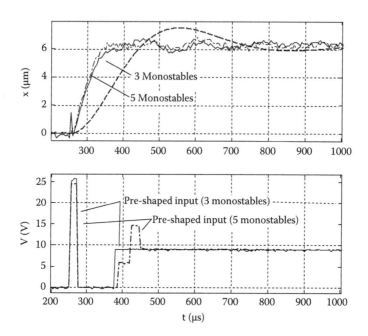

FIGURE 11.11
Preshaped open-loop response with and without the compensation of the higher modes (experimental results).

12 μs. The control action is given as

$$u_{ff} = (k/k_e)x_d u_{fb} + (1/N)\dot{u}_{fb} = K_P(x_d - x) + K_D(x)(\dot{x}_d - \dot{x}) \tag{11.16a}$$

$$u = u_{ff} + u_{fb} \qquad u \geq 0 \tag{11.16b}$$

$$V = \sqrt{u_{ff} + u_{fb}} \qquad 0 \leq V \leq 15V \tag{11.16c}$$

with $K_P = 3.2 \times 10^3$ and $K_D = 3.2 \times 10^7$.

The feed-forward gain ensures reaching the vicinity of the desired deflection. Proportional and derivative gains mitigate the remaining error, speed up the response and shape the signal, ensuring the aperiodic response. Experimental responses of the closed-loop system to the step input signal is shown in Figure 11.13. The open-loop response is also shown for comparison.

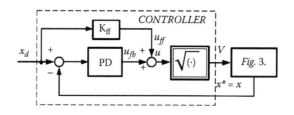

FIGURE 11.12
Controller structure.

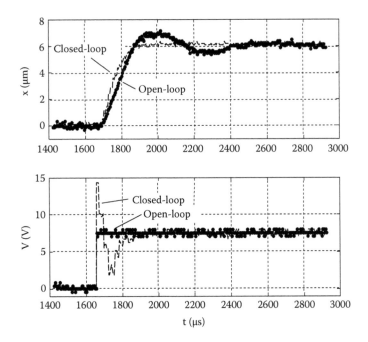

FIGURE 11.13
Open-loop and closed-loop responses and voltage corresponding to closed-loop response.

The rise time of the closed-loop system is around 170 μs, which is faster than the rise time of the open-loop step response (190 μs). However, it is slower than the rise time of the preshaped case (100 μs). The rise time in the closed-loop case is limited by the minimum achievable sampling time (12 μs). It is interesting to observe that there are no visible residual oscillations present in the closed-loop response. Typically, closed-loop control dramatically reduces the system's sensitivity to vibrations [21].

11.4.3 Discussion: Comparison and Issues

Having accomplished the comparison between the preshaped open-loop and closed-loop approaches, a meaningful discussion for choosing the driving approach can be presented. First, how the purpose of the device affects the choice is discussed. Next, the comparison in terms of parameter's sensitivity is given. Finally, the small scale feedback and sensing problems are addressed.

The light modulating device presented in this chapter can be used either as an optical switch or a variable optical attenuator. Optical switch mode requires switching the shutter position from fully closed to fully open and vice versa as quickly as possible, without an overshoot. As such, it does not require anything but a fast, aperiodic response. Therefore, the preshaped open-loop approach can be employed.

The results from Figure 11.14 show that fast response (100 μs) can be achieved with quite simple driving circuitry. The residual oscillations are not important if their amplitude is small enough, especially if it does not interfere with the light intensity.

On the other hand, the variable optical attenuator is typically used to condition the optical signal intensity after laser diodes, before fiber-optic amplifiers and photodetectors [15]. Therefore, in addition to fast, aperiodic response, the control of the light intensity requires

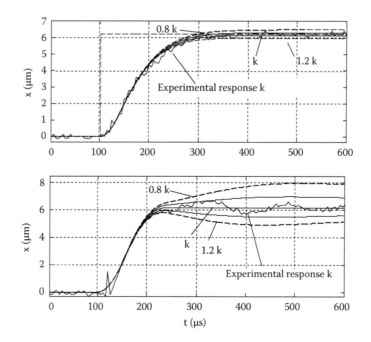

FIGURE 11.14
Sensitivity on change in stiffness, k represents the original stiffness of the device.

an accurate positioning of the modulating shutter. In this case, the closed-loop control is a better approach.

The next step is to compare the driving approaches in terms of the system's parameter sensitivity. A series of simulations was conducted, and the mass, damping, and stiffness were varied within their nominal values. Simulated results of nominal parameters' values and their variations are shown in Figures 11.14, 11.15, and 11.16. The experimental data for nominal values is also included for reference. It is obvious that the closed-loop driving renders much less sensitivity to the parameter changes.

Contemporary macroscale control systems are typically implemented with microcontrollers that execute a digital control algorithm. Sensors are relatively small in comparison to the controlled system and are quite easily implemented anywhere they need to be. Several factors make MEMS control systems unique. First, unlike macrosystems, MEMS systems are small and typically very fast. Second, the implementation of the sensor on the device can significantly change the size and dynamics of the device. Third, the whole control system should be integrated with the MEMS device and, therefore, should be as small as possible.

Response times of MEMS ranges from few microseconds for large DRIE fabricated thermal actuators to ranges for small surface electrostatic devices [17]. Conversion times for the standard D/A and A/D converters range from 1 to 10 μs. This excludes the application of the microcontroller for devices faster than 100 μs. The use of a microcontroller becomes questionable not only because of conversion times, but because it is too large to be integrated with MEMS devices. Due to the implementation size, the control algorithm should be kept as simple as possible. Control algorithms can be implemented as digital filters with sequential stages of multiplication and accumulation [21], or they can be implemented using analog techniques [7].

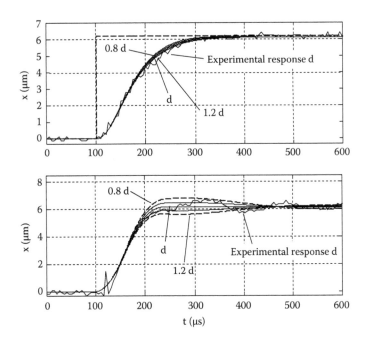

FIGURE 11.15
Sensitivity on change in damping, *d* represents the original damping of the device.

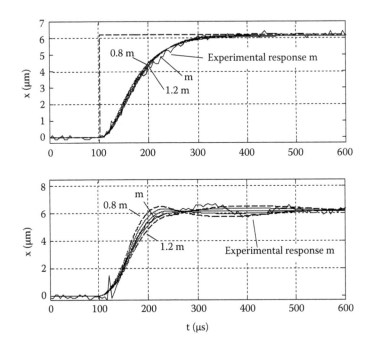

FIGURE 11.16
Sensitivity on change in mass; *m* represents the original mass of the variable optical attenuator.

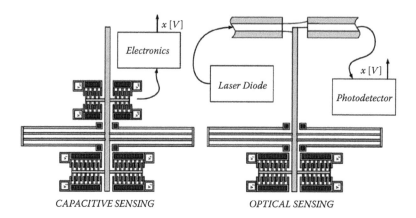

FIGURE 11.17
Difference between capacitive and optical sensing.

Position sensing makes the control system even more complicated. There are several ways to sense position. For the MEMS device described in this chapter, we implemented optical sensing, which, as a sensing approach, has several problems. First, the relationship between the actual position of the device and the output optical intensity is nonlinear. Moreover, small fiber misalignments can cause relatively large errors in the sensor output. Unstable light source can also add offset and noise to the measurement and influence it significantly.

Capacitive sensors are typically implemented as a differential capacitance [21]. As it is shown in Figure 11.17, the sensor becomes a part of the device. Electronic circuitry, converting the capacitance to a voltage and position, is attached to the sensor. A number of signal processing techniques have been developed [17–22] to extract the position from the measured capacitance. It is favorable to have as large of a capacitance as possible to get high resolution and high signal-to-noise ratio (SNR) [21]; however, the larger capacitor requires a larger area, which consequently increases the mass of the device slowing its response. Practically, however, the achievable values of microcapacitors are typically in the *fF–pF* range and, therefore, interfere with the value of the parasitic capacitance of the attached electronics (few *pF*) [21]. To mitigate the effects of these shunt capacitances it is desirable to integrate IC and MEMS devices together.

11.5 Summary

As a result of the analysis and experiments conducted for both open-loop and closed-loop control of MEMS, some conclusions regarding the performance of different control approaches can be drawn.

In terms of the complexity for the driving and sensing electronics, an open-loop approach has advantages over closed-loop control as it uses only driving circuits. On the other hand, open-loop driving is sensitive to parameter uncertainties and the shape of the input signal. The input voltage spikes have to be timed very precisely. For studied device, the accuracy of the triggering time is typically less than 0.1 μs. Faster responses and higher voltages require even higher precision. The closed-loop control approach is significantly less sensitive to changes in system parameters, and generates oscillation-free response.

In terms of application requirements, when a MEMS device is used for switching, only two signal levels are of interest, and the best way to drive it is using preshaped open-loop signals. However, if the actuator has to be accurately positioned between 0% and 100%, as in the case of variable optical attenuator, it is more suitable to use a closed-loop approach.

In conclusion, the choice of the control systems for MEMS depends on the available sensor, the size, and the speed of the device. The most difficult aspect of implementation is related to the hardware necessary for control, rather than the control algorithms. Finally, the control algorithms should be kept as simple as possible so they can be integrated directly in hardware with IC and optical components.

References

1. Bryzek, J., Abbott, E., Flannery, A., Cagle, D., and Maitan, J., Control issues for MEMS, *Proc. CDC, Maui, Hawaii, December 2003*, 2004.
2. Jensen, B.D., Mutlu, S., Miller, S., Kurabayashi, K., and Allen, J.J., Shaped comb fingers for tailored electromechanical restoring force, *J. Microelectromech. Syst.*, 12(3), 373, 2003.
3. Chen, C. and Lee, C., Design and modeling for comb drive actuator with enlarged static displacement, *Sens. Actuators A*, 115, 2–3, 177–627, 2004.
4. John, D.G., Jerman, H., and Thomas, W.K., Design of large deflection electrostatic actuators, *J. Microelectromech. Syst.*, 12(3), 335, 2003.
5. Popa, D.O., Wen, J.T., Stephanou, H.E., Skidmore, G., and Ellis, M., Dynamic modeling and input shaping for MEMS, *Tech. Proc. 2004 NSTI Nanotechnol. Conf. Trade Show (NANOTECH 2004)*, 2, 315, 2004.
6. Popa, D.O., Kang, B.H., Wen, J.T., Stephanou, H.E., Skidmore, G., and Geisberger, A., Dynamic modeling and input shaping of thermal bimorph actuators, *Proc. 2003 IEEE Internat. Conf. Robotics & Automation*, Taipei, Taiwan, 2003.
7. Borovic, B., Cai, H., Zhang, X.M., Liu, A.Q., and Lewis, F.L., Open vs. closed-loop control of the MEMS electrostatic comb drive, *13th Mediterranean Conf. Control and Automation, MED 2005*, Limassol, Cyprus, 2005.
8. Lu, M.S.-C. and Fedder, G.K., Position control of parallel-plate microactuators for probe-based data storage, *J. Microelectromech. Syst.*, 13(5), 759, 2004.
9. Borovic, B., Cai, H., Liu, A.Q., Xie, L., and Lewis, F.L., Control of a MEMS optical switch, *Int. Conf. Decision and Control, CDC 2004*, Nassau, Bahamas, 2004.
10. Sun, Y., Nelson, B.J., Potasek, D.P., and Enikov, E., A bulk microfabricated multi-axis capacitive cellular force sensor using transverse comb drives, *J. Micromech. Microeng.*, 12, 832, 2002.
11. Sun, Y., Piyabongkarn, D., Sezen, A., Nelson, B.J., and Rajamani, R., A high-aspect-ratio two-axis electrostatic microactuator with extended travel range, *Sens. Actuators A*, 102, 49, 2002.
12. Tang, W.C., Nguyen, T.H., and Howe, R.T., Laterally driven polysilicon resonant microstructures, *Sens. Actuators*, 20, 1–2, 25, 1989.
13. Liu, A.Q., Zhang, X.M., Lu, C., Wang, F., Lu, C., and Liu, Z.S., Optical and mechanical models for a variable optical attenuator using a micromirror drawbridge, *J. Micromech. Microeng.*, 13, 400, 2003.
14. Li, J., Zhang, Q.X., and Liu, A.Q., Advanced fiber optical switches using deep RIE (DRIE) fabrication, *Sens. Actuators A*, 102, 286–295, 2003.
15. Isamoto, K., Kato, K., Morosawa, A., Chong, C.H., Fujita, H., and Toshiyoshi, H., A 5-V-operated MEMS variable optical attenuator by SOI bulk micromachining, *IEEE J. Selected Top. Quantum Electron.*, 10(3), 570, 2004.
16. Klaasen, E., Petersen, K., Noworolski, J., Logan, J., Maluf, N., Brown, J., Storment, C., McCulley, W., and Kovacs, G., Silicon fusion bonding and deep reactive ion etching: a new technology for microstructures, *Sens. Actuators*, 52, 132, 1996.
17. Senturia, S.D., *Microsystem Design*, Kluwer Academic Publishers, Norwell, MA, 2001.

18. Elswenpoek, M. and Wiegerink, R., *Mechanical Microsensors*, Springer, Berlin, 2001.
19. Borovic, B., Liu, A.Q., Popa, D., Zhang, X.M., and Lewis, F.L., Lateral motion control of electrostatic comb drive: new methods in modeling and sensing, *Proc. 16th IASTED Int. Conf. Modeling and Simulation*, 301, 2005.
20. Lewis, F.L., Dawson, D.M., Lin, J., and Liu, K., Tank gun-pointing control with barrel flexibility effects, *ASME*, WAM, Atlanta, 1991.
21. Bryzek, J., Flannery, A., and Skurnik, D., Integrating microelectromechanical systems with integrated circuits, *Instrumentation & Measurement Magazine, IEEE*, 7(2), 51, 2004.
22. Seeger, J.I. and Boser, B.B., Charge control of parallel-plate, electrostatic actuators and the tip-in instability, *J. Microelectromech. Syst.*, 12(5), 656, 2003.

12

Control of Optical Devices

Bruno Borovic and Frank L. Lewis

CONTENTS

12.1   Introduction 447
12.2   MEMS VOA-Based Optical Waveform Generator 448
    12.2.1   Light Intensity Controller..449
    12.2.2   Experimental Results and Discussions 451
12.3   Electrostatic Comb-Drive Actuator—Lateral Instability................453
    12.3.1   Mechanical Model...454
    12.3.2   Electrostatic Model...455
    12.3.3   Extended Model for Lateral Sensor/Actuator........................457
    12.3.4   Control System Design... 461
12.4   Summary...464
References 464

Chapter 11 discussed in detail basic MEMS control and the feedback control systems that are used to improve dynamic performance of MEMS devices. This chapter focuses on two applications which take advantage of the feedback loop. The first one describes the way light intensity feedback improves the accuracy of the light output of the MEMS device. Applied this way, the light feedback allows designing a very accurate and fast variable optical attenuator-based optical waveform generator. The second example describes how the problem of lateral instability in electrostatic comb-drive actuators can be counteracted by application of the appropriate lateral feedback loop. Both examples are based on the optical switch and the variable optical attenuator, which was analyzed in details in Chapter 11.

12.1   Introduction

Ability of MEMS devices to modulate light has been successfully used to design various types of variable optical attenuators. The light attenuation mechanism is depending on its approach and applications. However, all of them require motion of the light-interacting part of the MEMS device. The simplest VOA is based on the shutter, which moves into the light path, therefore attenuating its intensity, just like the device described in Chapter 11. Motion of the MEMS device is controlled by the voltage applied on its electrodes. Going a step ahead, one can talk about optical waveform generator, a device where output light intensity follows, in the ideal case, voltage waveform applied at the electrodes. What prevents ideal light-tracking are dynamics of the MEMS device, unstable light source, and light-interacting imperfections of the shutter. In order to get around these issues, and improve the quality of the out-coming light intensity signal, a simple light intensity tracking controller can be designed. It ensures more accurate and much faster waveform tracking, therefore yielding the useful optical instrument.

Functionality of optical MEMS devices, such as the VOA and optical switch, requires large deflections. The second example describes how the traveling range of the electrostatic comb actuator is extended by the use of lateral position feedback. The comb drivers inherently suffer from electromechanical instability called *lateral pull-in, side pull-in,* or sometimes *lateral instability.* Although fabricated to be perfectly symmetrical, the actuator's comb structure is always unbalanced, causing adjacent finger electrodes to contact each other when voltage-deflection conditions are favorable. Therefore, lateral instability effectively decreases the active traveling range of the actuator. The pull-in problem can be successfully counteracted by introducing the active feedback steering of the lateral motion. The control system requires additional lateral actuation and sensing functionality.

12.2 MEMS VOA-Based Optical Waveform Generator

The VOAs are typically used to control optical signal intensity (power) after laser diodes and before fiber-optic amplifiers and photodetectors [1–3]. Various types of MEMS VOA have been developed and an excellent overview is given in Reference 2 Typical architectures are a shutter (blade) insertion type, a rotating or sliding mirror type, and interferometry-based types.

Advantages of the VOA used for controlling light intensity are various. The direct tuning range of the laser source is typically less than 20 dB. In contrast, the tuning range of the MEMS VOA can go up to 40 dB, depending on the quality of the device. Moreover, the wavelength of the light source is dependent on its power causing unwanted frequency shifts or, commonly refereed to as chirping. Frequency shifting is unacceptable in optical communications, especially in dense wavelength division multiplexed (DWDM) systems. This problem does not exist when MEMS VOA is used. Laser sources also exhibit a large noise when used at low power levels because a large part of the light comes from the spontaneous emission. Again, this phenomenon is unacceptable in optical communications. In addition to not suffering from such drawbacks, an MEMS VOA combines a fast response time, precise attenuation control, and long-term repeatability, and is less expensive than the existing products. Such devices are very flexible regarding any data rate or protocol, therefore eliminating the need for costly and lossy optical–electrical–optical conversion.

Due to its wavelength independency and excellent attenuation properties, the MEMS VOA, configured as a signal generator, can be used as a tool to simulate variable network losses and many scenarios of network events such as adding and dropping of users, breakdowns in the network, etc. The simple idea of practical implementation of the optical waveform generator is shown in Figure 12.1. The light coming from the light source is modulated by the MEMS VOA. The VOA is controlled using a microcontroller. Feedback, in terms of light intensity, is provided by photodetector attached to the optical-coupler. Interface electronics provides adjustments of signals being exchanged between microcontroller, photodetector, and VOA. The waveform generator can be integrated as a multichip package, containing MEMS chip, microcontroller, interface electronics, and optical connection.

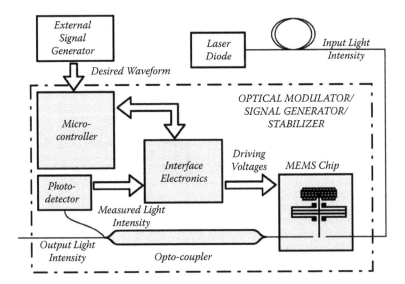

FIGURE 12.1
Light intensity waveform generator.

This light intensity waveform generator can be used to generate relatively accurate time-dependent signals with zero third and higher derivatives. It is an excellent tool to simulate variable network losses and many scenarios of network events such as adding or dropping of users, breakdowns in the network, etc.

The shutter-insertion type VOA is used as a platform for the design, implementation, and actual testing of a light intensity controller. Such feedback allows very accurate VOA's output signal that is independent on reconstruction of typically uncertain and complex relationship between position and light intensity. The purpose of the controller is two-pronged: to counteract the disturbances in light intensity, typically generated by the light source; and to make possible the generation of complex light intensity time-waveforms.

The experimental setup is shown in Figure 12.2. The voltages V_f and V_b are applied to the forward and backward actuators causing them to move. The movement of the actuator modulates the light generated by laser diode and the light is sensed by photodetector. The voltage from photodetector V_{PD} is processed to determine the deflection x and the light intensity P.

12.2.1 Light Intensity Controller

A hierarchical feedback controller, shown in Figure 12.3, is designed herein consisting of an outer, light intensity control loop and an inner, position control loop. The controller design is based on a simple but adequate mathematical model of the MEMS actuator as detailed in Chapter 11. For the purpose of analysis of the light intensity behavior, it is assumed that position loop is closed and its dynamics is described as the first order differential equation

$$\frac{dx}{dt} = (1/T_{pos})(K_{pos}x_d - x^*) \tag{12.1}$$

with unity gain and time constant being approximately 80 μs. The output of the system of Equation 12.1 is a voltage of the photodetector V_{PD}, as shown in Figure 12.2. It is generally

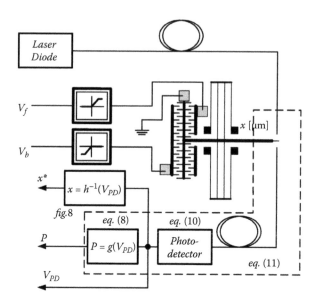

FIGURE 12.2
Experimental setup of the VOA.

nonlinear function of the shutter position $V_{PD} = h(x)$. This voltage is linearly related to the measured light intensity P.

To control the light intensity, a simple PID controller can be used

$$x_d = K_{LP}e + K_{LD}\frac{de}{dt} + K_{LI}\int^t e(\tau)d\tau \qquad (12.2)$$

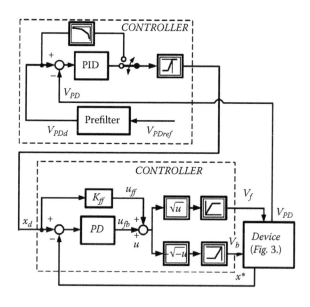

FIGURE. 12.3
Block scheme of the controller.

with light intensity error signal $e = V_{PDd} - V_{PD}$, where $V_{PDd} = V_{PDd}(t)$ is desired input waveform. This electrical waveform corresponds to the desired light intensity waveform.

Initial controller parameters are determined through any suitable linear controller design method. Here, a root locus method was used to determine the initial set of controller parameters. The parameters were further experimentally adjusted to meet requirements on the system dynamics. For the given example, the final set of controller parameters in Equation 12.2 are $K_{LP} = 0.475$, $K_{LD} = 17.6 \times 10^{-6} s$, and $K_{LI} = 17.2 \times 10^3 s^{-1}$.

It is often useful to filter the desired input signal to avoid high frequency components in the command signal path (i.e., V_{PDref}) of the tracking (servo) system [4]. The model, or pre-filter, helps ensure that the system is not driven too hard in response to command signal. Prefilter in a standard transfer function form [4] is given as

$$\frac{V_{PDd}(s)}{V_{PDref}(s)} = \frac{b_2 s^2 + b_1 s + b_0}{a_2 s^2 + a_1 s + a_0} \tag{12.3}$$

Parameters in Equation 12.3 are typically chosen so that Equation 12.3 is close to the desired closed system dynamics. Here, they were determined through simulation and their values are $a_0 = b_0 = 17.2 \times 10^3$, $b_1 = 0.475$, $a_1 = 1.475$, $b_2 = 1.7 \times 10^{-5}$, and $a_2 = 9.7 \times 10^{-5}$. The complete controller Equations 12.2 and 12.3 are shown in Figure 12.3.

12.2.2 Experimental Results and Discussions

The experimental analysis was done for two cases: with and without light intensity controller. The inner position controller was active in both cases, keeping the dynamics to be as shown in Equation 12.1. The light intensity of the light source was adjusted such that the maximal light intensity (i.e., 1.55 μW) corresponds to the maximal allowable opening of the VOA (i.e., -6.5 μm). Various desired light intensity waveforms were applied to the system in order to characterize it. Resulting waveforms were recorded. Results are represented in terms of light intensity P.

The prefiltered squared wave signal was applied to the system as a desired light intensity signal. The period of the signal wbas 1.4 μs and the desired light intensity varies between 0.82 μW and 1.25 μW. Desired and actual signals for controlled case are shown in top left part of the Figure 12.4, and the corresponding tracking error is shown right below. The uncontrolled signals and the tracking error are shown in the left bottom half of the Figure 12.4. The steady-state error is 10% in uncontrolled case and 0% in controlled case. Transitions errors are of the similar magnitude. Minimum achievable rise time 0%–90% is less than 200 μs. Light intensity control system significantly improves the accuracy of the waveform generator.

It is important to note that responses from the left part of Figure 12.4 correspond to the cases when only light intensity controller is either off or on. Position controller which speeds up the response of the system and mitigates overshoot and residual oscillations is turned on in both cases. Position control was discussed in Chapter 11 and is omitted here but reader should be aware that the open-loop response of the waveform generator, i.e., without either position or light intensity feedback loop, follows that in Figure 11.8. Improvements in speed and the shape of the response, as well as accuracy are significantly improved when light intensity is controlled.

The other example depicts the response of the system to the saw-like signal. The period of the signal is 45 μs. The desired signal alters between 0.85 μW and 1.22 μW. The results are shown on the right side in Figure 12.4. When the optical intensity controller is inactive, the average error is inside the 8% of the input light intensity range. When it is turned on, the tracking is perfect for the practical purpose.

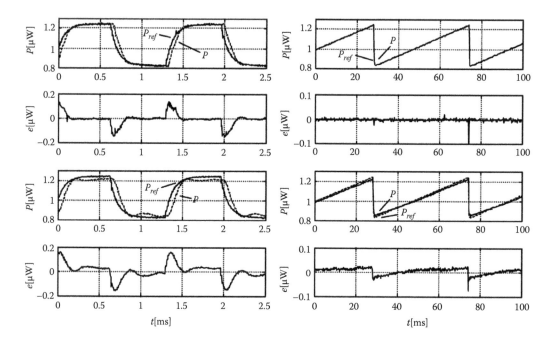

FIGURE 12.4
Desired and actual light intensity for squared signal (*left*), and desired and actual light intensity for saw-like signal of lower frequency.

Once designed, the mechanical structure of the device correlates damping, stiffness, and mass. Any change in the mechanical design affects all of them and there is no much space to adjust them. Ultimately, there exist optimal design and once it has been reached, there is no more space for improvements. The use of feedback provides additional freedom and ability to tune these parameters independently. It enables retuning the mass, stiffness, and damping in the artificial way. Consequently, much more can be achieved out of the given mechanical design.

This example depicts the simple and effective application of feedback control for MEMS device. An experimental setup and a practical system characterization of the light intensity control system for an optical waveform generator are given and the feedback control system is implemented on an actual MEMS VOA. The obtained results verify that the proposed control system significantly improves the dynamical behavior of the existing device. It can be assumed that the implementation of the light intensity feedback may improve accuracy of any VOA. Direct light feedback solves the problem of usually complex and uncertain relationship between position and light intensity. The price to be paid is implementation of the feedback itself. Additional information on the application of feedback with MEMS can be found in [5].

As it is, the light intensity controller used here is not perfect. Improved SISO nonlinear tracking controller based only on light intensity feedback enables a wider application range of the device.

The MEMS VOA with its excellent light modulation properties combined with the accuracy provided by controller, however, represents flexible and useful tool in analysis and testing of various optical networks.

12.3 Electrostatic Comb-Drive Actuator—Lateral Instability

One of the issues with the comb-drive design is achieving large deflections while mini-mizing the actuation voltage, resulting in a small deflection-to-size ratio of the actuator. These requirements are typically matched by balancing the design of actuator's suspen-sion and varying the size of the force-generating comb-drive structure. The comb driver, however, suffers from an electromechanical instability called lateral or side pull-in, or lat-eral instability. Electrostatic forces, perpendicular to the desired movement of the actuator, can get unbalanced causing the neighboring electrodes to contact each other when the voltage-deflection conditions are favorable. Weak suspensions and large forces, designed to achieve large traveling ranges, increase this problem even more. The example MEMS device actuated by comb-drive actuators is shown in Figure 12.5, and an example of the comb fingers in the state of lateral instability is shown in Figure 12.6.

The lateral instability occurs when the electrostatic stiffness transverse to the axial direc-tion of motion exceeds the transverse mechanical stiffness of the suspension. Therefore, the most common way to avoid it is by increasing the transverse stiffness of the suspension [6–10].

Another way to approach the lateral instability problem is use of feedback control. A requirement for doing so is to have a lateral motion sensing capability and an appropriate model of the device for the subsequent control system design. The introduction of the lat-eral feedback may impact the design of the comb drive, mitigating the requirements on the suspension, lowering the actuation voltage, and, therefore, decreasing the ratio between the size of the actuator and achievable deflection.

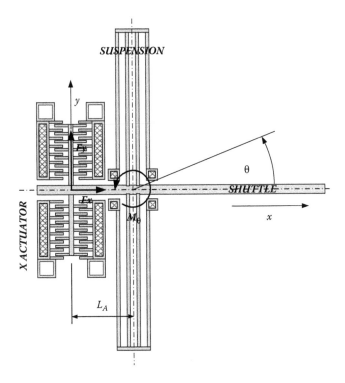

FIGURE 12.5
The structure of the combe drive with the shuttle. (From Borovic, B., *J. Micromech. Microeng.*, 15, 1917, 2005.)

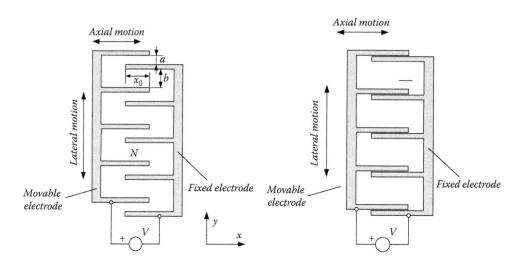

FIGURE 12.6
Electrostatic comb-drive actuator in the state of lateral instability.

To illustrate how lateral control system is implemented, the comb-drive model for the device from Figure 12.5, developed in Chapter 11, is extended to include the asymmetrical lateral DOF for lateral instability to be modeled. The device itself is then extended with both sensor and actuator functionality for generating lateral movement which is supposed to counteract lateral instability. Finally, these additional features are synchronized by the design of the controller for lateral motion which prevents lateral instability to occur.

12.3.1 Mechanical Model

The device shown in Figure 12.5 can be modeled as two second-order differential equations describing two degrees of freedom, axial displacement along x axis, and rotation around z axis. The dynamic model of the comb-drive actuator can be written as

$$m_x \frac{d^2x}{dt^2} + d_x \frac{dx}{dt} + k_x x = F_x \tag{12.4}$$

$$J_z \frac{d^2\theta}{dt^2} + d_\theta \frac{dx}{dt} + k_\theta \theta = M_z \tag{12.5}$$

where m_x and J_z are the effective moving mass along the x axis, and the effective moment of inertia around the z axes, respectively. Notations d_x and d_θ describe damping, and k_x and k_θ are the stiffnesses along the x axis and around the z axis, respectively.

It can be seen that θ is small during the operation of the device. In the other words, y direction movement of the comb structure is way smaller than the radius of rotation, i.e., $y \ll L_A$. Hence, we can approximate $tg\theta = y/L_A \approx \theta$. Therefore, Equation 12.5 is then expressed as

$$\frac{J_z}{L_A^2} \frac{d^2y}{dt^2} + \frac{d_\theta}{L_A^2} \frac{dy}{dt} + \frac{k_\theta}{L_A^2} y = -F_Y \tag{12.6}$$

Equations 12.4 and 12.6 represent the dynamics of the comb-drive actuator in two DOF.

Forces F_x and F_y in Equations 12.4 and 12.6, are determined as a contribution of all forces generated by comb drive as shown in Figure 12.6. As the structure in Figure 12.5 has four ($n = 4$) comb capacitances. Therefore,

$$F_x = \sum_n F_{xCn} = F_{xC1} + F_{xC2} + F_{xC3} + F_{xC4} \tag{12.7}$$

$$F_y = \sum_n F_{yCn} = F_{yC1} + F_{yC2} + F_{yC3} + F_{yC4} \tag{12.8}$$

where n is the number of comb capacitor units. Each force from Equations 12.7 and 12.8 can be determined by following the procedure described in Chapter 11.

For typical MEMS application, it is desirable that lateral y force from Equation 12.8 does not exist. However, because of various imperfections, it generally is not equal to zero. Lateral instability occurs if lateral force becomes large enough to overcome lateral stiffness.

12.3.2 Electrostatic Model

The structure of the comb-drive actuator is shown in Figure 12.7. One of the reasons for existence of the unsymmetrical lateral force is unsymmetrical geometry. The virtually symmetrical comb-drive actuator is made laterally unbalanced by introducing Δd. Figure 11.6 is recalled where the virtually symmetrical comb-drive actuator is modeled as unbalanced one by introducing Δd.

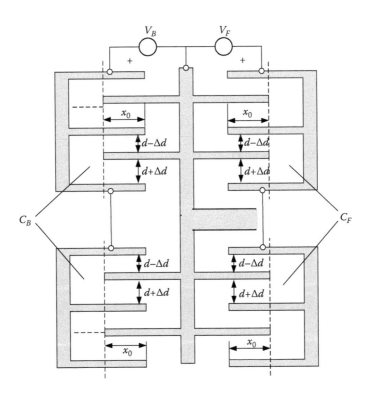

FIGURE 12.7
The comb drive with Δd used to model the unbalanced lateral geometry.

To determine total forces F_x and F_y from Equations 12.4 and 12.6, the similar procedure as the one from the previous chapter is recalled. The capacitances C_F and C_B are composed of two comb capacitance units as the one shown in Figure 12.6 and can be expressed as a function position x and y as:

$$C_F(x,y) = N\varepsilon_0 T(x+x_0)\left(\frac{1}{d-\Delta d-y}+\frac{1}{d+\Delta d+y}\right) \tag{12.9}$$

$$C_B(x,y) = N\varepsilon_0 T(-x+x_0)\left(\frac{1}{d-\Delta d-y}+\frac{1}{d+\Delta d+y}\right) \tag{12.10}$$

where C_F and C_B are forward- and backward-actuating capacitances, N is the number of finger electrodes of the each comb drive, ε_0 is the permittivity of the vacuum, T is the thickness of the structure, and x_0 is initial overlapping between the fingers. The force in x direction, F_x, is given as [11,12]

$$F_x = \frac{1}{2}\frac{\partial C_F}{\partial x}V_F^2 + \frac{1}{2}\frac{\partial C_B}{\partial x}V_B^2 \tag{12.11}$$

where V_F and V_B are the forward- and backward-driving voltages.

Partial derivatives $\frac{\partial C_F}{\partial x}$ and $\frac{\partial C_B}{\partial x}$ are calculated as

$$\frac{\partial C_F}{\partial x} = N\varepsilon_0 T\left(\frac{1}{d-\Delta d-y}+\frac{1}{d+\Delta d+y}\right) \tag{12.12}$$

$$\frac{\partial C_B}{\partial x} = -N\varepsilon_0 T\left(\frac{1}{d-\Delta d-y}+\frac{1}{d+\Delta d+y}\right) \tag{12.13}$$

And, finally, introducing Equations 12.12 and 12.13 into Equation 12.11, the resulting force in x direction as

$$F_x = \frac{1}{2}N\varepsilon_0 T\left(\frac{1}{d-\Delta d-y}+\frac{1}{d+\Delta d+y}\right)\left(V_F^2-V_B^2\right) \tag{12.14}$$

Similarly, the lateral force, F_y, is given as

$$F_y = \frac{1}{2}\frac{\partial C_F}{\partial y}V_F^2 + \frac{1}{2}\frac{\partial C_B}{\partial y}V_B^2 \tag{12.15}$$

Partial derivatives are

$$\frac{\partial C_F}{\partial y} = N\varepsilon_0 T(x_0+x)\left[\frac{1}{(d-\Delta d-y)^2}-\frac{1}{(d+\Delta d+y)^2}\right] \tag{12.16}$$

$$\frac{\partial C_F}{\partial y} = N\varepsilon_0 T(x_0-x)\left[\frac{1}{(d-\Delta d-y)^2}-\frac{1}{(d+\Delta d+y)^2}\right] \tag{12.17}$$

and, after substituting Equations 12.16 and 12.17 into Equation 12.15, the lateral force can be expressed as

$$F_y = \frac{1}{2}N\varepsilon_0 T\left[\frac{1}{(d-\Delta d-y)^2}-\frac{1}{(d+\Delta d+y)^2}\right]\left[x_0\left(V_F^2+V_B^2\right)+x\left(V_F^2-V_B^2\right)\right] \tag{12.18}$$

The lateral force exists only if y is not zero when $\Delta d = 0$. Equations 12.4, 12.6, 12.14, and 12.18 represent the dynamic model of the actuator in two DOF. The model is nonlinear and coupled through the generation of the electrostatic forces in Equations 12.14 and 12.18.

The model of the lateral pull-in behaviors are summarized as

$$m_x \frac{d^2x}{dt} + d_x \frac{dx}{dt} + k_x x = \eta F_x \tag{12.19}$$

$$\frac{J_z}{L_A^2} \frac{d^2y}{dt^2} + \frac{d_\theta}{L_A^2} \frac{dy}{dt} + \frac{k_\theta}{L_A^2} y = -\eta F_Y \tag{12.20}$$

where

$$F_x = \frac{1}{2} N \varepsilon_0 T \left(\frac{1}{d - \Delta d - y} + \frac{1}{d + \Delta d + y} \right) \left(V_F^2 - V_B^2 \right) \tag{12.21}$$

$$F_y = \frac{1}{2} N \varepsilon_0 T \left[\frac{1}{(d - \Delta d - y)^2} - \frac{1}{(d + \Delta d + y)^2} \right] \left[x_0 \left(V_F^2 + V_B^2 \right) + x \left(V_F^2 - V_B^2 \right) \right] \tag{12.22}$$

and the associated parameters are given as follows: $m_X = 7.75 \times 10^{-8}$ kg, $J_Z = 1.88 \times 10^{-15}$ kgm^2, $k_X = 1.08$ N/m, $k_\theta = 3.45 \times 10^{-6}$ Nm/rad, $L_A = 246$ μm, $d_X = 8 \times 10^{-5}$ kgs^{-1}, $d_\theta/(L_A)^2 = \eta = 2.07$, $\Delta d = 0.31$ μm, $N = 158$, $T = 75$ μm, $\varepsilon_0 = 8.854 \times 10^{-12}$, and $x_0 = 5$ μm.

The simulation results illustrating the dynamic behavior of the system at the edge of the lateral instability are shown in Figure 12.8. Notice that for transient conditions, the pull-in voltage may be slightly different from that for steady-state conditions.

12.3.3 Extended Model for Lateral Sensor/Actuator

With the model of the actual device developed in Section 12.3.2, the additional features, intended for lateral sensing and actuation, can be added to the device, as shown in Figure 12.9. For the lateral control analysis, these features are assumed not to have mass and damping. The detailed structure of both the lateral actuator and sensor are shown in Figure 12.10.

The lateral actuators contain top and bottom comb-drive structures designed to generate force in the y direction. These comb-drive structures are unbalanced with different gaps (a and b) between the electrodes, as shown enlarged in Figure 12.10. The maximum generated force occurs when the ratio between the smaller and the larger electrode gap is $a/b = 0.42$ [13]. Hence, the smaller gaps are defined by the minimum processing geometry, i.e., $a = 2.5$ μm. The larger gap is 6 μm wide.

The lateral sensor has the similar gap geometry to achieve maximum sensitivity. The number of fingers, N_s, and the initial electrode overlapping, x_{So}, may vary. Movable capacitors are connected to the bridge structure through serial capacitors C_s. Deflection in the y direction can be determined from the difference between voltages V_{sT} and V_{sB}. Because of the bridge structure of the sensor, any out-of-plane motion affects both sensing voltages equally, thereby canceling its influence with respect to lateral sensing. The structure of both the actuator and sensor ensures that no force is generated in x direction.

Referring to Figure 12.9, a layout of the MEMS device with comb actuator shown in Figure 12.5 with additional comb drives with actuator and sensor for lateral motion is shown. The lateral actuator is based on a four-element actuation structure that allows the generation of positive and negative forces in lateral direction. The lateral sensor is based on

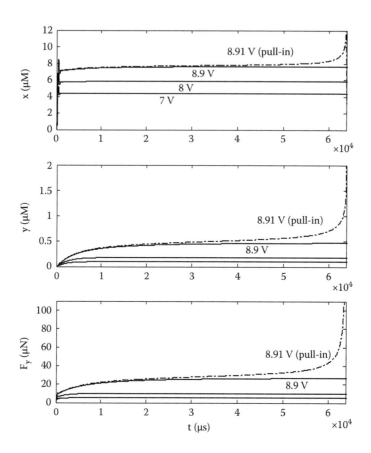

FIGURE 12.8
Verification of the model. Position response for different voltages applied. Simulations clearly show that the lateral pull-in occurs when 8.91 V is applied to the comb actuator. Maximal stable deflection is less than 8 μm.

a four-element bridge structure that allows the discrimination of positive and negative motion in lateral direction.

The lateral comb-like structures in Figure 12.9 includes both movable and fixed electrodes. Voltage V_{aT} applied to the actuating electrodes causes generation of the electrostatic forces and movement of the device. Motion is sensed by connecting the capacitive sensors to appropriately designed bridges and observing capacitance change on the fixed electrodes.

The lateral actuators include a top and bottom comb-drive structures designed to generate force in the y direction. These comb-drive structures are unbalanced with unequal gaps (a and b) between the electrodes, as shown in Figure 12.10. The maximum generated force per comb area occurs when the ratio between the smaller and the larger electrode gap, i.e., a and b is about 0.42. The smallest gaps are defined by the minimum processing geometry, e.g., if minimum processing gap is 2.5 μm, gap a is then about 2.5 μm wide and b is about 6 μm wide, although other dimensions can be used.

The lateral sensor has the similar gap geometry to achieve the maximum sensitivity. The number of fingers, N, and the initial electrode overlapping, x_0, may be different from the lateral actuator. Movable capacitors are connected to the bridge structure through serial capacitors. Deflection in the y direction can be determined from the difference between voltages. The structure of both the actuator and sensor ensures that no force is generated in x direction.

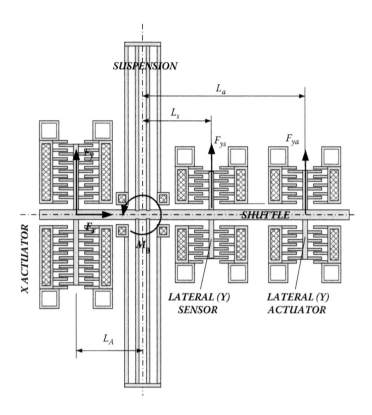

FIGURE 12.9
Hypothetical device for lateral DOF feedback control.

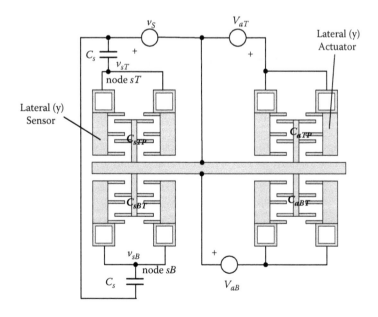

FIGURE 12.10
Schematics of lateral actuation and sensing.

The lateral actuator shown in Figure 12.9 is modeled using procedure, i.e., Equation 12.9 through Equation 12.18 that was used to model of the unbalanced comb actuator shown in Figure 12.5. Additionally, the force generated by lateral actuator and lateral sensor shown in Figure 12.9 may further be modeled in similar fashion. The capacitance of the two parallel top actuating capacitors, CTP, with respect to the x and y directions is given as

$$C_{TPa}(y) = 2\varepsilon_0 TN_a x_{a0}\left(\frac{1}{a-y}+\frac{1}{b+y}\right)$$ (12.23)

Similarly, the two parallel bottom capacitances, C_{BT}, are given as

$$C_{BTa}(y) = 2\varepsilon_0 TN_a x_{a0}\left(\frac{1}{a+y}+\frac{1}{b-y}\right)$$ (12.24)

Both capacitances in Equations 12.23 and 12.24 do not depend on x. Consequently, their contribution to the force along x axis does not exist. The unbalance coefficient Δd is omitted in Equations 12.23 and 12.24, because it is assumed that the sensing voltage is too low to influence lateral instability, and the actuation voltage is assumed to be the issue for the controller.

The total force in the lateral direction, F_{ya}, can be calculated using similar procedure as in Equations 12.18 and 12.22:

$$F_{ya} = \varepsilon_0 TN_a x_{a0}\left[\left(\frac{1}{(a-y)^2}-\frac{1}{(b+y)^2}\right)V_{aT}^2 +\left(\frac{1}{(b-y)^2}-\frac{1}{(a+y)^2}\right)V_{aB}^2\right]$$ (12.25)

Following a similar procedure, sensing comb capacitances and force generated by them are modeled as

$$C_{TPs}(y) = 2\varepsilon_0 TN_s x_{s0}\left(\frac{1}{a-y}+\frac{1}{b+y}\right)$$ (12.26)

$$C_{BTs}(y) = 2\varepsilon_0 TN_s x_{s0}\left(\frac{1}{a+y}+\frac{1}{b-y}\right)$$ (12.27)

$$F_{ys} = \varepsilon_0 TN_s x_{s0}\left[\left(\frac{1}{(a-y)^2}-\frac{1}{(b+y)^2}\right)V_{sT}^2 +\left(\frac{1}{(b-y)^2}-\frac{1}{(a+y)^2}\right)V_{sB}^2\right]$$ (12.28)

Now, with actuating and sensing forces in Equations 12.25 and 12.28 integrated into the model as expressed by Equations 12.4, 12.6, 12.14, and 12.18, for the device from Figure 12.9, summarized model is expressed as

$$\frac{J_z}{L_A^2}\frac{d^2y}{dt^2}+\frac{d_\theta}{L_A^2}\frac{dy}{dt}+\frac{k_\theta}{L_A^2}y = -F_Y+\frac{L_a}{L_A}F_{ya}+\frac{L_s}{L_A}F_{ys}$$ (12.29)

Additional parameters, necessary to conduct simulations and design of the controller, are: $a = 2.5\ \mu m$, $b = 6\ \mu m$, $N_A = N_S = 60$, $L_a = 246\ \mu m$, $L_s = 400\ \mu m$, and $x_{A0} = x_{S0} = 20\ \mu m$.

Additionally, the lateral sensor model shown in Figure 12.9 may further be modeled where the lateral position sensing shown in Figure 12.10 is based on the difference between capacitances C_{TPs} and C_{BTs} from Equations 12.26 and 12.27. As the purpose of the overall system is keeping lateral motion at 0 capacitance changes can be linearized around $y = 0$, and is given as

$$\Delta C_{Sense} = C_{TPs} - C_{BPs} = \left(\frac{\partial C_{TPs}}{\partial y}\right)_{y=0} \Delta y - \left(\frac{\partial C_{BPs}}{\partial y}\right)_{y=0} \Delta y \tag{12.30}$$

where

$$\left(\frac{\partial C_{TPs}}{\partial y}\right)_{y=0} = \left[2\varepsilon_0 TN_s x_{S0}\left(\frac{1}{(a-y)^2} - \frac{1}{(b+y)^2}\right)\right]_{y=0} = 2\varepsilon_0 TN_s x_{S0}\left(\frac{1}{a^2} - \frac{1}{b^2}\right) \tag{12.31}$$

and

$$\left(\frac{\partial C_{BTs}}{\partial y}\right)_{y=0} = \left[2\varepsilon_0 TN_s x_{S0}\left(\frac{1}{(b-y)^2} - \frac{1}{(a+y)^2}\right)\right]_{y=0} = 2\varepsilon_0 TN_s x_{S0}\left(\frac{1}{b^2} - \frac{1}{a^2}\right) \tag{12.32}$$

Introducing Equations 12.31 and 12.32 into Equation 12.30 yields

$$\Delta C_{Sense} = \frac{4\varepsilon_0 TN_s x_{S0}}{a^2 b^2}(b^2 - a^2)\Delta y = K_{C\Delta y}\Delta y \tag{12.33}$$

Equation 12.33 may be rewritten as $\Delta C_{Sense} = K_{C\Delta y}y$ assuming small y.

Differential change in capacitance can be sensed by properly designed voltage amplifier connected between V_{sT} and V_{sB} shown in Figure 12.10. Additionally, exactly the same sensor bridge configuration can be used with to measure by measuring the charge between sT and sB nodes with differential charge amplifier.

Sense bridge is designed in the way that all motions except the one in y direction (or around z axis) are rejected. Motion in x direction is cancelled by adding symmetrical structure pointing into negative direction of x. Equations 12.26 and 12.27 show that clearly. Any out-of-plane motion affects both top and bottom parts of the sensor in the same way therefore rejecting it. Sensitivity may change only for the case when Δy (or y) is relatively large and this does not happen because controller is assumed to keep lateral motion close to 0.

12.3.4 Control System Design

The purpose of the chapter is to describe a feedback approach that prevents lateral pull-in and extends the working range of the comb-drive actuator in the primary x direction DOF. The primary requirement for the controller is to keep $y = 0$. Additional shaping of the dynamics is also desirable. Detailed design of the controller is avoided due to available space, but the structure, parameters, and operational description are discussed.

The structure of the controller is shown in Figure 12.11. To simplify the analysis, a linear PID controller is implemented for the lateral DOF. We assume that the deflection y is measurable and available. When the lateral feedback loop is closed, the sensed value of y is compared to the referent $y = 0$, and the error signal is then passed through the controller. Saturation-type nonlinearities distribute voltages to the two channels leading to the left and right y direction comb-drive electrodes. The signal is then taken through the square root functions that take care of the electrostatic force being dependent on the squared value of the voltage.

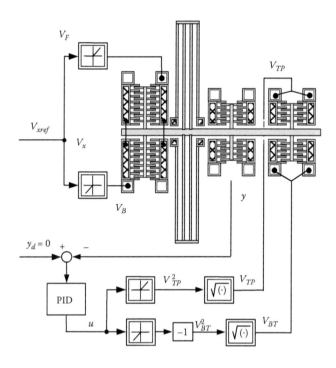

FIGURE 12.11
Lateral controller.

To keep the controller as simple as possible, the lateral degree of freedom as expressed by Equation 12.29 can be rewritten as

$$\frac{J_z}{L_A^2}\frac{d^2y}{dt^2} + \frac{d_\theta}{L_A^2}\frac{dy}{dt} + \frac{k_\theta}{L_A^2}y = \frac{L_a}{L_A}F_{ya} + F_D \tag{12.34}$$

where all lateral forces except the actuator one as a disturbance to the system, F_D are considered. Furthermore, for $y = 0$, Equation 12.25 becomes

$$F_{ya} = \varepsilon_0 T N_A x_{A0}\left(\frac{b^2-a^2}{a^2b^2}V_{aT}^2 + \frac{a^2-b^2}{a^2b^2}V_{aB}^2\right) \tag{12.35}$$

Merging Equations 12.34 and 12.35, and defining $k_{ab} = (b^2-a^2)/(ab)^2$, yields

$$\frac{J_z}{L_A^2}\frac{d^2y}{dt^2} + \frac{d_\theta}{L_A^2}\frac{dy}{dt} + \frac{k_\theta}{L_A^2}y = \frac{L_a}{L_A}\varepsilon_0 T N_a x_{a0}k_{ab}\left(V_{aT}^2 - V_{aB}^2\right) = k_L\left(V_{aT}^2 - V_{aB}^2\right) \tag{12.36}$$

where $k_L = \eta(L_a/L_A)\varepsilon_0 T N_a x_{a0}k_{ab}$.
  As noted above, the controller is assumed to be a PID type, and it is given in [4] as

$$u = -Py - D\frac{dy}{dt} - I\int_0^t y(\tau)d\tau \tag{12.37a}$$

$$V_{aT} = \sqrt{u} \quad u \geq 0, \quad V_{aB} = \sqrt{-u} \quad u < 0 \tag{12.37b}$$

Note that squared voltage term is cancelled by introduction of square root in Equation 12.37. It linearizes system expressed by Equation 12.36 and parameters of the controller in

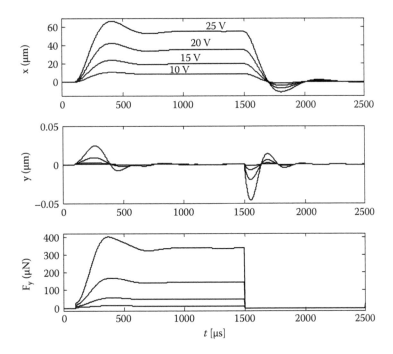

FIGURE 12.12
Dynamic behavior of the controlled system for the various voltage levels. In all cases, the achievable deflections are way greater than the maximum value for the uncontrolled case (8 μm).

Equation 12.37 can be determined by using any of the linear controller design methods. Here, following parameters have been calculated: $P = 7.5 \times 10^3$ V^2 m^{-1}, $D = 2.5 \times 10^{-1}$ sV^2 m^{-1}, and $I = 2.5 \times 10^8$ V^2 m^{-1}s^{-1}. The PID controller is shown in Figure 12.11.

A set of simulation results, shown in Figure 12.12, was conducted where different voltages greater than or equal to the pull-in value are applied to the actuator. It can be seen that the controller keeps the lateral motion at zero. Achievable deflections may depend on how much force the lateral steering actuators can provide, and their values are several times larger than the maximum deflection in the uncontrolled case (i.e., 8 μm).

The novel approach to counteract the lateral instability of the electrostatic comb-drive actuator was presented. The existing comb drive with its well-developed, experimentally verified mathematical model in one DOF [14] was extended with the lateral DOF model. The parameters of the lateral DOF model were determined through finite element analysis (FEA) and verified by static experimental results. This model was hypothetically extended with both sensor and actuator functions for lateral movement. Additional features were used to design the controller for lateral motion. Observations of the simulation results accomplished for the control system motivated several important conclusions.

The introduction of lateral feedback extends the range of the electrostatic comb-drive actuator. The amount of extension depends on how much force is generated with the lateral actuators. This approach can be easily combined with conventionally used suspension design methods extending the design possibilities.

Unfortunately, the utilization of feedback requires lateral sensor and actuator functionalities which have to be added to the MEMS device. These features can significantly change the inertia and, consequently, the dynamics of the device. Additional processing electronics for sensing and control should be integrated with MEMS device, as well. The sensing realization is not a problem as reliable techniques are available for high-resolution capacitance measurements [15,16]. Moreover, the lateral controller should be kept as simple as possible to prevent unnecessary consumption of valuable space.

Despite the disadvantages, the initial results are promising and may pave the way for more extensive use of feedback-based counteracting of lateral instability in practice.

12.4 Summary

This chapter provides two examples of optical MEMS devices where feedback loop is used to improve performance at device level. The first application discusses the implementation of light intensity feedback, introduced to improve the stability and accuracy of the light output from the optical MEMS device. The second example describes how deflection range of the optical MEMS device can be extended by means of counteracting lateral instability with the position feedback loop.

Light intensity control for VOA is a simple and effective application of feedback control for an MEMS device. The light intensity is used as a feedback and control signal is fed back to the MEMS devices. The obtained results verify that the proposed control system significantly improves overall dynamical behavior of the device. Taking a light intensity directly into the controller solves the problem of usually complex and uncertain relationships between position and light intensity. Existing nonlinearities are easily handled by cautiously designed linear controller. Such an MEMS VOA with excellent light modulation properties and accuracy provided by controller for a useful light source applicable in various optical networks.

The second example deals with improving the applicable deflection range of the electrostatic comb drive at the device level. The electrostatic comb drive can be used either as an optical switch or VOA. The extension of the device with both sensing and actuation functionality provides opportunity to control its lateral movement. The amount of the extension of useful range depends on how much force can be generated by the lateral actuators. This approach can be easily combined with conventionally used suspension design methods extending the design possibilities.

Unfortunately, for both examples, the utilization of feedback requires additional functionalities which have to be added to the device. Additional processing electronics for sensing and control, as well as additional features at device level have to be integrated with or added to the MEMS device. For that reason, the controller should be kept as simple as possible to prevent unnecessary consumption of valuable space. Therefore, depending on application and customer requirements, the trade-off between performance and available space determines whether feedback would be used.

References

1. Liu, A.Q., Zhang, X.M., Lu, C., Wang, F., and Liu, Z.S., Optical and mechanical models for a variable optical attenuator using a micromirror drawbridge, *J. Micromech. Microeng.*, 13, 400, 2003.
2. Isamoto, K., Kato, K., Morosawa, A., Chong, C., Fujita, H., and Toshiyoshi, H., A 5-V operated MEMS variable optical attenuator by SOI bulk micromachining, *IEEE J. Selected Top Quantum Electron.*, 10, 570, 2004.
3. Fujita, H. and Toshiyoshi, H., Optical MEMS, *IEICE Trans. Electron.*, E83-C, 9, 1427, 2000.

4. Astrom, K.J. and Wittenmark, B., *Computer-Controlled Systems: Theory and Design*, Prentice Hall, Upper Saddle River, NJ, 1997.
5. Bryzek, J., Abbott, E., Flannery, A., Cagle, D., and Maitan, J., Control issues for MEMS, *Proc. 42nd IEEE Conf. on Decision and Co,* 3, 9, Maui, HI, 2003.
6. Grade, J.D., Jerman, H., and Kenny, T.W., Design of large deflection electrostatic actuators, *J. Micromech. Microeng.,* 6, 320, 2003.
7. Legtenberg, R, Groeneveld, A.W., and Elwenspoek, M., Comb-drive actuators for large displacements, *J. Micromech. Microeng.,* 6, 320, 1996.
8. Elata, D., Bochobza-Degani, O., and Nemirovsky, Y., Analytical approach and numerical alpha-lines method for pull-in hyper-surface extraction of electrostatic actuators with multiple uncoupled voltage sources, *J. Microelectromech. Syst.,* 12, 681, 1996.
9. Chen, C. and Lee, C., Design and modeling for comb drive actuator with enlarged static displacement, *Sensors and Actuators A,* 115, 530, 2004.
10. Grade, J.D., Large deflection, high speed, electrostatic actuators for optical switching applications, *Mech. Eng. Stanford,* 1999.
11. Senturia, S.D., *Microsystem Design,* Kluwer Academic Publishers, Norwell, MA, 2000.
12. Elswenpoek, M. and Wiegerink, R., *Mechanical Microsensors,* Springer, Berlin, 2001.
13. Borovic, B., Liu, A.Q., Popa, D., Zhang, X.M., and Lewis, F.L., Lateral motion control of electrostatic comb drive: new methods in modeling and sensing, *16th IASTED International Conference on Modelling and Simulation,* Cancun, Mexico, May 18–20, 2005.
14. Borovic, B., Open-loop vs. closed-loop control of MEMS devices: choices and issues, *J. Micromech. Microeng.,* 15, 1917, 2005.
15. Bryzek, J., Flannery, A., and Skurnik, D., Integrating microelectromechanical systems with integrated circuits, *Instrumentation & Measurement Magazine, IEEE,* 7(2), 51–59, 2004.
16. Seeger, J.I. and Boser, B.B., Charge control of parallel-plate, electrostatic actuators and the tip-in instability, *J. Microelectromech. Syst.,* 12(5), 656, 2003.

Index

A

ABCD law, 103
Absorption length, 144
Acousto-optic switches, 3
Actuator
  analytical schematic, 73
  compliant structure, 72
  displacement, 53–54
  electromechanical optical switch design, 53–54
  stability, 53
  types, 29
Adjustable pivot, 276
Adjustable virtual pivot, 285
Air-bridge type PhC devices, 397
Air gap, 112
Airy disk, 147
All-optical network (AON), 2
All-optical switching
  output spectra, 340
  waveforms, 342
Aluminum, 12
  coating, 40
  highest reflection, 41
  highest reflection minimum thickness, 42
  incident angles relative reflectivity, 46
  polarization dependence reflection, 47
  reflectivity, 42
  relative reflection percentage *vs.* thickness, 40
  skin depth, 39
Amplification
  vs. input displacement, 85
  vs. vertical coordinates, 78
  x2, 76
Amplified spontaneous emission (ASE) spectrum, 306–308
Amplifier
  bifurcation, 82
  deformation, 84
Angular misalignment loss, 32
Antireflection (AR)
  coated gain chip, 280
  coated lensed SMF, 305
AON. *see* All-optical network (AON)
Aperture
  attenuation level, 202
  position, 220
  size, 202
  WDL, 220
Applied voltage *vs.* deformed gap, 69
AR. *see* Antireflection (AR)
ARDE. *see* Aspect-ratio-dependent etching (ARDE)
Argon fluoride lasers, 397

Argon plasma, 12
ASE. *see* Amplified spontaneous emission (ASE) spectrum
Aspect-ratio-dependent etching (ARDE), 379
Asymmetric shutter free-space MEMS VOA, 187–188
Attenuation
  elliptical mirror design, 185
  linearity, 185
  mirror displacement, 228
  simulated contour, 208
  tuning scheme, 200
Attenuation curve
  measured, 218, 219
  single-shutter tuning scheme, 217
Attenuation level, 185
  shutter position, 201
Attenuation model
  comparison, 198
  dual-shutter VOA, 199
  free-space single-shutter VOA, 191
  near-field condition, 196

B

Backcone, 431
Back reflection, 32
  free-space MEMS VOA specifications, 187–188
BARC. *see* Bottom antireflection coating (BARC)
Bcc. *see* Body-centered cubic (bcc)
Beams
  bending design, 213
  verticality *vs.* trench width, 377
Bessel function, 147
Bidirectional actuator, 119
Bifurcation, 80
  amplifier, 82
  compliant structure, 81
  effect self-latching, 79–82
  MEMS optical switches design, 79–82
Binary mask exposure system, 403
Black silicon, 417
Blazed grating and microlens
  continuously tunable lasers, 304–305
Blocking-wedge
  deflection free-space MEMS VOA, 187–188
  design, 182
Blue shift, 349
Body-centered cubic (bcc), 354
  cubic structure, 355
BOE. *see* Buffered oxide etchant (BOE)
Bottom antireflection coating (BARC), 394, 397, 402, 406, 409

Buffered oxide etchant (BOE), 383, 397, 424
Bulk photonic crystal lattice, 413

C

C, 21, 28, 182, 183, 186
  calculation, 231
  direct-coupling type MEMS VOA, 189
  dual-shutter VOA, 220
  EVOA, 232
  FVOA, 232
  metal-coated surface Fresnel reflection, 211
  minimization, 233
  optical switches, 21–22
  PVOA, 225
  reflection-type VOA, 190
  single-shutter VOA, 190
  temperature-dependent losses, 227–229
Cantilever beam, 289, 290
Capacitive *vs.* optical sensing, 444
CAR. *see* Chemically amplified resist (CAR)
Castigliano theory, 74
Cavity modes, 326
Cavity scheme configurations
  hybrid integration *vs.* single-chip integration, 268
  short cavity *vs.* long cavity, 267
CCL. *see* Coupled-cavity laser (CCL)
CD. *see* Critical dimension (CD)
Center wavelength shift *vs.* injected current
      square, 13
Cesium chloride, 354
  cubic structure, 355
CF. *see* Correction factor (CF)
Chemical etch, 358
Chemically amplified resist (CAR), 402, 406, 409, 423
  DUV, 394
Chemical mechanical polishing (CMP), 396
Closed-loop chiller, 154
Closed-loop control, 437, 442
  open-loop control, 438–440
Closed-loop response, 441
CMOS. *see* Complementary metal oxide
      semiconductor (CMOS)
CMP. *see* Chemical mechanical polishing (CMP)
CMT. *see* Coupled mode theory in time (CMT)
Coarse tuning, 219
Code V, 176
Coefficient of thermal expansion (CTE), 94
Coherent population trapping (CPT), 322–323
Comb drive, 435, 455
  depth parts, 62
  electromechanical optical switch design, 51–52
  lateral force, 51
  relative force, 68
  schematic, 48
  stability, 51–52
  structure, 453
Comb drive actuator, 10
  electromechanical optical switch design, 47, 49–50
  lateral-mode, 47

stability, 58
static actuation relationship, 264
Comb fingers
  DRIE, 49, 54
  MEMS optical switches design, 58–59
  native oxidation, 60
  oxide coverage, 61
  profile, 55
  relative displacement, 59
  relative electrostatic force, 57
  sloped angle, 57
  thickness, 54
Comb sets, 70
Comb structure generated force, 435
Complementary metal oxide semiconductor
      (CMOS), 189, 359
  DRIE, 189
Complex dielectric constant, 38
Compliant amplifier deformation, 84
Compliant mechanical mechanism dimension, 76
Compliant structure bifurcation, 81
Compound flexure design, 284
Concave grating, 281
Constitutive relations, 37
Constricted neck interfaces, 410
Continuous and discrete tuning transition
  discrete wavelength tuning, 260–261
  MEMS discretely tunable lasers, 260–261
Continuously tunable lasers, 304–305
Continuous shift, 239
Continuous wave (CW), 264, 322
  stabilizer signal, 322
Continuous wavelength tuning
  external reflector effective reflectivity, 298
  feedback model, 253–255
  gain-medium refractive index wavelength
      dependence, 299
  gain medium refractive index with grating
      rotation, 297
  ideal pivot position, 293–295
  MEMS continuously tunable lasers, 293–300
  mirror configuration wavelength-tuning range
      and stability, 157–256
  mirror translation laser gain and phase,
      250–257
  mirror translation wavelength, 252
  optimized design *vs.* conventional design, 300
  pivot position optimization, 296
Controlled system dynamic behavior, 463
Controller
  block scheme, 450
  structure, 440
Control signal, 340
Conventional sliding-block optomechanical
      VOA, 175
Coordinate system conversion model, 248
Corner mirror
  design, 183
  free-space MEMS VOA specifications,
      187–188

Correction factor (CF), 404
Coupled-cavity laser (CCL), 321
Coupled mode theoretical prediction, 139
Coupled mode theory in time (CMT), 136
Coupling cavity, 138
Coupling efficiency
    estimations, 240
    rotation angle, 249–250
    tuning behavior, 257
    variations, 241
Coupling loss
    vs. lateral misalignment, 34
    vs. misalignment angle, 34
    transmission distance, 33
CPT. *see* Coherent population trapping (CPT)
Critical dimension (CD), 362
    definition, 397
Cryogenic enhancement, 360
Crystals, 356
CTE. *see* Coefficient of thermal expansion (CTE)
Cubic photonic crystal holes array
        patterns, 410
Curved mirror
    configuration optical analysis, 244–246
    design, 242–243
    DRIE, 263
    micro-optical-coupling systems, 242–246
    SOI wafer, 263
CW. *see* Continuous wave (CW)

D

Damping, 443
Data signal, 340
DBR. *see* Distributed Bragg reflectors (DBR)
Deep-etched dual shutter VOAs
    device, 215
    linear tuning scheme, 217–219
    MEMS variable optical attenuators, 215–219
    single-shutter tuning scheme, 216
    wavelength- and polarization-dependent
        losses, 220
Deep-etched elliptical mirror VOA
    DRIE, 221
    micrographs, 221
Deep-etched parabolic mirror VOA, 224–225
Deep-etched trenches, 365
    fabrication process, 243
    platen power, 366
Deep etching fabrication process, 354–390
    beam verticality, 377
    cyclic etch/passivation cycles duration, 368
    design, 370
    DRIE chemistry and physics, 359–369
    dry release, 384–385
    dry silicon etching, 358
    etching process, 362–369
    etch rate along process, 375
    etch rate *vs.* trench width, 370–372
    etch rate *vs.* wafer exposed area, 373–374

etch rate *vs.* wafer trench position, 373
    gas flow with source power, 364–367
    loading effect, 370–376
    nonuniform beam release, 385
    notching effect, 377–389
    platen power, 363
    process pressure, 368–369
    silicon anisotropic wet etching, 358
    silicon crystal structure, 354–356
    silicon isotropic wet etching, 357
    silicon wet etching, 357–358
    SOI substrate DRIE fabrication, 390
    uniform beam release, 384
Deeply etched fiber grooves, 155, 156
Deep reactive ion etching (DRIE), 4, 11
    ARDE microloading effect, 379
    CMOS, 189
    comb fingers, 49, 54
    continuously tunable lasers, 301
    curved mirror, 263
    deep-etched dual-shutter VOA, 215
    development, 8
    eliminating notching effect, 381–382
    etched slanted trench schematic, 55
    fabrication tolerance, 69
    folded beam, 54
    lithographic processings, 416–420
    maximized, 362
    MEMS ILL, 335
    MEMS lasers, 239
    MEMS optical switch, 431
    notching effects, 288, 377
    parabolic mirror pair, 186
    plasma-based dry etch, 358
    prism, 109
    PVOA, 224
    SOI wafer, 83, 118, 122
    tolerances, 55
    tuning structure, 282
    undercut, 65, 364
    VOA, 180
Deep submicron photonic bandgap crystal
        fabrication processes, 394–425
    buried oxide interface RIE lag and
        notching, 422
    chemically amplified resist deactivation, 408
    deep reactive ion etching (DRIE), 416–419
    dense deep-submicrometer critical dimensions
        (CD) etching, 420–421
    etch postprocessing, 423–424
    fabrication process flow design, 394–396
    lithographic parameters manipulation, 401
    lithographic processings, 415–424
    lithography setup and methods, 400–414
    optical proximity error model, 402–404
    partial coherence and pattern density
        control, 406–407
    photonic crystal rods etch, 415
    photoresist trimming, 408–414
    subwavelength lithography, 397–399

Deep ultraviolet (DUV) lithography, 394
  CAR, 394, 408
  OL, 406
  theoretical resolution limits, 400
Deflection *vs.* minimum relative force, 68
Deformable membrane design, 179
Deformed gap *vs.* applied voltage, 69
Degrees of freedom, 199, 203, 401, 432–434, 434
  mathematical model, 463
Dense deep-submicrometer CD etching, 420–421
DenseLight (DL), 152
Dense wavelength division multiplexing (DWDM)
    systems, 8, 90, 174, 319, 448
Depth of focus, 400
Designed pattern *vs.* fabricated structure, 66
Design effect, 77
Developed MEMS continuously tunable
    lasers, 277
Deviant incident angle, 115
Device sidewalls postprocessing, 424
Device under test (DUT), 231
Diamond cubic structure, 355
Diffracted electrical fields, 210
Diffracting radiation schematic, 134
Diffraction beam patterns, 204
Diffraction type VOA, 190
  movable shutters, 180
Digital micromirror device (DMD), 185
Direct coupling design, 323
Direct-coupling MEMS VOA
  free-space, 187–188
  MEMS variable optical attenuators, 177
  PDL, 189
Discrete wavelength tuning
  continuous and discrete tuning transition,
      260–261
  discrete-dominant tuning hopping range, 262
  laser modes and frequency shift, 259
Distributed Bragg reflectors (DBR), 238
DL. *see* DenseLight (DL)
DMD. *see* Digital micromirror device (DMD)
DOF. *see* Degrees of freedom; Depth of focus
Doppler shift mode, 276
Double-clamped beam, 304
Double-clamped design, 287
Double cylindrical prism
  configuration, 109
  double cylindrical prism switch modeling and
      analysis, 112–116
  dynamic switching characterization, 124
  fabrication and experiment, 123, 124
  localized heating simulation, 116–117
  optical effects, 112–115
  static switching characterization, 123
  thermo-optic switch, 109–115
DRIE. *see* Deep reactive ion etching (DRIE)
Driving voltage *vs.* optical attenuation, 222
Drude formulations, 149
Drude model, 148, 149
Dry release

beams, 386
notching effect, 384–385
Dual-shutter VOA, 176
  attenuation model, 199–200
  attenuation modeling, 199
  linear tuning schemes, 201–202
  optical model and tuning schemes, 199–205
  tuning resolution, 203–205
DUT. *see* Device under test (DUT)
DUV. *see* Deep ultraviolet (DUV) lithography
DWDM. *see* Dense wavelength division multiplexing
    (DWDM) systems
Dynamic switching, 125
  light path MEMS subsystem, 21

E

ECP. *see* Electrochemical plating (ECP)
ECTL. *see* External cavity tunable laser (ECTL)
EDFA. *see* Erbium-doped fiber amplifier (EDFA)
EF. *see* Experiment factor (EF)
Effective external cavity length, 241
Effective injection power, 345
EH. *see* Electroholographic (EH) optical switches
Electrical field, 50
Electrochemical plating (ECP), 423
Electrode copper metallization, 424
Electrode design, 213
Electrode metallization integration, 423
Electroholographic (EH) optical switches, 2–3
  performance comparisons, 3
Electromagnetic (EM) fields, 326
Electromechanical optical switch design, 47–54
  actuator displacement, 53–54
  capacitor analysis, 49
  comb-drive actuator, 49–50
  comb-drive actuator lateral-mode, 47
  comb drive stability, 51–52
  lateral electrostatic force, 50
  parallel plate capacitor, 47–48
Electro-optic coefficient, 3
Electro-optic switch, 2–3
  performance comparisons, 3
Electrostatic actuator capacitance, 435
Electrostatic actuators mathematical modeling
  electrostatic model, 435
  mechanical model, 432–434
  optical intensity model, 436
Electrostatic comb-drive actuator, 431
  control system design, 461–463
  electrostatic model, 455–456
  lateral instability, 454–460
  lateral sensor/actuator extended model, 457–460
  mechanical model, 454
Electrostatic force comb sets, 70
Electrostatic MEMS devices control strategies, 430–443
  closed-loop control, 437–441
  comparison, 442
  electrostatic actuators mathematical modeling,
      432–436

electrostatic model, 435
mechanical model, 432–434
MEMS optical switch, 431
open-loop control, 437
optical intensity model, 436
Elliptical mirror VOA (EVOA), 206, 209, 227, 228
  attenuation linearity design, 185
  free-space MEMS, 187–188
  light, 210
  PDL, 223, 232
  polarization-dependent losses, 223
  SEM, 226
  specifications, 230
  TDL, 229
  WDL, 223, 230
EM. *see* Electromagnetic (EM) fields
Embedded resonator, 131
Erbium-doped fiber amplifier (EDFA), 10, 175
Etching
  cycle, 361
  depth trenches, 373–375
  openings, 417
  phases DRIE, 418
  postprocessing, 423–424
  profiles, 55–59
  selectivity platen power, 363, 367
  steps schematic, 362
Etch rate, 382
  platen power, 363
  trenches, 376
  vs. trench width, 376
Euler-Bernoulli hypothesis, 79
EVOA. *see* Elliptical mirror VOA (EVOA)
Experiment factor (EF), 404
Extended feedback model, 250
  wavelength, 256
  vs. week feedback model, 255
External cavity length, 257
External cavity tunable laser (ECTL), 274, 294

F

Fabrication
  actuator, 83
  dual-shutter VOA, 216
  etching isotropy failure, 417
  MEMS optical switches design, 55–69
  PhC devices, 141
  PhC process flow, 395
  photonic crystal lattice, 420
  structure *vs.* designed pattern, 66
  tolerance, 55–69, 69, 110
  VOA measured specifications, 224
Fabry-Pérot (FP)
  cavity, 130, 320
  cavity effect, 104, 106, 115, 130
  effect and polarization effect, 104–106
  etalon, 242
  filter, 179

filter configuration, 242
laser condition, 250
laser diode, 319
resonance, 278
semiconductor laser, 19
wavelength tuning, 250
Face-centered cubic (fcc), 354
  cubic structure, 355
Facet plane, 191
  amplitude and phase patterns, 194
Far-field beam
  model, 196
  profile laser, 307
FBG. *see* Fiber Bragg gratings (FBGs)
fcc. *see* Face-centered cubic (fcc)
FDTD. *see* Finite difference time domain (FDTD)
FEA. *see* Finite element analysis (FEA)
Feedback-coupling efficiencies, 248
Feedback proportional derivative loop, 438
Feed-forward (FF), 438
  and feedback PD loop, 438
FEM. *see* Finite element method; Focus exposure matrix
Ferroelectric liquid crystal (FLC) switches, 8
FF. *see* Feed-forward (FF)
Fiber Bragg gratings (FBGs), 8
Fiber coupling
  loss, 108–109, 115–116
  MEMS optical switches design, 29–31
  optical switch design, 29–31
Fiber grooves, 153
Fine tuning, 219
Finger gap, 207
Finite difference time domain (FDTD), 129, 132
  simulation, 133, 134, 164, 165, 169
  simulation results, 133, 165
Finite element analysis (FEA), 434, 463
Finite element method, 79
Fixed-wavelength lasers, 17
Flat mirror design, 183
  free-space MEMS VOA, 187–188
Flat mirror VOA (FVOA), 206, 209, 227, 228
  light, 210
  PDL, 232
  SEM, 226
  specifications, 230
  TDL, 229
  WDL, 230
FLC. *see* Ferroelectric liquid crystal (FLC) switches
FLS. *see* Fully locked state (FLS)
Focus exposure matrix
  antireflection coating, 402
  reduced-radius photonic crystal critical dimensions, 414
Folded beam
  DRIE, 54
  profile, 56–57, 58–59
  relative displacement, 59
  sloped angle, 59

Folded cantilever design, 284
Four-port optical intersection, 139
Four-port photonic crystal intersection
    device, 136
FP. *see* Fabry-Pérot (FP)
Free-space MEMS VOA, 187–188
Free spectral range (FSR), 161, 256, 258
Frequency, 250
    tuning, 261
Fresnel-Kirchhoff diffraction formula, 191, 192
Fresnel Loss, 32
    MEMS optical switches design, 32
    reflection, 108, 114–115
Frustrated total internal reflection (FTIR), 110, 113
F species
    consumption, 365
FSR. *see* Free spectral range (FSR)
FTIR. *see* Frustrated total internal reflection (FTIR)
Full-width half-maximum (FWHM),
    305, 306
Fully locked state (FLS), 321, 324, 325, 326,
    329–331, 341, 349
FVOA. *see* Flat mirror VOA (FVOA)
FWHM. *see* Full-width half-maximum (FWHM)

G

Gain-chip facet, 307
Gaussian beam, 6
    assumption, 190
    diffraction, 197
    divergence, 30, 102–104, 106
    Rayleigh range and waist radius, 195
    triangular prism switch modeling and analysis,
        102–104
    waist *vs.* transmission distance, 31
Gaussian optics
    ABCD law, 103
Generalized model (GM), 318, 327
    experimental data, 338
    photon numbers, 332
GM. *see* Generalized model (GM)
Gold
    highest reflection, 41
    highest reflection minimum thickness, 42
    incident angles relative reflectivity, 46
    polarization dependence reflection, 47
    reflectivity, 42
    skin depth, 39
Grating
    actuation voltage, 311
    characteristics, 305
    continuously tunable lasers, 304–305
    vs. driving voltage, 304
    ideal pivot position, 295
    output wavelength, 310
    profile, 281
    rotation angle, 295, 304, 310
    teeth position, 280
    threshold current, 311

H

Half-plane thin-shutter assumption, 190
HAR. *see* High aspect ratios (HAR)
Hard-mask oxide, 384
HDLP. *see* High-density low-pressure (HDLP)
    system
Hexamethyldisilazane (HMDS), 400
High aspect ratios (HAR), 157, 158, 163–164, 166, 394.
        see also Record-high aspect ratios
    PhC optical intersections, 167
    photonic crystal rods, 158
    SEM images, 158
High-density low-pressure (HDLP) system, 359
High intensity pump radiation, 146
High reflection mirror, 101
HMDS. *see* Hexamethyldisilazane (HMDS)
Hopping range, 239
Hybrid integration *vs.* single-chip integration, 268

I

IC. *see* Integrated circuit (IC)
ICP. *see* Inductively coupled plasma (ICP)
Ideal pivot position
    grating rotation angle, 295
    MEMS continuously tunable lasers,
        293–295
ILL. *see* Injection-locked laser (ILL)
Incident angle, 47
    output power, 107
    reflected light, 45
    reflection amplitude, 46
    relative reflectivity of metals, 46
    scattering power percentage difference, 36
Incident radiation, 146
Incident wave, 137
Inductively coupled plasma (ICP), 360
    DRIE, 369, 420
Information superhighway, 1
Initial capacitance, 53
Injected modes, 325
Injection light polarization direction, 349
Injection-locked laser (ILL)
    analytical model, 327
    multimode terms, 325
    semiconductor, 319
Injection power, 343, 346
    locking state hysteresis, 346
    photon numbers, 334
Injection seeding, 326
In-plane 8 × 8 switch matrix schematic, 6
Input displacement
    vs. amplification, 85
    vs. output displacement, 84
Input fibers misalignment, 31
Input point, 83
Input tunable laser source, 152
Input waveguides, 157
Insertion loss, 29–32

Instantaneous photonic crystal output modulation, 168
Integrated circuit (IC), 2, 6, 354
Integrated fiber grooves, 153
Integrated MEMS optical-switching OADM, 10
Integration scheme configurations
  hybrid integration *vs.* single-chip integration, 268
  short cavity *vs.* long cavity, 267
Intensity depth profile, 145
Interference type free-space MEMS VOA specifications, 187–188
Interference-type MEMS VOA, 178
Interjection locking, 321
Internet protocol (IP), 10
IP. *see* Internet protocol (IP)
IPA. *see* Isopropyl alcohol (IPA)
Isopropyl alcohol (IPA), 425
Isotropic oxide etch, 381
Isotropic silicon wet etching
  cross-sectional view, 357
  etchants, 357
ITU grids, 266

K

Knife-edge diffraction, 197
Krypton fluoride, 400

L

Lagrangian formulation, 80
Lamb's semiclassical laser theory, 320
Large-displacement actuator, 70–84
  bifurcation effect self-latching, 79–82
  design and theoretical analysis, 71–82
  experimental results, 83–84
  linear displacement amplification theoretical analysis, 72–78
  stable and large displacement actuator configuration, 71
Laser
  argon fluoride, 397
  far-field beam profile, 307
  gain chip, 306–308
  gain curve, 306
  light-current curve, 311
  output wavelengths, 265
  spectra, 265
Laser configurations design
  Littman configuration, 279
  Littrow configuration, 278
  nonstandard grating configuration, 279
Laser diode (LD), 241
Lateral actuation and sensing, 459
Lateral attack etch, 379
Lateral comb-like structures, 458
Lateral controller, 462
Lateral DOF, 461
  feedback control, 459

Lateral force
  comb drive, 51
  electrostatic, 50
Lateral instability, 448
Lateral misalignment
  vs. coupling loss, 34
  loss, 32
Lateral pull-in, 448, 457, 458
LC. *see* Liquid crystal (LC) switch
LD. *see* Laser diode (LD)
Leaked optical power, 103
LIGA. *see* Lithographic Galvano formumg Abformung (LIGA)
Light beam diffraction, 192
Light-current curve, 310, 311
Light intensity
  desired and actual, 452
  waveform generator, 449
Linear absorption coefficients, 143
Linear displacement ratio, 81
Linear superimposition, 326
Linear tuning scheme, 218
  deep-etched dual shutter VOAs, 217–219
Liquid crystal (LC) switch, 8
  performance comparisons, 3
Lithographic Galvano formumg Abformung (LIGA), 189
Lithographic parameters manipulation, 401
Lithographic processings
  buried oxide interface RIE lag and notching, 422
  dense deep-submicrometer CD etching, 420–421
  DRIE, 416–419
  etch postprocessing, 423–424
  photonic crystal rods etch, 417
Lithographic variations, 409
Lithography experiments
  critical dimension plot, 407
  patterns, 406
Lithography setup and methods, 400–414
Littman configuration, 275, 281, 283
  laser configurations design, 279
Littrow configuration, 275
  MEMS continuously tunable lasers, 278
Littrow external cavity laser, 312
  analytical model, 293
Loading effect
  position labels, 371
  process conditions, 370
Localized high intensity optical pumping, 170
Localized pump radiation, 145
Lockable spectrum range, 344
  experimental data, 345
  MEMS injection-locked lasers, 343
Locking depth, 346
  lateral misalignment, 347
Locking quality, 348
Longitude loss, 32
Lorentzian response, 165
L-shaped switch matrix configuration, 7
Lucent OXCs, 5

M

Mach-Zehnder (MZ)
  interferometer, 90
  switch, 10
  thermal switch, 91
  thermo-optic switch, 90
Magnetic permeability, 45
Main degree of freedom, 434
Mass
  change sensitivity, 443
Master, 318
Material equations, 37
Maximum mode, 343
Maximum static voltages, 436
Maxwell's equations, 36, 38
Mean free path (MFP), 144
Mechanical antireflection switch (MARS), 176
Medium feedback model, 250
MEMS
  external cavity tunable laser, 238
  fine tuning, 111
  micromachined double prism, 123
  optical devices, 390
  optical switch, 21
  rotary grating, 303
  tunable FBG, 10–13
  universal DRIE process flow, 390
MEMS continuously tunable lasers, 274–312
  blazed grating and microlens, 304–305
  comparison, 312
  continuous wavelength tuning, 293–300
  device, 301–303
  experimental studies, 301–312
  external reflector effective reflectivity, 298
  gain-medium refractive index wavelength
      dependence, 299
  gain medium refractive index with grating
      rotation, 297
  ideal pivot position, 293–295
  laser configurations design, 277–279
  laser gain chip, 306–308
  Littman configuration, 279
  Littrow configuration, 278
  MEMS virtual pivot designs, 282–284
  nonstandard grating configuration, 279
  optimized design *vs.* conventional
      design, 300
  pivot position optimization, 296
  real and virtual pivots, 287–290
  real-pivot designs, 285
  real pivot *vs.* virtual pivot, 291–292
  real pivot with double-camped beam, 287
  virtual and real pivot design, 282–284
  virtual pivot with cantilever beam, 288–290
  wavelength tunability, 309–311
MEMS discretely tunable lasers, 238–267
  cavity and integration scheme configurations,
      266–268
  characterization, 262–266

continuous and discrete tuning transition,
    260–261
continuous wavelength tuning, 250–257
curved-mirror configuration optical analysis,
    244–246
curved-mirror design, 242–243
device, 263
discrete-dominant tuning hopping range, 262
discrete wavelength tuning, 258–262
hybrid integration *vs.* single-chip
    integration, 268
laser modes and frequency shift, 259
micro-optical-coupling systems of external cavity,
    242–246
mirror configuration wavelength-tuning range,
    256–257
mirror translation laser gain and phase,
    250–257
mirror translation wavelength, 252
mode stability, 266
overview, 263
short cavity *vs.* long cavity, 267
stability, 256–257
three-dimensional optical-coupling system,
    242–243
wavelength tunability, 264–265
weak feedback *vs.* extended feedback model,
    253–255
MEMS injection-locked lasers, 318–347
  alignment tolerance, 346
  blazed grating, 336
  components, 336
  design, 323
  development, 321–322
  device, 335–336
  direct-coupling design, 323
  equations, 328–329
  experimental setup, 337
  experimental studies, 335–346
  GM *vs.* SSM, 331–333
  injection power influence, 343–345
  injection seeding model, 325–328
  lockable spectrum range, 343
  microlens, 336
  multiple external injection locking property,
      340–341
  multiple external injections locking
      property, 334
  polarization influence, 347–348
  rate equations, 319–320
  rotary comb drive, 336
  SEM, 335
  single external injection locking property, 337–339
  single injection locking property, 328–330
  technological origin, 318
  theoretical analysis, 333
  theoretical study, 324–333
  threshold injection power and critical
      detuning, 330
  weak injection locking property, 342–348

MEMS lasers
  DRIE, 239
  superimposed spectra, 309, 312
MEMS Littrow tunable laser
  overview, 302
  parameters, 296
MEMS optical switches, 2–23, 8
  actuator displacement, 53–54
  bifurcation effect self-latching, 79–82
  capacitor analysis, 49
  comb-drive actuator, 49–50
  comb-drive actuator lateral-mode, 47
  comb drive stability, 51–52
  comb fingers and folded beam profile, 58–59
  comb fingers native oxidation, 60
  comb fingers profile, 55
  design, 28–85
  design and theoretical analysis, 71–82
  dynamic response, 12
  electromechanical design, 47–54
  etch profiles, 55–59
  experimental results, 83–84
  fabrication tolerance effect, 55–69
  fiber coupling loss, 29–31
  folded beam profile, 56–57
  fresnel loss, 32
  insertion loss, 29–32
  large-displacement actuator, 70–84
  lateral electrostatic force, 50
  linear displacement amplification theoretical
    analysis, 72–78
  metal coating PDL, 41–46
  nonuniform gap, 67–68
  optical design, 29–46
  optical loss related to micromirror, 33–46
  optical switching systems, 6–22
  optical switch matrix, 6
  parallel plate capacitor, 47–48
  reconfigurable optical-switching OADM, 8
  stable and large displacement actuator
    configuration, 71
  surface material, 36–40
  surface roughness, 33–35
  switch types, 2–3
  tapered microstructures and oxidation, 60
  3-D, 5
  tolerances, 69
  tunable laser integrated OADM, 16–18
  2-D, 4
  undercut, 64–66
  uneven depth comb fingers, 61–63
MEMS subsystem
  dynamic light path switching, 21
  output wavelengths, 20
  tunable wavelength converter, 18
  wavelength-dependence loss, 22
MEMS thermo-optic switches, 90–125
  configuration, 100–101, 109
  design, 112
  double cylindrical prism, 109–124

  dynamic switching characterization, 124
  fabrication and experiment, 118–121, 122–124
  implementation, 118–124
  modeling and analysis, 112–116
  multimode interference thermo-optic switch, 91
  MZ interferometer thermo-optic switch, 90
  physics, 93–99
  silicon material thermo-optic effect, 94–96
  silicon thermo-optic effect and TIR, 98–99
  spatial modulation thermo-optic switch, 92
  static switching characterization, 123
  switch modeling and analysis, 102–107
  total internal reflection, 97
  triangular prism, 100–107, 118–121
MEMS tunable lasers
  configurations, 240
  integrated OADM configuration, 17
  optical-coupling systems, 243
MEMS variable optical attenuators, 174–234
  attenuation model, 199–200
  configurations, 177–185
  deep-etched dual shutter VOAs, 215–219
  deep-etched elliptical mirror VOA, 220–223
  deep-etched parabolic mirror VOA, 224–225
  device, 226
  diffraction, 179–181
  direction coupling, 177
  dual-shutter VOA optical model and tuning
    schemes, 199–205
  experimental results, 451–452
  experimental studies, 211–232
  far-field attenuation model, 193–195
  interfacing, 433
  interference-type MEMS VOA, 178
  light intensity controller, 449–450
  linear tuning schemes, 201–202
  near-field attenuation model, 196–198
  optical waveform generator, 449–450, 451–452
  PDL, 232
  polarization dependencies, 206–210
  polarization-dependent losses, 209–210, 230–232
  reflection, 183–185
  refraction, 182
  SEM micrograph, 213, 431
  single-shutter OVA optical attenuation model,
    190–198
  specifications, 186–199
  surface-micromachined single-shutter VOAs,
    211–214
  TDL, WDL, PDL, 226–230
  temperature-dependent losses, 206–207, 227–229
  tuning resolution, 203–205
  wavelength dependencies, 206–210, 230
Metals
  coated silicon, 40
  highest reflection, 41
  highest reflection minimum thickness, 42
  incident angles relative reflectivity, 46
  minimum thickness for highest reflectivity, 36–37
  polarization dependence reflection, 47

reflectivity, 40, 41, 42
skin depth, 39
MFP. *see* Mean free path (MFP)
Micro-electro-mechanical system (MEMS). *see* MEMS
Microlens
  characteristics, 305
  continuously tunable lasers, 304–305
Microloading effect, 61
Micromachined double prism, 123
Micromachined silicon prism, 119
Micro-optical-coupling systems of external cavity
  curved-mirror configuration optical analysis, 244–246
  curved-mirror design, 242–243
  laser facet coordinate system conversion, 245
  MEMS discretely tunable lasers, 242–246
  mode-coupling method feedback-coupling efficiency, 247–249
  ray transfer matrix, 245
  three-dimensional optical-coupling system, 242–243
Miller index, 354
MIML. *see* Multiple injections to multimode lasers (MIML)
Minimum mode, 343
Minimum relative force *vs.* deflection, 68
Mirror
  configuration wavelength-tuning range and stability, 256–257
  coupling efficiency, 245
  design, 213
  displacement, 20
  displacement *vs.* optical attenuation, 222
  optical switch, 41
  RMS roughness, 34
  rotation angle *vs.* measured attenuation, 225
  scattering loss, 109
  surface, 34
  wavelength, 20
Misalignment angle *vs.* coupling loss, 34
MMI. *see* Multimode interferometer (MMI) thermo-optic switch
Mode-hop-free wavelength tuning, 294
Mode hopping, 239
Mounting plates design, 213
Mueller matrix method, 231
Multimode interferometer (MMI) thermo-optic switch, 92
Multiple injections to multimode lasers (MIML), 324
Multiplexer (MUX), 174–175
Multishutter design, 181
MUX. *see* Multiplexer (MUX)
MZ. *see* Mach-Zehnder (MZ)

N

NA. *see* Numerical aperture (NA)
Nanoelectromechanical system (NEMS), 373
Nd:Yag pump beam
  application, 148

Near-field attenuation model, 214
  condition, 196
  single-shutter OVA, 196–198
NEMS. *see* Nanoelectromechanical system (NEMS)
Nickel
  highest reflection, 41
  highest reflection minimum thickness, 42
  incident angles relative reflectivity, 46
  polarization dependence reflection, 47
  reflectivity, 42
  skin depth, 39
Nondispersive mirror condition, 250
Nonideal folded beam, 58
Nonshifted axis illumination exposure system, 403
Nonstandard grating, 281
Nonuniform beams, 389
Normalized injection current *vs.* round-trip amplification factor, 309
Notching, 379
  charge accumulation, 378
  deep etching fabrication process, 377–389
  depth *vs.* trench width, 388
  DRIE, 288, 377, 381–382
  dry release, 384–385
  elimination, 380–383
  mechanisms, 377–379
  nonuniform beam release, 385
  rate *vs.* trench width, 389
  schematic, 378
  SEM, 378
  trenches, 380
  trenches SEM, 389
  uniform beam release, 384
Numerical aperture (NA), 398

O

OADM. *see* Optical add/drop multiplexers (OADM)
OAI. *see* Off-axis illumination (OAI)
OC. *see* Optical circulators (OCs)
Off-axis illumination (OAI), 397, 403
OL. *see* Optical lithography (OL)
On-axis illumination lithographic system schematic, 399
On-chip MEMS subsystem, 19
1 × 1 photonic crystal optical intersection, 135
OPC. *see* Optical proximity correction (OPC)
OPE, 403, 404, 406
Open-loop control, 437, 442
  closed-loop control, 438–440
Open-loop response, 438, 441
Optical absorption effects, 142
Optical add/drop multiplexers (OADM), 4, 8–16
  experimental results, 15
Optical attenuation
  vs. driving voltage, 222
  vs. mirror displacement, 222
Optical attenuators
  direct-coupling types, 177
  interference type, 178

Optical circulators (OCs), 8
Optical cross connects (OXCs), 4
  switch N × N schematic, 7
Optical design, 29–46
Optical devices control, 447–463
  control system design, 461–463
  electrostatic comb-drive actuator-lateral
      instability, 453–463
  electrostatic model, 455–456
  experimental results, 451–452
  lateral sensor/actuator extended model, 457–460
  light intensity controller, 449–450
  mechanical model, 454
  MEMS VOA-based optical waveform generator,
      448–452
Optical excitation thermal effects, 150
Optical experiment measurements, 164
Optical insertion loss, 116
Optical lithographic wavelengths, 398
Optical lithography (OL), 393, 408–409
Optical losses, 114–115
  related to micromirror, 33–46
  switching states, 108
  triangular prism switch modeling and analysis,
      104–108
Optically generated plasma
  photonic crystal device effects on optical
      modulation, 148–149
Optically induced modulation, 141
Optical MEMS switch, 2
  performance comparisons, 3
Optical microscope image, 157
Optical photonic crystal intersection, 157
Optical power
  deviant incident angle, 115
  polarization, 105
Optical proximity correction (OPC), 397
  model, 406
Optical proximity error model, 402–404
Optical pumping, 167
Optical spectrum analyzer (OSA), 156–157, 349
Optical spectrums
  single-line resonators, 165
Optical switch
  beam path schematic, 30
  matrix, 6
  MEMS system, 21
  micromachined silicon prism, 119
  performance comparisons, 3
  polarization-dependence loss, 22
  TIR state, 120
  transmission state, 120
Optical switch design
  fiber coupling loss, 29–31
  fresnel loss, 32
  insertion loss, 29–32
  metal coating PDL, 41–46
  optical loss related to micromirror, 33–46
  surface material, 36–40
  surface roughness, 33–35

Optical switching systems, 6–22
  configuration, 16–17
  construction, 9
  experimental results, 13–15
  experiment results, 18–23
  MEMS systems, 6–22
  reconfigurable optical-switching OADM, 8
  tunable FBG, 10–13
  tunable laser integrated OADM, 16–18
Optical testing structure (OTS), 396, 425
  PhC, 397
  postprocessing, 423
Optical *vs.* capacitive sensing, 444
Optical wavelength spatial resolution, 148
Optimized resist-trimming recipe, 411
Orthogonal component phase difference, 46
OSA. *see* Optical spectrum analyzer (OSA)
Oscillation modes, 326
OTS. *see* Optical testing structure (OTS)
Output fibers misalignment, 31
Output power incident angle, 107
Output *vs.* input displacement, 84
Output wavelengths
  grating rotation angle, 310
  laser spectra, 265
Overrreleased beams, 386
OXC. *see* Optical cross connects (OXCs)
Oxidized comb finger, 60

P

Packaged MEMS thermo-optic switch, 122
Packaged triangular prism
      optical switch, 118
Parabolic mirror, 186
  design, 184–186, 190, 321, 323–324
  DRIE, 186
Parabolic mirror VOA (PVOA), 174, 224–226, 233
  DRIE, 224
  free-space MEMS, 187–188
  PDL, 225
  specifications, 225
  WDL, 225
Parallel plate capacitor, 48
  electromechanical optical switch design, 47–48
Passivation
  cycle, 368
  DRIE, 418
  inappropriate reduction, 417
  polymer film deposition schematic, 361
PBG. *see* Photonic bandgap (PBG)
PD. *see* Proportional derivative (PD) loop
PDL. *see* Polarization-dependence loss (PDL)
PECVD. *see* Plasma enhanced chemical vapor
      deposition (PECVD)
Peeling
  electrode copper metallization, 424
Periscope design, 183
Phase compensation, 248
Phase shift incident angle, 46

Phase shift masking (PSM), 397, 402, 403, 406, 426
    technique, 398
PhC. *see* Photonic crystal (PhC)
Photon energies, 142
Photonic bandgap (PBG), 129–130, 416
    single-line-defect waveguide, 131
    SOI, 130
Photonic crystal (PhC), 394
    air bridge, 425
    array, 403
    defect structures, 403
    dense holes array, 425
    device effects on optical modulation, 148–150
    device fabrication, 141
    fabricated photomask, 405
    fabrication process flow, 395
    four ports PhC optical intersections, 132–135
    HAR and slab-type, 166
    holes lattice, 421
    input tunable laser source, 152
    intersection dynamic modulation, 166–170
    line defect waveguides, 162
    line intersections optical resonators, 136–140
    MERIE etching, 397
    OPC regression model, 405
    operating point, 138
    optical excitation thermal effects, 150
    optical intersection, 134, 138, 165
    optically generated plasma, 148–149
    planar optical, 393
    postprocessing, 423
    resonator rod size, 148
    rods, 405
    structure, 166
    trim result, 412
Photonic crystal (PhC) devices optical
        measurements
    silicon waveguides coupling and bending losses,
        158–162
    static photonic crystal device experiments,
        163–165
    test structures, calibrations, measurements,
        155–156
    waveguide propagation loss coefficients, 157
Photonic crystal (PhC) microresonators dynamic
        modulation devices, 129–171
    challenges, 144–147
    coupled model theory analytical formulations,
        136–140
    design, 141–150
    dynamic modulation, 166–170
    effects on optical modulation, 148–150
    four ports PhC optical intersections, 132–135
    intersections optical resonators, 132–140
    optical excitation thermal effects, 150
    optically generated plasma, 148–149
    optical measurements, 155–170
    silicon PhC optical modulations, 142–143
    silicon waveguides coupling and bending
        losses, 158–162

single-line defect waveguides with embedded
        microresonators, 130–131
static photonic crystal device experiments,
        163–165
test structures, calibrations, measurements,
        155–156
waveguide propagation loss coefficients, 157
Photonic crystal (PhC) optical intersection, 133,
        135, 166
    center wavelength shifts, 169
    critical dimensions, 414
    electric field modes, 135
    frequency response, 135
    resonator, 169
    resonator rod refractive index, 156
    schematic, 133
    SEM images, 163
    transmission spectrum measurement, 160
Photonic crystal (PhC) rods etch
    deep submicron photonic bandgap crystal
        fabrication processes, 415
    lithographic processings, 417
Photonic integrated circuit (PIC), 129
Photoresist strip, 397
Physical etch
    plasma-based dry etch, 358
Physical vapor deposition (PVD), 423
PIC. *see* Photonic integrated circuit (PIC)
PID controller, 450, 461–463
Piezoelectric actuators, 189, 203
Piezoelectric transducers (PZTs) tuning, 12
Pivot position, 276
    optimization, 296
Pivot shift, 283
Pivot structures, 282
Planar lightwave circuit (PLC), 2
Planar lightwave circuits, 6
Planck constant, 149
Plane wave, 42
Plane wave method (PWM), 129
Plasma-based dry etch, 358
Plasma enhanced chemical vapor deposition
        (PECVD), 370, 394
    notching effect dry release, 385
    oxide layer, 381, 384
    SEM, 383
    TEOS oxide, 388
    TEOS oxide deposition, 382
Platen power, 363
PLC. *see* Planar lightwave circuit (PLC)
Plutonium
    highest reflection, 41
    highest reflection minimum thickness, 42
    incident angles relative reflectivity, 46
    polarization dependence reflection, 47
    reflectivity, 42
    skin depth, 39
Polarization-dependence loss (PDL), 176
    deep-etched dual shutter VOAs, 220
    device, 226

EVOA, 223
MEMS VOA, 209–211
polarization-dependent losses, 230–232
wavelength-dependent losses, 230
Polarization dependence reflection, 47
Polarization external electric field, 94
Polymer cleaning, 397
Polymer deposition, 361
Position-optical intensity, 437
Potassium hydroxide, 358
Prefiltered squared wave signal, 451
Preshaped open-loop response, 439, 440
Prism
adjustment angle, 119
transient thermal response, 117
Prism surface scattering loss, 109
Probe needles, 153
Proportional derivative (PD) loop, 438
PSM. *see* Phase shift masking (PSM)
Pull-in problem, 283
Pump beam
absorption length, 144
intensity, 144, 167
radiation, 167
resonator, 154
PVD. *see* Physical vapor deposition (PVD)
PVOA. *see* Parabolic mirror VOA (PVOA)
PWM. *see* Plane wave method (PWM)
PZT. *see* Piezoelectric transducers (PZTs) tuning

Q

Q-switch Nd:YAG-based 532 nm direct-coupled
system, 154
Quasi-locking condition, 324

R

Radio frequency (RF), 1
Rapid etching, 362
Rayleigh criterion, 147
Rayleigh range and waist radius, 195
Rayleigh scattering, 233
Reactive ion etching (RIE), 61, 420, 422
lagging effect, 370
plasma-based dry etch, 358
SEM, 422
Real pivot, 282
designs, 285, 291
double-clamped beam, 286, 287
MEMS continuously tunable lasers,
287–290
position shift, 292
symmetric serpentine flexure, 286
Reconfigurable OADM, 14
Reconfigurable optical-switching OADM
MEMS optical switches and systems, 8
optical switching systems, 8
Record-high aspect ratios, 148, 161–167, 394–395,
397, 415, 426

Red shift
nominal detuning, 339
relationship with nominal detuning, 339
Reduced-radius photonic crystal critical
dimensions, 414
Reflected light
azimuth angle, 45
incident angle, 45
Reflected power temperature, 121
Reflection amplitude incident angle, 46
Reflection type MEMS VOA, 183–185
Reflectivity
aluminum, 42
gold, 42
measure, 44
metals, 42
mirror, 11
nickel, 42
Refraction-type MEMS VOA, 182
Refractive index
temperature change, 96
wavelength, 300
Relative driving force, 61
Relative electrostatic force, 57
Relative force, 68
Replicated holographic gratings, 281
Resolution enhancement techniques (RET), 397
Resonant frequency, 81
Resonant wavelength shifts, 168
Resonators
critical dimension measurements, 411
sizes, 165
RET. *see* Resolution enhancement techniques (RET)
RF. *see* Radio frequency (RF)
RIE. *see* Reactive ion etching (RIE)
RMS. *see* Root-mean-square (RMS)
Root-mean-square (RMS)
mirror surface, 34
prism, 109
PVOA, 224
roughness, 209
scattering power percentage, 35
Rotary comb drive, 336
Round-trip amplification factor *vs.* normalized
injection current, 309
Runaway etching, 417
SEM, 418

S

sc. *see* Simple cubic (sc)
Scalar Gaussian beam optics, 298
Scalar optical assumption, 190
Scallop, 419
Scanning electron micrograph (SEM), 335
EVOA, 226
fabricated actuator, 83
FVOA, 226
MEMS VOA, 213
packaged optical switch, 118

PECVD oxide, 383
photonic crystal holes lattice, 421
photonic crystal optical intersection, 163
RIE, 422
shutter-type VOA, 226
slab type photonic crystal intersection device, 159
SOI, 423
trimmed photonic crystal rods resist pattern, 415
2-D 2x2 MEMS optical switch, 11
Scattering power
    incident angle, 36
    percentage, 34, 36
    RMS roughness, 35
Scratch drive actuators (SDA), 29
SCS. *see* Single-crystal silicon (SCS)
SDA. *see* Scratch drive actuators (SDA)
SDH. *see* Synchronous digital hierarchy (SDH)
Secondary modes, 326
Seeded modes, 320, 325
Self-latched micromachined actuator
        schematic, 71
Self-latched micromachined mechanism, 82
SEM. *see* Scanning electron micrograph (SEM)
Semiconductor fabrication process, 3
Semiconductor laser, 319
Semiconductor optic amplifier (SOA), 2
    performance comparisons, 3
    switch, 2
Semitransparent silicon wedges, 182
Sense bridge, 461
Sensing, 444
Serial dynamic OADM, 9
SF. *see* Shape factor (SF)
Shape factor (SF), 404
Shift range, 239
Short external cavity length condition, 250
Short mirror translation condition
    wavelength tuning, 250
Shutter, 431
    amplitude and phase patterns, 194
    plane, 191, 194
    SEM, 226
    VOA, 179, 226
Shuttle, 431
Side modes, 325
Side-mode-suppression ratio (SMSR), 242,
        309, 321, 341
Side pull-in, 448
Signal-to-noise ratio (SNR), 444
Silicon
    planes, 356
    refractive index *vs.* wavelength, 99
    thermo-optic effect, 95
Silicon-air interface, 97
Silicon crystal structure, 354–356
Silicon etch, 361
    coil power, 367
Silicon isotropic wet etching, 357
Silicon material thermo-optic effect, 94–96
Silicon-on-insulator (SOI)

curved mirror, 263
deep-etched dual-shutter VOA, 215
DRIE, 83, 118, 122, 391
improved etching, 382
MEMS, 61
MEMS ILL, 335
parabolic mirror pair, 186
PBG, 130
PhC fabrication process flow, 396
PLC, 2
preparation, 394
SEM, 391, 423
tuning structure, 282
VOA, 180–183
Silicon PhC optical modulations, 142–143
Silicon prism thermo-optic switch design, 101
Silicon refractive index optical pumping, 150
Silicon waveguides, 156
    coupling and bending losses, 158–162
Simple cantilever design, 284
Simple cubic (sc), 354
    cubic structure, 355
Simplified lumped model switch, 433
Single-chip ILL, 335
Single-chip integration, 283
Single-crystal silicon (SCS), 93, 382
Single injection condition, 324
Single injection locking property
    equations, 328–329
    GM *vs.* SSM, 331–333
    threshold injection power and critical
        detuning, 330
Single injection single mode laser (SISL), 324
Single-line and resonator structures, 413
Single-line photonic crystal waveguides, 131
Single-line reduced-radius rods, 412
Single-line resonators
    optical spectrums, 165
    transmission spectrums, 132
Single-mode fiber (SMF), 29–31, 190, 306, 394
    AR coated lensed, 305
    coupled, 30
Single mode laser, 324
Single mode slave condition, 324
Single-shutter MEMS VOA
    drawbridge actuator, 212
    schematic, 212
Single-shutter OVA optical attenuation model
    far-field attenuation model, 193–195
    MEMS variable optical attenuators,
        190–198
    near-field attenuation model, 196–198
Single-shutter scheme, 203
Single-shutter VOA (SVOA), 206, 208, 209, 227
    PDL, 232
    specifications, 230
    TDL, 229
    WDL, 230
SISL. *see* Single injection single mode laser (SISL)
Slab type PhC devices, 163

intersection, 159
schematic, 160
structures, 161
Slave, 318
output spectra, 338
waveforms, 341
SLED. *see* Super-luminescent light emitting
diode (SLED)
Sloped comb fingers, 56
oxide, 62
Sloped folded beam schematic, 57
Small-signal condition, 324
Small-signal models (SSM), 318
photon numbers, 332
SMF. *see* Single-mode fiber (SMF)
SMSR. *see* Side-mode-suppression ratio (SMSR)
SMU. *see* Source measurement unit (SMU)
SNR. *see* signal-to-noise ratio (SNR)
SOA. *see* Semiconductor optic amplifier (SOA)
Sodium chloride, 354
SOI. *see* Silicon-on-insulator (SOI)
SOP. *see* State of polarization (SOP)
Source measurement unit (SMU), 123
Spacer-oxide removal, 381
Spatial modulation thermo-optic switch, 92
SSM. *see* Small-signal models (SSM)
Stability analysis schematic, 51
Stabilizer signal, 322
State of polarization (SOP), 230–231
Static actuation relationship, 264
Static photonic crystal device experiments, 163–165
Static switching, 124
Stiffness, 442
Straight photonic crystal waveguide, 161
STS. *see* Surface technology systems (STS)
Sublithographic wavelength exposure failures, 408
Subwavelength lithography, 397–399
Super-luminescent light emitting diode (SLED), 152
Surface material, 36–40
Surface-micromachined single-shutter VOAs
design, 211–213
experimental results, 214
Surface technology systems (STS), 360
ICP DRIE, 381
multiplex ICP process module, 360
Surviving photonic crystal rod array, 419
SVOA. *see* Single-shutter VOA (SVOA)
Switching states, 120
optical losses, 108
Symmetrical comb-drive actuator, 455
Symmetrical spring design, 285
Symmetric or single shutter free-space MEMS
VOA, 187–188
Symmetric spring design, 285
Synchronous digital hierarchy (SDH), 10

T

Targeted modes, 325
TDL. *see* Temperature-dependent losses (TDL)

TEC. *see* Thermo-electric cooler (TEC)
Temperature change *vs.* initial refractive angle, 99
Temperature-dependent losses (TDL), 206–208
device, 226
hot airflow, 208
MEMS VOA, 229
polarization-dependent losses, 230–232
temperature-dependent losses, 227–229
wavelength-dependent losses, 230
TEOS. *see* Tetraethyl silicate (TEOS)
Tetraethyl silicate (TEOS), 370
Tetramethyl ammonium hydroxide (TMAH), 358
Thermal effect
refractive index change, 117
schematic, 94
Thermally tuned FBG, 11–13
Thermal MEMS switch, 2
Thermal optical switch, 3
Thermo-electric cooler (TEC), 120, 228
currents, 121
Thermo-optic effect, 170
Thermo-optic modulation, 157
Thermo-optic switch, 110
Thermo-optic switching operations, 168
Three dimensional MEMS optical switches, 5
Three-dimensional optical-coupling
system, 242–243
3x3 photonic crystal optical intersection
resonator, 169
Tilting shutter VOA, 197
TIR. *see* Total internal reflection (TIR)
TLS. *see* Tunable laser system (TLS)
TMAH. *see* Tetramethyl ammonium hydroxide
(TMAH)
Torsional 2-D MEMS switch, 4
Total internal reflection (TIR), 93, 114, 118, 121
angle, 97–102, 107–110
cross talk, 122
interface, 93
prism, 109
required refractive index *vs.* refractive angle
shift, 98
state, 104, 120–122
reflection, 106
TPA, 142–143, 460
absorption coefficients, 143
Translation 2-D-mirror-based optical switches, 4
Transmission coefficient polarization light, 114
Transmissivity measure, 44
Transmitted power temperature, 121
Trench
deviation, 374
etch depth, 371
notching, 380, 387
rounded shape, 382
SEM, 371
sidewall schematic, 369
SOI wafer, 379, 380
width, 382
width etch rate, 372

width *vs.* beam verticality, 377
width *vs.* etch rate, 376
Triangular frame drawbridge actuator, 212
Triangular prism
    optical switch architecture, 100
    rotation, 101
Triangular prism switch modeling and analysis
    configuration, 100–102
    Fabry-Pérot effect and polarization effect, 104–106
    Gaussian beam divergence, 102–104
    modeling and analysis, 102–107
    optical loss sources, 104–108
Triangular thermo-optic switch
    MEMS thermo-optic switches, 100–108
    triangular prism thermo-optic switch
        configuration, 100–101
Trimmed photonic crystal rods resist
    pattern, 415
Tunable FBG, 8, 9
    dynamic response, 14
Tunable laser integrated OADM
    MEMS optical switches and systems, 16–18
    optical switching systems, 16–18
Tunable lasers, 254
Tunable laser system (TLS), 156
Tunable wavelength converter, 18
Tuning
    resolution contours, 205
    unstable region, 267
2-D digital mirror type free-space MEMS VOA,
    187–188
2-D photonic crystal device, 152
2-D switch configuration, 5
2-D 2x2 MEMS optical switch
    schematic, 28
    SEM micrograph, 11
2x2 MEMS optical switch, 8, 10–11

U

ULS. *see* Unlocked state (ULS)
Undercut
    DRIE, 364
    microbeam relative displacement, 66
    platen power, 363
    sloped comb-drive actuator displacement, 70
Undoped silicate glass (USG), 394, 396
Uneven comb drive, 65
Uneven-deep comb drive, 64
Uniform beam release, 384
Universal DRIE process flow, 390
Unlocked state (ULS), 324
Unoptimized process conditions, 401
USG. *see* Undoped silicate glass (USG)

V

Variable optical attenuators (VOA), 2
    applications, 175
    design and geometry, 432

DRIE, 180
    experimental setup, 450
    SOI, 180–183
    tilting shutter, 197
    wavelength dependence, 215
VCSEL. *see* Vertical-cavity side-emitting lasers
    (VCSELs)
Vertical-cavity side-emitting lasers (VCSELs),
    238–239, 322
Virtual pivots, 284
    cantilever beam, 289, 290
    design, 282–284, 291
    double-clamped beam, 304
    MEMS continuously tunable lasers, 282–284,
        287–290
    position shift, 292
VOA. *see* Variable optical attenuators (VOA)

W

Wafer flats, 356
Waveforms
    all-optical switching, 342
    slave, 341
Wavelength, 250
    converter, 17
    difference, 295
    mirror displacement, 20
    reflectivity, 41
    refractive index, 300
    tunability, 264–265, 309–311
Wavelength-dependent loss (WDL), 16, 21, 176, 178,
        179, 182, 183, 186, 231
    aperture position, 220
    deep-etched dual shutter VOAs, 220
    device, 226
    direct-coupling type MEMS VOA, 189
    dual-shutter VOA, 220
    EVOA, 223
    increases, 209
    MEMS subsystem, 22
    MEMS variable optical attenuators, 230
    physical mechanisms, 208
    polarization-dependent losses, 230–232
    PVOA, 225
    reduction, 224
    reflection-type VOA, 190
    temperature-dependent losses, 227–229
    wavelength-dependent losses, 230
Wavelength-division-multiplexing (WDM)
        combiner, 16
Wavelength tuning, 239
    coherence collapse, 257
    conventional *vs.* improved design, 301
    Doppler shift mode, 276
    Fabry-Pérot (FP) configuration, 240
    FP laser condition, 250
    measured and calculated relationships, 265
    stability, 257
WDL. *see* Wavelength-dependent loss (WDL)

WDM. *see* Wavelength-division-multiplexing
    (WDM) combiner
Weak feedback model, 250
    vs. extended feedback model, 255
    wavelength, 256
Wet silicon anisotropic etch, 358

X

X branch total internal reflection thermo-optic
    switch, 93

Y

Y branch digital thermo-optic switch
    schematics, 92
Young's modulus, 74, 79, 207

Z

ZEMAX, 176
Zinc blende, 354
    cubic structure, 355

Milton Keynes UK
Ingram Content Group UK Ltd.
UKHW052024071024
449327UK00027B/2414